Geometries in Interaction

GAFA special issue in honor of Mikhail Gromov

Edited by Y. Eliashberg
V. Milman
L. Polterovich
R. Schoen

Birkhäuser Verlag
Basel · Boston · Berlin

Reprint from GAFA (Geometric and Functional Analysis)
Volume 5 (1995), No. 2

Editorial Office:
School of Mathematical Sciences
Tel Aviv University
Tel Aviv 69978, Israel

Most of the contributors to this volume
participated in the Geometries in Interaction Workshop
at Tel Aviv University in December 1993.

The editors thank Miriam Hercberg for preparing
this book for publication.

A CIP catalogue record for this book is available from the Library of Congress, Washington D.C., USA

Die Deutsche Bibliothek - CIP-Einheitsaufnahme

Geometries in interaction: GAFA special issue in honor of
Mikhail Gromov / ed. by Y. Eliashberg ... - Basel ; Boston ;
Berlin : Birkhäuser, 1995
NE: Ēlî'ašberg, Ya'aqov [Hrsg.]; Gromov, Mikhail: Festschrift

Softcover reprint of the hardcover 1st edition 1995
Printed on acid-free paper produced of chlorine-free pulp ∞

ISBN-13: 978-3-0348-9907-9 e-ISBN-13: 978-3-0348-9102-8
DOI: 10.1007/978-3-0348-9102-8

9 8 7 6 5 4 3 2 1

Contents

The mathematical content of this book was first published in "Geometric And Functional Analysis" (GAFA), Volume 5:2 (1995). The original page numbering has been preserved to avoid confusion with references.

Photo E. Boubat Agence Top, Paris

Mikhail Gromov

Mikhail Gromov – Short Biography

Mikhail Gromov, recently celebrated his 50th birthday. Misha was born on 23 December, 1943, 200 miles from Leningrad, in the midst of the Second World War. Life during the war was very difficult and Misha decided not to speak until he was two and a half years old. By then the war was over, and Misha uttered his first words as whole sentences. At the age of six he annoyed his first grade teacher by solving a mathematical problem given to him by mistake and intended for third-graders. The teacher refused to believe that Misha had solved the problem without his parents' help. This especially hurt his feelings as his parents had indeed tried to solve the problem but failed. When Misha was ten his teacher told his mother that he would become a professor of mathematics. But at that time the future professor found more delight in playing with noxious chemicals than with theorems.

In 1960 the Mathematics Department of Leningrad University received a powerful boost when Professor V.A. Rokhlin began teaching there; Misha Gromov began studying as Rokhlin's student that same year.

After warming up on little problems during his undergraduate years (such as a problem of Banach on characterization of convex sets whose sections are mutually affine equivalent), Gromov succumbed to the charm of the Smale-Hirsh immersion theory. Blending their ideas with the Nash C^1-isometric immersion theorem in Riemannian geometry and the Oka-Grauert principle in several complex variables, he created powerful machinery for solving partial differential relations. This was the content of his PhD thesis in 1968. He found that many differential equations and inequalities arising in geometry enjoy the so-called *h-principle* which means, roughly speaking, that for the solvability of a differential relation it is sufficient to have a formal solution of the corresponding algebraic relation, where all derivatives are substituted by independent functions. For example he proved that a generic non-linear underdetermined (and sometimes even overdetermined, as in the case of isometric immersions) system of differential equations has many solutions if we are interested in the smooth solutions of the minimal class of differentiability. For instance, these solutions are C^0-dense in the space of all functions. Gromov developed several methods for establishing the h-principle: convex integration, the method of continuous sheaves and removal of singularities. These methods have been successfully applied to many concrete problems, especially in recent years, after Gromov's famous book *Partial Differential Relations* appeared in 1986. The h-principle was the subject of Gromov's invited talk at the 1970 International Congress of Mathematicians in Nice. This was the first of four talks he has given at ICMs (the others being in 1978, 1982 and 1986).

After the defence of his PhD thesis, Gromov obtained his first position at Leningrad University, where his main responsibility was learning conversational English as he was coached for the role of high school teacher in the Sudan. In 1972, after having learnt enough English, he defended his second thesis (Dr. Nauk) where he transplanted Nash's implicit function theorem to the ambience of continuous sheaves and developed a general theory of inducing geometric structures going far beyond Nash. By that time diplomatic relations between the USSR and the Sudan had been severed and, instead of to Africa, Misha moved first to Leningrad's Hydrometeorological Institute and then to the Pulp and Paper Institute. In 1974, disillusioned with "pulp and paper" he immigrated to the United States. From 1974 till 1981 he was a professor at the State University of New York at Stony Brook. In 1981 he moved to France, first as a professor at the University of Paris VI and then as a Permanent Fellow of IHES in Bures-sur-Yvette. Gromov is also a member of the Faculty of the University of Maryland and is Professor by Special Appointment in the School of Mathematical Sciences of Tel Aviv University.

Having escaped from the rigours of the Soviet regime, Gromov compensated by migrating to more rigid mathematics. One of the most intriguing problems on the borderline between soft and rigid mathematics resided in symplectic geometry. In his previous work on the h-principle, Gromov showed that symplectic structures enjoy unexpected flexibility, but there were indications (e.g. the Poincaré fixed point theorem, Arnold's conjectures) of certain rigidity. Accidentally, having failed to hack through the jungle of the quasi-analytic functions of Bers and Vekua, Gromov discovered their geometric counterpart – *pseudo-holomorphic curves in symplectic manifolds*. Their introduction, together with Gromov's compactness theorem for pseudoholomorphic curves, manifested symplectic rigidity and laid down the foundations of the new field, Symplectic topology.

In Riemannian geometry Gromov looked at the relaxed curvature equations, such as curvature $\leq \epsilon$, and found, in particular, a perturbative version of the Bieberbach theorem. This led to the new concept of *collapse* of Riemannian manifolds with a bound on curvature, which was developed jointly with Cheeger and Fukaya.

Together with Lawson, Gromov tracked the rigidity of the condition of positivity of scalar curvature to the analytic aspect of the Novikov conjecture on higher signatures.

He introduced the concept of convergence of Riemannian manifolds (*Gromov-Hausdorff convergence*) which he applied to groups of polynomial growth and showed that they are virtually nilpotent.

Gromov reanimated Dehn's original approach to infinite groups where combinatorics and geometry (especially that of curvature ≤ 0) were blended together into the *theory of hyperbolic groups*. Gromov found a homotopy theoretic definition of the volume of a hyperbolic manifold, as what is now called the *Gromov norm* on homology, which gave, in particular, a new topological proof of the Mostow rigidity theorem, and proved useful, for instance, in Thurston's theory of hyperbolic 3-manifolds.

Gromov proved that the positivity of sectional curvature restricts Betti numbers of a manifold and then, together with Burago and Perelman, started a new wave of synthetic geometry of singular spaces of non-negative curvature, which were originally introduced by A.D. Aleksandrov half a century ago.

He found restrictions on the transcendental part of the fundamental group of Kähler manifolds. Jointly with Schoen he proved the arithmeticity of lattices in $Sp(n,1)$ by using harmonic mappings of the corresponding symmetric spaces into Bruhat–Tits buildings.

Gromov has received a number of awards in recognition of his achievements, including the Moscow Mathematical Society Prize to Young Mathematicians (1971), the American Mathematical Society's Oswald Veblen Prize in Geometry (1981), the French Academy of Sciences' Elie Cartan Prize (1984), the Prix Union des Assurances de Paris (1989), and the Wolf Foundation Prize (1993). He is a foreign member of the National Academy of Science of the United States, the American Academy of Art and Sciences, and the French Academy of Science.

Misha Gromov has paved several routes, some of them highways, some avenues, and others breathtaking mountain paths. We follow these routes as will future generations.

Mikhail Gromov – Publications

1. On a geometrical hypothesis of Banach, Izv. A.N. SSSR 31 (1967), 1105-1114 (Russian).
2. Transversal mappings of foliations, Dokl. A.N. SSSR 182 (1968), 255-258 (Russian).
3. Maps of foliations into manifolds carrying additional structure, Vestnik Leningradskogo Univ. 23:19 (1968), 167 (Russian with English summary).
4. The number of simplexes of subdivisions of finite complexes, Matematicheskie Zametki 3 (1968), 511-522 (Russian).
5. Stable mappings of foliations into manifolds, Math. USSR-Izvestija 3 (1969), 671-694.
6. On simplexes inscribed in a hypersurface, Matematicheskie Zametki 5 (1969), 81-89 (Russian).
7. Imbeddings and immersions in Riemannian geometry (with V.A. Rokhlin), Russian Math. Surveys 25 (1970), 1-57.
8. Isometric immersions and embeddings, Soviet Math. Dokl. 11 (1970), 794-797.
9. A topological technique for the construction of solutions of differential equations and inequalities, Proceedings ICM (Nice 1970), vol. 2 (1971), 221-225.
10. Removal of singularities of smooth mappings (with Ya.M. Eliashberg), Math. USSR Izvestija 5 (1971), 615-639.
11. Construction of nonsingular isoperimetric surfaces (with Ya.M. Eliashberg), Steklov Institute Trudy 116 (1971) 18-331 (Russian).
12. Nonsingular mappings of Stein manifolds (with Ya.M. Eliashberg), Functional Analysis and its Applications 5 (1971), 156-157.
13. Smoothing and inversion of differential operators, Mat. USSR Sbornik 17 (1972), 382-435.
14. Isometric embeddings in multidimensional geometry, Leningrad University Thesis Abstract (1972).
15. Singular smooth maps, Mat. Zametki 14 (1973), 509-516 (Russian).
16. Convex integration of differential relations, I, Izv. A.N. SSSR 37 (1973), 329-343 (Russian).
17. Construction of a smooth mapping with prescribed Jacobian, I (with Ya.M. Eliashberg), Functional Analysis and its Applications 7:1 (1973), 27-32.
18. Topology of Riemannian manifolds with small curvature and diameter (preliminary report), Notices of A.M.S. 22:5 (1975), A-592.
19. Three remarks on geodesic dynamics and fundamental group, preprint SUNY 1976 (unpublished).
20. On the entropy of holomorphic maps, preprint 1977 (unpublished).
21. Homotopical effects of dilatation, Journal of Differential Geometry 13 (1978), 303-310.
22. Almost flat manifolds, Journal of Differential Geometry 13 (1978), 231-241.
23. Manifolds of negative curvature, Journal of Differential Geometry 13 (1978), 223-230.
24. Isometrical immersions, hyperbolic geometry and Burago's isoperimetric inequality, preprint SUNY 1979 (unpublished).

25. Synthetic geometry in Riemannian manifolds, Proceedings ICM 1978, vol. 1 (1980), 415-419.
26. Spin and scalar curvature in the presence of a fundamental group (with H.B. Lawson), Annals of Math. 111 (1980), 209-230.
27. The classification of simply connected manifolds of positive scalar curvature (with H.B. Lawson), Annals of Math. 111 (1980), 423-434.
28. Paul Levy's isoperimetric inequality, preprint IHES (1980), unpublished.
29. On the ergodicity of frame flows, (with M. Brin) Inv. Math. 60 (1980), 1-7.
30. Hyperbolic manifolds, groups and actions, Annals of Math. Studies 97 (1981), 183-215 (Princeton University Press).
31. Groups of polynomial growth and expanding maps, Publications Mathématiques IHES 53 (1981), 53-73.
32. Curvature, diameter and Betti numbers, Comm. Math. Helvetia 56 (1981), 179-195.
33. Hyperbolic manifolds according to Thurston and Jorgensen, Séminaire Bourbaki (1979/80), Springer Lecture Notes in Mathematics 842 (1981), 40-53.
34. Volume and bounded cohomology, Publications Mathématiques IHES 56 (1982), 5-99.
35. Finite propagation speed, kernel estimates, and the geometry of complete Riemannian manifolds (with J. Cheeger, M. Taylor), Journal of Differential Geometry 17 (1982), 15-53.
36. A topological application of the isoperimetric inequality (with V. Milman), Am. Journal of Mathematics 105 (1983), 843-853.
37. Filling Riemannian manifolds, Journal of Differential Geometry 18 (1983), 1-147.
38. Positive scalar curvature and the Dirac operator on complete Riemannian manifolds (with B. Lawson), Publications Mathématiques IHES 58 (1983), 83-196.
39. Infinite groups as geometric objects, Proceedings ICM (Warsaw 1983), vol.1,2 (1984), 385-392.
40. Brunn theorem and the concentration of volume phenomena for symmetric convex bodies (with V.D. Milman), GAFA Seminar Notes (1983/4), Tel Aviv University (1984).
41. Asymptotic geometry of homogeneous spaces, Proceedings of Conference on Differential Geometry and Homogeneous Spaces (Torino 1983), Rend. Sem. Mat. Univ. Politec. Special Issue (1984), 59-60.
42. Bounds on the Von Neumann dimension of L^2-cohomology and the Gauss-Bonnet theorem for open manifolds (with J. Cheeger), Journal of Differential Geometry 21 (1985), 1-34.
43. On the characteristic numbers of complete manifolds of bounded curvature and finite volume (with J. Cheeger), Differential Geometry and Complex Analysis, Rauch Memorial Volume, Springer-Verlag (1985), 115-154.
44. Manifolds of non-positive curvature (with W. Ballmann, V. Schroeder), Progress in Mathematics 61, Birkhäuser (1985).
45. Pseudo-holomorphic curves in symplectic manifolds, Inventiones Math. 82 (1985), 307-347.

46. Isometric immersions of Riemannian manifolds, In "The Mathematical Heritage of Élie Cartan" (Lyon 1984), Asterisque, Numero Hors Series (1985), 129-133.
47. Collapsing Riemannian manifolds while keeping the curvature bounded (with J. Cheeger), Journal of Differential Geometry 23 (1986), 309-346.
48. Partial Differential Relations, Springer-Verlag (1986).
49. L^2-cohomology and group cohomology, (with J. Cheeger), Topology 25:2 (1986), 189-215.
50. Large Riemannian manifolds, in "Curvature and topology of Riemannian manifolds" (Katata 1985), Springer Lecture Notes in Mathematics 1201 (1986), 108-121.
51. Isoperimetric inequalities in Riemannian manifolds, Springer-Verlag, Lecture Notes in Mathematics 1200 (1986), 114-130.
52. Structures métriques pour les variétés riemanniennes (J. Lafontaine and P. Pansu, eds.), Cedric Fernand Nathan, Paris (1987).
53. Pinching constants for hyperbolic manifolds, (with W.P. Thurston), Inventiones Math. 89:1 (1987), 1-12.
54. Generalization of the spherical isoperimetric inequality to uniformly convex Banach spaces (with V. Milman), Compositio Math. 62:3 (1987), 263-282.
55. Hyperbolic groups, Essays in Group Theory (S. Gersten, ed.), MSRI Publications Springer 8 (1987), 75-265.
56. Entropy, homology and semialgebraic geometry (after Yomdin), (Séminaire Bourbaki, 1985/6) Asterisque 145/6 (1987), 225-240.
57. Soft and hard symplectic geometry, Proceedings ICM (Berkeley 1986), AMS 1,2 (1987), 81-98.
58. Cauchy-Riemann equation in the Lagrange intersection theory, in "Periodic Orbits of Hamiltonian Systems and Related Topics" (Il Ciocco 1986), Adv. Sci. Inst. Ser. C: Math. Phys. Sci. 209 (1987), 175-176.
59. Monotonicity of the volume of intersection of balls, GAFA Seminar Notes 1985/6, Springer Lecture Notes in Mathematics 1267 (1987), 1-4.
60. Hyperbolic 4-manifolds and conformally flat 3-manifolds (with B. Lawson, W.P. Thurston), Publications Mathématiques IHES 68 (1988), 27-45.
61. Non-arithmetic groups in Lobachevsky spaces, (with I.I. Piatetski-Shapiro), Publications Mathématiques IHES 66 (1988), 93-103.
62. Rigid transformations groups, Géométrie Différentielle (Paris 1986), Travaux en Cours, 33, Hermann Paris (1988), 65-139.
63. Dimension, non-linear spectra and width, Springer-Verlag, Lecture Notes in Mathematics 1317 (1988), 132-184.
64. Width and related invariants of Riemannian manifolds, Astérisque 163-164 (1989), 93-109.
65. Sur le groupe fondamental d'une variété kählerienne, C.R. de Acad. Sci. Paris 308 (1989), 67-70.
66. Soft differential equations, in Proceedings of IX-th International Congress on Mathematical Physics (Swansea, 1988), 374-376, Hilger, Briston (1989).
67. Oka's principle for holomorphic sections of elliptic bundles, J. of American Mathematical Society 2:4 (1989), 851-897.

68. Estimates of Berstein width of Sobolev spaces (with J. Bourgain), GAFA Seminar Notes 1987/8, Springer Lecture Notes in Mathematics 1376 (1989), 176-185.
69. Cell division and hyperbolic geometry, In "Several Complex Variables and Complex Geometry", Part 2 (Santa Cruz, CA 1989), 135-162,
70. Convex sets and Kähler manifolds, in "Advances in Diff. Geom. and Topology", World Scientific Publishing, Teaneck, NJ (1990), 1-38.
71. Collapsing Riemannian manifolds while keeping their curvature bounded, II (with J. Cheeger), Journal Differential Geometry 32 (1990), 269-298.
72. Conjecture de Novikov et fibrés presque plats, (with A. Connes and H. Moscovici), C.R. Acad. Sci. Paris 310 série I:5 (1990), 273-277.
73. Chopping Riemannian manifolds (with J. Cheeger), in "Differential Geometry, Symposium in Honor of Manfredo de Carmo" (H.B. Lawson, K. Tenenblat, eds.), Pitman Monographs and Surveys in Pure and Applied Math. 52 (1991), 85-94.
74. Kähler hyperbolicity and L^2-Hodge theory, Journal Differential Geometry 33 (1991), 263-292.
75. Lectures on the transformation groups: geometry and dynamics (with G. d'Ambra), in "Surveys in Differential Geometry" (Cambridge, MA 1990) Lehigh Univ. Publ. (1991), 19-111.
76. Foliated Plateau problem, parts I, II, Geometric And Functional Analysis (GAFA) 1:1 (1991), 14-79; 1:3 (1991), 253-320.
77. Convex symplectic manifolds (with Ya. Eliashberg), Proc. Symp. Pure Math., AMS 52 (1991), Part 2, 135-162.
78. Rigidity of lattices : an introduction (with P. Pansu), in "Geometric Topology: Recent Developments" (Montecatini, Terme 1990), Springer Lecture Notes in Math. 1504 (1991), 39-137.
79. Nilpotent structures and invariant metric on collapsed manifolds (with J. Cheeger, K. Fukaya), Journal of Amer. Math. Soc. 5 (1992), 327-372.
80. Embeddings of Stein manifolds of dimension n into the affine space of dimension $3n/2 + 1$ (with Ya. Eliashberg), Annals of Math. 136 (1992), 123-135.
81. Stability and pinching, in "Geometry Seminars. Sessions in Topology and Geometry of Manifolds" (Bologna, 1990), 55-97, Univ. Stud. Bologna (1992).
82. Group cohomology with Lipschitz control and higher signatures (with A. Connes and H. Moscovici), GAFA 3:1 (1993), 1-78.
83. Sign and geometric meaning of curvature, Rend. Semm. Mat. Fis. Milano 61 (1991), 9-123 (1994).
84. Von Neumann spectra near zero (with M. Shubin), GAFA 1:4 (1991), 375-404.
85. The Riemann-Roch theorem for general elliptic operators (with M. Shubin), C.R. Acad. Sci. Paris Ser. I Math. 314 (1992), 363-367.
86. Harmonic maps into singular spaces and p-adic superrigidity for lattices in groups of rank one (with R. Schoen), Publications Math. de l'IHES 76 (1992), 165-246.
87. A.D. Aleksandrov's spaces with curvatures bounded below (with Yu. Burago and G. Perelman), Russian Math. Surveys 47 (1992), 1-58.
88. Spectral geometry of semi-algebraic sets, Ann. Inst. Fourier 42, 1-2 (1992), 249-274.

89. The Riemann-Roch theorem for elliptic operators (with M. Shubin), I.M. Gelfand Seminar 211-241 (S. Gelfand and S. Gindikin, eds.), Adv. Soviet Math. 16, Part 1, Amer. Math. Soc. (1993).
90. Near-cohomology of Hilbert complexes and topology of non-simply connected manifolds (with M. Shubin), in "Methodes Semi-classiques", Vol. 2, Colloque International (Nates, June 1991), Asterisque 210 (1993), 283-294.
91. Asymptotic invariants of infinite groups, in "Geometric Group Theory vol 2", London Math. Society Lecture Notes 182 (1993), 1-295.
92. Metric invariants of Kähler manifolds, Proc. of the Workshop on Differential Geometry and Topology, Algero, Italy 20-26 June, ed. Caddeo, Tricerri World Sci. (1993) 90-117.
93 The Riemann-Roch theorem for elliptic operators and solvability of elliptic equations with additional conditions on compact subsets (with M. Shubin), Journées "Équations aux Dérivées Partielles", École Polytechnique, Centre de Mathématiques, Palaiseau, 1993.
94 The Riemann-Roch theorem for elliptic operators and solvability of elliptic equations with additional conditions on compact subsets (with M. Shubin), Invent. Math. 117 (1994), 165-180.
95. Systoles and intersystolic inequalities, to appear in "Actes de la Table Ronde de Géométrie Différentielle en l'honneur de Marcel Berger" (Arthur L. Besse, ed.), Collection SMF Séminars and Congrès No. 1, 1994.
96. Carnot-Carathéodory spaces seen from within, preprint IHES M/94/06, 1994.
97. Geometric reflections on the Novikov conjecture, Lecture notes taken by J. Rosenberg, Oberwolfach (1993), to appear in Proc. of the Conference on the Novikov Conjecture.
98. Positive curvature, macroscopic dimension, spectral gaps and higher signatures, to appear in Gelfand's 80th Birthday volume.
99. L^2-holomorphic functions on pseudo-convex coverings (with G. Henkin and M. Shubin), to appear as IHES preprint.
100. Symplectic geometry of generating functions and pseudoisotopy (with Ya. Eliashberg), in preparation.

Geometric And Functional Analysis
Vol. 5, No. 2 (1995)
1016-443X/95/0200105-36$1.50+0.20/0

ASPECTS OF LONG TIME BEHAVIOUR OF SOLUTIONS OF NONLINEAR HAMILTONIAN EVOLUTION EQUATIONS

J. Bourgain

0. Introduction

In this paper we will be mainly concerned with the behaviour of solutions of (space periodic) nonlinear wave equations

$$u_{tt} = \Delta u + p(u; t, x) \qquad (x \in \mathbb{T}^d) \tag{1}$$

and nonlinear Schrödinger equations

$$-iu_t = -\Delta u + V(x)u + \frac{\partial}{\partial \overline{u}} G(u, \overline{u}; t, x) \qquad (x \in \mathbb{T}^d)\ . \tag{2}$$

Most of the techniques used have a wider range of applicability however.

We are interested in nonlinear Hamiltonian PDE in a general (nonintegrable or close to integrable) context where one does not expect KAM results. Assuming local existence of solutions established, there are many issues to be addressed about their behaviour for $t \to \infty$ and the properties of the flow S_t in phase space, such as

- global existence, blowup behaviour (3)
- asymptotic stability (4)
- behaviour of higher Sobolev norms of smooth solutions when $t \to \infty$; spreading of energy to higher modes (5)
- recurrence properties in time (6)

Global wellposedness may often be derived from local wellposedness and conservation laws, without much further insight into the behaviour of the solutions. On the other hand, the Hamiltonian property of the equation permits exploiting invariants of the flow in an appropriate phase space. For instance, Liouville's theorem leads to invariant Gibbs measures that permit establishing Poincaré recurrence (in appropriate topology). More recently, other invariants were discovered, such as symplectic capacities and applied to Hamiltonian mechanics by various authors. Symplectic capacities enable proving certain nonsqueezing properties of S_t, which are of relevance for (4) and (5). In a recent work ([K2]), S. Kuksin adjusted the finite dimensional theory to an infinite dimensional phase space setting, provided the map S_t is of the form

$$S_t = \text{linear operator} + \text{compact smooth operator} \tag{7}$$

(essentially speaking). In this statement (7), the phase space is well defined, due to the finite dimensional normalization. For instance, for equation (1), the "symplectic Hilbert space" is $H^{1/2}(\mathbf{T}^d) \times H^{1/2}(\mathbf{T}^d)$, while for equation (2) it is $L^2(\mathbf{T}^d)$. Examples of results obtained along these lines in [K2] is the nonsqueezing of balls in cylinders defined with respect to a Darboux basis, in the case of the nonlinear string equation

$$u_{tt} = u_{xx} + p(u;t,x) \qquad (x \in \mathbf{T}) \tag{8}$$

where p is a smooth function which has at most polynomial growth as $|u| \to \infty$ and the quadratic nonlinear wave equation

$$u_{tt} = \Delta u + a(t,x)u + b(t,x)u^2 \qquad (x \in \mathbf{T}^2)\ . \tag{9}$$

These are special cases of (1).

Thus the squeezing theorem states that

$$S_t(B_R) \subset \mathbf{T}_r^{(k)} \Rightarrow R \leq r\ . \tag{10}$$

Here B_R denotes an R-ball in the symplectic Hilbert space Z (not necessarily centered at 0) and $\mathbf{T}_r^{(k)}$ stands for a translate of the cylinder $\{\Sigma p_j \varphi_j^+ + q_j \varphi_j^- \mid p_k^2 + q_k^2 < r^2\}$ where $\{\varphi_j^\pm\}$ is a Darboux basis of Z.

The squeezing theorem implies in particular that if B_ρ is a ball centered at some initial point, then the diameter of the set $S_t(B_\rho)$ cannot tend to zero. This fact is referred to in [K2] as the failure of "uniform asymptotic stability" as $t \to \infty$ of bounded solutions of the equation.

Another consequence of the squeezing theorem relates to spreading of the energy to higher frequencies. It follows indeed from (10) that for any time t and mode k, one cannot have

$$p_k^2 + q_k^2 < (\rho - \varepsilon)^2 \tag{11}$$

for all members of $S_t(B_\rho)$.

In this paper, we will continue some of these investigations by extending Kuksin's results to other equations, requiring more PDE analysis or with flow map not of the form (7).

PROPOSITION 1. *The nonsqueezing theorem* (10) *in* $H^{1/2} \times H^{1/2}$ *holds for the wave equation* (1) *with at most degree 4 nonlinearity (resp. at most quadratic nonlinearity) in dimension 2 (resp. dimension* 3, 4). *In fact, the flow map is of the form* (7) *in these cases.*

The symplectic Hilbert space of the NLSE (2) is the space $L^2(\mathbb{T}^d)$ of complex functions. We restrict the discussion to $d = 1$. In the interesting cases, such as for instance the cubic NLSE $iu_t + u_{xx} \pm u|u|^2 = 0$, the map S_t is not of form (7). However, in [B3], we proved the nonsqueezing theorem for equations of the form

$$iu_t + u_{xx} + a(x,t)u + b(x,t)u|u|^2 = 0 \tag{12}$$

where a, b are sufficiently smooth real functions, both periodic in x. The property is derived from a direct approximation argument by finite dimensional models

$$-iU_t = -U_{xx} + P_N \frac{\partial}{\partial \overline{U}} G(U, \overline{U}; t, x) \qquad U = P_N U \tag{13}$$

where P_N is the usual Dirichlet projection. An L^2-analysis for the 1-dimensional NLSE with L^2-local wellposedness theorem seems presently only available for cubic nonlinearity, i.e. $G(u, \overline{u}; t, x)$ is a polynomial of degree ≤ 4 in $u, \overline{u}$. The argument in [B4] depends moreover on the conservation of the L^2-norm $\int_{\mathbb{T}} |u|^2 dx$ which holds for G of the form $G(|u|^2; t, x)$ as in (12). We consider here the more general case and prove the following.

Proposition 2. *Consider the NLSE*

$$iu_t + u_{xx} + \frac{\partial}{\partial \overline{u}} G(u, \overline{u}; t, x) = 0 \tag{14}$$

where G is a real polynomial in $u, \overline{u}$ of degree ≤ 4. Then bounded solutions of (14) are not uniformly asymptotically stable in L^2 for $t \to \infty$. The same statement holds without degree restriction on G in H^s, $s > \frac{1}{2}$.

One may compare solutions of (13),(14) for data $\varphi = P_N \varphi$ which "tail" Fourier coefficients are sufficiently small and invoke a result of Ekeland and Hofer [EH] according to which in $2n$-dimensional phase space the product $B^2(r) \times \cdots \times B^2(r)$ (n copies) cannot be symplectically embedded in a ball $B^{2n}(\rho)$ with $\rho < \sqrt{n} r$.

In the last section, we exhibit smooth global solutions of Hamiltonian NLSE with smooth nonlinear term which develop large $\|u(t)\|_{H^s}$-norm for $t \to \infty$. Here the Sobolev exponent s_0 is numerical. The construction used leads to equations which are close to the linear Schrödinger equation. Similar arguments apply for other evolution equations as well.

Proposition 3. (1) *There is a Hamiltonian NLSE with smooth nonlinear term depending on u and projections Pu of u on the trigonometric system, such that for smooth data the solution u satisfies $\overline{\lim}_{t\to\infty} \|u(t)\|_{H^{s_0}} = \infty$ for some fixed exponent s_0.*

(2) *There is a Hamiltonian NLSE with smooth and local nonlinearity*

such that $S_t(B^s(\delta))$, $t > 0$ *is not a bounded subset of* H^{s_0}, *for any* $s < \infty$ *and* $\delta > 0$. *Here* $B^s(\delta)$ *denotes* $\{\varphi \in H^s \mid \|\varphi\|_s < \delta\}$.

In previous constructions, we use the fact that solutions of the linear equation $iu_t + u_{xx} = 0$ are periodic in time (for periodic boundary conditions). The last example discussed deals with perturbations of a linear Schrödinger operator $iu_t + u_{xx} + V(x)u$ where $V(x)$ is a real smooth potential with nearly resonant spectrum, in the sense that for some n_0 and infinite sequence $\{n_j\}$ one has $\text{dist}(\lambda_{n_j}, \mathbf{Z}\lambda_{n_0}) \to 0$ rapidly for $j \to \infty$. In this context we construct a Hamiltonian perturbation $\Gamma(u) = \frac{\partial}{\partial \bar{u}}G$ such that if $u_{\varepsilon,q}$ denotes the solution of the IVP

$$\begin{cases} -iu_t = -u_{xx} + V(x)u + \varepsilon\Gamma(u) \\ u(0) = q \end{cases} \tag{15}$$

then

$$\inf_{q\in\gamma} \sup_t \|u_{\varepsilon,q}(t)\|_{H^{s_0}} \to \infty \quad \text{for} \quad \varepsilon \to 0 \tag{16}$$

where γ is some open subset of H^{s_0}.

This example is related to infinite dimensional versions of Lyapounov's theorem in KAM theory on persistence of invariant tori and the work of Kuksin, Craig-Wayne and the author on small Hamiltonian perturbations of integrable equations (see [K1], [CW], [B5]).

1. Nonlinear Wave Equations

We consider the nonlinear wave equation

$$u_{tt} = \Delta u - u - f(u; t, x) \tag{1}$$

where $x \in \mathbf{T}^d$ and f is a polynomial in u with smooth coefficients in x, t.

Denoting B the operator $(-\Delta + 1)^{1/2}$ we may write (1) in Hamiltonian form as

$$\begin{cases} u_t = -Bv \\ v_t = Bu + B^{-1}f(u; t, x) \, . \end{cases} \tag{2}$$

The sumplectic Hilbert space here is $H^{1/2}(\mathbf{T}^d) \times H^{1/2}(\mathbf{T}^d)$. We study the Cauchy problem for (2) with data in $H^s(\mathbf{T}^d)$ (s close to $\frac{1}{2}$) local in time. From the integral equation we get

$$u(t) = \sum_\xi \left[\frac{\widehat{u(0)}(\xi)+i\widehat{v(0)}(\xi)}{2} e^{i(\langle x,\xi\rangle+(1+|\xi|^2)^{1/2}t)} + \frac{\widehat{u(0)}(\xi)-i\widehat{v(0)}(\xi)}{2} e^{i(\langle x,\xi\rangle-(1+|\xi|^2)^{1/2}t)}\right] \tag{3}$$

$$+\sum_\xi \int d\lambda \frac{\widehat{f}(\xi,\lambda)}{\lambda^2-1-|\xi|^2} e^{i\langle x,\xi\rangle}$$

$$\left[e^{i\lambda t}-\frac{1}{2}\left(1+\frac{\lambda}{(1+|\xi|^2)^{1/2}}\right)e^{i(1+|\xi|^2)^{1/2}t}-\frac{1}{2}\left(1-\frac{\lambda}{(1+|\xi|^2)^{1/2}}\right)e^{-i(1+|\xi|^2)^{1/2}t}\right] \tag{4}$$

where $\widehat{f}$ denotes the Fourier transform of $f(u;t,x)$ on $\mathbf{T}^d\times I$. The expression for $v(t)$ is obtained as $-B^{-1}u_t$.

The assumptions on dimension d and f are the following

$$d=2 \qquad f \text{ of the form } a_1(x,t)u+a_2(x,t)u^2+a_3(x,t)u^3+a_4(x,t)u^4 \tag{5}$$

$$d=3,4 \quad f \text{ of the form } a_1(x,t)u+a_2(x,t)u^2\ . \tag{6}$$

Consider following norm for functions $A(x,t)=\Sigma_\xi \int d\lambda \widehat{A}(\xi,\lambda)e^{i(\langle x,\xi\rangle+\lambda t)}$ on $\mathbf{T}^d\times I$

$$|||A|||_s=\left(\sum_\xi \int d\lambda (1+|\xi|)^{2s}(1+||\lambda|-|\xi||)^{2\rho}|\widehat{A}(\xi,\lambda)|^2\right)^{1/2} \tag{7}$$

(to be understood as a restriction norm with respect to $\mathbf{T}^d\times I$).

Here ρ is chosen a bit larger than $\frac{1}{2}$. Observe that

$$\|A(t)\|_{H^s}\le c|||A|||_s \quad \text{for} \quad t\in I\ . \tag{8}$$

The expression (3) defines a linear operator of $u(0),v(0)$; observe that

$$|||(3)|||_s+|||B^{-1}(3)_t|||_s\le c\big(\|u(0)\|_{H^s}+\|v(0)\|_{H^s}\big)\ . \tag{9}$$

Our purpose is to show that for some $s_1<\frac{1}{2}<s_2$

$$|||(4)|||_{s_2}+|||B^{-1}(4)_t|||_{s_2}\le C\sigma^c(1+|||u|||_{s_1}^3)|||u|||_{s_1} \tag{10}$$

where $\sigma=|I|$, for some constants $0<c,C<\infty$. Replacing s_2 by s_1 in the left member of (10), one gets as a first consequence, letting σ be sufficiently small,

$$|||u|||_{s_1}+|||v|||_{s_1}\le c\big(\|u(0)\|_{H^{s_1}}+\|v(0)\|_{H_{s_1}}\big)\ . \tag{11}$$

Hence, from (8),(10),(11), it follows that for $t\in I$

$$\|(4)\|_{H^{s_2}}+\|B^{-1}(4)_t\|_{H^{s_2}}\le\|u(0)\|_{H^{s_1}}+\|v(0)\|_{H^{s_1}} \tag{12}$$

and hence the nonlinear part of the flow map acts boundedly from $H^{s_1}\times H^{s_1}$ to $H^{s_2}\times H^{s_2}$ ($s_1<\frac{1}{2}, s_2>\frac{1}{2}$) which is the required condition to make the results from [K2] applicable.

The $a(x,t)$-coefficients will play little role in the verification of (10) and we ignore them for simplicity sake. In fact the relevant calculation appears for $f(u;t,x)=u^4$ in $d=2$ and $f(u;t,x)=u^2$ in $d=3$, $d=4$.

Consider the expression (4) with $||\lambda| - |\xi|| < 10$. One easily verifies because t is local (multiply the expression with a localizing function $\varphi(t)$) that the corresponding contribution to (7) is bounded by

$$\Big[\sum_\xi (1+|\xi|)^{2(s-1)} \int_{||\lambda|-|\xi||<10} \big|\widehat{f}(\xi,\lambda)\big|^2\Big]^{1/2} . \tag{13}$$

Hence $|||(4)|||_{s_2}$ may be estimated by the sum of

$$\Big[\sum_\xi |\xi|^{2s_2} \int d\lambda \frac{|\widehat{f}(\xi,\lambda)|^2}{(|\xi|+|\lambda|+1)^2(||\xi|-|\lambda||+1)^{2(1-\rho)}}\Big]^{1/2} \tag{14}$$

$$\Big[\sum_\xi |\xi|^{2(s_2-1)} \Big(\int d\lambda \frac{|\widehat{f}(\xi,\lambda)|}{||\xi|-|\lambda||+1}\Big)^2\Big]^{1/2} \tag{15}$$

and hence, from Hölder's inequality ($\rho > \frac{1}{2}$)

$$\Big[\sum_\xi \int d\lambda \frac{|\widehat{f}(\xi,\lambda)|^2}{(1+|\xi|)^{2(1-s_2)}(1+||\xi|-|\lambda||)^{2(1-\rho)}}\Big]^{1/2} . \tag{16}$$

To estimate $|||B^{-1}(4)_t|||_{s_2}$, we need to introduce an extra factor $\frac{\lambda^2}{1+|\xi|^2}$ in (14). Hence (16) is an estimate on both $|||(4)|||_{s_2}$ and $|||B^{-1}(4)_t|||_{s_2}$.

Consider first the case $d=2$. Letting $f = u^4$, one has $\widehat{f} = \widehat{u} * \widehat{u} * \widehat{u} * \widehat{u}$. According to (7), define

$$c(\xi,\lambda) = (1+|\xi|)^{s_1}(1+||\lambda|-|\xi||)^{\rho}|\widehat{u}(\xi,\lambda)| \tag{17}$$

hence

$$\|c\|_2 = |||u|||_{s_1} . \tag{18}$$

In the sequel, we will write $|\ldots|$ instead of $1+|\ldots|$.

By duality (16) may be estimated by

$$\sum_{\substack{\xi=\xi_1+\xi_2+\xi_3+\xi_4 \\ \lambda=\lambda_1+\lambda_2+\lambda_3+\lambda_4}} \int \prod_{i=1}^4 \frac{c(\xi_i,\lambda_i)}{|\xi_i|^{s_1}||\lambda_i|-|\xi_i||^{\rho}} \frac{d(\xi,\lambda)}{|\xi|^{1-s_2}||\lambda|-|\xi||^{1-\rho}} \tag{19}$$

where $d(\xi,\lambda) \geq 0$, $\|d\|_2 \leq 1$.

Observe that in the problem of estimating (19), the discrete character of the summation plays clearly no role and we may as well replace $\Sigma_{\xi=\xi_1+\xi_2+\xi_3+\xi_4}$ by $\int_{\xi=\xi_1+\xi_2+\xi_3+\xi_4}$ (all denominators are taken > 1).

At this stage, we invoke Strichart's inequality on the Fourier transform of an L^2-density carried by a cone in $\mathbf{R}^{d+1}$ (see [S]).

Let $q = \frac{2(d+1)}{d-1}$. Then

$$\Big\| \int_{|\xi|\sim R} a(\xi) e^{i(\langle x,\xi\rangle + t|\xi|)} d\xi \Big\|_{L^q(dx dt)} \le C R^{1/2} \Big(\int |a(\xi)|^2 d\xi \Big)^{1/2} . \tag{20}$$

In our case $d = 3$, $q = 6$. It follows from (20) and Hölder's inequality that

$$\Big\| \int_{|\xi|\sim R} d\xi \int d\lambda \frac{c(\xi,\lambda)}{||\lambda| - |\xi||^{\frac{1}{2}+}} e^{i(\langle x,\xi\rangle + \lambda t)} \Big\|_{L^6(dx dt)}$$
$$\le C R^{1/2} \Big(\iint |c(\xi,\lambda)|^2 d\xi\, d\lambda \Big)^{1/2} \tag{21}$$

and hence, interpolating with the obvious (Parseval) L^2-inequality

$$\Big\| \int d\xi \int d\lambda\, c(\xi,\lambda) e^{i(\langle x,\xi\rangle + \lambda t)} \Big\|_{L^2(dx dt)} \le \Big(\iint |c(\xi,\lambda)|^2 d\xi\, d\lambda \Big)^{1/2} \tag{22}$$

we get the inequality

$$\Big\| \int_{|\xi|\sim R} d\xi \int d\lambda \frac{c(\xi,\lambda)}{||\lambda| - |\xi||^{\frac{9}{20}+}} e^{i(\langle x,\xi\rangle + \lambda t)} \Big\|_{L^5(dx dt)}$$
$$\le C R^{9/20} \Big(\iint |c(\xi,\lambda)|^2 d\xi\, d\lambda \Big)^{1/2} \tag{23}$$

which is used to bound (19). Restricting ξ_i, ξ to dyadic regions

$$\begin{cases} |\xi_i| \sim R_i \quad (i = 1,2,3,4) \\ |\xi| \sim R \end{cases} \tag{24}$$

one gets

$$(R_1 R_2 R_3 R_4)^{-s_1} R^{-(1-s_2)} \sum_{\substack{\xi = \Sigma\xi_i; \lambda = \Sigma\lambda_i \\ |\xi_i|\sim R_i, |\xi|\sim R}} \int \prod_{i=1}^4 \frac{c(\xi_i,\lambda_i)}{||\xi_i| - |\lambda_i||^{\rho}} \frac{d(\xi,\lambda)}{||\xi| - |\lambda||^{q-\rho}} . \tag{25}$$

Define $F_i = F_i(x,t)$, $G = G(x,t)$ letting $\widehat{F}_i(\xi,\lambda) = \frac{c(\xi,\lambda)}{||\xi|-|\lambda||^{\rho}} \chi_{[|\xi|\sim R_i]}(\xi)$, $\widehat{G}(\xi,\lambda) = \frac{d(\xi,\lambda)}{||\xi|-|\lambda||^{1-\rho}} \chi_{[|\xi|\sim R]}(\xi)$. Thus (25) equals

$$(R_1 R_2 R_3 R_4)^{-s_1} R^{-(1-s_2)} \int \prod_{i=1}^4 F_i \cdot G\, dx\, dt$$
$$\le (R_1 R_2 R_3 R_4)^{-s_1} R^{-(1-s_2)} \prod_{i=1}^4 \|F_i\|_5 \cdot \|G\|_5 . \tag{26}$$

Assume $s_1 < \frac{1}{2} < s_2$ chosen such that $s_1, 1 - s_2 > \frac{9}{20}$ and $\rho, 1-\rho > \frac{9}{20}$. It follows from (23) that $\|F_i\|_5 \le C R_i^{9/20} \|c\|_2 (1 \le i \le 4)$ and $\|G\|_5 \le C R^{9/20}$, so that by (18)

$$(26) \le (R_1R_2R_3R_4)^{\frac{9}{20}-s_1} R^{-\frac{11}{20}+s_2} |||u|||^4_{s_1} \tag{27}$$

which is summable for dyadic values of R_i, R.

Considering a small time interval I, $|I| = \sigma$, there is an extra saving of σ^c, for some $c > 0$, which is inequality (10). Consider functions u which are supported on a 2σ neighborhood of 0. It follows from the definition of the norm (7), in particular the $||\lambda| - |\xi||^\rho$-multiplier, that localizing (3),(4) to I will affect the $|||\ |||_{s_2}$-norm by a factor $\left(\frac{1}{\sigma}\right)^{\rho-\frac{1}{2}}$ $\left(\rho > \frac{1}{2}\right)$. On the other hand, repeating the previous L^5-estimate, one gets factors

$$\begin{aligned} &R^{9/20} \left\| \frac{c(\xi,\lambda)}{||\xi| - |\lambda||^{\rho - \frac{9}{20}}} \chi_{|\xi|\sim R} \right\|_{L^2_{\xi,\lambda}} \\ &\quad = R^{9/20+s_1} \left\| ||\xi| - |\lambda||^{9/20} \, |\widehat{u}(\xi,\lambda)| \chi_{|\xi|\sim R} \right\|_{L^2_{\xi,\lambda}} \end{aligned}$$

which by interpolation are bounded by

$$\begin{aligned} &R^{\frac{9}{20}+s_1} \|\widehat{u}\chi_{|\xi|\sim R}\|_2^{1-\frac{9}{20\rho}} \left\| ||\xi| - |\lambda||^\rho |\widehat{u}(\xi,\lambda)| \chi_{|\xi|\sim R} \right\|_2^{\frac{9}{20\rho}} \\ &\quad \le R^{\frac{9}{20}+(1-\frac{9}{20\rho})s_1} \cdot \left(\|\widehat{u}(\xi)\|_{L^2_{|\xi|\sim R} L^2_t} \right)^{1-\frac{9}{20\rho}} \cdot \|c\|_2^{\frac{9}{20\rho}} . \end{aligned} \tag{28}$$

Since $\operatorname{supp} u \subset \mathbb{T}^d \times I$, $\|\widehat{u}(\xi)\|_{L^2_t} \le \sigma^{1/2} \|\widehat{u}(\xi)\|_{L^\infty_t} \le \sigma^{1/2} \| \, ||\lambda| - |\xi||^\rho |\widehat{u}(\xi,\lambda)| \, \|_{L^2_\lambda}$ by Hölder's inequality and (28) $\le R^{\frac{9}{20}} \sigma^{\frac{1}{2}(1-\frac{9}{20\rho})} \|c\|_2 \le R^{\frac{9}{20}} \sigma^{\frac{1}{20}} |||u|||_{s_1}$. For $\rho > \frac{1}{2}$ close enough to $\frac{1}{2}$, this clearly implies inequality (10) with the σ^c-factor. From the earlier discussion, Proposition I follows in case (5).

The proof of (6) is completely analogous. For the argument to work, one needs the exponent $q = \frac{2(d+1)}{d-1}$ from Strichart's inequality to fulfil the condition $q > k+1$, where k is the degree of $f(u;t,x)$ in u. Thus for $d \ge 3$, $\frac{d+3}{d-1} > k \ge 2$ only permits (6).

2. Behaviour of Solutions of Nonlinear Schrödinger Equations with Nonlinear Term of Degree At Most 3

As pointed out in [K2], the symplectic Hilbert space for the NLSE

$$iu_t = -\Delta u + \frac{\partial}{\partial \overline{u}} G(u, \overline{u}; t, x) \tag{1}$$

where G is a real valued smooth function periodic in $x \in \mathbb{T}^d$ is the space $L^2(\mathbb{T}^d)$. Consider dimension $d = 1$, G a polynomial in $u, \overline{u}$. At this point, there is only a local wellposedness theory in $L^2(\mathbb{T})$ for degree ≤ 4 (see [B1]) and this will be our assumption here. Thus G is a sum of following terms

$$au + \overline{au} \qquad \text{(linear)}$$

$$\begin{cases} au^2 + \overline{au}^2 \\ a_r|u|^2 \end{cases} \qquad \text{(degree 2)}$$

$$\begin{cases} au^3 + \overline{au}^3 \\ (au + \overline{au})|u|^2 \end{cases} \qquad \text{(degree 3)}$$

$$\begin{cases} au^4 + \overline{au}^4 \\ (au^2 + \overline{au}^2)|u|^2 \\ a_r|u|^4 \end{cases} \qquad \text{(degree 4)}$$

where a (resp. a_r) is a smooth (resp. real) function of x, t, periodic in x. In the case G has the form $f(|u|^2, x, t)$, the equation preserves the L^2-norm and the local wellposedness property is global.

In [B3], we considered equations of the form

$$iu_t + u_{xx} + a(x,t)u + b(x,t)u|u|^2 = 0 \tag{2}$$

and proved nonsqueezing of balls in cylinders, i.e.

$$S_t(B_r) \subset \mathbf{T}_R^{(k)} \Rightarrow R \geq r\ . \tag{3}$$

Here B_r is a translate of the L^2-ball of radius r

$$\{u \in L^2(\mathbf{T}) \mid \|u\|_2 < r\}$$

and $\mathbf{T}_R^{(k)}$ a translate of the cylinder defined with respect to the kth-element of the (Darboux) basis $\varphi_k(x) = \sqrt{2}\cos kx$ or $\sqrt{2}\sin kx$

$$\{u \in L^2(\mathbf{T}) \mid |\langle u, \varphi_k\rangle| < R\}\ .$$

S^t denotes the flow map of (1),(2). Recall that canonical coordinates are $\operatorname{Re} u$, $\operatorname{Im} u$ here.

The proof uses a certain (uniform) approximation property of solutions of (1), (2) by solutions of a "truncated" equations

$$iU_t + U_{xx} + P_N\left[\frac{\partial}{\partial\overline{U}} G(U, \overline{U}; t, x)\right] = 0 \tag{4}$$

where $U = P_N U$ and P_N stands for the usual Dirichlet projection on the space $[e^{ikx} \mid -N \leq k \leq N]$. The phase space for equation (4) is finite dimensional and the squeezing property follows from the symplectic capacity theory. To obtain previous approximation result, the conservation of $\int_{\mathbf{T}} |u|^2 dx$ is used in an essential way (besides getting a priori L^2-bounds) in the case of equation (2). This conservation fails if we allow a general degree ≤ 4 expression for G as described earlier. Our main purpose here is to study these more general equations. Essentially speaking, we show that there is an L^2-approximation of the solutions of the IVP's

$$\begin{cases} iu_t + u_{xx} + \frac{\partial}{\partial \overline{u}} G(u, \overline{u}, t, x) = 0 \\ u(0) = \phi \end{cases} \tag{5}$$

$$\begin{cases} iU_t + U_{xx} + P_N \left[\frac{\partial}{\partial \overline{U}} G(U, \overline{U}; t, x)\right] = 0 \\ U(0) = \phi \end{cases} \tag{6}$$

($U = P_N U$, $\phi = P_N \phi$) on a time interval $[0, T]$, provided there is no blowup and the data ϕ satisfies a condition on Fourier coefficient size

$$|\widehat{\phi}(k)| < \delta \qquad (N_0 < |k| < N) \tag{7}$$

where N depends on N_0, $\|u(t)\|_2$ for $t < T$ and the approximation ε

$$\|u(t) - U(t)\|_2 < \varepsilon \qquad (0 < t < T) \tag{8}$$

and δ will be any power N^{-c}, $c > 0$.

This fact will imply the absence of "uniform asymptotical stability" for $t \to \infty$, in the sense that for any ρ-ball B_ρ the diameter of the set $S_t(B_\rho)$ cannot tend to zero. We may derive this from the result of [EH] about nonsqueezing of a translate of the set

$$\left(\frac{\rho}{\sqrt{N}} B^2\right) \times \cdots \times \left(\frac{\rho}{\sqrt{N}} B^2\right) \qquad (N \text{ copies})$$

in a ball of radius ρ, considering here the (Hamiltonian) flow $S_N(t)$ corresponding to (6).[1]

We first recall some facts on the analysis of (1) ($n = 1$) with cubic nonlinearity (see [B1] for details). Consider an equation of the form

$$iu_t = u_{xx} + F(u, \overline{u}, x, t) \tag{9}$$

where F is smooth in x, t, periodic in x and a polynomial in $u, \overline{u}$ of degree ≤ 3 (not necessarily Hamiltonian). Consider a sufficiently small time interval $[0, \sigma]$, where σ depends on the L^2-norm size $\|\phi\|_2$ of the data $u(0) = \phi$. A wellposedness result on $[0, \sigma]$ may then be proved using Picard's contraction principle and the equivalent integral equation

$$u(t) = S(t)\phi + i \int_0^t S(t - \tau) w(\tau) d\tau \ ; \quad w(\tau) = F\big(u(\tau), \overline{u}(\tau), x, \tau\big) \tag{10}$$

where $S(t)$ is the unitary group solving the linear equation $iu_t = u_{xx}$. The fixpoint argument is applied in the space

[1] This property becomes in fact already evident in the discussion of solutions of (5) assuming (7).

$$|||u||| = \Big(\sum_n \int d\lambda \big(|\lambda - n^2| + 1\big)^{2\rho} \big|\widehat{u}(n,\lambda)\big|^2\Big)^{1/2} \tag{11}$$

assuming

$$u(x,t) = \sum_n \int d\lambda \widehat{u}(n,\lambda) e^{i(nx+\lambda t)} \quad \text{on} \quad \mathbf{T} \times [0,\sigma] .$$

In (11), the exponent ρ is chosen slightly larger than $\frac{1}{2}$, so that

$$|||u||| \geq \big\|u(t)\big\|_2 \quad \text{for} \quad t < \sigma \tag{12}$$

(by Hölder's inequality). The main estimate used in the analysis of the $||| \ |||$-norm of the nonlinear term in (10) is following L^4-inequality

$$\|u\|_{L^4(\mathbf{T}\times[0,1])} \leq c \Big(\sum \big(|\lambda - n^2| + 1\big)^{3/4} \big|\widehat{u}(n,\lambda)\big|^2\Big)^{1/2} \tag{13}$$

which is general (and sharp).

Write F as a sum of monomials and consider say the cubic terms $a\, u_1\, u_2\, u_3$ where $u_i = u$ or $\overline{u}$. Writing using Fourier transform

$$\int_0^t S(t-\tau)(au_1u_2u_3)(\tau)d\tau$$

$$= \sum_n e^{inx} \int d\lambda \widehat{(au_1u_2u_3)}(n,\lambda) \frac{e^{i\lambda t} - e^{in^2 t}}{\lambda - n^2} \tag{14}$$

the $||| \ |||$-norm of (14) may be estimated by

$$\Big[\sum_n \int d\lambda \frac{|\widehat{(au_1u_2u_3)}(n,\lambda)|^2}{|\lambda - n^2|^{2(1-\rho)}}\Big]^{1/2} \tag{15}$$

(denominators $|\lambda - n^2|$ will mean $|\lambda - n^2| + 1$ in the sequel). Define

$$c(n,\lambda) = \big(1 + |\lambda - n^2|\big)^{\rho} \cdot \big|\widehat{u}(n,\lambda)\big| . \tag{16}$$

Since $\widehat{(au_1u_2u_3)} = \widehat{a} * \widehat{u}_1 * \widehat{u}_2 * \widehat{u}_3$ (convolution), one may estimate by duality (15), taking $d(n,\lambda) \geq 0$, $\Sigma_n \int d\lambda\, d(n,\lambda)^2 \leq 1$

$$\sum_{\substack{n = n_0 + \epsilon_1 n_1 + \epsilon_2 n_2 + \epsilon_3 n_3 \\ \lambda = \lambda_0 + \epsilon_1 \lambda_1 + \epsilon_2 \lambda_2 + \epsilon_3 \lambda_3}} \int |\widehat{a}(n_0,\lambda_0)|$$

$$\frac{c_1(n_1,\lambda_1)}{|\lambda_1 - n_1^2|^{\rho}} \frac{c(n_2,\lambda_2)}{|\lambda_2 - n_2^2|^{\rho}} \frac{c(n_3,\lambda_3)}{|\lambda_3 - n_3^2|^{\rho}} \frac{d(n,\lambda)}{|\lambda - n^2|^{1-\rho}} . \tag{17}$$

Because a is smooth

$$\big|\widehat{a}(n_0,\lambda_0)\big| < \big(|n_0| + |\lambda_0|\big)^{-c} \tag{18}$$

where c is an arbitrary exponent.

Consider for fixed n_0, λ_0 the expression

$$\sum_{\substack{n=n_0+\varepsilon_1 n_1+\varepsilon_2 n_2+\varepsilon_3 n_3 \\ \lambda=\lambda_0+\varepsilon_1\lambda_1+\varepsilon_2\lambda_2+\varepsilon_3\lambda_3}} \int \frac{c(n_1,\lambda_1)}{|\lambda_1-n_1^2|^{3/8}} \frac{c(n_2,\lambda_2)}{|\lambda_2-n_2^2|^{3/8}} \frac{c(n_3,\lambda_3)}{|\lambda_3-n_3^2|^{3/8}} \frac{d(n,\lambda)}{|\lambda-n^2|^{3/8}} \tag{19}$$

and define functions $F_i = F_i(x,t)$, $G = G(x,t)$ letting

$$\widehat{F}_i(n,\lambda) = \frac{c_i(\varepsilon_i\, n, \varepsilon_i\lambda)}{|\lambda-n^2|^{3/8}} \quad \text{and } \widehat{G}(n,\lambda) = \frac{d(n,\lambda)}{|\lambda-n^2|^{3/8}} \, .$$

Then (19) $= \int (F_1F_2F_3G)dx\,dt$, bounded by $\|F_1\|_4\|F_2\|_4\|F_3\|_4\|G\|_4$. Hence, (13), (16) yield an estimate by $c|||u|||^3$.

Since the exponents $\rho, 1-\rho > \frac{3}{8}$, it follows from (18) that (17) $\leq c|||u|||^3$. In fact, since $\rho > \frac{3}{8}$, a choice of a time interval of small size σ will give an extra saving of a factor σ^c, thus

$$(17) < c\sigma^{c_1}|||u|||^3 \tag{20}$$

for some fixed constant $c_1 > 0$. Indeed, consider functions u supported by a 2σ-neighborhood of 0. Observe from the multiplier in (11) that multiplying a function (here given by (14)) with a localizing function in the t-variable to a σ-neighborhood will increase at most the $|||\ |||$-norm by a $\left(\frac{1}{\sigma}\right)^{\rho-\frac{1}{2}}$ factor.

On the other hand, repeating previous L^4-estimate, one gets factors

$$\left\| \frac{c(n,\lambda)}{|\lambda-n^2|^{\rho-\frac{3}{8}}} \right\|_{\ell^2_n L^2_\lambda} = \left\| |\lambda-n^2|^{3/8}|\widehat{u}(n,\lambda)| \right\|_{\ell^2_n L^2_\lambda}$$

which by interpolation and (13) is bounded by

$$\|\widehat{u}\|_2^{1-\frac{3}{8\rho}} \big\| |\lambda-n^2|^\rho |\widehat{u}(n,\lambda)| \big\|_2^{\frac{3}{8\rho}} \leq \sigma^{\frac{1}{4}\left(1-\frac{3}{8\rho}\right)} \|u\|_4^{1-\frac{3}{8\rho}} |||u|||^{\frac{3}{8\rho}} \leq \sigma^{\frac{1}{4}\left(1-\frac{3}{8\rho}\right)} |||u||| \, .$$

Similarly, considering difference expressions for functions u, v, previous reasoning leads to estimates of the form $c\sigma^{c_1}(|||u|||+|||v|||+1)^2|||u-v|||$, from where the contraction principle may be derived for sufficiently small σ.

These estimates appear in [B1] in the context of the cubic NLS $iu_t + u_{xx} \pm u|u|^2 = 0$.

The purpose of the next considerations is to verify which systems of frequencies (n_0, n_1, n_2, n_3) considering $\widehat{a}(n_0)\widehat{u}_1(n_1)\widehat{u}_2(n_2)\widehat{u}_3(n_3)$ will be significant in the $|||\ |||$-estimate of $\int_0^t S(t-\tau)(au_1u_2u_3)(\tau)d\tau$.

Further Analysis. Observe that since in (17) the exponents $\rho, 1-\rho > \frac{3}{8}$, there will be an extra saving of a factor B^{-c_2}, for some constant $c_2 > 0$, unless each of the factors $\lambda_1 - n_1^2, \lambda_2 - n_2^2, \lambda_3 - n_3^2, \lambda - n^2$ is bounded by B, hence

$$\left|(n_0+\varepsilon_1 n_1+\varepsilon_2 n_2+\varepsilon_3 n_3)^2 - \varepsilon_1 n_1^2 - \varepsilon_2 n_2^2 - \varepsilon_3 n_3^2\right| \leq |\lambda_0| + n_0^2 + B \, . \tag{21}$$

Consider a partial summation

$$\widetilde{F} = \sum_{(n_0,n_1,n_2,n_3)\in E} e^{i(n_0x+\lambda_0 t)}\widehat{u}_1(n_1)\widehat{u}_2(n_2)\widehat{u}_3(n_3) \tag{22}$$

where E is some index set with properties to be specified.

It follows from the preceding that $|||\int_0^t S(t-\tau)\widetilde{F}(\tau)d\tau|||$ may be bounded by an expression

$$B^4 \sum_{\substack{(n_0,n_1,n_2,n_3)\in E\\ (21)\text{ holds}}}$$

$$\bar{c}_1(n_1)\bar{c}_2(n_2)\bar{c}_3(n_3)\bar{d}(n_0+\varepsilon_1 n_1+\varepsilon_2 n_2+\varepsilon_3 n_3) + CB^{-c_2}|||u|||^3 \tag{23}$$

letting $\bar{c}_i(n_i) = c(n_i,\lambda_i)$ and $\bar{d}(n) = d(n,\lambda)$ for some λ_i (resp. λ) satisfying $|\lambda_i - n_i^2| < B$ (resp. $|\lambda - n^2| < B$). Our purpose is to exploit a small size property of the $\bar{c}_i(n_i)$, hence $|\widehat{u}(n,\lambda)|$, in estimating the first term in (23). In fact we estimate

$$\sum_{\substack{(n_0,n_1,n_2,n_3)\in E\\ (21)\text{ holds}}} \bar{c}_1(n_1)\,\bar{c}_2(n_2)\bar{c}_3(n_3)\ . \tag{24}$$

Assume as first condition on E the absence of indices (n_0,n_1,n_2,n_3) with

$$n_0 = 0 \tag{25}$$

and one of following cases

$$\varepsilon_1 = 1,\ \varepsilon_2 = 1,\ \varepsilon_3 = -1\ ;\ n_3 = n_1 \text{ or } n_2 \tag{26}$$

$$\varepsilon_1 = 1,\ \varepsilon_2 = -1,\ \varepsilon_3 = 1\ ;\ n_2 = n_1 \text{ or } n_3 \tag{27}$$

$$\varepsilon_1 = -1,\ \varepsilon_2 = 1,\ \varepsilon_3 = 1\ ;\ n_1 = n_2 \text{ or } n_3 \tag{28}$$

(unless $n_1 = n_2 = n_3$).

Fix n_3 say and estimate the number of pairs (n_1,n_2) satisfying (21). Thus we consider following quadratic expression in n_1, n_2

$$\begin{aligned}&(1-\varepsilon_1)n_1^2 + (1-\varepsilon_2)n_2^2 + 2\varepsilon_1\varepsilon_2 n_1 n_2 +\\ &2(\varepsilon_3 n_3 + n_0)(\varepsilon_1 n_1 + \varepsilon_2 n_2) + (\varepsilon_3 n_3 + n_0)^2 - \varepsilon_3 n_3^2\ .\end{aligned} \tag{29}$$

Assume all indices n_1,n_2,n_3 are bounded by N. We denote N^ε an arbitrary small power of N (appearing here as $\exp\frac{\log N}{\log\log N}$ from divisor numbers).

CASE $\varepsilon_1 = -1$, $\varepsilon_2 = -1$. This case reduces to lattice point counting on an ellipse $x^2+3y^2 = A$, $A \le N^2 + |\lambda_0| + n_0^2 + B$, and hence is a bound by N^ε (we assume N much larger than n_0, λ_0, B).

CASE $\varepsilon_1 = 1$, $\varepsilon_2 = 1$. (29) yields then

$$2(n_1+\varepsilon_3 n_3+n_0)(n_2+\varepsilon_3 n_3+n_0) - (\varepsilon_3 n_3 + n_0)^2 - \varepsilon_3 n_3^2\ . \tag{30}$$

Considering divisors, there is again a bound by N^ε unless $n_1+\varepsilon_3 n_3+n_0 = 0$

or $n_2+\varepsilon_3 n_3+n_0=0$ and $|(\varepsilon_3 n_3+n_0)^2+\varepsilon_3 n_3^2|\le|\lambda_0|+n_0^2+B$. Hence either

$$|n_3|\le|\lambda_0|+n_0^2+B \tag{31}$$

or

$$n_0=0\ ,\ \varepsilon_3=1 \quad \text{hence} \quad n_1=n_3 \text{ or } n_2=n_3$$

which is the excluded case (25), (26).

CASE $\varepsilon_1=1$, $\varepsilon_2=-1$. ($\varepsilon_1=-1$, $\varepsilon_2=1$ similar) (29) becomes then

$$2(n_1-n_2)(\varepsilon_3 n_3+n_0-n_2)+(\varepsilon_3 n_3+n_0)^2-\varepsilon_3 n_3^2\ . \tag{32}$$

Thus there is again a bound by N^ε unless $n_1=n_2$ or $n_2=n_0+\varepsilon_3 n_3$ and either (31) or $\varepsilon_3=1$, $n_0=0$. Since (27) is excluded, (31) only is possible.

The conclusion is that if E excludes (25)+(26), (25)+(27), (25)+(28) and

$$|n_3|\ge|\lambda_0|+n_0^2+B \tag{33}$$

is given, the number of pairs (n_1,n_2) satisfying (21) is at most N^ε.

Consider the terms in (24) with $|n_1|,|n_2|,|n_3|\ge|\lambda_0|+n_0^2+B$. Estimate by Hölder's inequality as

$$\|c\|_2\Big[\sum_{n_3(33)}\Big(\sum_{(n_0,n_1,n_2,n_3)\in E;(21)}\bar c_1(n_1)\bar c_2(n_2)\Big)^2\Big]^{1/2}\ . \tag{34}$$

Since for fixed n_3 the number of pairs (n_1,n_2) is at most N^ε

$$(34)\ll N^\varepsilon\|c\|_2\Big[\sum_{(n_0,n_1,n_2,n_3)\in E;(21)}\bar c_1(n_1)^2\bar c_2(n_2)^2\Big]^{1/2}\ . \tag{35}$$

Similarly, for fixed n_2 there are at most N^ε pairs (n_1,n_3) in the summation, so that

$$(35)\ll N^\varepsilon\|c\|_2^2\max_n \bar c_1(n)\ . \tag{36}$$

Consequently

$$(24)\ll N^\varepsilon\Big(\min_{i=1,2,3}\max_n \bar c_i(n)\Big)\|c\|_2^2\ . \tag{37}$$

Suppose now

$$\min\big(|n_1|,|n_2|,|n_3|\big)\le|\lambda_0|+n_0^2+B \tag{38}$$

say $|n_3|<|\lambda_0|+n_0^2+B$. Rewrite (21) as

$$\begin{aligned}(1-\varepsilon_1)n_1^2+(1-\varepsilon_2)n_2^2+2\varepsilon_1\varepsilon_2 n_1 n_2+2(n_0+\varepsilon_3 n_3)(\varepsilon_1 n_1+\varepsilon_2 n_2)\\ \le\big(|\lambda_0|+n_0^2+B\big)^2\ .\end{aligned} \tag{39}$$

CASE $\varepsilon_1=-1$, $\varepsilon_2=-1$. Clearly

$$|n_1|,|n_2|\le|\lambda_0|+n_0^2+B\ . \tag{40}$$

CASE $\varepsilon_1 = 1$, $\varepsilon_2 = 1$. Then (39) gives

$$|n_0 + n_1 + \varepsilon_3 n_3|\,|n_0 + n_2 + \varepsilon_3 n_3| \leq \left(|\lambda_0| + n_0^2 + B\right)^2$$

hence one of the following cases

$$(40)$$

$$n_0 + \varepsilon_1 n_1 + \varepsilon_3 n_3 = 0 \tag{41}$$

$$n_0 + \varepsilon_2 n_2 + \varepsilon_3 n_3 = 0\ . \tag{42}$$

CASE $\varepsilon_1 = 1$, $\varepsilon_2 = -1$. Then

$$|n_1 - n_2|\,|n_0 + \varepsilon_2 n_2 + \varepsilon_3 n_3| \leq \left(|\lambda_0| + n_0^2 + B\right)^2$$

and hence (40) or

$$n_1 = n_2 \tag{43}$$

or

$$(42)\ .$$

Considering these remaining cases (40)-(43), inequality (37) will be valid provided E excludes moreover following cases

$$\begin{cases} |n_1|, |n_2| \leq |\lambda_0| + n_0^2 + B \\ n_0 + \varepsilon_1 n_1 + \varepsilon_2 n_2 = 0 \\ \varepsilon_3 = 1 \end{cases} \tag{44}$$

and cyclic

$$\begin{cases} |n_1| \leq |\lambda_0| + n_0^2 + B \\ n_2 = n_3\ . \end{cases} \tag{45}$$

and cyclic.

Thus the cases to be avoided in E are
(25)+(26), (25)+(27), (25)+(28)
(44) etc.
(45) etc.

Considering an expression $F = a u_1 u_2 u_3$, a satisfying (18), we may as a consequence of the preceding write F as a sum of certain terms of the form

$$\Big(\int_{\mathbb{T}} a\, dx\Big)\Big(\int_{\mathbb{T}} |u|^2\, dx\Big) u \tag{46}$$

$$a_0 \left(\int |u|^2\right) P_K(u) \tag{47}$$

$$a \left(\int |u|^2\right) P_K \overline{u} \tag{48}$$

$$\left(\int a_0 |P_K\ u|^2\right) u \tag{49}$$

$$\left(\int a(P_K u)^2\right) u \tag{50}$$

$$\left(\int a(P_K \overline{u})^2\right) u \tag{51}$$

and an expression $\widetilde{F}$ that will satisfy

$$\left|\left|\left| \int_0^t S(t-\tau)\widetilde{F}(\tau)d\tau \right|\right|\right|$$
$$< B^4 N^\varepsilon \Big(\min_i \max_{n,\lambda} |\widehat{u}_i(n,\lambda)|\Big) |||u|||^2 + CB^{-c_2} |||u|||^3 \, . \tag{52}$$

Here $a_0 = a - \int a\, dx$ and K in the projection P_K depends on B and a. We assume $|n| < N$.

Approximation of solutions of Hamiltonian NLS with at most cubic nonlinearity. Consider

$$F(u, \overline{u}; t, x) = \frac{\partial}{\partial \overline{u}} G(u, \overline{u}; t, x) \tag{53}$$

where G is a sum of degree $1, 2, 3, 4$ terms in $u, \overline{u}$ as described earlier in this section. The corresponding terms for F are as follows

$$au + \overline{a}\overline{u} \longrightarrow \overline{a} \tag{54}$$

$$\begin{cases} au^2 + \overline{a}\overline{u}^2 \\ a_r |u|^2 \end{cases} \qquad \begin{cases} \overline{a}\overline{u} \\ a_r u \end{cases} \tag{55}$$

$$\begin{cases} au^3 + \overline{a}\overline{u}^3 \\ (au + \overline{a}\overline{u})|u|^2 \end{cases} \qquad \begin{cases} \overline{a}\overline{u}^2 \\ au^2 + 2\overline{a}\overline{u}u \end{cases} \tag{56}$$

$$\begin{cases} au^4 + \overline{a}\overline{u}^4 \\ au^3\overline{u} + \overline{a}\overline{u}^3 u \\ a_r |u|^4 \end{cases} \qquad \begin{cases} \overline{a}\overline{u}^3 \\ au^3 + 3\overline{a}\overline{u}^2 u \\ a_r |u|^2 u \, . \end{cases} \tag{57}$$

Recall that a_r stands for a real function. Fix B and choose K depending on the smooth coefficients a and B as above. The discussion done above for cubic expressions $F = a u_1 u_2 u_3$ may be done for lower degree too, so that (52) will hold after removal of certain specific parts of the expression. Looking at the formulas (55)-(57), we isolate the following contributions

$$\left(\int a_r\right) u \tag{58}$$

$$\left[\int (aP_K u + \overline{a} P_k \overline{u})\right] u \tag{59}$$

$$\overline{a}\int |u|^2 \tag{60}$$

$$\left[\int a(P_K\, u)^2 + \overline{a}(P_k\overline{u})^2\right]u \tag{61}$$

$$\left(\int a_r\right)\left(\int |u|^2\right)u \tag{62}$$

$$\left(\int a_r\cdot P_k u\cdot P_k\overline{u}\right)u \tag{63}$$

$$\overline{a}\left(\int |u|^2\right)\cdot P_k\overline{u} \tag{64}$$

$$a_r\left(\int |u|^2\right)\cdot P_k u\ . \tag{65}$$

Let $M > K$ and add to the list (58)-(65) the additional contributions of $F(P_M u, P_M\overline{u}; t, x)$. This clearly yields an (algebraic) decomposition

$$F(u,\overline{u};t,x) = \Omega_{M,u}(t)\cdot u + F_{M,u} + \widetilde{F}_{M,u} \tag{66}$$

where $\Omega_{M,u}(t)$ is a real function of t (depending on u), $F_M = P_{3M}F_M$ and $\widetilde{F}_M$ satisfies the estimate

$$\left|\left|\left|\int_0^t S(t-\tau)\widetilde{F}_M(\tau)d\tau\right|\right|\right|$$
$$< B^4N^{\varepsilon}\Big(\sup_{|n|>M}|n|^{\varepsilon}|\widehat{u}(n)|\Big)\big(1+|||u|||\big)^2 + CB^{-c_2}\big(1+|||u|||\big)^3\ . \tag{67}$$

Consider the Cauchy problem (5)

$$\begin{cases} iu_t + u_{xx} + F(u) = 0\\ u(0) = \phi\end{cases} \tag{68}$$

where $\phi = P_N\phi$ and satisfies (7), thus $|\widehat{\phi}(k)| < \delta$ ($N_0 < |k| < N$).

Assume $\|u(t)\|_2$ bounded on $[0,T]$, T given. We compare the solutions u, v of (68) and

$$\begin{cases} iv_t + v_{xx} + \Omega_{10M,v}(t)v + F_{M,v} = 0\\ v(0) = \phi\end{cases} \tag{69}$$

where

$$F(v,\overline{v};t,x) = \Omega_{10M,v}(t)v + F_{10M,v} + \widetilde{F}_{10M,v} \tag{70}$$

corresponds to the decomposition (66). Hence

$$\Omega_{10M,v}(t)\cdot v + F_{M,v}(v) = F(v) + F_{M,v} - F_{10M,v} - \widetilde{F}_{10M,v}. \tag{71}$$

Write the integral equations

$$u(t) = S(t)\phi + i\int_0^t S(t-\tau)F(u(\tau), \overline{u}(\tau); \tau, x)d\tau \tag{72}$$

and

$$v(t) = S(t)\phi + i\int_0^t S(t-\tau)F(v(\tau), \overline{v}(\tau); \tau, x)d\tau \tag{73}$$

$$+ i\int_0^t S(t-\tau)(F_{M,v} - F_{10M,v})(\tau)d\tau \tag{74}$$

$$- i\int_0^t S(t-\tau)\widetilde{F}_{10M,v}(\tau)d\tau\ . \tag{75}$$

Consider a time interval $[0,\sigma]$, σ depending on $\|u(t)\|_2$ $(0 < t < T)$. Substract (72) and (73) and estimate (74), (75). We use the $|||\ |||$-estimates given earlier leading to (20) and inequality (67). Thus clearly on $[0,\sigma]$

$$|||u-v||| \leq c\sigma^{c_1}|||u-v|||(1+|||u|||+|||v|||)^2 \tag{76}$$

$$+ c\sigma^{c_1}|||P_{10M}v - P_M v|||(1+|||v|||)^2 \tag{77}$$

$$+ B^4\Big(\sup_{|n|>10M} |n|^\varepsilon|\widehat{v}(n)|\Big)(1+|||v|||)^2$$

$$+ CB^{-c_2}(1+|||v|||)^3\ . \tag{78}$$

Write

$$|||P_{10M}v - P_M v||| \leq |||P_{10M}u - P_M u||| + 2|||u-v|||$$

and observe that by considering the (finite) sequence M_0, $10M_0$, $10^2M_0, \dots,$ $10^s M_0$, $M_0 = \max(N_0, K)$, one may clearly get an M-value at most $10^s M_0$ such that

$$|||P_{10M}u - P_M u||| < c\frac{|||u|||}{s^{1/2}}\ . \tag{79}$$

Assume $|n| > 10M$. Since $F_{M,v} = P_{3M}(F_{M,v})$, the n-th-Fourier coefficient $\widehat{v}(n) = \widehat{v}(n)(t)$ satisfies the equation

$$\begin{cases} i\frac{d\widehat{v(n)}}{dt} - n^2\widehat{v}(n) + \Omega_{10M,v}(t)\widehat{v}(n) = 0 \\ \widehat{v}(n)(0) = \widehat{\phi}(n)\ . \end{cases} \tag{80}$$

Thus, because in particular $M > N_0$, $|\widehat{\phi}(n)| < \delta$ by hypothesis. The fact that $\Omega_{10M,v}(t)$ is real implies that $|\widehat{v}(n)|^2$ is constant in time, hence

$$|\widehat{v}(n)| < \delta \quad \text{for} \quad |n| > 10M$$

$$\widehat{v}(n) = 0 \quad \text{for} \quad |n| > N\ . \tag{81}$$

Substituting (79), (81) in inequality (76)-(78) gives

$$|||u-v||| < c\sigma^{c_1}|||u-v|||(1+|||u|||+|||v|||)^2 + \left(\frac{1}{s^{1/2}} + \delta B^4 N^\varepsilon + B^{-c_2}\right)(1+|||u|||+|||v|||)^3 \,. \quad (82)$$

Recall that $\delta = N^{-c}$ for some $c > 0$. Choosing s and B suitably large, the coefficient of the second term in (82) may be made small. Thus (82) yields an estimate on $[0,\sigma]$

$$\|u(t)-v(t)\|_2 \leq c|||u-v||| < \kappa \quad (83)$$

(σ has to be taken small enough depending on $|||u|||$, hence $\|u(t)\|_2$, $t < T$). The number M will depend on N_0, κ and $N \gg M$.

Instead of (5), consider now the Cauchy problem (6), i.e.

$$\begin{cases} iU_t + U_{xx} + P_N F(U) = 0 \\ U(0) = \phi \end{cases} \quad (84)$$

where $U = P_N U$. The integral equation for U is now

$$U(t) = S(t)\phi + i\int_0^t S(t-\tau)P_N F\big(U(\tau),\overline{U}(\tau);\tau,x\big)d\tau \,. \quad (85)$$

Since $v = P_N v$, it also follows from (73)-(75) that

$$\begin{aligned} v(t) = S(t)\phi &+ i\int_0^t S(t-\tau)P_N F\big(v(\tau),\overline{v}(\tau);\tau,x\big)\, d\tau \\ &+ i\int_0^t S(t-\tau)(F_{M,v} - F_{10M,v})(\tau)d\tau \\ &- i\int_0^t S(t-\tau)P_M\widetilde{F}_{10M,v}(\tau)d\tau \,. \end{aligned} \quad (86)$$

The same argument as for u now permits by subtraction of (85),(86), establishing the approximation on $[0,\sigma]$

$$\|U(t)-v(t)\|_2 < \kappa \,. \quad (87)$$

Hence, from (83), (87)

$$\|u(t)-U(t)\|_2 < \kappa \quad \text{for} \quad 0 < t < \sigma \,. \quad (88)$$

To cover the whole interval $[0,T]$, we partition in $T\sigma^{-1}$ intervals of length σ and repeat previous approximation. At the second step, we redefine N_0 as $N_1 = 10M$ and replace $u(t_1)$, $U(t_1)$ by $\phi_1 = v(t_1)$, v given by equation (69), which is now the new data. The perturbation is at most κ, from (87). By (81), one has again

$$|\widehat{\phi}_1(n)| < \delta \quad \text{for} \quad |n| > N_1 \,. \quad (89)$$

The continuation of the process is clear. For sufficiently good approximations at each stage, one ensures (8) at the end. The final size condition on N will relate to N_0, T, ε and $\|u(t)\|_2$ for $t < T$ (determining σ).

As a conclusion of the preceding, we get following

PROPOSITION 1. *Consider the Cauchy problems (with G as above) on time interval $[0,T]$*

$$\begin{cases} iu_t + u_{xx} + \frac{\partial}{\partial \bar u} G(u, \bar u; t, x) = 0 \\ u(0) = \phi \end{cases} \tag{90}$$

$$\begin{cases} iU_t + U_{xx} + P_N \left[\frac{\partial}{\partial \bar U} G(U, \bar U; t, x)\right] = 0 \\ U(0) = \phi \end{cases} \tag{91}$$

where $\phi = P_N \phi$ and satisfies

$$\|\phi\|_2 < C \quad \text{and} \quad |\widehat{\phi}(n)| < N^{-c} \quad \text{for} \quad |n| > N_0 \tag{92}$$

for some $C, c > 0$. Assume the solution u of (90) does not blowup on $[0,T]$ and

$$\|u(t)\|_2 < A \quad \text{for} \quad 0 \le t \le T\,. \tag{93}$$

Let $\varepsilon > 0$ and assume

$$N > N(N_0, A, \varepsilon, T)\,. \tag{94}$$

Then the solution U of (91) is an ε-approximation of u on $[0,T]$, i.e.

$$\big\|u(t) - U(t)\big\|_2 < \varepsilon \quad \text{for} \quad 0 \le t \le T\,. \tag{95}$$

(In fact, (94) *is a condition on the ratio $\frac{N}{N_0}$ to be sufficiently large depending on A, ε, T).*

COROLLARY 1. *Assume* (90) *wellposed for ϕ in an L^2-ball B_ρ of radius $\rho > 0$, for $0 \le t \le T$. Denote S_t the flowmap. Then*

$$\operatorname{diam} S_t(B_\rho) > \rho \quad \text{for} \quad 0 \le t \le T. \tag{96}$$

Proof: Denote ϕ_0 the center of B_ρ and assume $\operatorname{supp} \widehat{\phi_0} \subset [-N_0, N_0]$ (by approximation). Define the sets of trigonometric polynomials

$$R_{N,\kappa} = \left\{ \phi \mid \operatorname{supp} \widehat{\phi} \subset [-N, N] \quad \text{and} \quad |\widehat{\phi}(n)| < \frac{\kappa}{\sqrt{2N+1}} \right\}\,. \tag{97}$$

Thus $R_{N,\kappa} \subset B(0,\kappa)$ for each N. By the results of Ekeland and Hofer (see [EH]), there is no symplectic embedding in $2(2N+1)$-dimensional phase space of $R_{N,\kappa}$ in $B_{2(2N+1)}(0,\kappa')$ for $\kappa > \kappa'$ (balls are defined wrt ℓ^2-norm). The same statement holds for translates of these sets.

Choose $\varepsilon > 0$ and N satisfying (94). Here (93) results from the wellposedness hypothesis on B_ρ. Observe that $\phi_0 + R_{N,\rho} \subset B_\rho$. Let $0 < t < T$ and denote S_t^N the flowmap corresponding to the truncated equation (91). Since $\phi \in \phi_0 + R_{N,\rho}$ satisfies (92), with $c = \frac{1}{2}$, and $S_t \phi$ satisfies (93), it follows that $\|S_t\phi - S_t^N \phi\|_2 < \varepsilon$ $(t < T)$ uniformly for $\phi_0 + R_{N,\rho}$. The (finite dimensional) symplectic nonsqueezing property mentioned above implies that $S_t^N(\phi_0 + R_{N,\rho}) \not\subset B(S_t\phi_0, \rho-)$. Hence there is $\phi \in \phi_0 + R_{N,\rho}$ such that $\|S_t\phi - S_t\phi_0\|_2 > \|S_t^N\phi - S_t\phi_0\|_2 - \varepsilon > \rho - \varepsilon$, implying $\operatorname{diam} S_t(B_\rho) > \rho - \varepsilon$. The claim follows letting $\varepsilon \to 0$.

COROLLARY 2. *Assume the solution of (90) exists for all time. Then* $\operatorname{diam} S_t(B_\rho) \not\to 0$ *for* $t \to \infty$, *for any neighborhood* $B_\rho = B(\phi, \rho)$ *of the initial data* ϕ *(failure of uniform asymptotic stability for* $t \to \infty$*).*

Proof: In case the statement fails, there is T_0 such that $\operatorname{diam} S_t(B_\rho) < 1$ for $t > T_0$. Hence, by the L^2-wellposedness of NLS with cubic nonlinearity in L^2, there is some $0 < \rho' < \rho$ such that (90) is globally wellposed on $B_{\rho'} = \phi + B(0, \rho')$. It follows thus from Corollary 1 that $\operatorname{diam} S_t(B_\rho) > \operatorname{diam} S_t(B_{\rho'}) \geq \rho' > 0$ for all t.

Remarks: (1) Assume $G(u, \overline{u}; t, x)$ does not contain monomials of the form $a\,\overline{u}$, $a|u|^2\overline{u}$. Consider the IVP

$$\begin{cases} iu_t + u_{xx} + \frac{\partial}{\partial \overline{u}} G = 0 \\ u(0) = \phi \end{cases} \tag{98}$$

and assume $\phi = P_N\phi$, $\|\widehat{\phi}\|_\infty < N^{-c}$ for some $c > 0$. Then, for $N > N(T, \varepsilon)$, (98) is wellposed on $[0, T]$ and the solution $u(t)$ satisfies

$$\Big(\sum \big||\widehat{u(t)}(n)| - |\widehat{\phi}(n)|\big|^2\Big)^{1/2} < \varepsilon\,. \tag{99}$$

Let

$$\frac{\partial}{\partial \overline{u}} G = \Omega_{M,u}(t)u + F_{M,u} + \widetilde{F}_{M,u} \tag{100}$$

be the decomposition considered earlier and compare u with the solution v of

$$\begin{cases} iv_t + v_{xx} + \Omega_{M,v}(t)v = 0 \\ v(0) = \phi\,. \end{cases} \tag{101}$$

Recall that since $\Omega_{M,v}(t)$ is real, $|\widehat{v}(n)| = |\widehat{\phi}(n)|$ for all n.

It follows from (98), (100), (101) that (locally)

$$\begin{aligned} |||u - v||| \leq & \left|\left|\left| \int_0^t S(t-\tau)F_{M,v}(\tau)d\tau \right|\right|\right| \\ & + \left|\left|\left| \int_0^t S(t-\tau)\widetilde{F}_{M,v}(\tau)d\tau \right|\right|\right| + c\sigma|||u - v||| \end{aligned} \tag{102}$$

and for the hypothesis on G and (67)

$$\begin{aligned}|||u-v||| &\le |||P_M v||| + M^4 N^\varepsilon \sup_n |\widehat{v}(n)| + M^{-c_2}\\ &\le |||P_M\phi|||_2 + M^4N^{\varepsilon-c} + M^{-c_2}\\ &< M^{1/2}N^{-c} + M^4N^{\varepsilon-c} + M^{-c}\\ &< N^{-c'}\end{aligned}$$

for appropriate choice of M. Hence also

$$\Big(\sum \big(|\widehat{u(t)}(n)| - |\widehat{\phi}(n)|\big)^2\Big)^{1/2} = \Big(\sum \big(|\widehat{u(t)}(n)| - |\widehat{v}(t)(n)|\big)^2\Big)^{1/2} \le \|u(t)-v(t)\|_2 < N^{-c'} . \tag{103}$$

Since in particular $\|u(t)\|_2$ remains bounded, the statement follows.

(2) The limitation to $G(u,\overline{u};t,x)$ of degree ≤ 4 in $u,\overline{u}$ is due to the fact that an L^2-analysis n only developed for NLS with degree ≤ 3 nonlinearity (cf. [B1]). On the other hand, if L^2 is replaced by the space H^s, $s>\frac{1}{2}$, one has a local wellposedness theorem for any $F(u,\overline{u};t,x)$ which is a polynomial in $u,\overline{u}$ say. One considers now the norm

$$|||u|||_s = \Big[\sum_n (1+|n|^2)^s \int d\lambda (1+|\lambda-n^2|)^{2\rho} |\widehat{u}(n,\lambda)|^2\Big]^{1/2} . \tag{104}$$

Letting again

$$c(n,\lambda) = (1+|n|)^s \cdot (1+|\lambda-n^2|^\rho)|\widehat{u}(n,\lambda)| \tag{105}$$

the estimation of $\big|\big|\big| \int_0^t S(t-\tau)\, F(\tau)d\tau \big|\big|\big|$ reduces to bounding the expression

$$\sum_{\substack{n=n_0+\Sigma\varepsilon_i n_i\\ \lambda=\lambda_0+\Sigma\varepsilon_i\lambda_i}} \int \frac{|n|^s d(n,\lambda)}{|\lambda-n^2|^{1-\rho}} \prod_i \frac{c(n_i,\lambda_i)}{|n_i|^s|\lambda_i-n_i^2|^\rho} \quad (\|d\|_2\le 1) . \tag{106}$$

Assume $|n_1| = \max|n_i|$, hence $|n_1|\ge c|n|$. Replacing (106) by

$$\sum_{\substack{n=n_0+\Sigma\varepsilon_i n_i\\ \lambda=\lambda_0+\Sigma\varepsilon_i\lambda_i}} d(n,\lambda)c(n_1,\lambda_1) \prod_{i\ge 2} \frac{c(n_i,\lambda_i)}{|n_i|^s|\lambda_i-n_i^2|^\rho} \tag{107}$$

one immediately gets an estimate by $\Pi_i\|c_i\|_2$. In fact, there is clearly a saving if

$$\max_{i\ge 2}|n_i| \tag{108}$$

or

$$\max\big(|\lambda-n^2|, |\lambda_i-n_i^2|\big) \tag{109}$$

is large. Thus up to a small additional contribution B^{-c}, we may assume

$$|n|\approx|n_1| , \qquad \max_{i\ge 2}|n_i| < B \tag{110}$$

$$\left|\left(\varepsilon_1 n_1 + \sum_{i\geq 2} \varepsilon_i n_i + n_0\right)^2 - \varepsilon_1 n_1^2 - \sum_{i\geq 2} \varepsilon_i n_i^2\right| < B \tag{111}$$

and thus, from (110),(111)

$$|n_1| \leq B \tag{112}$$

or

$$\varepsilon_1 = 1\,, \qquad \sum_{i\geq 2} \varepsilon_i n_i + n_0 = 0\,. \tag{113}$$

This means that the contributing part of $F(u,\overline{u};t,x) = \frac{\partial}{\partial \overline{u}} G(u,\overline{u};t,x)$ in evaluating $\left|\left|\left| \int_0^t S(t-\tau)F(\tau)d\tau \right|\right|\right|$ comes from

$$\left[\int \frac{\partial^2}{\partial u\, \partial \overline{u}} G(P_B u, P_B \overline{u}; t, x) dx\right] u \tag{114}$$

$$F(P_B u, P_B \overline{u}; t, x)\,. \tag{115}$$

Hence, there is a good approximation of u by the solution of

$$\begin{cases} iv_t + v_{xx} + \Omega_{M,v}(t)v + P_M F(P_M v, P_M \overline{v}; t, x) = 0 \\ v(0) = u(0) \end{cases} \tag{116}$$

on $[0,T]$, provided $\|u(t)\|_{H^s}$ remains controlled on $[0,T]$. Here $\Omega_{M,v}(t)$ is the real function appearing in (114) and M depends on $\|u(t)\|_s$, T and the approximation. (There is no smallness hypothesis on the Fourier coefficients of $u(0)$ here.)

From (116), the failure of uniform asymptotic stability in H^s, $s > \frac{1}{2}$ is obtained directly.

PROPOSITION 2. *Bounded solutions in H^s $\left(s > \frac{1}{2}\right)$ of a Hamiltonian NLSE $iu_t + u_{xx} + \frac{\partial}{\partial \overline{u}} G(u,\overline{u};t,x) = 0$, G a real polynomial in $u, \overline{u}$, are not uniformly asymptotically stable for $t \to \infty$.*

3. On the Behaviour of Higher Sobolev Norms for Large Time in Nonlinear Hamiltonian PDE's

The main purpose of this section is to exhibit smooth global solutions of smooth Hamiltonian PDE's with periodic boundary conditions (on $\mathbf{T}$) that develop large $\|u(t)\|_{H^{s_0}(\mathbf{T})}$ norm for $t \to \infty$. Here the exponent s_0 will be some fixed constant. Thus the Hamiltonian property does not imply necessarily bounds on higher order derivatives. The basic construction used is a general (small) perturbation argument of linear equations with resonant or almost resonant spectrum (under appropriate nonresonance conditions, the KAM results would make this impossible). Such an argument was worked out in detail in [B4] for certain KdV-type equations

$$u_t + u_{xxx} + u_x f(u) = 0 \ . \tag{1}$$

The examples considered here are nonlinear Schrödinger equations (NLSE), obtained as small Hamiltonian perturbation of a linear Schrödinger equation

$$iu_t = u_{xx} + V(x)u \tag{2}$$

with V a smooth real periodic potential. In fact the NLS case is easier and will be used as a model to illustrate certain phenomena.

Consider a nonlinear equation

$$-iu_t = -u_{xx} + V(x)u + \varepsilon\Gamma(u) \ . \tag{3}$$

Here V is a real smooth periodic potential. Denote (λ_n) the periodic spectrum of $-\frac{d^2}{dx^2} + V(x)$ and (φ_n) the corresponding normalized eigenfunctions basis (taking into account possible multiplicity), $\varepsilon > 0$ is a small parameter. We consider a (not necessarily local) nonlinear map $\Gamma : L^2(\mathbf{T}) \to L^2(\mathbf{T})$, such that Γ has bounded Lipschitz constant, thus

$$\big\|\Gamma(\phi)\big\|_2 \leq c(1 + \|\phi\|_2) \quad \text{and} \quad \big\|\Gamma(\phi) - \Gamma(\psi)\big\|_2 \leq c\|\phi - \psi\|_2 \ . \tag{4}$$

More specifically Γ will be of the form

$$\Gamma(u) = \frac{\partial}{\partial\, \overline{u}}\Big\{\sum_j \int_{\mathbf{T}} F_j(P_j u, P_j \overline{u})\Big\} \tag{5}$$

where the P_j are projections on parts of the eigenfunction basis (φ_n) and the F_j real smooth functions.

Denoting $S(t)\phi = \Sigma\langle\phi, \varphi_n\rangle\varphi_n e^{i\lambda_n t}$ the solution operator to the linear problem

$$-iu_t = -u_{xx} + V(x)u \ ; \quad u(0) = \phi \ . \tag{6}$$

We may rewrite (3) in integral equation form

$$u(t) = S(t)\phi + i\varepsilon \int_0^t S(t-\tau)\Gamma(u)(\tau)d\tau \ . \tag{7}$$

Since the group $S(t)$ are L^2-isometries, if follows from (4), letting $\varepsilon|t| \ll 1$,

$$\big\|u(t) - S(t)\phi\big\|_2 \leq \varepsilon \int_0^t \big\|\Gamma(u(\tau))\big\|_2 d\tau \leq C\varepsilon t\Big(1 + \max_{0<\tau<t} \|u(\tau)\|_2\Big) \tag{8}$$

$$\big\|u(t) - S(t)\phi\big\|_2 \leq C\varepsilon t(1 + \|\phi\|_2) \leq C\varepsilon t \ . \tag{9}$$

Writing

$$u(t) = S(t)\phi + i\varepsilon \int_0^t S(t-\tau)\Gamma\big(S(\tau)\phi\big)d\tau$$
$$+ i\varepsilon \int_0^t S(t-\tau)\big[\Gamma(u(\tau)) - \Gamma(S(\tau)\phi)\big]d\tau \tag{10}$$

it follows thus from (4) and (9) that the last term in (10) is at most $C(\varepsilon t)^2$ in L^2. Hence, in L^2

$$u(t) = S(t)\phi + i\varepsilon \int_0^t S(t-\tau)\Gamma\big(S(\tau)\phi\big)d\tau + 0(\varepsilon^2 t^2)\ . \tag{11}$$

Projecting on a basis element φ_{n_0} gives

$$\big|\langle u(t) - \phi, \varphi_{n_0}\rangle\big| = \varepsilon \left| \int_{\mathbf{T}} \int_0^t \Gamma\big(S(\tau)\phi\big)\varphi_{n_0}(x)e^{-i\lambda_{n_0}\tau}dx\, d\tau\right| + 0(\varepsilon^2\, t^2)\ . \tag{12}$$

Assume $V = 0$. Then $\lambda_n = n^2$, $\varphi_n = \sqrt{2}\cos nx$, $\sqrt{2}\sin nx$ and $(S(t)\phi)(x)$ is periodic in x and t. Hence the integral term in (12) equals

$$t \iint_{\mathbf{T}^2} \Gamma\big(S(\tau)\phi\big)e^{i(nx-n^2\tau)}dx\, d\tau + 0(1)\ . \tag{13}$$

Assume

$$\frac{1}{t} \ll \int_{\mathbf{T}^2} \Gamma\big(S(\tau)\phi\big)e^{i(nx-n^2\tau)}dx\, d\tau \sim \varepsilon t\ . \tag{14}$$

From (12),(13) it follows now

$$\big|\widehat{u(t)}(n_0) - \widehat{\phi}(n_0)\big| \sim \left[\int_{\mathbf{T}^2} \Gamma(S(\tau)\phi)e^{i(n_0x-n_0^2\tau)}dx\, d\tau\right]^2\ . \tag{15}$$

Choose $\Gamma(u)$ of the form

$$\frac{\partial}{\partial\overline{u}}\left\{\frac{1}{K^2}\int_{\mathbf{T}} \sin KF(Pu, P\overline{u})\right\} = -\frac{1}{K}P\left\{(\cos KF)\cdot \frac{\partial F}{\partial \overline{U}}(U, \overline{U})\right\}\ , \tag{16}$$
$$U = Pu$$

where F is a real smooth function, P a projection on the eigenfunction basis and K a suitably chosen large number. Assume $\widehat{P}(n_0) = 1$. Writing

$$\cos KF = \tfrac{1}{2}(e^{iKF} + e^{-iKF})$$

the integral in (15) will lead to oscillatory integrals

$$\int_{\mathbf{T}^2} e^{i(\pm KF(S(\tau)P\phi, \overline{S(\tau)P\phi}) + n_0x - n_0^2\tau)} \frac{\partial F}{\partial\overline{U}}\big(S(\tau)P\phi, \overline{S(\tau)P\phi}\big)dx\, d\tau\ . \tag{17}$$

The main idea now is to exploit critical points of the phase function. Assume

$$F\big(S(\tau)P\phi(x), \overline{S(\tau)P\phi(x)}\big)$$

as a function of $(x,\tau) \in \mathbf{T}^2$ has a nondegenerate critical point. For K

sufficiently large (mainly depending on n_0), there will be a solution $(\overline{x}, \overline{t})$ of the equation

$$\nabla F(S(\tau)P\phi, \overline{S(\tau)P\phi}) = \left(\mp\frac{n_0}{K}, \pm\frac{n_0^2}{K}\right) \tag{18}$$

and one expects a contribution $0(\frac{1}{K})$ of this stationary point $(\overline{x}, \overline{t})$ to the integral (17). If this contribution is not cancelled out by the contributions of other possible stationary points, the expression (15) will yield $0(K^{-4})$. This amounts to

$$\left|\widehat{u(t)}(n_0) - \widehat{\phi}(n_0)\right| \geq |n_0|^{-8} \quad \text{hence} \quad \left|\widehat{u(t)}(n_0)\right| \geq |n_0|^{-8} \tag{19}$$

assuming n_0 large.

In the preceding K is chosen depending on n_0. One then chooses ε such that the perturbation $\varepsilon\Gamma$ in (3) has bounded (or small) derivatives up to any specified order (assuming F smooth). Finally, one picks a time t such that (14) is satisfied. Observe also that one expects the preceding to be true for data ϕ which are "generic".

The conservation of the Hamiltonian

$$\int_{\mathbf{T}} |u_x|^2 + \frac{\varepsilon}{K^2}\int_{\mathbf{T}} \sin KF(Pu, \overline{Pu}) \tag{20}$$

implies an a priori bound on $\|u(t)\|_{H^1}$. Analyzing the IVP, it follows that for data $\phi \in H^s(\mathbf{T})$, $s \geq 1$, there is global wellposedness and the solution u satisfies $u(t) \in H^s$ for all time, with bounds possibly depending on t.

Next, we make the previous discussion more precise. Let f be a smooth real function and denote Φ its primitive, $\Phi'(0) = 0$. Since $\frac{\partial}{\partial \overline{u}} \int_{\mathbf{T}} \Phi(\text{Re}Pu) = Pf(\text{Re}Pu)$, we take $\Gamma(u) = Pf(\text{Re}Pu)$ as Hamiltonian perturbation in (3), provided (4) holds. Let $f(t) = \frac{1}{K}\sin Kt$ and P a projection on an eigenspace containing $\{e^{\pm im_0x}, e^{\pm in_0x}\}$. Take for the initial data

$$\phi = (a + ib)\cos m_0 x \tag{21}$$

hence

$$S(\tau)P\phi = (a + ib)\cos m_0x \cdot e^{im_0^2\tau}$$

and

$$\text{Re}\, S(\tau)P\phi = \cos m_0x(a\cos m_0^2\tau - b\sin m_0^2\tau)\ . \tag{22}$$

Substituting (22), the integral (17) becomes

$$\int_{\mathbf{T}^2} e^{i[\pm K\cos m_0x(a\cos m_0^2\tau - b\sin m_0^2\tau) + n_0x - n_0^2\tau]}dx\, d\tau\ . \tag{23}$$

Assume $n_0 = n \cdot m_0$. By a change of variable $x' = m_0x$, $\tau' = m_0^2\tau$, (23) yields

$$\int_{\mathbf{T}^2} e^{i[\pm K\cos x(a\cos\tau - b\sin\tau) + nx - n^2\tau]}dx\, d\tau\ . \tag{24}$$

Write $a\cos\tau - b\sin\tau = (a^2+b^2)^{1/2}\cos(\tau+\varphi)$, $\varphi = \varphi(a,b)$. Another change of variable gives

$$e^{in^2\varphi}\int_{\mathbf{T}^2} e^{i[\pm K\sqrt{a^2+b^2}\cos x\cdot\cos\tau + nx - n^2\tau]}dx\,d\tau\ . \tag{25}$$

The stationary points are given by

$$\begin{cases} \sin x\cdot\cos\tau = \pm\frac{n}{K\sqrt{a^2+b^2}} \\ \cos x\cdot\sin\tau = \mp\frac{n^2}{K\sqrt{a^2+b^2}}\ . \end{cases} \tag{26}$$

Assuming K large with respect to n^2 (hence n_0^2), solutions of (26) will be close to multiples of $\frac{\pi}{2}$. Assuming n a multiple of 4, (26) will therefore essentially be

$$e^{in^2\varphi}\int_{\mathbf{T}^2} e^{\pm iK\sqrt{a^2+b^2}\cos x\cdot\cos\tau}dx\,d\tau \tag{27}$$

$$= e^{in^2\varphi}\Big\{\int_{\mathbf{T}^2} e^{\pm i\frac{K}{2}\sqrt{a^2+b^2}\cdot\cos\theta}d\theta\Big\}^2 = 0(K^{-1})\ . \tag{28}$$

Thus the solution u of

$$\begin{cases} -iu_t = -u_{xx} + \varepsilon\Gamma(u) \\ u(0) = \phi = (a+ib)\cos m_0 x \end{cases} \tag{29}$$

letting

$$\Gamma(u)=\frac{1}{K}\sin K\mathrm{Re}\,u \text{ or } \Gamma(u)=\frac{1}{K}P\big[\sin K\mathrm{Re}(Pu)\big]\ ,\ \widehat{P}(m_0)=1=\widehat{P}(n_0) \tag{30}$$

will satisfy for some time t

$$\big|\widehat{u(t)}(n_0)\big| > |n_0|^{-C} \tag{31}$$

where C is some numerical constant. The construction permits moreover to make the data ϕ and the perturbation $\varepsilon\Gamma(u)$ arbitrarily small and smooth. This exhibits the behaviour of solutions of certain NLS mentioned in the beginning of this section. The Sobolev exponent s_0 depends on C in (31) and is thus fixed.

Take $\Gamma(u) = \frac{1}{K}P[\sin K\mathrm{Re}(Pu)]$ in (30).

Observe then that by construction $\operatorname{supp}\widehat{u(t)} \subset \operatorname{supp}\hat{p}$ for the solution u of (29).

Then one may clearly piece together such examples considering a sequence of projections $\{p_j\}$ on disjoint subsets of the eigenfunction basis. By previous observation, the corresponding IVP's are decoupled and one gets a Hamiltonian perturbation $\Gamma(u)$ of the form (5) and smooth data $\phi = \Sigma\,\delta_j\cos m_j x$ such that the solution u of

$$\begin{cases} -iu_t = -u_{xx} + \Gamma(u) \\ u(0) = \phi \end{cases} \tag{32}$$

satisfies

$$\overline{\lim_{t\to\infty}} \|u(t)\|_{H^{s_0}} = \infty \,. \tag{33}$$

This solution u is obtained as $u = \Sigma u^j$ where $u^j = P_j u^j$ solves

$$\begin{cases} -iu_t^j = -u_{xx}^j + \varepsilon_j \Gamma_j(u^j) \\ u^j(0) = \phi_j = \delta_j \cos m_j x \end{cases} \tag{34}$$

as in (29)-(31), thus

$$\left|\widehat{u^j(t_j)}(n_j)\right| > |n_j|^{-C} \,. \tag{35}$$

The nonlinear term $\Gamma(u) = \Sigma \varepsilon_j \Gamma_j(u)$ is smooth.

COROLLARY 1. *There is a Hamiltonian NLS* $-iu_t = -u_{xx} + \Gamma(u)$ *with* $\Gamma(u)$ *smooth of the form (5) such that for some smooth data* ϕ, *the (global and smooth) solution* u *of the IVP (32) satisfies (33), for some Sobolev exponent* s_0.

In the next construction, we will exhibit a local nonlinearity $\Gamma(u)$, such that the flow map of $-iu_t = -u_{xx} + \Gamma(u)$, denoted $U(t)$, has the following property

$$\overline{\lim_{\delta\to 0}} \sup_{t>0} \|U(t)(\delta \cdot \cos x)\|_{H^{s_0}} = \infty \,. \tag{36}$$

Consider first (29)

$$\begin{cases} -iv_t^j = -v_{xx}^j + \varepsilon_j \Gamma_j(v^j) \\ v^j(0) = \delta_j = \delta_j \cos x \end{cases} \tag{37}$$

where $\Gamma_j(v)$ is now defined by

$$\Gamma_j(v) = \frac{1}{K_j^2} (\cos K_j \mathrm{Re}\, v) \chi_j(\mathrm{Re}\, v) \tag{38}$$

where $0 \leq \chi_j \leq 1$ is 0 on a neighborhood of 0 of size $\frac{1}{K_j^2}$, is 1 outside a neighborhood of 0 of size $\sim \frac{1}{K_j^2}$ and $|\chi_j'| < cK_j^2$. Thus Γ_j fulfils (4). For the integral (17) one gets instead of (23)

$$\int_{\mathbf{T}^2} e^{i[\pm K_j \delta_j \cos x \cos \tau + n_j x - n_j^2 \tau]} \chi_j(\delta_j \cos x \cos \tau) dx\, d\tau \,. \tag{39}$$

The evaluation $0(K_j^{-1})$ in (28) will be affected by an error bounded by

$$\mathrm{mes}\left[(x,\tau) \in \mathbf{T}^2 \,\Big|\, |\cos x \cdot \cos \tau| < \frac{1}{K_j^2 \delta_j}\right] \ll \frac{1}{K_j} \tag{40}$$

for K_j chosen large enough with respect to δ_j. Hence we keep the lower estimate $0(K_j^{-1})$ for (39). The $\frac{1}{K_j^2}$-factor in (38) yields now

$$\left|\widehat{v^j(t_j)}(n_j)\right| \geq K_j^{-3}$$

hence still

$$\left|\widehat{v^j(t_j)}(n_j)\right| > |n_j|^{-C} \tag{41}$$

for some fixed constant C. Define

$$\Gamma(u) = \sum \varepsilon_j \frac{1}{K_j^2} (\cos K_j \mathrm{Re}\, v) \chi_j(\mathrm{Re}\, v) \ . \tag{42}$$

Fix j and analyze the effect of a replacement of Γ_j in (37) by the sum (42) on the solution v^j for $0 \leq t \leq t_j$. Observe that from the conservation of the Hamiltonian

$$\int_{\mathbf{T}} |v_x^j|^2 + \frac{\varepsilon_j}{K_j^2} \int_{\mathbf{T}} D^{-1} \left[\cos K_j t \cdot \chi_j(t)\right] (\mathrm{Re}\, v^j) \tag{43}$$

there is the a priori bound

$$\left\| v - \widehat{v}(0) \right\|_\infty \leq \left\| v - \widehat{v}(0) \right\|_{H^1} < \delta_j + \varepsilon_j^{1/2} \ . \tag{44}$$

Also, since $\Gamma_j(v)$ is real in (37), $\frac{d}{dt}\left(\int_{\mathbf{T}} \mathrm{Re}\, v\right) = 0$, hence $\mathrm{Re}\, \widehat{v(t)}(0) = 0 = \widehat{v(0)}(0)$. Consequently $\|\mathrm{Re}\, v^j\|_\infty < \delta_j + \varepsilon_j^{1/2} \ll \frac{1}{K_{j'}^2}$ for $j' < j$ (by construction). Hence, for $j' < j$, $\chi_{j'}(\mathrm{Re}\, v^j) = 0$ and the addition of the $j' < j$ terms in (42) does not affect the solution of (37). Now, for $j' > j$, the $\varepsilon_{j'}$ may clearly be taken small enough so that on $[0, t_j]$ the solution v^j is only slightly perturbed. The conclusion of previous considerations is that the inductive construction may be done such that

$$\begin{cases} -iv_t^j = -v_{xx}^j + \varepsilon_j \Gamma_j(v^j) \\ v^j(0) = \delta_j \cos x \end{cases} \quad \text{and} \quad \begin{cases} -iu_t^j = -u_{xx}^j + \Gamma(u^j) \\ u^j(0) = \delta_j \cos x \end{cases}$$

have approximately the same solutions on $[0, t_j]$ and in particular, by (41)

$$\left| \widehat{(U(t_j)(\delta_j \cos x))}(n_j) \right| = \left| \widehat{u^j(t_j)}(n_j) \right| > |n_j|^{-C} \ . \tag{45}$$

Thus (36) follows. As corollary, one has

COROLLARY 2. *There is a Hamiltonian NLS* $-iu_t = -u_{xx} + \Gamma(u)$ *with* $\Gamma(u)$ *smooth and local, such that for all exponents* $s > 0$ *and all* $\delta > 0$, *the set*

$$\left\{ U(t)\phi \mid \|\phi\|_{H^s} < \delta, t > 0 \right\}$$

is unbounded in H^{s_0}, *for some fixed exponent* s_0.

The next construction is more KAM related and shows in particular the failure of Lyapounov's theorem on the persistency of 1-dimensional tori in ε-perturbed infinite dimensional Hamiltonian systems corresponding to an equation of the form (3), when the spectrum $\{\lambda_n\}$ of $-\frac{d^2}{dx^2} + V(x)$ satisfies an almost resonance assumption

$$|\lambda_{n_j} - m_j \lambda_{n_0}| < n_j^{-c_1} \tag{46}$$

and also some diophantine property

$$|\xi_1 \lambda_{n_0} + \xi_2 \lambda_{n_j}| \geq \kappa_j \big(|\xi_1| + |\xi_2|\big)^{-c_2} \quad \text{for all} \quad \xi_1, \xi_2 \in \mathbf{Z} \tag{47}$$

for some fixed n_0 and infinite sequence $\{n_j\}$, where c_1, c_2 are constants and $\kappa_j > 0$.

Remarks: (1) KAM theory on the perturbation of invariant tori may be developed in infinite dimensional systems (for finite dimensional tori). A general Melnikov-type theory with application to quasiperiodic solutions of PDE appears in [K1], [B5]. For an infinite dimensional version of Lyapounov's theorem on the persistency of periodic solutions, see [CW].

(2) There is an asymptotic formula (cf. [PT])

$$\lambda_n = \pi^2 n^2 + \sum_{j=0}^{j_1} c_j(V) n^{-j} + 0(n^{-j_1 - 1}) \tag{48}$$

for the periodic spectrum of a sufficiently smooth periodic potential V (depending on j_1). Thus an assumption such as (46),(47) is realistic, assuming say $\lambda_{n_0} = \pi^2$ and $c_j(V) = 0$ for j up to c_1. This fact was pointed out by S. Kuksin to the author.

We will prove the following fact.

PROPOSITION 3. *Assuming* (46), (47), *one may construct a smooth Hamiltonian perturbation* $\Gamma(u)$ *of the form* (5), *the smoothness depending on the exponent* c_1 *in* (46), *and a fixed Sobolev exponent* s_0 *such that*

$$\inf_{q \in \gamma} \sup_t \big\| u_{\varepsilon, q}(t) \big\|_{H^{s_0}} \to \infty \quad \text{for} \quad \varepsilon \to 0\ . \tag{49}$$

Here $u_{\varepsilon,q}$ *denotes the solution of the IVP*

$$\begin{cases} -iu_t = -u_{xx} + V(x) u + \varepsilon \Gamma(u) \\ u(0) = q \end{cases} \tag{50}$$

and γ *denotes a bounded set in* H^{s_0} *such that* $|\langle q, \varphi_{n_0} \rangle| > c$ *for* $q \in \gamma$.

In particular, there is no invariant torus in an H^{s_0}-neighborhood of the periodic solution $u_{0,\varphi_{n_0}} = a\varphi_{n_0}(x) e^{i\lambda_{n_0} t}$ of the linear equation, no matter how small ε.

The equation in (50) appears as

$$-iu_t = \frac{\partial}{\partial \bar{u}} H_\varepsilon(u)$$

where $H_\varepsilon(u) = \int_{\mathbf{T}} |u_x|^2 + \int_{\mathbf{T}} V(x)|u|^2 + \Sigma \delta_j \int_{\mathbf{T}} D^{-1} F_j(P_j \mathrm{Re}\, u)$, thus

$$\Gamma(u) = \sum \delta_j \Gamma_j(u) \quad \text{with} \quad \Gamma_j(u) = P_j F_j(\mathrm{Re}\, P_j u) \tag{51}$$

and we will again take F_j of the form $F_j(t) = \frac{1}{K_j} \cos K_j t$.

In particular, Γ will fulfil the Lipschitz estimate (4).

Fix j_0 and consider $\varepsilon > 0$ and times t_1, t all depending on j_0 (to be specified). We may from (11) write

$$u(t_1) = S(t_1)\phi + i\varepsilon \int_0^{t_1} S(t_1 - \tau)\Gamma\big(S(\tau)\phi\big)d\tau + 0\big((\varepsilon t_1)^2\big) \tag{52}$$

for the solution u of the IVP

$$\begin{cases} -iu_t = -u_{xx} + V(x)u + \varepsilon\Gamma(u) \\ u(0) = \phi\,. \end{cases} \tag{53}$$

Write

$$\int_0^{t_1} S(t_1-\tau)\Gamma(S(\tau)\phi)d\tau = \sum_{j<j_0} \delta_j \int_0^{t_1} S(t_1-\tau)\Gamma_j\big(S(\tau)\phi\big)d\tau$$
$$+ \sum_{j\geq j_0} \delta_j \int_0^{t_1} S(t_1-\tau)\Gamma_j\big(S(\tau)\phi\big)d\tau\,. \tag{54}$$

Let P_j be the projection on the eigenfunctions $\varphi_{n_0}, \varphi_{n_j}$. From the form (51) of Γ_j, it follows that (assuming ϕ real to simply the formulas)

$$\int_0^{t_1} S(t_1-\tau)\Gamma_j(S(\tau)\phi)d\tau =$$
$$\frac{1}{K_j}\varphi_{n_0}(x)e^{i\lambda_{n_0}t_1}\Big\{\int_{\mathbf{T}}\int_0^{t_1} \cos K_j\big[\langle\phi,\varphi_{n_0}\rangle\varphi_{n_0}(x)\cos\lambda_{n_0}\tau$$
$$+ \langle\phi,\varphi_{n_j}\rangle\varphi_{n_j}(x)\cos\lambda_{n_j}\tau\big]\cdot\varphi_{n_0}(x)e^{-i\lambda_{n_0}\tau}d\tau\, dx\Big\} \tag{55}$$
$$+ \frac{1}{K_j}\varphi_{n_j}(x)e^{i\lambda_{n_j}t_1}\Big\{\int_{\mathbf{T}}\int_0^{t_1} \cos K_j\big[\langle\phi,\varphi_{n_0}\rangle\varphi_{n_0}(x)\cos\lambda_{n_0}\tau$$
$$+ \langle\phi,\varphi_{n_j}\rangle\varphi_{n_j}(x)\cos\lambda_{n_j}\tau\big]\cdot\varphi_{n_j}(x)e^{-i\lambda_{n_j}\tau}d\tau\, dx\Big\}\,. \tag{56}$$

We will show first that for $j < j_0$ these terms are small (compared with t_1), because of the assumption (47). We will use that for $t_1 = t_1(j_0)$ sufficiently large, $\lambda_{n_0}\tau$ and $\lambda_{n_j}\tau$ behave almost like independent variables on $[0, t_1]$. We use following simple lemma proved by Fourier expansion.

LEMMA. *Consider a smooth function $f = f(\theta, \dots, \theta_k)$ on $\mathbf{T}^k$ and let $\lambda_1, \dots, \lambda_k \in \mathbf{R}$ be rationally independent numbers such that*

$$|\Sigma\xi_j\lambda_j| > \kappa\big(\Sigma|\xi_j|\big)^{-M} \tag{57}$$

for all $\xi_j \in \mathbf{Z}$. Here M is some constant and $\kappa > 0$ some small number. Then

$$\left|\frac{1}{T}\int_0^T f(\lambda_1\tau,\dots,\lambda_k\tau)d\tau - \int_{\mathbf{T}^k} f(\theta_1,\dots,\theta_k)d\theta\right| < \|f\|_{C^{M+k+1}}(T\kappa)^{-1} . \tag{58}$$

In the application to (55) say, we let $k=2$, $\lambda_1=\lambda_{n_0}$, $\lambda_2=\lambda_{n_j}$. Hence $\kappa=\kappa_j$, $M=C_2$ by (47). The function $f=f(\theta_1,\theta_2)$ is given by

$$f(\theta_1,\theta_2)=\cos K_j\big[\langle\phi,\varphi_{n_0}\rangle\varphi_{n_0}(x)\cos\theta_1+\langle\phi,\varphi_{n_j}\rangle\varphi_{n_j}(x)\cos\theta_2\big]e^{-i\theta_1} \tag{59}$$

(where x is fixed). Application of (58) for the τ-integration gives

$$\begin{aligned}&\int_0^{t_1}\cos K_j\big[\langle\phi,\varphi_{n_0}\rangle\varphi_{n_0}(x)\cos\lambda_{n_0}\tau+\langle\phi,\varphi_{n_j}\rangle\varphi_{n_j}(x)\cos\lambda_{n_j}\tau\big]e^{-i\lambda_{n_0}\tau}d\tau\\ &\quad=t_1\int_{\mathbf{T}^2}f(\theta_1,\theta_2)d\theta_1\,d\theta_2+0(K_j^{C_2+3}\kappa_j^{-1})\\ &\quad=0(K_j^{C_2+3}\kappa_j^{-1})\end{aligned} \tag{60}$$

since the $\mathbf{T}^2$-integral vanishes, as is clear from (59), making a variable change $(\theta_1,\theta_2)\mapsto(\theta_1+\pi,\theta_2+\pi)$. In case of complex data ϕ, the conclusion would be the same (there would be additional terms $\sin\theta_1,\sin\theta_2$). Hence (55) is bounded by $K_j^{C_2+2}\kappa_j^{-1}$ and so is (56). Hence the summation $\sum_{j<j_0}$ in (54) is at most $\sum_{j<j_0}\delta_jK_j^{C_2+3}\kappa_j^{-1}<K_{j_0-1}^{C_2+2}\kappa_{j_0-1}^{-1}$ (assuming the κ_j decreasing). Thus

$$\int_0^{t_1}S(t_1-\tau)\Gamma\big(S(\tau)\phi\big)d\tau=$$

$$\sum_{j\ge j_0}\delta_j\int_0^{t_1}S(t_1-\tau)\;\Gamma_j\big(S(\tau)\phi\big)d\tau+0(K_{j_0-1}^{C_2+2}\kappa_{j_0-1}^{-1}) \tag{61}$$

and the idea is to have the error term small compared with t_1. Denote

$$\Gamma^{j_0}=\sum_{j\ge j_0}\delta_j\Gamma_j\ . \tag{62}$$

We will compare the solution u of (53) with the solution v of the IVP

$$\begin{cases}-iv_t=-v_{xx}+V(x)v+\varepsilon\Gamma^{j_0}(v)\\ v(0)=\psi\end{cases} \tag{63}$$

on $[0,t_1]$.

From (52), (61)

$$u(t_1)=$$

$$S(t_1)\phi+i\varepsilon\int_0^{t_1}S(t_1-\tau)\Gamma^{j_0}\big(S(\tau)\phi\big)d\tau+0\big(\varepsilon K_{j_0-1}^{C_2+2}\kappa_{j_0-1}^{-1}+(\varepsilon t_1)^2\big) \tag{64}$$

and from (63)

$$v(t_1) = S(t_1)\psi + i\varepsilon \int_0^{t_1} S(t_1 - \tau)\Gamma^{j_0}\big(S(\tau)\psi\big)d\tau + 0\big((\varepsilon\, t_1)^2\big)\,. \tag{65}$$

Hence, by (64), (65), (62) and the Lipschitz bounds on the Γ_j

$$\big\|\Gamma^{j_0}(\phi_1) - \Gamma^{j_0}(\phi_2)\big\|_2 \leq \delta_{j_0}\|\phi_1 - \phi_2\|_2 \tag{66}$$

$$\big\|u(t_1) - v(t_1)\big\|_2 \leq (1 + C\varepsilon t_1\delta_{j_0})\|\phi - \psi\|_2 + 0\big(\varepsilon K_{j_0-1}^{C_2+2}\kappa_{j_0-1}^{-1} + (\varepsilon t_1)^2\big)\,. \tag{67}$$

Consider the interval $[0, t]$ now and subdivide in intervals of length t_1. Iterating the estimate (67) clearly yields the following comparison, if the data for $t = 0$ are both ϕ

$$\big\|u(t) - v(t)\big\|_2 \leq C\frac{t}{t_1}(1 + C\varepsilon\delta_{j_0}t_1)^{\frac{t}{t_1}}\big(\varepsilon K_{j_0-1}^{C_2+2}\kappa_{j_0-1}^{-1} + (\varepsilon t_1)^2\big)\,. \tag{68}$$

Assume $t = t(j_0)$ satisfies

$$\varepsilon\delta_{j_0}t < 1 \tag{69}$$

one gets from (68)

$$\big\|u(t) - v(t)\big\|_2 < \left\{\frac{K_{j_0-1}^{C_2+2}\kappa_{j_0-1}^{-1}}{\delta_{j_0}t_1} + \frac{\varepsilon t_1}{\delta_{j_0}}\right\}\varepsilon\delta_{j_0}t \tag{70}$$

and we denote γ the expression $\{\ \}$ in (70), which will be taken small. From (66), (11), the solution v of

$$\begin{cases} -iv_t = -v_{xx} + V(x)v + \varepsilon\Gamma^{j_0}(v) \\ v(0) = \phi \end{cases} \tag{71}$$

satisfies

$$v(t) = S(t)\phi + i\varepsilon \int_0^{t} S(t - \tau)\Gamma^{j_0}\big(S(\tau)\phi\big)d\tau + 0\big((\delta_{j_0}\varepsilon t)^2\big)\,. \tag{72}$$

It follows from (70), (72)

$$\langle u(t), \varphi_{n_{j_0}}\rangle - \langle S(t)\phi, \varphi_{n_0}\rangle =$$

$$i\varepsilon\delta_{j_0}\Big\langle \int_0^t S(t - \tau)\Gamma^{j_0}\big(S(\tau)\phi\big)d\tau, \varphi_{n_{j_0}}\Big\rangle + 0\big((\gamma + \delta_{j_0}\varepsilon t)\delta_{j_0}\varepsilon t\big) \tag{73}$$

where the integral is the expression (56), thus

$$\frac{1}{K_{j_0}}e^{i\lambda_{n_{j_0}}t}\int_{\mathbf{T}}\int_0^t \cos K_{j_0}\big[\langle\phi, \varphi_{n_0}\rangle\varphi_{n_0}(x)\cos\lambda_{n_0}\tau$$
$$+ \langle\phi, \varphi_{n_{j_0}}\rangle\varphi_{n_{j_0}}(x)\cos\lambda_{n_{j_0}}\tau\big]\cdot\varphi_{n_{j_0}}(x)e^{-i\lambda_{n_{j_0}}\tau}d\tau\, dx\,. \tag{74}$$

Recall that by (46)

$$|\lambda_{n_{j_0}} - m_{j_0}\lambda_{n_0}| < n_{j_0}^{-C_1} \tag{75}$$

which permits in (74) to replace $\lambda_{n_{j_0}}$ by $m_{j_0}\lambda_{n_0}$ provided say

$$n_{j_0}^{-C_1} t < K_{j_0}^{-10} \tag{76}$$

the error being $tK_{j_0}^{-10}$.

After this replacement, the $\frac{2\pi}{\lambda_{n_0}}$-periodicity in τ yields the expression

$$\frac{1}{K_{j_0}} e^{i\lambda_{n_{j_0}} t}(\varepsilon\delta_{j_0} t) \int_{\mathbf{T}^2} \cos K_{j_0}\big[\langle\phi, \varphi_{n_0}\rangle\varphi_{n_0}(x)\cos\tau$$
$$+ \langle\phi, \varphi_{n_{j_0}}\rangle\varphi_{n_{j_0}}(x)\cos m_{j_0}\tau\big]\cdot\varphi_{n_{j_0}}(x)e^{-im_{j_0}\tau} d\tau\, dx \tag{77}$$

for the second term in (73), up to an error term $(\varepsilon\delta_{j_0} t)\cdot K_{j_0}^{-10}$.

Since ϕ is the function q in Proposition 3 controlled in H^{s_0}-norm, one has

$$|\langle\phi, \varphi_{n_{j_0}}\rangle| \le n_{j_0}^{-s_0} \tag{78}$$

and the second term between $[\cdots]$ in (77) may be deleted, up to an error term $(\varepsilon\delta_{j_0} t)n_{j_0}^{-s_0}$. Thus the second term in (73) becomes

$$\frac{1}{K_{j_0}} e^{i\lambda_{n_{j_0}} t}(\varepsilon\delta_{j_0} t) \int_{\mathbf{T}_2} \cos\big(K_{j_0}\langle\phi, \varphi_{n_0}\rangle\varphi_{n_0}(x)\cos\tau\big)\cdot\ \varphi_{n_{j_0}}(x)e^{-im_{j_0}\tau} d\tau\, dx$$
$$+ \varepsilon\delta_{j_0} t 0(K_{j_0}^{-10} + n_{j_0}^{-s_0})\ . \tag{79}$$

Assuming K_{j_0} sufficiently large with respect to n_{j_0}, say

$$K_{j_0} > n_{j_0}^{c} \tag{80}$$

the integral in (79) will essentially be

$$\sum_{(x_c,\tau_c)} \varphi_{n_{j_0}}(x_c)e^{-im_{j_0}\tau_c} \int_{(x,\tau)\approx(x_c,\tau_c)} \cos(K_{j_0}\langle\phi, \varphi_{n_0}\rangle\varphi_{n_0}(x)\cos\tau) dx\, d\tau \tag{81}$$

where the summation extends to the critical points

$$\nabla\big[\varphi_{n_0}(x)\cos\tau\big](x_c, \tau_c) = 0 \tag{82}$$

and hence one expects a typical size $0(K_{j_0}^{-1})$ again, since $|\langle\phi, \varphi_{n_0}\rangle| > c$. This gives for the first term in (79) size $\frac{1}{K_{j_0}^2}(\varepsilon\delta_{j_0} t)$ and hence the right side of (73) has size

$$(\varepsilon\delta_{j_0} t)\left(\frac{1}{K_{j_0}^2} + 0(K_{j_0}^{-10} + n_{j_0}^{-s_0} + \gamma + \varepsilon\delta_{j_0} t)\right)\ . \tag{83}$$

Choosing

$$\varepsilon\delta_{j_0}t \sim \frac{1}{K_{j_0}^2} \tag{84}$$

one gets a lower estimate $K_{j_0}^{-4} = n_{j_0}^{-C}$, provided

$$n_{j_0}^{-s_0} < K_{j_0}^{-10}, \qquad \gamma < K_{j_0}^{-10}. \tag{85}$$

By (73), this leads to an estimate

$$\left|\langle u(t), \varphi_{n_{j_0}}\rangle\right| > n_{j_0}^{-C} \tag{86}$$

where $u(t) = u_{\varepsilon,\phi}(t)$, $\varepsilon = \varepsilon_{j_0}$, $t = t_{j_0}$. This last fact permits obtaining large H^{s_0}-norm for $u(t)$, leading to (49).

It remains to keep track of the conditions on the various parameters.

From (70),(76),(80),(84),(85), we get

$$\delta_{j_0} = K_{j_0}^{-C_3} \tag{87}$$

(c_3 determines the smoothness of Γ)

$$\gamma = \frac{K_{j_0-1}^{C_2+2}}{\kappa_{j_0-1}\delta_{j_0}t_1} + \frac{\varepsilon t_1}{\delta_{j_0}} \tag{88}$$

$$n_{j_0}^{-C_1}t < K_{j_0}^{-10} \tag{89}$$

$$K_{j_0} > n_{j_0}^{C} \tag{90}$$

$$\varepsilon\delta_{j_0}t \sim \frac{1}{K_{j_0}^2} \tag{91}$$

$$n_{j_0}^{-s_0} < K_{j_0}^{-10} \tag{92}$$

$$\gamma < K_{j_0}^{-10}. \tag{93}$$

Assume the construction performed up to stage $j_0 - 1$. Choosing n_{j_0} large, depending on $K_{j_0-1}, \kappa_{j_0-1}$, let $K_{j_0} = n_{j_0}^{C}$, $\delta_{j_0} = n_{j_0}^{-CC_3}$, $s_0 > 10C$. Condition (88)-(93) leads to a choice of $t_1 > n_{j_0}^{CC_3+11C}$, $\varepsilon < n_{j_0}^{-2CC_3-21C}$ and the condition on t given by (89) is compatible with (91) for C_1 large enough (depending on C_3).

This concludes the proof of Proposition 3.

References

[B1] J. BOURGAIN, Fourier restriction phenomena for certain lattice subsets and applications to nonlinear evolution equations, Part I, Geometric and Functional Analysis 3 (1993), 107-156.

[B2] J. BOURGAIN, Fourier restriction phenomena for certain lattice subsets and applications to nonlinear evolution equations, Part II, Geometric and Functional Analysis 3 (1993), 209-262.

[B3] J. BOURGAIN, Approximation of solutions of the cubic NLSE by finite dimensional equations and non-squeezing properties, International Math. Res. Notices 2 (1994), 79-90.

[B4] J. BOURGAIN, On the Cauchy problem for periodic KdV type equations, Preprint IHES (1993), J. of Fourier Analysis (Kahane issue), to appear.

[B5] J. BOURGAIN, Constructions of Quasi-Periodic solutions for Hamiltonian Perturbations of Linear Equations and Applications to Nonlinear PDE, IMRN 11 (1994), 475-497.

[CW] W. CRAIG, E. WAYNE, Newton's method and periodic solutions of nonlinear wave equations, Preprint (1992).

[EH] J. EKELAND, H. HOFER, Symplectic topology and Hamiltonian dynamics II, Math. Z. 203 (1990), 553-567.

[K1] S. KUKSIN, Nearly Integrable Infinite-dimensional Hamiltonian Systems, Springer Lect. Notes in Math. 1556 (1993).

[K2] S. KUKSIN, Infinite-dimensional symplectic capacities and a squeezing theorem for Hamiltonian PDE's, Preprint ETH (1993).

[PT] J. PÖSCHEL, E. TRUBOWITZ, Inverse Spectral Theory, Academic Press, Boston (1987).

[S] R. STRICHART, Restriction of Fourier transform to quasi surfaces and decay of solutions to the wave equation, Duke Math. J. 44 (1977), 705-714.

Jean Bourgain
Institute of Advanced Study
Olden Lane
Princeton, NJ 08540
USA

Submitted: April 1994

Geometric And Functional Analysis
Vol. 5, No. 2 (1995)

1016-443X/95/0200141-23$1.50+0.20/0

COLLAPSED RIEMANNIAN MANIFOLDS WITH BOUNDED DIAMETER AND BOUNDED COVERING GEOMETRY

J. CHEEGER AND X. RONG

Abstract

We study the class of n-Riemannian manifolds in the title such that the torsion elements in the fundamental group have a definite bound on their orders. Our main result asserts the existence of a kind of generalized Seifert fiber structure on M^n, for which the fundamental group of fibers injects into that of M^n. This provides a necessary and sufficient topological condition for a manifold to admit a sufficiently collapsed metric in our class. Among other consequences we obtain a strengthened version of the "gap conjecture" in this context.

0. Introduction

For each $n \geq 2$, $D > 0$, let $\mathcal{M}_D^n$ denote the class of the Riemannian n-manifolds whose diameter and sectional curvature satisfy, $\mathrm{diam}(M) \leq D$, $|K_M| \leq 1$. The non-collapsed situation, where one also assumes $\mathrm{Vol}(M) > v > 0$, is well understood ([C],[GroLP]; see also [GWu],[N],[Pe],[SSh]). Thus our interest is the collapsed case in which $\mathrm{Vol}(M)$ is sufficiently small.

The cornerstone of the theory in the collapsed case is Gromov's theorem on almost flat manifolds ([Gro1],[Ru],[BuK]). It asserts that there exists $\epsilon(n) > 0$ such that any n-manifold for which $\mathrm{diam}(M) \leq \epsilon(n)$ is an infranil manifold.

In [F1–3], an equivariant version and a parameterized version of Gromov's result were proved and were shown to apply to the frame bundle of all collapsed manifolds in $\mathcal{M}_D^n$. In [CFGro], the same was done for a version which is (simultaneously) equivariant *and* parameterized. The version in [CFGro] is the one required in this paper.

Collapsed manifolds were also studied from a somewhat different (but closely related) standpoint in [CGro1,2]. In each of the above mentioned works, a certain symmetry structure was shown to exist on a sufficiently collapsed manifold.

The work of the first author is partially supported by NSF Grant DMS 9303999. The work of the second author is supported by MSRI through NSF grant DMS 9022140 and partially supported by NSF Grant DMS 9204095.

Consider a Riemannian n-manifold M. The Riemannian connection on M gives rise to a canonical metric on the frame bundle, FM, of M, so that $O(n)$ acts on FM by isometries. A (second) fibration structure on FM is called *$O(n)$-invariant* if the $O(n)$-action on FM preserves its fibers and structural group.

Suppose that FM admits an $O(n)$-invariant fiber bundle structure, $\tilde{\eta}: FM \to B_{\tilde{\eta}}$, whose fibers are equipped with smoothly varying affine structures, affine isomorphic to a nilmanifold, $\Gamma\backslash N$, with its canonical connection, i.e. the one for which all left invariant vector fields are parallel. Note that the structural group of affine automorphisms of a nilmanifold $\Gamma\backslash N$ has a continuous part generated by the right translations. Thus, the bundle $\tilde{\eta}$ has no natural flat structure and need not have any flat structure at all.

The $O(n)$-action on FM defines a $O(n)$-action on $B_{\tilde{\eta}}$ by isometries and the projection, $\tilde{\eta}$, descends to a map, $\eta : M \to B_\eta = B_{\tilde{\eta}}/O(n)$, such that the following diagram commutes.

$$\begin{array}{ccc} FM & \xrightarrow{\ \tilde{\eta}\ } & B_{\tilde{\eta}} \\ \Big\downarrow{\scriptstyle \pi} & & \Big\downarrow{\scriptstyle \tilde{\pi}} \\ M & \xrightarrow{\ \eta\ } & B_\eta \end{array}$$

The centers of the fibers of $\tilde{\eta}$ determine a second $O(n)$-invariant fiber bundle structure on FM, $\tilde{f} : FM \to B_{\tilde{f}}$, with fiber a torus and affine structural group. We shall call $\tilde{f}$ the *canonical torus bundle of* $\tilde{\eta}$. As above, $\tilde{f}$ determines a map $f : M \to B_f = B_{\tilde{f}}/O(n)$ such that the diagram below commutes.

$$\begin{array}{ccc} FM & \xrightarrow{\ \tilde{f}\ } & B_{\tilde{f}} \\ \Big\downarrow{\scriptstyle \pi} & & \Big\downarrow{\scriptstyle \tilde{\pi}} \\ M & \xrightarrow{\ f\ } & B_f \end{array}$$

In this paper our primary concern is with $\tilde{f}$ and its subfibrations (see below).

There is a canonical sheaf on FM whose local sections restrict to local right invariant vector fields on the fibers of $\tilde{\eta}$; see [CFGro, page 335]. For the fibration $\tilde{f}$, the corresponding point is explained at the beginning of §2. This is all that we require here.

The metric, g, on M is said to be *invariant* if local sections of the above mentioned sheaf are local Killing fields for g. In this case, in particular, the restriction of g to each fiber of $\tilde{\eta}$ (and $\tilde{f}$) is a left invariant metric. Moreover, $\tilde{\eta}$ and $\tilde{f}$ are Riemannian submersions for unique quotient metrics on $B_{\tilde{\eta}}$, $B_{\tilde{f}}$.

By a *subfibration* of $\tilde{f}$, we mean a torus bundle such that each of its fibers is a totally geodesic submanifold of some fiber of $\tilde{f}$.

Remark 0.1: Any subfibration of $\tilde{f}$ is automatically $O(n)$-invariant. This is because the $O(n)$-action defines continuously varying affine automorphisms while compact totally geodesic submanifolds of a torus are rigid, i.e. cannot be varied continuously. Note that the same is true for (analogously defined) subfibrations of $\tilde{\eta}$ (though this will not be used here).

THEOREM 0.2 ([F1,2], [CFGro]). *For each $n \geq 2$, $D > 0$, there exists a constant, $v = v(n, D) > 0$, such that for any $M \in \mathcal{M}_D^n$, if the volume of M satisfies* $\mathrm{Vol}(M) < v$, *then*

(0.2.1) *FM admits an $O(n)$-invariant fiber bundle, $\tilde{\eta} : FM \to B_{\tilde{\eta}}$ with fiber a nilmanifold and affine structural group;*

(0.2.2) *There exist constants, $\rho = \rho(n)$ and $C = C(n)$, such that the injectivity radius of $B_{\tilde{\eta}}$ is $\geq \rho$ and the second fundamental form of all fibers of $\tilde{\eta}$ are bounded by C.*

(0.2.3) *For each $\epsilon > 0$, M admits an invariant metric, g_ϵ, such that*
$$e^{-\epsilon} g < g_\epsilon < e^{\epsilon} g \ , \quad |\nabla^g - \nabla^{g_\epsilon}| < \epsilon \ , \quad \left|(\nabla^{g_\epsilon})^i R_{g_\epsilon}\right| < C(n, i, \epsilon) \ ,$$
where $C(n, 0, \epsilon) = 1$.

It is important to note that for the induced possibly singular fibration, η, of M, all fibers have positive dimension (see Proposition A1.14 of [CFGro]).

Recall that if a compact manifold, M^{4k}, has nonzero signature, then for any metric with $|K| \leq 1$, the volume of M^{4k} is at least $c(4k) > 0$. In particular, if $\epsilon < c(4k)$, then M^{4k} admits no ϵ-volume collapsed metric in $\mathcal{M}_D^{4k}$ (for any D). However, there do exist such M^{4k} admitting singular fibration structures satisfying (0.2.1)–(0.2.3). The metrics may even be taken to be arbitrarily injectivity radius collapsed with $|K| \leq 1$, provided that the diameters are correspondingly large, so that the condition, $\mathrm{Vol}(M^{4k}) \geq c(4k)$, is not violated see [CGro1].

Since not every singular fibration structure satisfying (0.2.1)–(0.2.3) arises from a sufficiently collapsed manifold in $\mathcal{M}_D^n$, it is natural to expect that those structures which do so will have additional properties.

Here are two general problems.

Let $M \in \mathcal{M}_D^n$, $\mathrm{Vol}(M) < v$.

1) Determine topological constraints on M imposed by the existence of a collapsed metric with bounded diameter.

2) In particular, find constraints on the singular fibration structure or more precisely, on its holonomy group and singular set (see §2).

A more specific question concerns the existence of a "gap".

3) (Cheeger-Gromov) Given $n \geq 2$ and $D > 0$, does there exist a constant, $v = v(n, D) > 0$, such that for $M \in \mathcal{M}_D^n$, if $\mathrm{Vol}(M) < v$, then M admits a sequence of metrics with bounded diameter for which the volumes converge to zero?

Remark 0.3: If in Problem 3) one restricts attention to simply connected manifolds, then the answer is affirmative. This follows from Theorem 0.2, the fact that a *pure F-structure* (see below) on a simply connected manifold is actually defined by a global torus action and the collapsing construction of [CGro1].

A Riemannian manifold, M, with $|K| \leq 1$ is said to have *bounded covering geometry* (BCG) if the pullback metric on the universal covering space, $\widetilde{M}$, has injectivity radius ≥ 1; compare [CGro3]. Note that if instead $\mathrm{injrad}_{\widetilde{M}} \geq \delta > 0$, then $\mathrm{injrad}_{\widetilde{M}} \geq 1$ can be achieved by multiplying the metric by δ^{-2}. Thus, the choice of number, 1, is just a convenient normalization.

In this paper, we will concentrate on an important special case of our problems, that of sufficiently collapsed manifolds in $\mathcal{M}_D^n$ which have BCG and for which there is an a priori bound on the order of a torsion element of $\pi_1(M)$. In particular, this includes the study of collapsing sequences with BCG on a fixed manifold; compare [CGro2], [R1,2]. We emphasize that in the BCG case, stronger results hold than are true for arbitrary manifolds in $\mathcal{M}_D^n$. In [CR], we will study 1)–3) in full generality.

Before stating our main result, we need some terminology. In what follows, we will call the two (singular) fibrations, η and f of M, *a pure positive rank N-structure* and *its canonical F-structure*, respectively. We use the symbols, $\mathcal{N}$ and $\mathcal{F}$ to denote the N-structure and the canonical F-structure, respectively. Similarly, we call the $O(n)$-invariant fiber bundles, $\tilde{\eta}$ and $\tilde{f}$, the *lifting of η and f to FM*, and denote them by $\widetilde{\mathcal{N}}$ and $\widetilde{\mathcal{F}}$ respectively; compare [CFGro].

Given a pure positive rank N-structure on M, $\mathcal{N}$. An orbit of $\mathcal{N}$ (i.e. the fiber of η) is said to be *regular* if it has the highest dimension among all orbits in some open neighborhood of itself. All other fibers are called *singular*. An F-structure is said to be *polarized* if all its orbits are regular.

DEFINITION 0.4: Given a positive rank F-structure, an orbit is *injective* if the homomorphism of fundamental groups induced by the natural inclusion of the orbit into the manifold is injective. A positive rank F-structure is *injective* if all orbits are injective.

Obviously, an injective F-structure is polarized, but the converse is not true. Note that *a pure F-structure is injective if and only if its lifting to FM is injective.* This follows easily from the homotopy sequence of the fibration, $O(n) \to FM \to M$.

The main result in this paper is the following.

THEOREM 0.5. *For $n \geq 2, D > 0$ and $m \geq 1$, there exists a constant,*

$v = v(n, D, m) > 0$, such that for $M \in \mathcal{M}_D^n$, if
(0.5.1) $\gamma \in \pi_1(M)$, ord$(\gamma) < \infty$ implies ord$(\gamma) \leq m$,
(0.5.2) M has BCG,
(0.5.3) $\mathrm{Vol}(M) < v$,
then M admits a pure injective F-structure.

The injective F-structure on M in Theorem 0.5 is actually a substructure of the canonical F-structure in Theorem 0.2. However, the canonical F-structure on M may not even be polarized; see Example 6.1.

Theorem 0.5 is sharp in the sense that there are counterexamples if any part of the hypothesis is removed and no further assumptions are added (see Examples 6.2 and 6.3).

A direct consequence of Theorem 0.5 is

COROLLARY 0.6. *Given the assumptions of Theorem 0.5, let k be the dimension of the orbits of the injective F-structure on M. Then, the fundamental group $\pi_1(M)$ has a normal free abelian subgroup of rank k.*

The following result is proved in §4.

COROLLARY 0.7. *Given the assumptions of Theorem 0.5, assume in addition that $\pi_1(M)$ is torsion-free (i.e. $m = 1$). Then either*
(0.7.1) $\pi_1(M)$ has a normal free abelian subgroup of rank ≥ 2,
or
(0.7.2) $\pi_1(M)$ has an infinite cyclic center, any two cyclic subgroups have non-trivial intersection. In particular, if $\pi_1, (M)$ is not cyclic, then the quotient by the commutator group is a finite group.

We remark that unless $\pi_1(M)$ is cyclic, (0.7.2) provides a strong constraint even from the point of view of group theory; see Example 6.6. It seems possible that for fixed D, such a noncyclic group cannot occur as the fundamental group of a sufficiently collapsed manifold in $\mathcal{M}_D^n$.

Using the center theorem in [GWu] and [LaY], we immediately get

COROLLARY 0.8. *Given the assumptions of Theorem 0.5, assume in addition that M admits a metric with non-positive sectional curvature. Then $\pi_1(M)$ has a free normal abelian subgroup of rank ≥ 2.*

Note that in Corollary 0.8, no conditions on the diameter and injectivity radius of the metric of non-positive curvature are assumed.

Remark 0.9: A special case of Theorem 0.5 in which M is assumed to be an aspherical manifold was proved in [F4] and in [CGro1] (for the more general case of so called essential manifolds). Actually, in [F4] and [CGro1] the canonical fibration itself is shown to be injective. As previously mentioned, in our situation this need not hold; see Example 6.1. Corollary 0.8 also partly overlaps the result of [Buy], stating that the fundamental group of a sufficiently collapsed manifold with $-1 \leq K \leq 0$ has a free abelian subgroup of rank ≥ 2.

Using the collapsing construction in [CGro1] (see also [R1]) we obtain from Theorem 0.5

COROLLARY 0.10. *Given the assumptions of Theorem 0.5, there exists a sequence of invariant metrics on M with bounded diameter and bounded covering geometry for which the volumes converge to zero.*

In particular, Corollary 0.10 provides a positive answer to Problem 3) above in this particular case.

As noted above, for those manifolds in Theorem 0.2 which are of dimension $4n$ and are orientable, the top Pontrjagin numbers vanish. In the case that the pure polarized F-structure in Theorem 0.5 has higher rank, by applying the collapsing construction in [CGro1] we get

COROLLARY 0.11. *Given the assumptions of Theorem 0.5, suppose that the rank of the injective F-structure is $k \geq 4$. Then all elements of the real Pontrjagin ring of degree $\geq \frac{4n-k}{4}$ vanish.*

In low dimensions, we have the following result which is proved in §5.

COROLLARY 0.12. *Given the assumptions of Theorem 0.5, then,*

i) *For $n = 3$, M is homeomorphic to a Seifert manifold of infinite fundamental group or doubly covered by a solvmanifold.*
ii) *For $n = 4$, let $\mathcal{F}$ denote the canonical singular fibration as in Theorem 0.2. If $\mathcal{F}$ is not injective, then its holonomy group is the identity.*
iii) *M admits a sequence of invariant metrics, g_i, with bounded covering geometry such that (M, g_i) converges to B_f in the Gromov-Hausdorff sense.*
iv) *The fundamental group of M has either a free abelian subgroup of rank ≥ 2 or an infinite cyclic subgroup of finite index.*

Lastly, we mention that Theorem 0.5 and Corollary 0.10 can be used to prove the rationality conjecture of Cheeger-Gromov on limiting η-invariants for manifolds as in Theorem 0.5; see [CGro3], [R1,2].

The remainder of this paper is divided into six sections as follows. 1. Algebraic preliminaries; 2. Application to singular fibrations; 3. BCG and the null subfibration; 4. Proof of Corollary 0.7; 5. Proof of Corollary 0.12; 6. Examples; 7. A refinement of Theorem 0.5.

1. Algebraic Preliminaries

In this section we summarize some elementary facts which when taken together, provide an algebraic model of our geometric situation. The specific results which will be used in the sequel are Propositions 1.2 and 1.3.

Let F be a smooth manifold equipped with an affine connection, ∇, such that $(F,\nabla) \simeq (T^k, \nabla^{\text{can}})$, where $T^k = \mathbf{Z}^k\backslash\mathbf{R}^k$ and ∇^{can} is the canonical flat connection (we write $\mathbf{Z}^k\backslash\mathbf{R}^k$ rather than $\mathbf{R}^k/\mathbf{Z}^k$ in order to maintain consistency with convention of [CFGro]).

Let V^k denote the space of the parallel vector fields with respect to ∇. For any $x \in F$, let $\exp_x : T_xF \to F$ denote the exponential map at x. Put

$$\Lambda^k = \left\{\lambda \in V^k \mid \exp_x \lambda(x) = x, \text{for all } x \in F\right\} .$$

Then, Λ^k is a *lattice* in V^k. Note that $F \simeq \Lambda^k\backslash V^k$. The isomorphism is canonical up to a choice of base point in F.

There is a canonical exact sequence,

$$e \to \Lambda^k\backslash V^k \to \text{Aff}(F) \to \text{Aut}(\Lambda^k) \to e ,$$

where $\Lambda^k\backslash V^k = \text{Aff}_0(F)$, the identity component of $\text{Aff}(F)$. Indeed, there is an (non-canonical) isomorphism, $\text{Aff}(F) \simeq \Lambda^k\backslash V^k \propto \text{Aut}(\Lambda^k)$. Again, such isomorphisms are in 1-1 correspondence with base points, i.e. with points of F.

Subtori, $\Lambda^r\backslash\Sigma^r$, of $\Lambda^k\backslash V^k$ are in 1-1 correspondence with subgroups, $\Lambda^r \subseteq \Lambda^k$, such that there is a subspace, Σ^r, of V^k for which $\Lambda^r = \Sigma^r \cap \Lambda$. By abuse of language, we simply call Λ^r a subtorus.

Every totally geodesic submanifold of F (closed or not) is the fiber of a unique fibration of F by mutually parallel totally geodesic submanifolds. Such fibrations are in 1-1 correspondence with subspaces of V^k. Those for which the fibers are closed (and thus are tori) are in 1-1 correspondence with subtori $\Lambda^r \subseteq \Lambda^k$.

Consider $(F_1, V_1), (F_2, V_2)$ and $A : F_1 \to F_2$. Then A is affine if and only if it maps parallel fields to parallel fields. In this case write $A \in \text{Aff}(F_1, F_2)$ and let $A^\#$ denote the induced linear map, $A^\# : V_1 \to V_2$. Note that in fact, $A^\# : \Lambda_1 \to \Lambda_2$. If $B_1 \in \text{Aff}_0(F_1)$, $B_2 \in \text{Aff}_0(F_2)$, then $(AB_1)^\# = (B_2A)^\# = A^\#$.

Now suppose F has an invariant Riemannian metric, i.e. V^k has an inner product. Let $|v|$ denote the norm of $v \in V^k$. By a standard procedure one can find a basis $v_1, ..., v_k$, for Λ^k, such that $|v_i| \leq |v_{i+1}|$ and the angle between v_i and the subspace spanned by $v_1, ..., \hat{v}_i, ..., v_n$ is at least $\theta_k > 0$; see [GroP]. This basis is called a *canonical basis*. Note that a canonical basis is unique up to finitely many possibilities (see [GroP]).

LEMMA 1.1. *There is a constant, $C_k > 0$, such that if $v_1, \ldots, v_k$ is a canonical basis and $v = \sum_{i=1}^{k}, a_i v_i$, then for all j*

$$|v| \geq C_k |a_j v_j| .$$

Proof: Let Σ_j denote the subspace of V generated by $v_1, \ldots, \hat{v}_j, \ldots, v_n$. Since the angle between v_j and Σ_j is at least $\theta_k > 0$, the angle between v_j and the normal vector, n_j, to Σ_j is at most $\pi/2 - \theta_k < \pi/2$.

Thus,

$$\begin{aligned} |v| \geq |\langle v, n_j \rangle| &= \left| \left\langle \sum_{i=1}^{k} a_i v_i, n_j \right\rangle \right| = \left| \langle a_j v_j, n_j \rangle \right| \\ &\geq |a_j| |v_j| \cos(\pi/2 - \theta_k) = C_k |a_j v_j| . \end{aligned}$$ □

An *automorphism* of $\Lambda^k \backslash V^k$ is uniquely determined by a linear map, $A : V^k \to V^k$, which preserves Λ^k. A subtorus, Λ^r, is called *A-invariant* by $A \in \mathrm{Aut}(\Lambda^k)$ if A preserves Λ^r.

The norm of a linear map, A, of V^k, is defined as usual by

$$\|A\| = \max_{v \in V^k - \{0\}} \frac{|A(v)|}{|v|} .$$

Given any real number $R \geq 1$. A subtorus, Λ^r, is called *R-round* if Λ^r has a canonical basis, $v_1, \ldots, v_r$, which satisfies the condition,

$$\frac{|v_{i+1}|}{|v_i|} \leq R .$$

Consider a flat torus with a canonical basis, $v_1, \ldots, v_k$, and fix $R \geq 1$. Assume that $|v_{i+1}|/|v_i| \leq R$ for $i \leq r-1$ and $|v_{r+1}|/|v_r| > R$. The subtorus, generated by $v_1, \ldots, v_r$, is called a *canonical R-round subtorus*.

PROPOSITION 1.2. *Let $A \in \mathrm{Aut}(\Lambda^k)$ with $\|A\| \leq C_k R$. Then, a canonical R-round subtorus, Λ^r, is A-invariant.*

Proof: Take a canonical basis $v_1, \ldots, v_k$ for Λ^k. Assume that $v_1, \ldots, v_r$ generate a canonical R-round subtorus, say Λ^r. Put $A = (a_{ij})$, where $a_{i,j} \in \mathbf{Z}$. We shall show that $a_{ij} = 0$ for all $1 \leq j \leq r$ and $i > r$, i.e. Λ^r is invariant by A.

Suppose that, for some $1 \leq j \leq r$, $i_0 > r$, we have $0 \neq a_{i_0 j} \in \mathbf{Z}$ and $a_{ij} = 0$ for all $i > i_0$. Then by Lemma 1.1, we derive

$$\begin{aligned} |A(v_j)| &= \left| \sum_{i=1}^{i_0} a_{ij} v_i \right| \geq C_k |a_{i_0 j} v_{i_0}| \geq C_k |v_{i_0}| \\ &= C_k |v_{i_0}| \cdot |v_j|^{-1} \cdot |v_j| > C_k R |v_j| , \end{aligned}$$

which contradicts $\|A\| \leq C_k R$. □

Given a subtorus, Λ^r, put

$$\lambda(\Lambda^r) = \frac{\min_{v' \in \Lambda^r - \{0\}} |v'|}{\min_{v \in \Lambda^k - \{0\}} |v|} .$$

Obviously, $\lambda(\Lambda^r)$ is invariant under scaling. The number, $\min_{v' \in \Lambda^r - \{0\}} |v'|$ is the injectivity radius of the flat torus $\Lambda^r \backslash \Sigma^r$.

An R-round invariant metric on a torus, $\Lambda^k \backslash V^k$, is called *normal* if the R-round canonical basis $v_1, \dots, v_k$ for Λ^k, satisfies $|v_1| = 1$. Any R-round invariant metric can be normalized by rescaling.

Two subtori, Λ^r and Λ^t, are said to be *transversal* to each other if $\Sigma(\Lambda^r) \cap \Sigma(\Lambda^t) = \{0\}$.

PROPOSITION 1.3. *Given $R \geq 1$ and $C \geq 1$, there exists $l = l(R, C, k) < \infty$ with the following property. Let $\Lambda^k \backslash V^k$ be an R-round torus, and let $\mathcal{A} \subset \mathrm{Aut}(\Lambda^k \backslash V^k)$ satisfy $\|A\| \leq C$ for all $A \in \mathcal{A}$. If Λ^r is a proper $\mathcal{A}$-invariant subtorus with $\lambda(\Lambda^r) \geq l$, there is a proper $\mathcal{A}$-invariant subtorus transversal to Λ^r.*

Any choice of a basis for Λ^k determines an isomorphism of the group of automorphisms, $\mathrm{Aut}(\Lambda^k)$, with $SL(k, \mathbb{Z})$. A *canonical isomorphism*, $\mathrm{Aut}(\Lambda^k) \simeq SL(k, \mathbb{Z})$, is by definition, one which is induced by a choice of a canonical basis.

LEMMA 1.4. *There is a constant $N(k, R, C) < \infty$, such that if $\Lambda^k \backslash V^k$ is an R-round torus and $A \in \mathrm{Aut}(\Lambda^k)$ has norm $\leq C$, then the image of A under a canonical isomorphism is a matrix whose entries are bounded in absolute value by N. In particular there are only finitely many such A and such matrices.*

Proof: This is an immediate consequence of Lemma 1.1. □

Proof of Proposition 1.3: Fix $\mathcal{A} \subset SL(k, \mathbb{Z})$. In view of Lemma 1.4 it clearly suffices to prove that there is a constant, $\lambda(\mathcal{A})$, such that if $\Lambda \subset \mathbb{Z}^k$ is an $\mathcal{A}$-invariant subtorus and $\lambda(\Lambda) > \lambda(\mathcal{A})$, then there is a nontrivial $\mathcal{A}$-invariant subtorus transversal to Λ. Since if $\Lambda^k \subset V^k$ is R-round, then a canonical isomorphism is a $C(R, k)$-quasi-isometry for some constant $C(R, k)$, this clearly implies our assertion. Actually it would suffice to assume that $\mathcal{A}$ is a finite set, but it is not necessary to do so.

If the assertion is false, then there is a sequence of tori, $\Lambda_i \subset \mathbb{Z}^k$, with $\lambda(\Lambda_i) \to \infty$, such that no Λ_i has an $\mathcal{A}$-invariant transversal torus.

Note that if Λ, Λ' are tori, then either $\Lambda \cap \Lambda' = \Lambda$, or $\Lambda \cap \Lambda'$ is a torus of rank strictly smaller than that of Λ. Thus, either the sequence, $\Lambda_1, \Lambda_1 \cap \Lambda_2, \dots$ is eventually constant or there exists m such that $\bigcap_{j=1}^m \Lambda_j \neq$

0, $\bigcap_{j=1}^{m+1} \Lambda_j = 0$. The former possibility contradicts $\lambda(\Lambda_i) \to \infty$. In the later case, $\bigcap_{j=1}^{m} \Lambda_j \neq 0$ is an $\mathcal{A}$-invariant torus transversal to Λ_{m+1}, also a contradiction. □

The result of Proposition 1.3 requires a priori bounds, $\lambda(\mathcal{A}), R$ and C. We conclude this section by giving simple examples which show that Proposition 1.3 fails if one relaxes any of these bounds.

EXAMPLE 1.5: Consider $\mathbf{R}^2$ with its standard metric. The torus, $\mathbf{Z}^2 \subset \mathbf{R}^2$ is 1-round. The subtorus, Λ_1, generated by e_1 is a $\mathcal{A}$-invariant, for $\mathcal{A} = \{A\}$ where $A = \left(\begin{smallmatrix} 1 & 1 \\ 0 & 1 \end{smallmatrix}\right)$. But there is no $\mathcal{A}$-invariant subtorus transversal to Λ_1 (since $\lambda(\Lambda_1)$ is *not* sufficiently large).

EXAMPLE 1.6: Consider again the standard two torus, $\mathbf{Z}^2 \subset \mathbf{R}^2$. Let g_0 denote the standard metric on $\mathbf{R}^2$. Choose a sequence of basis for $\mathbf{Z}^2\backslash\mathbf{R}^2$, v_1^i, v_2^i, such that $g_0(v_1^i, v_1^i) \geq i$. For each i, define an automorphism of $\mathbf{Z}^2$ by $\phi_i(e_1, e_2) = (v_1^i, v_2^i)$, where e_1, e_2 is the standard basis for $\mathbf{Z}^2$. Put $g_i = \phi_i^* g_0$.

Let $A = \left(\begin{smallmatrix} 1 & 1 \\ 0 & 1 \end{smallmatrix}\right) \in SL(2, \mathbf{Z})$. Then, A has a *unique* invariant subtorus, Λ_1, generated by e_1. Thus, there is no A-invariant subtorus transversal to Λ_1. It is obvious that $\mathbf{Z}^2$ is 1-round with respect to each g_i and $\lambda(\Lambda_1, g_i) = g_0(v_1^i, v_1^i) \geq i$. By Proposition 1.3, $\{\|A\|_{g_i}\}$ is *not* bounded (as can be checked directly).

EXAMPLE 1.7: Let g_i denote the invariant metric on $\mathbf{R}^2$ defined by

$$g_i(e_1, e_1) = i^2 \,, \quad g_i(e_1, e_2) = 0 \,, \quad g_i(e_2, e_2) = 1 \,.$$

Then,

$$2 \geq \|A\|_{g_i} \to 1 \,, \quad \lambda(\Lambda_1, g_i) = i \to \infty \,, \quad i \to \infty \,.$$

Given any $R \geq 1$, it is obvious that $(\mathbf{Z}^2, g_i)$ is *not* R-round for $R \leq i$. Again, the conclusion of Proposition 1.3 fails.

2. Applications to Singular Fibrations

In this section we prove Theorem 0.5 modulo one result, Proposition 2.10. The proof is obtained by reducing our study of the geometric situation to the algebraic one considered in the previous section.

The proof of Proposition 2.10 will be given in §3. Only there do the assumptions concerning BCG and the bound on the order of torsion elements of the fundamental group enter our discussion.

a) The associated flat bundle $\mathcal{V}$. Consider a fiber bundle, $p : E \to B$, with fiber $F \simeq \mathbf{Z}^k\backslash\mathbf{R}^k$ and structural group $\mathrm{Aff}(\mathbf{Z}^k\backslash\mathbf{R}^k)$. Thus, each fiber carries a canonical flat affine structure isomorphic to $\mathbf{Z}^k\backslash\mathbf{R}^k$ and this structure varies continuously from fiber to fiber.

Associated to E is the vector bundle $\mathcal{V} \to B$ whose fiber, $\mathcal{V}_x$, at x is the space of parallel fields on E_x, the fiber of E at x. We can also consider the associated torus bundle, $\mathcal{E}$ with fiber, $\Lambda_x \backslash \mathcal{V}_x$.

Let $\{g_{\alpha\beta}\}$ be the 1-cocycle defining E, with values in $\mathrm{Aff}(\mathbf{Z}^k \backslash \mathbf{R}^k)$. Thus, $g_{\alpha\beta} : U_\alpha \cap U_\beta \to \mathrm{Aff}(\mathbf{Z}^k \backslash \mathbf{R}^k)$, where with no loss of generality we can assume that $U_\alpha \cap U_\beta$ is connected. For any fixed $x_0 \in U_\alpha \cap U_\beta$, we can write $g_{\alpha\beta} = B(x) g_{\alpha\beta}(x_0)$, where $B(x) \in \mathrm{Aff}_0(\mathbf{Z}^k \backslash \mathbf{R}^k)$, $B(x_0) = Id$. Thus, $\{g^{\#}_{\alpha\beta}\}$, the cocycle which defines $\mathcal{V}$ is given by $g^{\#}_{\alpha\beta}(x) = (B(x) g_{\alpha\beta}(x_0))^{\#} = g^{\#}_{\alpha\beta}(x_0)$. Hence, $\mathcal{V}$ has a canonical flat structure, with structural group contained in $\mathrm{Aut}(\mathbf{Z}^k) = SL(k, \mathbf{Z})$. Therefore, $\mathcal{E}$ has such a flat structure as well.

From the above description it is clear that the parallel translation, $\mathcal{P}$, associated to the flat structure on $\mathcal{V}$, is defined as follows. Given $c : [0,1] \to B$, let $A(t) \in \mathrm{Aff}(E_{c(0)}, E_{c(t)})$ be continuous in t (in the obvious sense) with $A(0) = Id_{c(0)}$. Then $\mathcal{P}_c : \mathcal{V}_{c(0)} \to \mathcal{V}_{c(1)}$ is given by $\mathcal{P}_c = A^{\#}(1)$. Let $\nabla^{\mathcal{P}}$ be the corresponding flat connection.

As pointed out in the introduction (see also §1) there is a 1-1 correspondence between subfibrations of E for which each fiber is a totally geodesic torus in some E_x and subbundles $\mathcal{E}'$ of $\mathcal{E}$ (necessarily with torus as fiber). Note that the rigidity of such subtori guarantees that the corresponding subbundle $\mathcal{V}'$ of $\mathcal{V}$ is flat. Moreover, the *fiber* $\mathcal{V}'_x$ *is generated by a unique subtorus* $\Lambda'_x \subset \Lambda_x$. If we fix $x \in B$, then we can restate this as

LEMMA 2.1. *Subfibrations of E correspond to subtori of Λ_x which are invariant under the holonomy of the parallel translation (flat connection) $\mathcal{P}$.*

If $U \subset B$ is contractible and $x \in U$, then $v \in \mathcal{V}_x$ determines a vector field, $\hat{V}$, on $p^{-1}(U)$ such that $\hat{V}|E_y$ is parallel, for all $y \in U$ and such that the local section of $\mathcal{V}$ corresponding to $\hat{V}$ is parallel with respect to $\mathcal{P}$.

A Riemannian metric, g, on (the total space) E is called *invariant*, if each vector field $\hat{V}$ is a locally defined Killing field for g, i.e. $L_{\hat{V}} g = 0$. By the argument of [CFGro, §8], such metrics exist. In this case, the restriction of g to any fiber is invariant, so that in particular g induces a metric (also denoted by g) on $\mathcal{V}$. Moreover, the horizontal subbundle $H \subset TE$, the orthogonal complement to the tangent bundle to the fibers, is invariant under any $\hat{V}$. Thus, there is a unique metric on B for which $p : E \to B$ is a Riemannian submersion.

The horizontal distribution, H, defines a parallel translation, P, on E, or equivalently, a connection. Thus, if $c : [0,1] \to B$, the collection of all horizontal lifts, $\tilde{c}$, of c determines a map, $P_c : E_{c(0)} \to E_{c(1)}$.

Typically this connection is not flat. However, since H is invariant under the flow of any $\hat{V}$, it follows that $P^{\#}_{c(t)}(\hat{V}|E_{c(0)}) = \hat{V}|E_{c(t)}$. Thus,

$\mathcal{P}_{c(t)} : E_{c(0)} \to E_{c(t)}$ is affine (and $\mathcal{P}_{c(0)} = Id_{c(0)}$). Hence, by the description of $\mathcal{P}$ given above, $\mathcal{P}$ is given in terms of P by $P_c^{\#} = \mathcal{P}_c$.

Now we can compute $\nabla^{\mathcal{P}} g$ on $\mathcal{V}$. Let V_1, V_2 be local sections of $\mathcal{V}$ with $\nabla^{\mathcal{P}} V_1 = 0, \nabla^{\mathcal{P}} V_2 = 0$ and let $\hat{V}_1, \hat{V}_2$ denote the corresponding local Killing fields on E. Then if $c : [0,1] \to B$ with V_1, V_2 defined near $c(0)$, it follows that

$$\begin{aligned}(\nabla^{\mathcal{P}}_{c'} g)(V_1, V_2) &= \tilde{c}'(g(\hat{V}_1, \hat{V}_2)) \\ &= 2g(II_{E_x}(\hat{V}_1, \hat{V}_2), \tilde{c}') \ ,\end{aligned}$$

where II_{E_x} denotes the second fundamental form of E_x. Therefore, by a standard argument we get

Proposition 2.2. *If the norms of the second fundamental forms of the fibers of E are bounded by a constant C, then the norm of $\mathcal{P}_c$ is bounded by $e^{2Cl(c(t))}$, where $l(c(t))$ is the length of $c(t)$.*

We now apply Proposition 2.2 to $M \in \mathcal{M}^n_D$. Assume that $\mathrm{Vol}(M) < v$, where $v = v(n, D)$ is the constant given in Theorem 0.2. Then, we obtain an $O(n)$-invariant fiber bundle $\tilde{\eta} : FM \to B_{\tilde{\eta}}$ such that the second fundamental forms of the fibers are bounded by $C(n)$. Since any fiber of the canonical fiber bundle $\tilde{f} : FM \to B_{\tilde{f}}$ is a totally geodesic submanifold of some fiber of $\tilde{\eta}$, the second fundamental forms of the fibers of $\tilde{f}$ are also bounded by $C(n)$.

Since $B_{\tilde{f}}$ has diameter less than a constant depending on D, $\pi_1(B_{\tilde{f}})$ can be generated by loops whose lengths are less than a constant depending only on D. Applying Proposition 2.2 we obtain

Corollary 2.3. *Let M satisfy the assumptions of Theorem 0.2. Let $\tilde{f} : FM \to B_{\tilde{f}}$ be the canonical bundle on FM. Then, there exists $R = R(n, D) < \infty$, such that the holonomy group of the associated flat bundle $\mathcal{V}FM$ can be generated by the affine automorphisms of norm less than R.*

b) The *R*-round subfibration. Consider $p : E \to B$ equipped with an invariant metric. Thus, the induced metrics on fibers are flat.

Definition 2.4: Let $p : E \to B$ have an invariant metric. Given any real $R \geq 1$, the fibration E is called *R-round* if there exists at least one fiber which is an R-round flat torus.

Note that an R-round bundle may have fibers which are not R-round.

Corollary 2.5. *Given the assumptions of Corollary 2.3, $\tilde{f} : FM \to B_{\tilde{f}}$ has an R'-round subfibration, $\tilde{f}_{R'} : FM \to B_{R'}$, where $R' = R/C_k$ and R and C_k are given by Corollary 2.3 and Lemma 1.1.*

Proof: Consider the associated bundle $\mathcal{V}FM$. Let $x \in B_{\tilde{f}}$. By Lemma 2.1, it suffices to show that a canonical R'-round subtorus of $\mathcal{V}FM_x$ is invariant under the holonomy group of $\mathcal{V}FM$. The proof is then finished by applying Corollary 2.3 and Proposition 1.2. □

By applying Lemma 1.4, we obtain the following result (which is not used in the remainder of the present paper).

COROLLARY 2.6. *There exists $\mathcal{A}(n,D) = \{A_1, \dots, A_s\} \subset SL(k, \mathbf{Z})$, such that, given the assumptions of Corollary 2.3, the image under the canonical representation of the holonomy group of $\mathcal{V}FM_x$ can be generated by matrices in $\mathcal{A}(n,D)$.*

c) BCG and the null subfibrations. For any $z \in E$, consider the map, $i_* : \pi_1(E_{p(z)}, z) \to \pi_1(E, z)$, induced by the inclusion, $i : (E_{p(z)}, z) \to (E, z)$.

DEFINITION 2.7: The image, $i_*(\pi_1(E_{p(z)}, z))$, is called the *fiber subgroup of $E_{p(z)}$.*

By the homotopy exact sequence, the fiber subgroup of $E_{p(z)}$ is a normal abelian subgroup of $\pi_1(E, z)$.

LEMMA 2.8. *The fibration, $p : E \to B$, has a subfibration, $p_u : E \to B_u$, such that the fiber subgroup of p_u is the torsion subgroup of the fiber subgroup of p. In particular, if the fiber subgroup of p is neither $\mathbf{Z}^k$ nor finite, then p_u is a proper subfibration.*

Proof: For each fiber, E_x, consider the fibration of E_x into the maximal totally geodesic submanifolds in which any loop has a multiple which is homotopically trivial in E. Clearly, this collection of fibrations defines a subfibration of E. □

We will call the subfibration, p_u, the *null subfibration of p.*

Let $p_i : E \to B$, $i = 1, 2$, be subfibrations of $p : E \to B$. The subfibrations, p_1, p_2 are called *transversal* if their fibers at one (and hence all) $x \in E$ are transversal to each other, in the sense of §1.

The importance of the null subfibration is the following obvious fact.

COROLLARY 2.9. *A subfibration of $p : E \to B$ is injective if and only if it is transversal to the null subfibration.*

We now turn to the canonical fibration of FM as in Theorem 0.2, where $M \in \mathcal{M}^n_D$ such that $\mathrm{Vol}(M) < v$. Let $\tilde{f} : FM \to B_{\tilde{f}}$ denote the canonical fibration and $\tilde{f}_u : FM \to B_u$ the null subfibration. It turns out that in the presence of BCG, the fibers of $\tilde{f}_u$ cannot be collapsed.

PROPOSITION 2.10. *Let the assumptions be as in Theorem 0.5. Then, with respect to the induced metric, the injectivity radius of any fiber of $\tilde{f}_u$ is greater than $\rho(m) > 0$.*

The proof of Proposition 2.10 will be given in §3.

d) Proof of Theorem 0.5 modulo Proposition 2.10.

Proof of Theorem 0.5: Let $M \in \mathcal{M}_D^n$ with $\mathrm{Vol}(M) < v(n, D)$. First, by Theorem 0.2 the frame bundle of M admits a canonical bundle, $\tilde{f} : FM \to B_{\tilde{f}}$. With no loss of generality, we can assume that the canonical metric on FM is invariant, see (0.3.2).

Recall that a subfibration of FM is injective if and only if the corresponding fibration of M is injective. Also, by Remark 0.1, a subfibration of $\tilde{f}$ is automatically $O(n)$-invariant. Thus, under the stronger assumption $\mathrm{Vol}(M^n) < v(n, D, m)$, it will suffice to construct an injective subfibration of $\tilde{f}$.

By Corollary 2.3, we obtain an R'-round subfibration $\tilde{f}_{R'} : FM \to B_{R'}$, of $\tilde{f}$. If the null subfibration of $\tilde{f}_{R'}$ is trivial, then $\tilde{f}_{R'}$ is injective. Otherwise, we will show the existence of a nontrivial subfibration of $\tilde{f}_{R'}$ which is transversal to $\tilde{f}_{R',u}$; see Corollary 2.9.

Let $\mathcal{V}'FM$ denote the flat subbundle of $\mathcal{V}FM$ corresponding to $\tilde{f}_{R'}$. Let $\mathcal{V}'FM_x$ be R'-round and let $(\Lambda'_u)_x$ denote the subtorus corresponding to $\tilde{f}_{R',u}$. By Lemma 2.1, Proposition 1.3 and Corollary 2.3, it suffices to check that $\lambda((\Lambda'_u)_x) \geq l(n, D)$, for suitable $l(n, D) > 0$. But if $\mathrm{Vol}(M) < v(n, D, m)$, for suitable $v(n, D, m) > 0$, then this follows from Proposition 2.10. □

3. BCG and The Null Subfibration

In this section, we will prove Proposition 2.10.

First we introduce some terminology. Fix any fiber, $\tilde{\eta}_x$, of the $O(n)$-equivariant fiber bundle $\tilde{\eta} : FM \to B_{\tilde{\eta}}$. Consider the restriction of the exponential map at x on the normal subspace of $\tilde{\eta}_x$. The *normal injectivity radius of* $\tilde{\eta}_x$ is by definition the largest number r such that the exponential map is embedding in the open ball of radius r in the normal subspace. Since the metric is invariant, the normal injectivity radius of $\tilde{f}_x$ does not depend on $x \in \tilde{\eta}_x$. The infimum of the normal injectivity of all fibers of $\tilde{\eta}$ is called the *normal injectivity radius of* $\tilde{\eta}$.

We can restate (0.2.3) as

(3.1.1) The normal injectivity radius of $\tilde{\eta}$ is greater than $\delta(n) > 0$.

The following is also clear.

(3.1.2) If M has BCG then after suitable normalization, the canonical metric on FM has BCG as well.

Proof of Proposition 2.10: Fix $x \in FM$. Let $B_\rho(x)$ denote the metric ball at x of radius ρ. By (3.1.2) we conclude that there exists $C(n) > 0$ such that for $\rho \leq C(n)$, any loop at x of length $l < 2\rho$, which is homotopically nontrivial in $B_\rho(x)$, is homotopically nontrivial in FM.

Consider any geodesic loop, γ, in the fibre $\tilde{f}_{u,x}$ of $\tilde{f}_u$ at x. By the definition of $\tilde{f}_u$, the loop, $m\gamma$, is homotopically trivial in FM. But by (3.1.1), γ has infinite order in $B_\rho(x)$, provided $\rho + \mathrm{diam}(\tilde{f}_{u,x}) < \delta(n)$. Clearly, the proposition follows. □

4. Proof of Corollary 0.7

Consider a manifold, M, as in Corollary 0.7. By Theorem 0.5, M admits a pure polarized F-structure, $\mathcal{F}$, of rank k. Moreover, the fiber subgroup of some principle fiber is a normal free abelian subgroup of rank $k \geq 1$. Thus, to obtain Corollary 0.7, it is only necessary to check the case $k = 1$. Note that the fiber subgroup is an infinite cyclic normal subgroup of $\pi_1(M)$. Since $\pi_1(M)$ is torsion free, Corollary 0.7 is an immediate consequence of the following elementary algebraic fact.

PROPOSITION 4.1. *Let G be a finitely generated group with generators, $g_0, g_1, \ldots, g_s$. Suppose that G satisfies the following conditions:*
(4.1.1) G has no torsion elements;
(4.1.2) G has no free abelian subgroup of rank ≥ 2;
(4.1.3) The infinite cyclic subgroup generated by g_0 is normal in G.
Then, any two cyclic subgroups of G have non-trivial intersection. In particular, if G is not cyclic, $G/[G,G]$ is a finite group.

In what follows, we shall use $(\alpha_1, \ldots, \alpha_s)$ to denote the subgroup generated by $\alpha_1, \ldots, \alpha_s$.

LEMMA 4.2. *Under the same assumptions as in Proposition 4.1, (g_0) is in the center of G. In particular, any element in G has a power in (g_0).*

Proof: If the statement fails, there exists $g \in G$, such that $g^{-1}g_0g = g_0^{-1}$. Then $g^{-2}g_0g^2 = g_0$, i.e. (g_0, g^2) is an abelian subgroup. Since (4.1.1) and (4.1.2) also apply to the abelian group, (g_0, g^2), this group must be cyclic. Thus, there are positive integers k, m such that $g^{2k} = g_0^m$. From $g^{-2}g_0g^2 = g_0$, we deduce

$$g^{2k} = g^{-1}g^{2k}g = g^{-1}g_0^m g = (g^{-1}g_0g)^m = g_0^{-m} = g^{-2k} .$$

Consequently, $g^{4k} = 1$. This contradicts the assumption that G is torsion free. The second statement is obvious. □

Recall that a group is abelian if and only if its commutator group is trivial. Combining Lemma 4.1 and Lemma 4.3, we obtain

COROLLARY 4.3. *Under the assumptions of Proposition 4.1, if* $(g_0) \cap [G,G] = 1$, *then* G *is infinite cyclic.*

Proof of Proposition 4.1: Assume that G is not infinite cyclic. By Corollary 4.3, $(g_0) \cap [G,G] \neq 1$. For all i, (g_0, g_i) is abelian and satisfies (4.1.1), (4.1.2). Thus it is cyclic. In particular, that $(g_0) \cap [G,G] \neq 1$ implies that $(g_i) \cap [G,G] \neq 1$. Therefore, $G/[G,G]$ is a finite group. □

5. Proof of Corollary 0.12

Proof of Corollary 0.12: i) Let N denote a 3-manifold which satisfies the conditions of Theorem 0.5. Let $f : N \to B_f$ denote an injective fibration on N. If the fiber of f is a circle, then N is a Seifert manifold (see [O]). If the fiber of f has dimension two, then N is the total space of a bundle over a circle with fiber either a torus or a Klein bottle. Clearly, either N or its a double cover is a solvmanifold. If the fiber of f has dimension three, then N is finitely covered by a nilmanifold (see [Gro1]).

Note that in any case either $\pi_1(N)$ has a free abelian subgroup of rank ≥ 2 or $\pi_1(N)$ is cyclic with index ≤ 2.

ii) Let M denote any 4-manifold which satisfies the conditions of Theorem 0.5. Let $f : M \to B_f$ denote the canonical singular fibration on M as in Theorem 0.2. Assume that f is not injective. By Theorem 0.5 and the result in [Gro1], a regular fiber of f has dimension $k = 2$ or 3. If $k = 3$, then B_f is homeomorphic to either a circle or a closed interval. Since f is not injective, B_f is a closed interval. Thus, the holonomy group of f is trivial.

Assume $k = 2$. If the holonomy group of f is not trivial, then for some $x \in M$, there is a loop, γ, in B_f at some $f(x)$, such that the holonomy matrix, A_γ, associated to γ is not the identity matrix. We will show that this implies that the fiber subgroup $i_*\pi_1(f_x, x)$ is finite; a contradiction to Theorem 0.5.

Clearly, we can choose γ so that all fibers over $\gamma(t)$ are regular. Without essential loss of generality, we assume that a regular fiber is a torus. Consider the bundle, $f^{-1}(\gamma(t))$. Identify $f_x \simeq T^2$, and let h_1, h_2 denote two generators of $\pi_1(f_x, x)$. Then,

$$\gamma^{-1} h_1 \gamma = A_\gamma(h_1) = h_1^a h_2^b \,, \quad \gamma^{-1} h_2 \gamma = A_\gamma(h_2) = h_1^c h_2^d \,, \quad ac - bd = 1 \,.$$

Since the fiber subgroup of f_x is not trivial, there is at least one more relation,

$$(h_1^p h_2^q)^m = 1 \,, \quad p,\ q \text{ are coprime integers.}$$

We now choose a new basis, (k_1, k_2), for f_x, where, $k_1 = h_1^p h_2^q$ and $k_2 = h_1^u h_2^v$, $pu - qv = 1$. The new basis satisfies

$$\gamma^{-1}k_1\gamma = k_1^{a'}k_2^{b'}\ , \quad \gamma^{-1}k_2\gamma = k_1^{c'}k_2^{d'}\ , \quad k_1^m = 1\ , \quad a'c' - b'd' = 1\ .$$

We now view k_i as elements in $\pi_1(M, x)$ subject to at least the above three relations. Clearly, since A_γ is not trivial, these relations imply that k_2 also has finite order.

iii) and iv) follow easily from i) and ii). □

6. Examples

Our first example shows that the canonical F-structure on a sufficiently collapsed manifold which satisfies the conditions of Theorem 0.5 need not to be polarized.

EXAMPLE 6.1: Consider the standard T^2-action on $S^2 \times S^1$. Choose any 1-dimensional dense subgroup of T^2, and the product metric of the standard metrics on S^2 and S^1. By scaling the metric on the orbits of the subgroup while leaving it unchanged on the orthogonal direction, one obtains a sequence of metrics, g_δ, on $S^2 \times S^1$, with uniformly bounded curvature and diameter. It is easy to check that the injectivity radius of the pullback metrics on the universal covering space of $S^2 \times S^1$ has a uniform lower bound.

Since the orbit of the 1-dimensional subgroup is dense in T^2, the sequence, $(S^2 \times S^1, g_\delta)$, converges in the Gromov-Hausdorff sense [GroLP], to a closed interval as $\delta \to 0$. Clearly, for δ sufficiently small, the canonical F-structure on $S^2 \times S^1$ as in Theorem 0.2 coincides with the T^2-action. Thus, it is not polarized.

Next, we will give two simple examples showing that Theorem 0.5 fails if one removes any part of the hypothesis without adding other restrictions.

EXAMPLE 6.2: Given any small $v > 0$, let N denote any lens space with natural metric of constant curvature $\equiv 1$ and $\mathrm{Vol}(N) < v$. Note that the smaller the value v, the larger the size $|\pi_1(N)|$. Since $\pi_1(N)$ is finite, N does not admit any pure injective F-structure. This gives a counterexample if assumption (0.5.1) is removed.

EXAMPLE 6.3: Any sufficiently collapsed simply connected manifold with bounded sectional curvature and diameter will serve as a counterexample if one removes (0.5.2). A typical example is Berger's collapse of S^3.

It is easy to see that the canonical pure F-structure on any manifold in the above mentioned examples does have a pure polarized substructure (though not an injective one). It is natural to ask if this is always true.

This question is closely related to problem 3 of the introduction. Below, a slightly more general question is raised.

Consider a canonical fibration, $\tilde{f} : FM \to B_{\tilde{f}}$ and its associated flat bundle $\mathcal{V}FM$ (see §2). Recall that subfibrations of $\tilde{f}$ are in one-to-one correspondence with subtori which are invariant under the holonomy group of $\mathcal{V}FM$. Suppose that $\mathcal{V}FM$ has an $O(n)$-invariant subbundle which is also invariant under the holonomy group. Here we do not assume that each fiber of this bundle contains a cocompact sublattice. The corresponding decomposition of FM into possibly open totally geodesic submanifolds of the fibers of $\tilde{f}$ induces a decomposition of M, whose orbits need not be closed. If all such orbits have the same dimension, this decomposition is called a *pure polarization*. In case the orbits are closed, the notion of pure polarization coincides with that of pure polarized substructure as defined previously.

Consider a manifold which satisfies the conditions of Theorem 0.2. Let $f : M \to B_f$ denote the canonical singular fibration on M.

QUESTION: Does the canonical singular fibration f always have a pure polarization, provided the volume of M is sufficiently small ?

Note that an affirmative answer would imply the minimal diameter-volume conjecture of Cheeger-Gromov (i.e. problem 3). However, the following example shows that the answer is negative.

EXAMPLE 6.4: We will construct a collapsing sequence of compact 4-manifolds, $\{M_i\}$, with uniform bounds on sectional curvature and diameter, such that, for no member of the sequence does the canonical singular fibration have a pure polarization.

We start by constructing two collapsing sequences of metrics with uniformly bounded curvature and diameter, on two fixed compact 4-manifolds with boundary. At every stage, there exists an isometry between the boundaries, but when viewed as maps of the underlying manifolds, these isometries become increasingly twisted, the further out one goes in the sequence. By successively gluing together the boundaries by each member of the sequence of isometries, we obtain a collapsing sequence of closed 4-manifolds (with varying topology).

Step 1: Let Σ^2 be a closed surface with $\partial\Sigma^2 = S^1$. Let M_1^4 be the total space of a T^2-bundle over Σ^2, with holonomy in $SL(2, \mathbf{Z})$, whose restriction to the boundary is trivial i.e. isomorphic to $T^2 \times (1/2, 1] \times S^1$ where $\partial\Sigma^2 = 1 \times S^1$. Then M_1^4 has an obvious T^2-structure whose restriction to a neighborhood of ∂M_1^4 is the obvious T^2-action on $T^2 \times (1/2, 1] \times S^1$.

Let D^2 denote the 2-disk and let $S^1 \times D^2$ have the T^2 action given by rotation in S^1 and rotation in D^2. Extend this action trivially to $M_2^4 =$

$S^1 \times D^2 \times S^1$.

Again, near ∂M_2^4 the action looks like the standard T^2 action on $T^2 \times (1/2, 1] \times S^1 = S^1 \times S^1 \times (1/2, 1] \times S^1$, where in this case, we regard a neighborhood of ∂D^2 as $S^1 \times (1/2, 1]$ and $\partial D^2 = S^1 \times 1$.

Let $\phi \in SL(2, \mathbf{Z})$. Identify ∂M_2^4 with ∂M_1^4 by $\phi \times Id_{1 \times S^1}$. The resulting closed manifold M_ϕ^4 has an obvious T^2 structure, $\mathcal{F}_\phi$.

Step 2: It is easy to choose M_1^4 such that the T^2-structure on M_1^4 has no 1-dimensional polarizations. On the other hand, since the T^2-action on M_2^4 has a 1-dimensional orbit, the corresponding T^2 structure has only 1-dimensional polarizations. Thus, $\mathcal{F}_\phi$ has no pure polarization whatsoever.

Step 3: Choose an invariant metric on M_1^4. By scaling the metric on the fibers while leaving it unchanged in the orthogonal direction we obtain a sequence of metrics, $g_{1,\delta}$, such that

(6.4.1) $|K_{g_{1,\delta}}| \leq 1$;

(6.4.2) $\mathrm{diam}(M_1, g_{1,\delta}) \leq D_1$ for some constant $D_1 > 0$;

(6.4.3) $\mathrm{Vol}(M_1, g_{1,\delta}) \to 0$ as $\delta \to 0$.

Near the boundary we can assume that $g_{1,\delta} = \delta^2 \tilde{g} + dr^2 + k$, where $\tilde{g}$ is some flat metric on T^2, k is the standard metric on the unit circle, S^1, and $dr^2 + k$ is the product metric on $(1/2, 1] \times S^1$. Clearly, the constant D_1 in (6.4.2) can be chosen uniformly, for any collection of metrics, $\tilde{g}$, which are uniformly quasi-isometric to the product metric, $k + k$, on $S^1 \times S^1$. Below, we will consider a sequence of choices from within such a class.

Pick a rotationally invariant metric $\hat{g}$ on D^2 which on a neighborhood, $S^1 \times (1/2, 1]$ of ∂D^2, is isometric to $k + dr^2$. Define a metric g_2 on M_2^4 by $g_2 = k + \hat{g} + k$. Thus, near the boundary $g_2 = k + k + dr^2 + k$.

Choose a 1-dimensional dense subgroup of T^2. Let $g_{2,\delta}$ denote the sequence of invariant metrics obtained by scaling $\hat{g}$ on the orbits of the subgroup while leaving it unchanged in the orthogonal direction. Clearly, $g_{2,\delta}$ also satisfies bounds corresponding to (6.4.1)–(6.4.2).

Step 4: By a well known elementary argument, there is a sequence, $\eta_i \to 0$, such that the following holds. There is a sequence of metrics, $\tilde{g}_i$, on T^2 such that for all i, g_i is uniformly quasi-isometric (via the identity map) to the standard metric, $k + k$. Moreover, near ∂M_2^4, for some sequence $\phi_i \in SL(2, \mathbf{Z})$, we have $g_{2,\eta_i} = \epsilon_i^2 \phi_i^*(\tilde{g}_i) + dr^2 + k$ where $\epsilon_i \to 0$.

Step 5: For $i = 1, 2, \ldots$, consider the metric, g_{1,δ_i} obtained by choosing $\delta_i = \epsilon_i$, $\tilde{g} = \tilde{g}_i$. Clearly, for $M_{\phi_i}^4$ as above, the sequence $\{M_{\phi_i}^4\}$ collapses, with bounded curvature and diameter. Let $f_i : M_{\phi_i}^4 \to B_{f_i}$ denote the canonical singular fibration on $M_{\phi_i}^4$ for i sufficiently large. From the above construction it is clear that f_i coincides with the obvious pure T^2-structure $\mathcal{F}_{\phi_i}$, on M_{ϕ_i}. Thus, $\mathcal{F}_{\phi_i}$ has no 1-dimensional pure polarizations.

Remark 6.5: Recall that previous constructions of collapsing sequence of manifolds with bounded diameter use a (globally defined) *pure* polarization; see [CGro1], [F1–4]. The new method of Example 6.4 suggests the likelihood of a negative answer to problem 3 in the introduction. We intend to discuss this further elsewhere.

Our last example concerns Corollary 0.7. Assume that M admits an injective S^1-fibration and $\pi_1(M)$ is torsion free (one can even assume that M is $K(\pi,1)$-space). A purely topological question is the following.

QUESTION: Does $\pi_1(M)$ of rank one imply that $\pi_1(M)$ is cyclic ?

The following example is constructed in [A].

EXAMPLE 6.6: Given an integer, $m \geq 1$, and odd $n \geq 665$. Let $A(m,n)$ denote the group given by generators,

$$g_1, \ldots, g_m, g_0 ,$$

and the system of defining relations,

$$g_i g_0 = g_0 g_i , \quad W^n = g_0 ,$$

where W denotes any elementary word. In [A], the following properties of $A(m,n)$ were proved:

(6.6.1) $A(m,n)$ is torsion free;

(6.6.2) $A(m,n)$ has rank one;

(6.6.3) The quotient group of $A(m,n)$ by its center (a cyclic group) is finitely generated infinite torsion group.

Since any finitely presented group can be realized as the fundamental group of a 4-manifold, Example 6.6 shows that the question above is non-trivial.

7. A Refinement of Theorem 0.5

In this section, we state a refined version, of Theorem 0.5 in which the full N-structure plays a role. Since for the most part, the proof is just a technical extension of that of Theorem 0.5, we will omit it. As an example given at the end of this section illustrates, the refined version, Theorem 7.3, actually does provide new information, beyond that in Theorem 0.5.

Let N be a closed nilmanifold, $N \simeq \tilde{N}/\Gamma$, where $\tilde{N}$ is a simply connected nilpotent Lie group and Γ a lattice. The group, $\tilde{N}$, has derived series

$$\tilde{N}_r \subset \tilde{N}_{r-1} \subset \cdots \subset \tilde{N}_1 = \tilde{N} ,$$

where

$$\tilde{N}_1 = [\tilde{N}, \tilde{N}] , \ \tilde{N}_2 = [\tilde{N}_1, \tilde{N}], \ldots, \tilde{N}_r = [\tilde{N}_{r-1}, \tilde{N}] , \ \tilde{N}_{r+1} = [\tilde{N}_r, \tilde{N}] = 1 .$$

For each $1 \le i \le r$, put $N_i = \widetilde{N}_i/\widetilde{N}_i \cap \Gamma$. Clearly, N_i is a closed submanifold of N and

$$N_r \subset N_{r-1} \subset \cdots \subset N_1 \subset N .$$

In addition, $N_i/N_{i+1} \simeq T^{k_i}$ is a torus of positive dimension. Finally, any affine isomorphism of $\widetilde{N}$ which is Γ-invariant preserves the derived series of N.

We now consider a bundle, $p : E \to B$, with fiber a closed nilmanifold N and affine structural group (see §2). Each manifold N_i as above determines a subbundle of E with fiber N_i. We will denote this subbundle by $p_i : E \to B_i$.

Associated to the sequence $\{P_i\}$ is a sequence of quotient bundles where for each i, the bundle, $\hat{p}_i : B_{i+1} \to B_i$, is a torus bundle with fiber $T^{k_i} \simeq N_i/N_{i+1}$.

Let $p_s : E \to B_s$ be a subbundle of p. Observe that for each i, the intersection of p_s with p_i is a subbundle of p_i. In an obvious way, this subbundle determines a subbundle, $\hat{p}_{s,i}$, of $\hat{p}_i$.

DEFINITION 7.1: Let $p : E \to B$ be a fiber bundle with fiber a closed nilmanifold and affine structural group. Let $p_s : E \to B_s$ be a subbundle of p. We call p_s a *strongly maximal injective* subbundle of p if for each $1 \le i \le r$, $\hat{p}_{s,i}$ is a nontrivial maximal injective subbundle of $\hat{p}_i$, i.e. $\hat{p}_{s,i}$ is injective, and is not a proper subbundle of any injective subbundle of $\hat{p}_i$.

Remark 7.2: Note that the condition "strongly maximal injective subbundle" in the above is stronger than the condition "maximal injective" since in the former case each quotient subbundle is required to be nontrivial.

The natural N-structure on a closed nilmanifold is itself strongly maximal injective. Note also that a strongly maximal injective subbundle can be a proper subbundle (see Example 7.4). On the other hand, a given bundle may have no strongly maximal injective subbundle. For instance, if such a subbundle exists, then the fundamental group of E is infinite.

Let M be a collapsed manifold in $\mathcal{M}^n_D$, and $\tilde{\eta} : FM \to B_{\tilde{\eta}}$, the canonical nilpotent fibration of the frame bundle. Assume that $\tilde{\eta}$ has a strongly maximal injective subbundle. We will call the substructure of the canonical N-structure on M, defined by the projection of the maximal injective subbundle, a *strongly maximal injective nilpotent substructure.*

THEOREM 7.3. *Let the assumptions be as in Theorem 0.5. Then, the nilpotent fibration on the frame bundle, has a strongly maximal injective subbundle. Equivalently, the canonical N-structure on M has a pure strongly maximal injective nilpotent substructure.*

EXAMPLE 7.4: Let $M = S^2 \times S^1 \times N^3$, where N^3 is a closed nilpotent 3-manifold. We will first construct a sequence of collapsing metrics on M with bounded diameter and bounded covering geometry. Then we will compare Theorem 7.3 with Theorem 0.5 in the context of this sequence.

Let $g_{1,\delta}$ denote a sequence of collapsed metrics on $S^2 \times S^1$ as in Example 6.1. In particular, $g_{1,\delta}$ ($\delta \to 0$) has bounded diameter and bounded covering geometry. Let $g_{2,\delta}$ ($\delta \to 0$) denote a standard sequence of collapsed metrics on N^3 such that the limit space is a point. It is clear that $g_{2,\delta}$ also has bounded covering geometry. Thus, for each δ, the product, $g_\delta = g_{1,\delta} \times g_{2,\delta}$, is a sequence of collapsed metrics on M with bounded diameter and bounded covering geometry.

From the way g_δ is constructed, it is clear that the canonical N-structure of a sufficiently collapsed g_δ is independent of δ, i.e. it is a fixed N-structure. This canonical N-structure, $\mathcal{N}$, on M is defined by the product of the T^2-action on $S^2 \times S^1$ and the canonical N-structure on N^3. The canonical F-structure, $\mathcal{F}$, is defined by the product of the T^2-action on $S^2 \times S^1$ and the S^1-action on N^3 coming from its center.

It is clear that any injective substructure of $\mathcal{F}$ (by Theorem 0.5) has one of the following forms:

(7.4.1) A S^1-action on $S^2 \times S^1$ without fixed points.

(7.4.2) The product of a S^1-action on $S^2 \times S^1$ without fixed point and the S^1-action on N^3 by its center.

It is also obvious that any strongly maximal injective nilpotent substructure of $\mathcal{N}$ is the product of a S^1-action on $S^2 \times S^1$ without fixed point and the standard N-structure on N^3.

References

[A] S.I. ADIAN, The Burnside problem and identities in groups, Ergebn. d. Math Springer 95 (1979), 169-176.

[B] G. BREDON, Introduction to Compact Transformation Groups, Academic Press (1972).

[BuK] P. BUSER, H. KARCHER, Gromov's almost flat manifolds, Asterisque 81 (1981), 1-148.

[Buy] V. BUYDO, Volume and the fundamental group of a manifold of non-positive curvature, Math. USSR Sobornik 50 (1985), 137–150.

[C] J. CHEEGER, Finiteness theorems for Riemannian manifolds, Am. J. Math 92 (1970), 61–75.

[CFGro] J. CHEEGER, K. FUKAYA, M. GROMOV, Nilpotent structures and invariant metrics on collapsed manifolds, J. A.M.S. 5 (1992), 327–372.

[CGro1] J. CHEEGER, M. GROMOV, Collapsing Riemannian manifolds while keeping their curvature bound I, J. Diff. Geom. 23 (1986), 309–364.

[CGro2] J. CHEEGER, M. GROMOV, Collapsing Riemannian manifolds while keeping their curvature bound II, J. Differential Geom. 32 (1990), 269–298.

[CGro3] J. CHEEGER, M. GROMOV, Bounds on the von Neumann dimension of L^2-cohomology and the Gauss-Bonnet theorem for open manifolds, J. Differential Geom. 21 (1985), 1–34.

[CR] J. CHEEGER, X. RONG, In preparation.

[F1] K. Fukaya, Collapsing Riemannian manifolds to ones of lower dimensions, J. Diff. Geom. 25 (1987), 139–156.
[F2] K. Fukaya, A boundary of the set of Riemannian manifolds with bounded curvatures and diameters, J. Diff. Geom. 28 (1988), 1–21.
[F3] K. Fukaya, Collapsing Riemannian manifolds to ones of lower dimensions II, J. Math. Soc. Japan 41 (1989), 333–356.
[F4] K. Fukaya, Hausdorff convergence of Riemannian manifolds and its applications , Recent Topics in Differential and Analytic Geometry (T. Ochiai, ed), Kinokuniya, Tokyo (1990).
[GWu] R. Greene, H. Wu, Lipschitz convergence of Riemannian manifolds, Pacific J. Math. 131 (1988), 119–141.
[GrW] D. Gromoll, J. Wolf, Some relations between the metric structure and the algebraic structure of the fundamental group in manifolds of nonpositive curvature, Bull. Amer. Math. Soc. 77 (1971), 545–552.
[Gro1] M. Gromov, Almost flat manifolds, J. Diff. Geom. 13 (1978), 231–241.
[Gro2] M. Gromov, Volume and bounded cohomology, I.H.E.S. Publ. Math. 56 (1983), 213–307.
[GroLP] M. Gromov, J. Lafontaine, P. Pansu, Structures Metriques pour les Varietes Riemannienes, Cedic-Fernand Paris (1981).
[GroP] M. Gromov, P. Pansu, Rigidity of lattices: An introduction, preprint (1991).
[LaY] B. Lawson, S-T. Yau, Compact manifolds of nonpositive curvature, J. Diff. Geom. 7 (1972), 211-228.
[N] I.G. Nikolaev, Parallel translation and smoothness of the metric of spaces of bounded curvature, Soviet Math. Dokl. 21 (1980), 263–265.
[O] P. Orlik, Seifert Manifolds, Lecture Note in math., Springer Berlin 291 (1972).
[Pe] S. Peters, Convergence of Riemannian manifolds, Comp. Math. 62 (1987), 3–16.
[R1] X. Rong The limiting eta invariant of collapsed 3-manifolds, J. Diff. Geom. 37 (1993), 535-568.
[R2] X. Rong, Rationality of geometric signatures of complete 4-manifolds, Invent. Math (to appear).
[Ru] E. Ruh, Almost flat manifolds, J. Diff. Geom. 17 (1982), 1-14.
[SSh] I. Sabitov, S. Shefel, The connection between the order of smoothness of a surface and its metric, Sibirsk. Matem. Zh. 17 (1976), 916–925.

Jeff Cheeger
Courant Institute of Math. Sci.
251 Mercer Street
New York, NY 10012
USA

Xiachun Rong
Dept. Math.
Columbia University
New York, NY 10027
USA
Current address:
Dept. of Math.
University of Chicago
Chicago, IL 60637
USA

Submitted: August 1994

Geometric And Functional Analysis
Vol. 5, No. 2 (1995)
1016-443X/95/0200164-10$1.50+0.20/0

UNITARITY AND FUNCTORIALITY

J.W. COGDELL AND I.I. PIATETSKI-SHAPIRO

Dedicated to M. Gromov on the occasion of his 50th birthday

Let G be a reductive algebraic group defined over a global field k. Denote by $A(G)$ the set of all automorphic representations of $G(\mathbb{A})$. Denote by $A^s(G)$ the set of all unitary automorphic representations which occur in Langlands' spectral decomposition of $L^2(G(k)\backslash G(\mathbb{A}))$ and $A^d(G)$ those which occur discretely. Throughout this paper, cuspidal automorphic representations will mean unitary cuspidal automorphic representations.

Let G and H be split reductive groups over k. Assume that there exists a homomorphism of L-groups

$$\rho : {}^LH \to {}^LG .$$

According to Langlands' functoriality conjecture ρ induces a map, which we denote by $\hat{\rho}$,

$$\hat{\rho} : A(H) \to A(G) .$$

If $\pi \in A(H)$ we will call its image under $\hat{\rho}$ a *Langlands lift* of π corresponding to ρ. We also have compatible local versions $\hat{\rho}_v$ of this lifting. We will say the $\Pi \in A(G)$ is a *weak Langlands lifting* (or simply a weak lifting) of π if for all places v of k where π_v is unramified we have Π_v is the local Langlands lifting of π_v. The difference between a weak lifting and Langlands lifting is that for weak lifting we do not require local compatibility at the archimedean places or at the finite ramified places.

In this paper we would like to investigate the behavior of $A^s(H)$ under Langlands liftings, in particular, to investigate conditions under which $\hat{\rho}(\pi) \subset A^s(G)$ for $\pi \in A^s(H)$.

The first author was supported in part by NSA grants MDA904-91-H-0040 and MDA904-93-H-3028. The second author was supported in part by NSF grants DMS-8807336 and DMS-9302732.

1. Sufficient Conditions

We begin by giving sufficient conditions on $\pi \in A^s(H)$ to ensure that $\hat{\rho}(\pi) \subset A^s(G)$.

THEOREM 1. *Let $\pi \in A^s(H)$. Assume that there exists a place v such that π_v is a tempered unramified representation. Then $\hat{\rho}(\pi)$ will be contained in $A^s(G)$.*

Proof: Let $\Pi \in \hat{\rho}(\pi)$ be a lifting (which we assume to exist). From the definition of Langlands functoriality it follows that $\Pi_v = \hat{\rho}(\pi_v)$ will also be a tempered unramified representation. In other words, Π_v will be a constituent of a representation induced from a unitary unramified character the Borel group.

According to [L] there exist cuspidal automorphic representations $\sigma_1, \ldots, \sigma_r$ and real numbers $t_1, \ldots, t_r$ such that

$$\Pi \subset \mathrm{Ind}_P^G\, \sigma_1 |\det|^{t_1} \otimes \ldots \otimes \sigma_r |\det|^{t_r}$$

where P is a parabolic subgroup of G. The meaning of this formula is that Π is a constituent of the induced representation.

In order to prove our theorem it is enough to prove that $t_1 = \cdots = t_r = 0$.

Denote by $\lambda_1, \ldots, \lambda_n$ the eigenvalues of the Satake conjugacy class in ${}^L H$ which corresponds to π_v. Since π is tempered $|\lambda_i| = 1$ for all i. Denote by Λ_i, $i = 1, \ldots, N$ the eigenvalues of the Satake class in ${}^L G$ which corresponds to Π_v. It is easy to see that we will again have $|\Lambda_i| = 1$.

Since Π_v is unramified at v the representations $\sigma_{i,v}$ are also unramified and

$$\Pi_v \subset \mathrm{Ind}_P^G\, \sigma_{1,v} |\det|^{t_1} \otimes \cdots \otimes \sigma_{r,v} |\det|^{t_r} .$$

Denote by $\mu_{\ell,1}, \ldots, \mu_{\ell,m_\ell}$ the eigenvalues of the Satake class for $\sigma_{\ell,v}$. Since σ_ℓ is a cuspidal representation and hence unitary, we have

$$\left| \prod_{i=1}^{m_\ell} \mu_{\ell,i} \right| = 1 .$$

Since each $\mu_{\ell,i} q_v^{t_\ell}$ is one of the Λ_j, this implies $t_\ell = 0$. Hence Π is unitarily induced from cuspidal representation and $\Pi \in A^s(G)$. □

DEFINITION. *Let π be an automorphic representation of the group H. We will call π weakly Ramanujan if there exists an infinite sequence of places v_m such that*

(i) *the local components π_{v_m} are unramified with Satake eigenvalues $\{\lambda_{v_m,i}\}$ and*

(ii) *for every $\varepsilon > 0$ we have $\max_i\{|\lambda_{v_m,i}|, |\lambda_{v_m,i}^{-1}|\} = O(q_{v_m}^{\varepsilon})$.*

From the result of [JS] it follows that any cuspidal automorphic representation of $GL(3)$ is weakly Ramanujan.

The following theorem can be proved along the lines of Theorem 1.

THEOREM 1′. *Assume that $\pi \in A^s(H)$ is weakly Ramanujan. Then $\hat{\rho}(\pi)$ will be contained in $A^s(G)$.*

Proof: Let $\Pi \in \hat{\rho}(\pi)$ be some Langlands lift of π. As in Theorem 1, we can realize Π as a constituent of an induced representation

$$\Pi \subset \mathrm{Ind}_P^G\, \sigma_1|\det|^{t_1} \otimes \cdots \otimes \sigma_r|\det|^{t_r}$$

with the σ_i cuspidal and we need to show that each $t_i = 0$.

For each place v where the local component π_v is unramified we let $\lambda_{v,1}, \ldots, \lambda_{v,n}$ denote the eigenvalues of the Satake class associated to π_v. At these places Π_v will also be unramified and we let $\Lambda_{v,1}, \ldots, \Lambda_{v,N}$ denote the eigenvalues of the Satake class associated to Π_v. Since π is weakly Ramanujan we have a sequence of places $\{v_m\}$ such that for every $\varepsilon > 0$ we have a constant c_ε such that

$$c_\varepsilon^{-1} q_{v_m}^{-\varepsilon} < |\lambda_{v_m,i}| < c_\varepsilon q_{v_m}^{\varepsilon}$$

for each $i = 1, \ldots, n$. It is easy to see from finite dimensional representation theory that there will be a constant $c(\rho)$ depending only on ρ and for each $\varepsilon > 0$ a constant $c_{\varepsilon,\rho}$ such that

$$c_{\varepsilon,\rho}^{-1} q_{v_m}^{-c(\rho)\varepsilon} < |\Lambda_{v_m,i}| < c_{\varepsilon,\rho} q_{v_m}^{c(\rho)\varepsilon}$$

for $i = 1, \ldots, N$.

For the places where Π_v is unramified, we must have that the local components $\sigma_{\ell,v}$ of the cuspidal representations σ_ℓ are unramified. Let $\mu_{\ell,v,1}, \ldots, \mu_{\ell,v,m_\ell}$ denote the Satake parameters for $\sigma_{\ell,v}$. Since each $\mu_{\ell,v,i} q_v^{t_\ell}$ must be a $\Lambda_{v,j}$, then at the places where π is weakly Ramanujan we have an estimate

$$c_{\varepsilon,\rho}^{-1} q_{v_m}^{-c(\rho)\varepsilon} < |\mu_{\ell,v_m,i} q_{v_m}^{t_\ell}| < c_{\varepsilon,\rho} q_{v_m}^{c(\rho)\varepsilon} .$$

Again, since each σ_{ℓ,v_m} is cuspidal we know

$$\left| \prod_{i=1}^{m_\ell} \mu_{\ell,v_m,i} \right| = 1 .$$

This gives us

$$c_{\varepsilon,\rho}^{-1} q_{v_m}^{-c(\rho)\varepsilon} < q_{v_m}^{t_\ell} < c_{\varepsilon,\rho} q_{v_m}^{c(\rho)\varepsilon}$$

or

$$-c(\rho)\varepsilon - \frac{\log(c_{\varepsilon,\rho})}{\log(q_{v_m})} < t_\ell < c(\rho)\varepsilon + \frac{\log(c_{\varepsilon,\rho})}{\log(q_{v_m})} .$$

If we now let m go to infinity, we obtain

$$-c(\rho)\varepsilon < t_\ell < c(\rho)\varepsilon$$

for each $\varepsilon > 0$, and hence $t_\ell = 0$. □

Now we restrict our attention to general linear groups. Consider the Langlands lifting which corresponds to a homomorphism $\rho : GL(n, \mathbb{C}) \to GL(N, \mathbb{C})$. It is enough to consider the case when ρ is irreducible. Our aim is to formulate sufficient conditions which imply that

$$\hat{\rho}(A^s(GL(n))) \subset A^s(GL(N)) \ .$$

Let us assume that any cuspidal automorphic representation of $GL(m)$, $m \le n$, is weakly Ramanujan.

Assume first that the representation $\pi \in A^s(GL(n))$ is generic. In this case we can get π by unitary induction from cuspidal automorphic representations $\sigma_1, \ldots, \sigma_m$ of the linear groups $GL(n_1), \ldots, GL(n_m)$, where $n = n_1 + \cdots + n_m$. According to the usual formalism of functoriality, to analyze the lift we have to restrict ρ to

$$GL(n_1, \mathbb{C}) \times \cdots \times GL(n_m, \mathbb{C}) \subset GL(n, \mathbb{C}) \ .$$

Then ρ becomes reducible and we have

$$\rho = \bigoplus_k \rho_1^k \otimes \cdots \otimes \rho_m^k \ ,$$

where ρ_i^k will be a homomorphism of $GL(n_i, \mathbb{C})$ into $GL(N_i^k, \mathbb{C})$.

We now assume that for any ρ_i^k there exists a corresponding Langlands lifting. Now we can apply Theorem 1′ to each σ_i. From Theorem 1′ we obtain that the Langlands lifting Σ_i^k of σ_i which corresponds to the homomorphism ρ_i^k is lying in the spectral decomposition of $GL(N_i^k)$. Let $N^{(k)} = \prod N_i^k$ and let Σ^k denote the Langlands lifting of $\Sigma_1^k \otimes \cdots \otimes \Sigma_m^k$ to $GL(N^{(k)})$ associated to the natural embedding of L-groups $GL(N_1^k, \mathbb{C}) \otimes \cdots \otimes GL(N_m^k, \mathbb{C}) \hookrightarrow GL(N^{(k)}, \mathbb{C})$. Arguing by Satake parameters as in the proof of Theorem 1′ we see that $\Sigma^k \in A^s(GL(N^{(k)}))$. Now we get Π by unitary induction of the representation $\Sigma = \otimes_k \Sigma^k$ from the standard parabolic associated to the partition $N = \sum N^{(k)}$ to $GL(N)$. Thus $\Pi \in A^s(GL(N))$.

We next consider the case when π is not generic. It is clear that it is enough to consider the case when π is a representation of the discrete spectrum but not cuspidal, since any representation lying in Langlands' spectrum decomposition we can get by unitary induction from representations of the discrete spectrum. In this case, we have to use the formalism which was suggested by J. Arthur ([AG]). Now we restrict ρ to

$$GL(m) \times \alpha_\ell(SL(2)) \ ,$$

where $n = m\ell$, and α_ℓ is the irreducible ℓ-dimensional representation of $SL(2)$. In order to explain how to attach m and ℓ to a given representation π, we recall the following. It is possible to get any representation of the discrete spectrum of $GL(n)$ in the following way. Put $n = m\ell$. Denote by σ a cuspidal automorphic representation of $GL(m)$. Then all the discrete non-cuspidal spectrum is obtained as the unique irreducible quotient of the induced representation $\mathrm{Ind}_P^{GL(n)}\ \sigma|\det|^{(\ell-1)/2}\otimes\cdots\otimes\sigma|\det|^{-(\ell-1)/2}$, where P is the standard parabolic associated to the partition $n = m + \cdots + m$. We now put

$$\rho(GL(m) \times \alpha_\ell(SL(2))) = \bigoplus_j \rho_j \otimes \alpha_{\ell_j} \ .$$

Now we can apply the result of Theorem 1$'$ as in the generic case.

Combining these two cases, we obtain the following theorem.

THEOREM 2. *Assume that every cuspidal representation of $GL(m)$, $m \leq n$, is weakly Ramanujan and that all Langlands liftings from the various $GL(n_i)$ to $GL(N_i^k)$ needed above exist. Then for any homomorphism $\rho : GL(n,\mathbb{C}) \to GL(N,\mathbb{C})$ we have $\hat\rho(A^s(GL(n)) \subset A^s(GL(N))$.*

2. An Example

If the target group G is not a $GL(N)$ then it is no longer true that under Langlands lifting $A^s(H)$ necessarily goes into $A^s(G)$. In fact, there are cases where certain representations in $A^s(H)$ do not even have weak Langlands lifts which lie in $A^s(G)$. The example presented below is the first example of this phenomenon.

Our example is the following. Put $H = SO(3)\times SO(3)$, and $G = SO(5)$, both groups split. Then ${}^LH = SL(2,\mathbb{C}) \times SL(2,\mathbb{C})$, and ${}^LG = Sp(4,\mathbb{C})$. It is obvious that there exists an embedding

$$\rho : {}^LH \to {}^LG \ .$$

Denote by σ a cuspidal automorphic representation of $PGL(2) \simeq SO(3)$ and by $\mathbf{1}$ the trivial representation. We will prove that if $L(\sigma\otimes\chi,\frac{1}{2}) \equiv 0$ for every quadratic character χ, then $\hat\rho(\sigma\otimes\mathbf{1}) \notin A^s(G)$. In fact, we will prove $\sigma\otimes\mathbf{1}$ does not have any weak lifting in $A^s(G)$. It is not difficult to show, that if $L(\sigma\otimes\chi,\frac{1}{2}) \not\equiv 0$, then $\hat\rho(\sigma\otimes\mathbf{1}) \in A^s(G)$. Even more, $\hat\rho(\sigma\otimes\mathbf{1})$ is lying discretely in $L^2(G(k)\backslash G(\mathbb{A}))$. If $L(\sigma,\frac{1}{2}) = 0$ and $L(\sigma\otimes\chi,\frac{1}{2}) \not\equiv 0$, then $\hat\rho(\sigma\otimes\mathbf{1})$ will be a cuspidal automorphic representation of G.

The existence of automorphic cuspidal representations σ of $PGL(2)$, such that

$$L(\sigma\otimes\chi,\tfrac{1}{2}) \equiv 0 \ ,$$

where χ is an arbitrary quadratic character, was proved by Waldspurger ([W]). He also proved that the image of such a representation is always 0 under the θ-correspondence

$$\widetilde{SL(2)} \leftrightarrow PGL(2) ,$$

for an arbitrary choice of additive character ψ. At the same time the image of the local correspondence will always be not trivial. The reason for this is that according to Waldspurger, if $L(\sigma \otimes \chi, \frac{1}{2}) \equiv 0$, then any local component σ_v of σ will be a principal series representation. Interesting examples of such representations over $\mathbf{Q}$ are given in [R].

THEOREM 3. *If σ is a cuspidal automorphic representation of $PGL(2)$ such that for all quadratic characters χ we have $L(\sigma \otimes \chi, \frac{1}{2}) \equiv 0$ then $\hat{\rho}(\sigma \otimes \mathbf{1}) \notin A^s(G)$. In fact, no weak Langlands lift of $\sigma \otimes \mathbf{1}$ can lie in $A^s(G)$.*

Before we begin the proof of Theorem 3, let us recall the following idea ([P-S1]). Denote by P the maximal parabolic subgroup of G which has an abelian radical. Denote by Q the maximal parabolic of G with nonabelian radical.

DEFINITION. *Let S be the abelian radical of P. Let k_v be a local field, and let (π, V) be a representation of G_v. Let ψ_T be a nondegenerate character of S_v. We say that (π, V) has the U-property (uniqueness) if, whenever a linear functional ℓ_T satisfies*

$$\ell_T\big(\pi(s)v\big) = \psi_T(s)\ell_T(v)$$

($s \in S_v$, $v \in V$, T nondegenerate), then

$$\ell\big(\pi(\delta)v\big) = \ell_T(v) \quad \text{for all } \delta \in O_T^\circ,$$

where O_T° is the connected component of the stabilizer of the character ψ_T in the Levi subgroup of P_v.

Proof of Theorem 3: Let us assume that Π is a weak Langlands lift of $\sigma \otimes \mathbf{1}$ which lies in $A^s(G)$. Put $\Pi = \bigotimes_v \Pi_v$, where $\Pi_v = \hat{\rho}(\sigma_v \otimes \mathbf{1}_v)$ if σ_v is unramified. Then at these unramified places

$$L(\Pi_v, s) = L(\sigma_v, s)(1 - q_v^{-s-\frac{1}{2}})^{-1}(1 - q_v^{-s+\frac{1}{2}})^{-1}.$$

From [P-S1] it follows that Π_v has the U-property at these places.

Now we are going to prove that Π has the global U-property. This means that if

$$\varphi_T(g) = \int_{S(k)\backslash S(\mathbb{A})} \varphi(xg)\psi_T^{-1}(x)dx \tag{1}$$

is a nondegenerate Fourier coefficient, then

$$\varphi_T(\delta g) = \varphi_T(g) \quad \text{for all} \quad \delta \in O_T^\circ(\mathbb{A}) \, . \tag{2}$$

This follows immediately from weak approximation since we already know that (2) is true for all $\delta \in O_T^\circ(k)$ (because φ is automorphic) and for all $\delta \in O_T^\circ(k_v)$, for almost all v (all v for which σ_v is unramified). This implies ([P-S1]) that Π_v has the U-property for all v.

From the fact that Π_v satisfies the U-property at all places and our assumption that Π_v is unitary, we obtain that each Π_v is a θ-lifting from $\widetilde{SL}(2)$. Hence we get that

$$\Pi = \otimes \theta_v(\nu_v) \, ,$$

where θ_v means a local θ-lifting. It remains to prove that there exists an automorphic representation ν of $\widetilde{SL}(2)$ such that Π will be a global theta lifting of ν. In order to prove this we will use the following interpretation of ν. Let us recall that according to [P-S1] the representation Π remains irreducible after restriction to $Q(\mathbb{A})$ and it can be written in the form

$$\Pi = \operatorname{Ind}_D^Q \nu \otimes \beta_\psi$$

where β_ψ is the Weil representation of the unipotent radical of Q, which is a Heisenberg group. This shows that we can choose ν to be an automorphic representation. (Π is an automorphic representation!) Hence Π is lying in the image of the so-called Saito-Kurokawa lifting [P-S2].

Let τ denote the theta lifting of ν to PGL_2. As was proved in [P-S2] or [W] this implies that $L\left(\tau \otimes \chi, \frac{1}{2}\right) \neq 0$ for some choice of quadratic χ. A consequence of the fact that Π is the theta lift of ν is then that

$$L^S(\Pi, s) = L^S(\tau, s)\zeta^S\left(s + \tfrac{1}{2}\right)\zeta^S\left(s - \tfrac{1}{2}\right) \, ,$$

where S is a sufficiently large finite set of places such that S includes all archimedean places and Π_v is unramified at all places outside of S. However, from the definition of Π it follows that

$$L^S(\Pi, s) = L^S(\sigma, s)\zeta^S\left(s + \tfrac{1}{2}\right)\zeta^S\left(s - \tfrac{1}{2}\right) \, .$$

By strong multiplicity one for PGL_2 this gives $\sigma = \tau$. This contradicts our assumption that $L(\sigma \otimes \chi, \frac{1}{2}) \equiv 0$. □

Waldspurger proved that if $L(\sigma \otimes \chi, \frac{1}{2}) \equiv 0$ for any quadratic character χ, then any local component σ_v will be a principal series representation. Using this fact it is easy to check that representation $\hat{\rho}(\sigma \otimes 1)$ will be a constituent of some induced representation. It is not difficult to check that each component of this representation will be equal to a local θ-lifting. Hence this representation will be abstractly unitarizable. It is also not difficult to check that there exists a lift of this representation to $GL(4)$, which will be lying in $A^s(GL(4))$.

3. Conjectures

To end we would like to formulate the following conjectures.

CONJECTURE 1. *Let H and G be reductive groups over a global field k and assume that G is split. Consider the weak Langlands lifting associated to a homomorphism $\rho : {}^LH \to {}^LG$. If $\pi \in A^d(H)$ then there always exists a weak lifting $\Pi \in A(G)$ which is abstractly unitarizable.*

CONJECTURE 2. *Let H be a reductive group over a global field k and let $G = GL(n)$. Consider the weak Langlands lifting associated to a homomorphism $\rho : {}^LH \to {}^LG$. Then for any $\pi \in A^s(H)$ there exists a weak lifting $\Pi \in A^s(G)$.*

To establish the second conjecture it would of course be enough to prove it for $\pi \in A^d(H)$.

In order to explain why these conjectures are in terms of weak Langlands lifts rather than Langlands lifts, we consider the following example. Let $H = GSp(4)$ and $G = GL(4)$ over the field $k = \mathbb{Q}$. Consider the Langlands lifting associated to the natural embedding of L–groups

$$\rho : {}^LH = GSp(4, \mathbb{C}) \to {}^LG = GL(4, \mathbb{C}) .$$

We will give an example of a $\pi \in A^d(H)$ such that any Langlands lift Π^L of π will not lie in $A^s(G)$ but there will be a weak lift $\Pi^w \in A^s(G)$.

For π we take the cuspidal automorphic representation associated to a holomorphic Siegel modular form of level 1, of sufficiently high weight so that π_∞ is a discrete series and such that π lies in the image of the Saito-Kurokawa lift. This is Kurokawa's original example ([K]). Either from Kurokawa or as in the proof of Theorem 3, we have associated to π a cuspidal representation τ of $PGL(2)$ which is again unramified at all finite places and such that

$$L_f(\pi, s) = L_f(\tau, s)\zeta(s + \tfrac{1}{2})\zeta(s - \tfrac{1}{2}) .$$

As a weak lifting of π we may take

$$\Pi^w = \mathrm{Ind}_{P_{2,2}}^G \tau \otimes \mathbf{1}$$

where $P_{2,2}$ is the parabolic subgroup of $GL(4)$ with Levi $GL(2) \times GL(2)$ and we have put τ on the first $GL(2)$ and the trivial representation $\mathbf{1}$ on the second $GL(2)$. Π^w then lies in $A^s(G)$.

Any Langlands lift Π^L of π must agree with Π^w at all finite places, but Π^L_∞ must be tempered by the Langlands classification over $\mathbb{R}$. Therefore we cannot have $\Pi^L_\infty = \mathrm{Ind}\ \tau_\infty \otimes \mathbf{1}_\infty$ since the latter representation is not tempered. Therefore Π^L cannot lie in $A^s(G)$ since we have strong multiplicity one in this space.

In fact,

$$\Pi_\infty^L = \mathrm{Ind}_{P_{2,2}}^G \ \tau_\infty \otimes St_\infty$$

where by St we mean the Steinberg representation of $PGL(2)$. Both of the representations Π^w and Π^L are constituents of

$$\mathrm{Ind}_{P_{2,1,1}}^G \ \tau \otimes |\ |^{1/2} \otimes |\ |^{-1/2} .$$

This shows that both of these representations are indeed automorphic. Π^w lies as a proper automorphic representation while Π^L is only a sub-quotient of the space of automorphic forms.

Let us introduce a notion of *totally unitary* representations for groups over local fields. Let π_v a be representation of a group G_v over the local field k_v. We will call π_v totally unitary if it is unitary and remains unitary under any Langlands lifting to a split group. With this definition, Conjecture 1 would have the following corollary.

COROLLARY TO CONJECTURE 1. *Let G be a reductive group over a global field k and let $\pi \in A^d(G)$. Then every local component π_v of π is totally unitary.*

This corollary greatly restricts the local components of elements of $A^d(G)$, and in particular of cuspidal representations.

Let us consider now the Arthur group ([AG]). For the unramified case, this is just the product $\mathbb{Z} \times SL(2,\mathbb{C})$. For the archimedean case it will be $W \times SL(2,\mathbb{C})$ where W is the Weil group of the archimedean field. If we modify the Arthur group by replacing $SL(2,\mathbb{C})$ by $SU(2)$, then it is clear that representations of the Arthur group and representations of this modified group which you get by restriction will be the same. To unitary representations of the modified Arthur group correspond totally unitary representations of G_v, but not conversely. The correspondence between representations of G_v and homomorphisms of the Arthur group into ${}^LG_v$ is described in [AG].

CONJECTURE 3. *Let G be a reductive group over a global field k. Let $\pi \in A^d(G)$. For each place v such that π_v is unramified we have an associated unitary homomorphism of the modified Arthur group $\rho_v : \mathbb{Z} \times SU(2) \to {}^LG_v$ such that each of the maps ρ_v are of the same type, i.e. the restriction $\rho_v : SU(2) \to {}^LG^o$ is independent of the place v.*

Conjecture 3 is the correct generalization of the Ramanujan conjecture to arbitrary groups. It says that the deviation from the usual Ramanujan conjecture is uniform for all unramified components of a given cuspidal representation. Conjecture 3 follows from Conjecture 2 and the Ramanujan conjecture for $GL(n)$.

Acknowledgement. We would like to thank J. Arthur, P. Deligne, R. Howe, and R. Langlands for many fruitful discussions on this topic.

References

[AG] J. Arthur, S. Gelbart, Lectures on automorphic L-functions, in "L-functions and Arithmetic: Proceedings of the Durham Symposium, July, 1989" London Mathematical Society Lecture Notes 153, Cambridge University Press, Cambridge (1991), 1–59.

[JS] H. Jacquet, J. Shalika, On Euler products and the classification of automorphic representations, I, Am. J. Math. 103 (1981), 499–558; II, 103 (1981), 107–120.

[K] N. Kurokawa, Examples of eigenvalues of Hecke operators on Siegel cusp forms of degree two, Inv. Math. 49 (1978), 149–165.

[L] R. Langlands, On the notion of an automorphic representation, Proc. Symp. Pure Math. 33, part 1 (1979), 203–207.

[P-S1] I. Piatetski-Shapiro, Special automorphic forms on $PGSp_4$, in "Arithmetic and Geometry" I, 309–325, Progress in Mathematics 35, Birkhäuser Verlag, Basel (1983).

[P-S2] I. Piatetski-Shapiro, On the Saito-Kurokawa lifting, Inv. Math. 71 (1983), 309–338.

[R] D. Rohrlich, Nonvanishing of L-functions and structure of Mordell-Weil groups, J. reine angew. Math. 417 (1991), 1–26.

[W] J.-L. Waldspurger, Correspondances de Shimura et Quaternions, Forum Math. 3 (1991), 219–307.

J.W. Cogdell
Department of Mathematics
Oklahoma State University
Stillwater, OK 74078
USA

I.I. Piatetski-Shapiro
Department of Mathematics
Yale University
New Haven, CT 06520
USA

Submitted: October 1994

Geometric And Functional Analysis
Vol. 5, No. 2 (1995)
1016-443X/95/0200174-70$1.50+0.20/0

THE LOCAL INDEX FORMULA IN NONCOMMUTATIVE GEOMETRY

A. Connes and H. Moscovici

We dedicate this paper to Misha Gromov

Abstract

In noncommutative geometry a geometric space is described from a spectral vantage point, as a triple $(\mathcal{A}, \mathcal{H}, D)$ consisting of a $*$-algebra $\mathcal{A}$ represented in a Hilbert space $\mathcal{H}$ together with an unbounded selfadjoint operator D, with compact resolvent, which interacts with the algebra in a bounded fashion. This paper contributes to the advancement of this point of view in two significant ways: (1) by showing that any pseudogroup of transformations of a manifold gives rise to such a spectral triple of finite summability degree, and (2) by proving a general, in some sense universal, local index formula for arbitrary spectral triples of finite summability degree, in terms of the Dixmier trace and its residue-type extension.

Introduction

Many of the tools of differential calculus acquire their full power only when formulated at the level of variational calculus, where the original space X one is dealing with is replaced by a functional space $\mathcal{F}(X)$ of functions or fields on X. The space X itself is involved only indirectly in $\mathcal{F}(X)$, for instance to write down the right hand side $F(\varphi)$ of a nonlinear evolution equation,

$$\frac{d\varphi}{dt} = F(\varphi) , \qquad \varphi \in \mathcal{F}(X) ,$$

with the right hand side usually involving the pointwise product of functions φ on X and partial differentiation.

The essence of noncommutative geometry is the existence of many situations in which $\mathcal{F}(X)$ makes perfectly good sense, while X itself is no longer an ordinary space, described set-theoretically by means of points $p \in X$ and coordinates. When X is given by a set, the basic structure on the space $\mathcal{F}(X)$ of (real or complex valued) functions on a set X is the pointwise product of functions. Given two functions f_1, f_2 one forms a new function $f_1 f_2$ by:

$$(f_1 f_2)(p) = f_1(p) f_2(p) , \qquad \forall p \in X . \tag{1}$$

In noncommutative geometry one still has a product on $\mathcal{F}(X)$ but the commutativity property of (1)

$$f_1 f_2 = f_2 f_1 \,, \qquad \forall\, f_j \in \mathcal{F}(X) \,. \tag{2}$$

is dropped. It is precisely this commutativity property which signals that X is an ordinary set. When dropped, one no longer deals with just a set X, but essentially with a set endowed with relations between different points. For instance, if one considers a set Y consisting of two points $\{1,2\}$ and the relation which identifies 1 and 2, then $\mathcal{F}(Y,\mathrm{rel})$ is the space $M_2(\mathbb{C})$ of 2×2 complex matrices with the product

$$(f_1 f_2)(i,j) = \sum f_1(i,k) f_2(k,j) \,, \tag{3}$$

i.e. the usual product of matrices.

In this simple example the ordinary space $\{1,2\}$, given by the two points without any relation, is described by the subalgebra of diagonal matrices. It is the "off-diagonal" matrices, such as $e_{12} = \left[\begin{smallmatrix} 0 & 1 \\ 0 & 0 \end{smallmatrix}\right]$ or $e_{21} = \left[\begin{smallmatrix} 0 & 0 \\ 1 & 0 \end{smallmatrix}\right]$, which describe the relation. This type of construction of an algebra $\mathcal{F}(X)$ extends easily to a pseudogroup of transformations of a manifold and also to the holonomy pseudogroup of a foliation (see [C1]). The resulting noncommutative algebra encodes the structure of the "space with relations". We shall later discuss in detail the case of a smooth manifold together with its full diffeomorphism group.

As another simple example we can consider the case of a single point divided by a discrete group Γ. Then the corresponding algebra $\mathcal{F}$ is the group ring attached to Γ, whose elements f are functions (with finite support) on Γ,

$$g \to f_g \in \mathbb{C} \,, \tag{4}$$

with the product given by linearization of the group law $g_1, g_2 \to g_1 g_2$ in Γ:

$$(f_1 f_2)_g = \sum_{g_1 g_2 = g} f_{1,g_1} f_{2,g_2} \,. \tag{5}$$

So far, in describing the functional space $\mathcal{F}(X)$ associated to an ordinary space X we have ignored the degree of regularity of the elements $f \in \mathcal{F}(X)$ as functions of $p \in X$. To various degrees of regularity correspond various branches of the general theory of noncommutative associative algebras. The latter are assumed to be algebras over $\mathbb{C}$, which moreover are involutive, i.e. endowed with an antilinear involution

$$f \to f^* \,, \qquad (f_1 f_2)^* = f_2^* f_1^* \,. \tag{6}$$

The two kinds of regularity assumptions for which the corresponding algebraic theory is satisfactory are:

measurability, which corresponds to the theory of von Neumann algebras;

continuity, which corresponds to the theory of C^*-algebras.

In both theories the Hilbert space plays a key role. Indeed, both types of algebras are faithfully representable as algebras of operators in Hilbert space with a suitable closure hypothesis. One can trace the role of the Hilbert space to the simple fact that *positive* complex numbers are those of the form

$$\lambda = z^* z\,. \tag{7}$$

In any of the above algebras, functional analysis provides the existence, via Hahn-Banach arguments, of sufficiently many linear functionals L which are positive

$$L(f^* f) \geq 0\,. \tag{8}$$

From such an L, one easily constructs a Hilbert space together with a representation, by left multiplication, of the original algebra.

Next, many of the tools of *differential topology*, such as the de Rham theory of differential forms and currents, the Chern character etc. are well captured (see [C1]) by cyclic cohomology applied to *pre C^*-algebras*, i.e. to dense subalgebras of C^*-algebras which are stable under the holomorphic functional calculus:

$$f \to h(f) = \frac{1}{2i\pi} \int \frac{h(z)}{f - z}\, dz\,, \tag{9}$$

where h is holomorphic in a neighbourhood of Spec(f). The prototype of such an algebra is the algebra $C^\infty(X)$ of smooth functions on a manifold X. The cyclic cohomology construction then recovers the ordinary differential forms, the de Rham complex of currents and so on. More significantly, this construction also applies to the highly noncommutative example of group rings, in which case the group cocycles give rise to cyclic cocycles with direct application to the Novikov conjecture on the homotopy invariance of the higher signatures of non-simply connected manifolds with given fundamental group. (For a more thorough discussion, see [C1]).

If one wants to go beyond differential topology and reach the geometric structure itself, including the metric and the real analytic aspects, it turns out that the most fruitful point of view is that of *spectral geometry*. More precisely, while our measure theoretic understanding of the space X was encoded by a (von Neumann) algebra of operators $\mathcal{A}$ acting in the Hilbert space $\mathcal{H}$, the *geometric* understanding of the space X will be encoded, not by a suitable subalgebra of $\mathcal{A}$, but by an operator in Hilbert space:

$$D = D^*\,, \quad \text{selfadjoint unbounded operator in } \mathcal{H}\,. \tag{10}$$

In the compact case, i.e. X compact, the operator D will have discrete spectrum, with (real) eigenvalues λ_n, $|\lambda_n| \to \infty$, when $n \to \infty$.

Formulating the precise conditions to which the triples $(\mathcal{A}, \mathcal{H}, D)$ should be subjected is tantamount to devising the axioms of noncommutative geometry. If we let F and $|D|$ be the elements of the polar decomposition of D,

$$D = F|D| \ , \quad |D|^2 = D^2 \ , \quad F = \operatorname{Sign} D \ , \tag{11}$$

then the operators F and $|D|$ play a similar role to the measurements of angles and, respectively, of length in Hilbert's axioms of geometry. In particular the operator $F = \operatorname{Sign} D$ captures the conformal aspect while D describes the full geometric situation.

Considering F alone, the quantized calculus was developed (cf. [C1]) based on the following dictionary.

Classical	*Quantum*
Complex variable	Bounded operator in Hilbert space $\mathcal{H}$
Real variable	Selfadjoint operator
Infinitesimal	Compact operator
Infinitesimal of order $\alpha > 0$	Compact operator whose characteristic values μ_n satisfy $\mu_n = O(n^{-\alpha})$
Differential	$dT = [F, T] = FT - TF$
Integral of infinitesimal of order 1	Dixmier trace $\operatorname{Tr}_\omega(T)$

We refer to Appendix A for a thorough treatment of the Dixmier trace. For a host of applications of the quantized calculus, including Julia sets, the quantum Hall effect and the analysis of group rings, the reader is referred to [C1]. A further application, namely the construction of a 4-dimensional conformal invariant analogue of the 2-dimensional Polyakov action, is discussed in [C3].

Our goal in the present paper is to use the quantized calculus to develop geometry from a spectral point of view. In more precise terms, our initial datum will be a triple $(\mathcal{A}, \mathcal{H}, D)$ where $\mathcal{A}$ is an involutive algebra represented in the Hilbert space $\mathcal{H}$ and D is a selfadjoint operator in $\mathcal{H}$ with compact resolvent, which almost commutes with any $a \in \mathcal{A}$, to the extent that

$$[D, a] \text{ is bounded for any } a \in \mathcal{A} \ .$$

The basic example of such a triple is provided by the Dirac operator on a closed Riemannian (Spin) manifold. In that case, $\mathcal{H}$ is the Hilbert space of L^2 spinors on the manifold M, $\mathcal{A}$ is the algebra of (smooth) functions acting in $\mathcal{H}$ by multiplication operators and D is the (selfadjoint) Dirac operator. One can easily check that no information has been lost in trading the geometric space M for the spectral triple $(\mathcal{A}, \mathcal{H}, D)$. Indeed (see [C1]), one recovers (*i*) the space M, as the spectrum Spec($\overline{\mathcal{A}}$), of the norm closure of the algebra $\mathcal{A}$ of operators in $\mathcal{H}$;

(*ii*) the geodesic distance d on M, from the formula:

$$d(p,q) = \mathrm{Sup}\left\{|f(p) - f(q)| \; ; \; \|[D,f]\| \leq 1\right\} , \qquad \forall p,q \in M .$$

The right hand side of the above formula continues to make sense in general and the simplest non-Riemannian example where it applies is the 0-dimensional situation in which the geometric space is finite. In that case both the algebra $\mathcal{A}$ and the Hilbert space $\mathcal{H}$ are finite dimensional, so that D is a selfadjoint matrix. For instance, for a two-point space, one lets $\mathcal{A} = \mathbf{C} \oplus \mathbf{C}$ act in the 2-dimensional Hilbert space $\mathcal{H}$ by

$$f \in \mathcal{A} \to \begin{bmatrix} f(a) & 0 \\ 0 & f(b) \end{bmatrix} ,$$

and one takes $D = \begin{bmatrix} 0 & \mu \\ \mu & 0 \end{bmatrix}$. The above formula gives $d(a,b) = 1/\mu$.

As a slightly more involved 0-dimensional example, one can consider the algebraic structure provided by the elementary Fermions, i.e. the three families of quarks. Thus, one lets $\mathcal{H}$ be the finite dimensional Hilbert space with orthonormal basis labelled by the left-handed and right-handed elementary quarks such as $u_L^r, u_R^b, \ldots$. The algebra $\mathcal{A}$ is $\mathbf{C} \oplus \mathbf{H}$, where the complex number λ in $(\lambda, q) \in \mathcal{A}$ acts on the right-handed part by λ on "up" particles and $\overline{\lambda}$ on "down" particles. The isodoublet structure of the left-handed (up, down) pairs allows the quaternion q to act on them by the matrix

$$\begin{bmatrix} \alpha & \beta \\ -\overline{\beta} & \overline{\alpha} \end{bmatrix} \qquad q = \alpha + \beta j \; ; \; \alpha, \beta \in \mathbf{C} .$$

Then the Yukawa coupling matrix of the standard model provides the self-adjoint matrix D.

In [CL] the theory of matter fields was developed in the above framework, under the finite-dimensionality hypothesis that the characteristic values of D^{-1} are $O(n^{-1/d})$, for some finite d. This makes it possible to define the action functional of Quantum Electrodynamics at the same level of generality (cf. [C1]). The striking fact there is that if one replaces the usual picture of space-time by its product by the above 0-dimensional example, the QED action functional gives the Glashow-Weinberg-Salam standard model Lagrangian with its Higgs fields and symmetry breaking mechanism. In the development of this theory, the tools of the quantized calculus, in particular the Dixmier trace as the substitute for the Lebesgue integral, played an essential role.

The matter field Lagrangian involves the metric $g_{\mu\nu}$ but does not involve any derivative of $g_{\mu\nu}$. This indicates that the difficulty involved in developing the analogue of gravity in the above context is of a different scale. In order to overcome it, one needs both a good list of examples of spectrally defined spaces and a difficult mathematical problem to solve. By a spectrally defined space we mean a triple $(\mathcal{A}, \mathcal{H}, D)$ as above; the involutive algebra $\mathcal{A}$ is not necessarily commutative. We shall also refer to them as *spectral triples*.

Before running through the list of examples, let us state the mathematical problem:

compute by a local formula the cyclic cohomology Chern character of $(\mathcal{A}, \mathcal{H}, D)$.

More specifically, the representation of $\mathcal{A}$ in $\mathcal{H}$ together with the operator D allows setting up an index problem:

$$\mathrm{Ind}_D : K_j(\mathcal{A}) \to \mathbb{Z}$$

where $j = 0$ in the $\mathbb{Z}/2$-graded (or even case) and $j = 1$ otherwise. The index map turns out to be polynomial and given, in the above generality, by the pairing of $K_j(\mathcal{A})$ with the following cyclic cocycle

$$\tau(a^0, a^1, \ldots, a^n) = \mathrm{Trace}\big(a^0[F, a^1] \ldots [F, a^n]\big) , \qquad \forall a^j \in \mathcal{A}$$

where n has the same parity as j and $n > d - 1$. In the even case, one replaces the trace by the supertrace, i.e. one uses the $\mathbb{Z}/2$-grading γ of $\mathcal{H}$ to write

$$\tau(a^0, a^1, \ldots, a^n) = \mathrm{Trace}\big(\gamma a^0[F, a^1] \ldots [F, a^n]\big) , \qquad \forall a^j \in \mathcal{A} .$$

The class of τ in the cyclic cohomology $HC^n(\mathcal{A})$ is the Chern character of $(\mathcal{A}, \mathcal{H}, D)$. We refer to [C1] for more details as well as for the appropriate normalizations.

The general problem is to compute the class of τ by a *local formula*. A partial answer to this problem was already obtained in [C1], by means of a general local formula for the *Hochschild class* of τ as the Hochschild n-cocycle:

$$\varphi(a^0, \ldots, a^n) = \mathrm{Tr}_\omega\big(a^0[D, a^1] \ldots [D, a^n]|D|^{-n}\big) , \qquad \forall a^j \in \mathcal{A} , \qquad (12)$$

where n is as above and, in the even case, with γ inserted in front of a^0 .

In the above formula Tr_ω is the Dixmier trace, which when evaluated on a given operator T only depends upon the asymptotic behavior of its eigenvalues. More precisely, for $T \geq 0$, with $\mu_n(T)$ the nth eigenvalue of T

in decreasing order, one has (cf. Appendix A):

$$\mathrm{Tr}_\omega(T) = \lim_\omega \frac{1}{\log N} \sum_0^N \mu_n(T) \ ;$$

this is insensitive to the perturbation of μ_n by any sequence $\varepsilon_n = o\left(\frac{1}{n}\right)$, i.e. such that $n\varepsilon_n \to 0$, $n \to \infty$.

For a classical pseudodifferential operator P with distributional kernel $k(x,y)$ the Dixmier trace is given by the Wodzicki residue

$$\mathrm{Tr}_\omega(T) = \int a(x)$$

where $k(x,y)$ has an asymptotic expansion near the diagonal of the form

$$k(x,y) = a(x)V \log\big(d(x,y)\big) + b(x,y) \ ,$$

with b bounded.

In particular, when one evaluates Tr_ω on a product $T_1 \cdots T_n$ of such operators the result is expressed as an integral *in a single variable* x of a local quantity. This is in sharp contrast with what happens for the ordinary trace, which when evaluated on $T_1 \cdots T_n$ involves a multiple integral, of the form

$$\int k_1(x_1,x_2)k_2(x_2,x_3)\ldots k_n(x_n,x_1) \ ,$$

where the x_j's vary arbitrarily in the manifold.

While the expression (12) of the Hochschild cocycle φ is local in full generality, it only accounts for the Hochschild class of the Chern character of $(\mathcal{A},\mathcal{H},D)$, which is not sufficient to recover the index map. In the manifold case for instance, it only gives the index of D with coefficients in the Bott K-theory class supported by an arbitrarily small disk.

In section II of this paper we shall obtain a general local formula for all the components of the cyclic cocycle τ. This will be achieved by adapting the Wodzicki residue, the unique extension of the Dixmier trace to pseudo-differential operators of arbitrary order, to all our examples. For spectrally defined spaces $(\mathcal{A},\mathcal{H},D)$, we shall see that the usual notion of dimension is replaced by a *dimension spectrum*, which is a subset of $\mathbb{C}$. Under the assumption of simple discrete dimension spectrum, the Wodzicki residue makes sense and defines a trace on the algebra of the pseudodifferential operators of $(\mathcal{A},\mathcal{H},D)$. The latter algebra is obtained by analysing the one-parameter group $\sigma_t = |D|^{it} \cdot |D|^{-it}$ in a manner very similar to Tomita's analysis of the modular automorphism group of von Neumann algebras. When the dimension spectrum is discrete but not simple, the analogue of the Wodzicki residue is no longer a trace; it satisfies, however, cohomological identities which relate it to higher residues.

Under the sole hypothesis of discreteness of the dimension spectrum, we shall obtain a *general local formula for the Chern character* of a spectral triple $(\mathcal{A}, \mathcal{H}, D)$, expressing the components of the Chern character in terms of finite linear combinations, with *rational coefficients*, of higher residues applied to products of iterated commutators of D^2 with $[D, a^j]$, $a^j \in \mathcal{A}$. A noteworthy feature of the proof is the use of renormalization group techniques to remove the transcendental coefficients which arise when the dimension spectrum has multiplicity. In the manifold case, this formula reduces, of course, to the classical local index formula. In general however it is necessarily more intricate, in several respects, because of its large domain of applicability, which encompasses for instance the diffeomorphisms-equivariant situation described in section I.

We conclude the introduction with a list of spectral triples corresponding to geometric or group-theoretic spaces.

1. *Riemannian manifolds* (with some variations allowing for Finsler metrics and also for the replacement of $|D|$ by $|D|^\alpha$, $\alpha \in]0,1]$).
2. *Manifolds with singularities*. For this, the work of J. Cheeger on conical singularities is very relevant. In fact, the spectral triples are stable under the operation of "coning", which is easy to formulate algebraically.
3. *Discrete spaces and their product with manifolds* (as in the discussion in [C1] of the standard model). The spectral triples are of course stable under products.
4. *Cantor sets*. Their importance lies in the fact that they provide examples of dimension spectra which contain complex numbers (cf. section II).
5. *Nilpotent discrete groups*. The algebra $\mathcal{A}$ is the group ring of the discrete group Γ, and the nilpotency of Γ is required to ensure the finite-summability condition $D^{-1} \in \mathcal{L}^{(p,\infty)}$. We refer to [C1] for the construction of the triple for subgroups of Lie groups.
6. *Transverse structure for foliations*. This example, or rather the intimately related example of the *Diff*-equivariant structure of a manifold will be treated in detail in section I of this paper.

I. Diffeomorphism Invariant Geometry

1. Triangular structure of the frame bundle. Let W be a smooth manifold and Diff(W) its group of diffeomorphisms. Given an arbitrary subgroup Γ of Diff(W) one can form the crossed product of the algebra of functions on W by the action of Γ (cf. [C1]) and describe in this way the measure theory and topology of the noncommutative space encoding the identifications of points in W by the action of Γ.

The basic idea which we developed in [C2], in order to obtain invariants of K-theory of the C^*-algebra $C_0(W) \rtimes \Gamma$, is to relate the general case of an arbitrary Γ acting on W to a "type II" situation, in which the action of Γ preserves a certain G-structure which we shall describe and use at great length below.

In general, and for instance if we take $\Gamma = \mathrm{Diff}(W)$, the action of Γ on W preserves no structure at all. Thus, if we take $W = S^1$ there is no Γ-invariant measure in the Lebesgue measure class, and even at the measure theory level the crossed product $L^\infty(S^1) \rtimes \Gamma$ is of type III. The basic structure theory of type III factors, as crossed products of type II by an action of $\mathbb{R}_+^*$, is easy to interpret in this example (cf. e.g. [C1]) as the replacement of the manifold $W = S^1$ by the total space P of the $\mathbb{R}_+^*$-principal bundle of (positive) 1-densities on S^1.

On this total space P the group $\mathrm{Diff}(W)$ is still acting and now there is on P a tautological invariant measure for the action of $\mathrm{Diff}(W)$, so that the crossed product of P by Γ is of type II. Furthermore the group $\mathbb{R}_+^*$ acts on this crossed product and gives back (up to Morita equivalence) the original crossed product of W by Γ.

Even though Γ acting on P preserves a natural density, it does not preserve a Riemannian geometry, since it would then have to be contained in the Lie group of isometries of that geometry. Let us describe the natural geometric structure on P preserved by the action of $\mathrm{Diff}(S^1)$. By construction, P is an $\mathbb{R}_+^*$ principal bundle over S^1 and we let $\pi : P \to S^1$ be the canonical projection. Given $x \in S^1$, a point $p \in \pi^{-1}\{x\}$ is the same thing as a unit of length in the tangent space $T_x(S^1)$. Moreover, the canonical action of $\mathbb{R}_+^*$ on P also specifies a unit of length in $T_p(\pi^{-1}\{x\})$, given by the vertical vector field $\frac{d}{dt}(e^t p)$.

We thus have a natural integrable subbundle V of the tangent bundle of P together with Euclidean metrics on both V and $N = T/V$, where at any $p \in P$ we use the identification of N_p with $T_x(S^1)$, $x = \pi(p)$, to define the metric at p.

Since this construction is completely canonical it is invariant under the action of $\mathrm{Diff}(S^1)$.

If we make the (non canonical) choice of section $d\theta$ of P we can label the points of P by (θ, λ), $\theta \in S^1$, $\lambda \in \mathbb{R}_+^*$ or equivalently (θ, s), $s \in \mathbb{R}$, $\lambda = e^s$. In these coordinates, the vertical metric is ds^2 and the transverse one is $(e^s d\theta)^2$. A diffeomorphism φ of S^1 acts in the obvious way, namely:

$$\varphi(\theta, s) = \big(\varphi(\theta), s + \log \varphi'(\theta)\big) .$$

If we perform the measure theory construction of the principal $\mathbb{R}_+^*$ bundle in higher dimension, we obtain the correct description of the corresponding type II algebra but since we use only the (absolute value of the) determinant of the Jacobian matrix of φ we only control the volume distorsion by φ but not the geometric distorsion.

To describe the latter, we take for P the bundle over W whose fiber P_x at each $x \in W$ is the space of all Euclidean metrics on the vector space $T_x(W)$. Thus, a point p of P is given by a point $x \in W$ together with a non-degenerate quadratic form, $g_{\mu\nu}\, dx^\mu\, dx^\nu$ in local coordinates, on $T_x(W)$. Equivalently, we can describe P as the quotient of the frame bundle of W, whose fiber at $x \in W$ is the space of linear isomorphisms $\mathbb{R}^n \to T_xW$, by the action of the subgroup $O(n) \subset GL(n, \mathbb{R})$. On the symmetric space $GL_n/O(n)$ we use the natural invariant Riemannian metric, which on the tangent space at the unit matrix, identified with the space of symmetric matrices, is given by the Hilbert-Schmidt norm. Once transported to the fiber P_x, $x \in W$, this metric gives an Euclidean structure on the vertical bundle $V \subset TP$. Given a vertical path $p(t)$, $p(0) = p$ its square length at $t = 0$ is simply the trace of $(p^{-1}\dot{p})^2$.

Also, exactly as above, we can identify the transverse space $N_p = T_pP/V_p$ with the tangent space $T_x(W)$, $x = \pi(p)$, so that the quadratic form p provides us with a natural Euclidean structure on N_p. In order to have a convenient terminology, we introduce the following definition:

DEFINITION I.1. *By a triangular structure on a manifold M we mean an integrable subbundle V of the tangent bundle TM together with Euclidean metrics on both V and $N = T/V$.*

We can summarize the above discussion as follows:

PROPOSITION I.1. *Given a manifold W, the space P of all metrics on W, defined above, has a canonical triangular structure, invariant under the action of* Diff(W).

This construction was used in [C2] to prove analytic properties of cyclic cocycles such as the transverse fundamental class or Gelfand-Fuchs cohomology classes. We refer to [C2] for purely geometric corollaries of this technique. To obtain them, it was crucial to relate the K-theory of the C^*-algebras obtained (from crossed products by subgroups Γ of Diff W) using W and using P. This followed from the "dual Dirac" construction of a bivariant Kasparov class (cf. [C2]). In [HSk], M. Hilsum and G. Skandalis went further and constructed the transverse fundamental class in K-homology for the space P, using hypoelliptic operators. They did this at the level of homotopy classes of Fredholm modules and the central theme of the first part of this paper will be to refine their construction in order to describe the geometry of the crossed products (of P by Γ) by a spectral triple $(\mathcal{A}, \mathcal{H}, D)$ satisfying all the conditions of our general spectral geometry.

Thus, we shall now free ourselves from the particular features of the example P above and discuss, in the context of triangular manifolds, the construction of the corresponding spectral triple.

One of the new features will be the *quartic* aspect of the discussion, as opposed to the quadratic feature of Riemannian geometry.

Finally one should keep in mind that the crossed product of P by a given subgroup Γ of Diff(W) is only the "type II" counterpart of the crossed product of W by Γ. To obtain the latter from the former, one needs to take a crossed product by "$GL_n/O(n)$", operation in which the non amenability of the Lie group GL_n, $n > 1$ comes into play. In this paper we shall content ourselves with the type II discussion.

2. The spectral triple $(\mathcal{A}, \mathcal{H}, D)$ of a triangular structure. We let M be a smooth (not necessarily compact) manifold together with an integrable subbundle V of its tangent bundle. We let $N = TM/V$ be the transverse bundle and assume that both N and V are oriented Euclidean even dimensional vector bundles.

Our first aim is to construct a hypoelliptic differential operator Q corresponding to the signature of M, which modulo lower order only depends upon the Euclidean structures of both V and N but not upon a choice of Riemannian metric on M. This will be done by combining a longitudinal signature operator of order 2 with the usual signature operator in the transverse direction. Then, using Q, we shall define a first-order operator D by the equation

$$Q = D|D| \,. \tag{1}$$

Let us first see how one can replace the usual signature operator by an equivalent operator of order 2.

α) Second order signature operator. Let V be a smooth (and, for simplicity, compact) oriented even-dimensional Riemannian manifold. On the bundle $\wedge T^*_{\mathbb{C}}$ of exterior differential forms on V one has a natural $\mathbb{Z}/2$-grading γ, $\gamma^2 = 1$, $\gamma = \gamma^*$, given by the $*$-operation, such that

$$d^* = -\gamma \, d\gamma \; ; \tag{2}$$

thus, the signature operator $d + d^*$ anticommutes with γ. Let

$$\Delta = (d + d^*)^2 = dd^* + d^*d \,.$$

It commutes with both d and d^* and we can thus consider, for $\lambda \in [0,1]$, the operators

$$U_\lambda = \Delta^{1/2} + \lambda(d - d^*) \,, \quad U^*_\lambda = \Delta^{1/2} + \lambda(d^* - d) \,. \tag{3}$$

One has

$$U_\lambda U_\lambda^* = \Delta - \lambda^2(d-d^*)^2 = (1+\lambda^2)\Delta$$

and similarly

$$U_\lambda^* U_\lambda = (1+\lambda^2)\Delta \ .$$

LEMMA I.1. 1) *U_λ commutes with the $\mathbb{Z}/2$-grading γ.*

2) *$U_\lambda(d+d^*)U_\lambda^* = 2\Delta^{1/2}(dd^* - d^*d)$ for $\lambda = 1$.*

Proof: 1) Both Δ and $d - d^* = d + \gamma d\gamma$ commute with γ.

One has

$$\left(\Delta^{1/2} + (d-d^*)\right)(d+d^*)\left(\Delta^{1/2} - (d-d^*)\right) =$$

$$\Delta(d+d^*)+(d-d^*)(d+d^*)\Delta^{1/2}+\Delta^{1/2}(d+d^*)(d^*-d)-(d-d^*)(d+d^*)(d-d^*) \, .$$

On the other hand,

$$(d-d^*)(d+d^*) = -d^*d + dd^* \ , \quad (d+d^*)(d^*-d) = -d^*d + dd^*$$

and

$$-(d-d^*)(d+d^*)(d-d^*) = (dd^*-d^*d)(d^*-d) = -dd^*d-d^*dd^* = -\Delta(d+d^*) \, .$$

□

It follows, ignoring finite rank operators and using the operators $U_\lambda\Delta^{-1/2}$ which are bounded, that one gets a homotopy between the signature operator and the operator $\Delta^{-1/2}(dd^* - d^*d)$ with the same $\mathbb{Z}/2$ grading. This latter operator is an elliptic pseudodifferential operator of order 1 defined by the equation:

$$D|D| = dd^* - d^*d \ . \tag{4}$$

The second order operator $dd^* - d^*d$, with the $\mathbb{Z}/2$-grading γ, thus represents the signature class on M.

Let us now combine it with $d + d^*$ in the above context.

β) THE MIXED SIGNATURE OPERATOR. We let M, $V \subset TM$ and $N = TM/V$ be as above. We consider over M the hermitian vector bundle E with fiber

$$E = \wedge^* \, V_{\mathbb{C}}^* \otimes \wedge^* \, N_{\mathbb{C}}^* \ . \tag{5}$$

Its metric comes from the metrics of V and N, together with the orientations these yield $\mathbb{Z}/2$-grading operators γ_V and γ_N. It also yields a natural volume element, i.e. a section of

$$\wedge^v \, V^* \otimes \wedge^n \, N^* = \wedge^d \, T^*M$$

where $v = \dim V$, $n = \dim N$, $d = \dim M = v + n$.

Thus the Hilbert space $L^2(M,E)$ of sections of E has a natural inner product, independent of any additional choice.

Using the canonical *flat* connection of the restriction of the bundle N to the leaves of the foliation by V, we can define the longitudinal differential d_L, as an operator of degree $(1,0)$ with respect to the obvious bigrading, satisfying

$$d_L^2 = 0 \ . \tag{6}$$

The operator d_L^* is, by definition, the adjoint of d_L. It is of the form

$$d_L^* = -\gamma_V \ d_L \ \gamma_V + \mathrm{Order}\, 0 \tag{7}$$

with the additional term of order 0 uniquely prescribed without any extra choice.

This means that the following operator is a well defined longitudinal elliptic operator:

$$Q_L = d_L d_L^* - d_L^* d_L \ . \tag{8}$$

By the discussion of section α) this operator describes at the K-theory level the longitudinal signature class. To obtain the full signature of M we need to combine it with a transverse signature operator, which is of order 1 as a differential operator.

Our next step will thus be to define the operator $d_H + d_H^*$, where d_H is of degree $(0,1)$ in the bigrading of E and corresponds to transverse differentiation.

This step will require an additional choice of a (non-integrable) subbundle H of TM transverse to V, $\dim H = n$. It is crucial that such a choice does not affect the principal symbol of the operator as a hypoelliptic operator (see below).

The choice of H provides a natural isomorphism

$$j_H : \wedge V_x^* \otimes \wedge N_x^* \to \wedge T_x^* \ , \qquad \forall x \in M \ , \tag{9}$$

and for $\omega \in C^\infty(M, \wedge^r V^* \otimes \wedge^s N^*)$ we define $d_H(\omega)$ as the component of bidegree $(r, s+1)$ of

$$j_H^{-1} d(j_H \omega) \ . \tag{10}$$

To understand the ambiguity in the choice of H we consider locally a function f which is leafwise constant, i.e. $d_L f = 0$. Then $d_H f$ is independent of H and given as a section of N^*. We can then define the transverse symbol of d_H using its commutation with such f:

$$d_H(f\omega) - f d_H(\omega) = df \wedge \omega \qquad \forall f, d_L \ f = 0 \ ; \tag{11}$$

in the right hand side we use the natural algebra structure for $\wedge V^* \otimes \wedge N^*$.

Thus, (11) means that the transverse symbol is independent of the choice of H. We let:

$$Q_H = d_H + d_H^* \tag{12}$$

where the $*$ is taken relative to the inner product in $L^2(M,E)$. We now combine Q_L and Q_H, using the parity $(-1)^{\partial_N}$ in the transverse direction which commutes with Q_L and anticommutes with Q_H, to define

$$Q = Q_L\,(-1)^{\partial_N} + Q_H\,. \tag{13}$$

We should remark that one can use $(-1)^{\partial_N}$ instead of $1 \otimes \gamma_N$, without changing the homotopy class of the operators, since these two gradings are homotopic among operators which anticommute with Q_H.

The selfadjoint operator D is now uniquely defined by the equation:

$$D|D| = Q\,. \tag{14}$$

Note that Q is formally selfadjoint by construction. We shall not discuss here the problem of selfadjointness of Q in the non-compact case. This issue will need to be adressed eventually, in connection with our main example, i.e. the total space P in I.1.

The following theorem shows that the operator D constructed above gives rise to a spectral triple $(\mathcal{A}, \mathcal{H}, D)$ for the crossed product

$$\mathcal{A} = C_c^\infty(M) \rtimes \Gamma\,,$$

where Γ is any group of diffeomorphisms preserving the triangular structure.

THEOREM I.1. 1) *$[D,f]$ is bounded for any $f \in C_c^\infty(M)$, and both f and $[D,f]$ belong to $\cap_{n\geq 1}$ Dom δ^n, where $\delta = [|D|,\cdot]$.*

2) *If M is compact D has compact resolvent; in all cases*

$$f(D-\lambda)^{-1} \quad \text{is compact} \quad f \in C_c^\infty(M)\,, \qquad \lambda \notin \mathbb{R}\,.$$

3) *Changing the choice of H only affects D and $|D|$ by bounded operators (locally bounded, in the non-compact case).*

4) *Let $\varphi \in \mathrm{Diff}(M)$ preserve the foliation V and be isometric on both V and N. Let U_φ be the corresponding unitary in $\mathcal{H}$ then $[D, fU_\varphi]$ is bounded, and both fU_φ and $[D, fU_\varphi]$ belong to $\cap_{n\geq 1}\mathrm{Dom}\,\delta^n$ for all $f \in C_c^\infty(M)$.*

We shall also give the precise summability in 2) by showing that $f(D-\lambda)^{-1} \in \mathcal{L}^{(p,\infty)}$, $p = v + 2n$, and compute the Dixmier trace of the product $f|D|^{-p}$.

3. Preliminaries on $\Psi DO'$ calculus. As a technical tool in the proof of Theorem I.1, we shall describe the pseudodifferential calculus which is adapted to the situation. It is just a special case of the pseudodifferential calculus on Heisenberg manifolds (cf. [BG]), which is however sufficiently different from the ordinary ΨDO calculus to deserve a careful treatment. The reader familiar with [BG] can skip this section.

Recall that M is foliated by the integrable subbundle V. We shall only use charts, i.e. local coordinates x^i, which are foliation charts, that is

$$V \quad \text{is generated by} \quad \frac{\partial}{\partial x^j} = \partial_j \, , \qquad j = 1, \dots, v \, . \tag{15}$$

Thus the plaques, i.e. the leaves of the restriction of the foliation, are $\mathbf{R}^v \times \mathrm{pt}$. In such coordinates we shall use the ordinary formula to pass from a symbol $\sigma(x,\xi)$ to the corresponding operator:

$$P_\sigma = (2\pi)^{-m} \int e^{i\langle (x-y),\xi\rangle} \sigma(x,\xi) d^m\xi \qquad m = v + n \, . \tag{16}$$

One has $\xi \in (\mathbf{R}^v \times \mathbf{R}^n)^* = \mathbf{R}_v \times \mathbf{R}_n$ and one defines a (coordinate dependent) notion of homogeneity of symbols using

$$\lambda \cdot \xi = (\lambda \xi_v, \lambda^2 \xi_n) \quad \text{for} \quad \xi = (\xi_v, \xi_n) \, , \quad \lambda \in \mathbf{R}_+^* \, . \tag{17}$$

The natural length for which ξ is homogeneous of degree 1 is

$$\|\xi\|' = \left(\|\xi_v\|^4 + \|\xi_n\|^2\right)^{1/4} \, . \tag{18}$$

Let us start with a symbol σ, smooth on $\mathbf{R}^m \times \mathbf{R}_m \backslash \{0\}$ and homogeneous of degree q for the dilations (17), i.e.

$$\sigma(x, \lambda \cdot \xi) = \lambda^q \sigma(x,\xi) \, . \tag{19}$$

In order to control the operator P_σ defined in (16), one needs to control the partial derivatives

$$\partial_x^\alpha \partial_\xi^\beta \sigma(x,\xi) \, .$$

When we apply ∂_x^α, the homogeneity property (19) is preserved. The action of $\partial_{\xi_v}^i$, $i = 1, \dots, v$, lowers q by 1 while the action of $\partial_{\xi_n}^i$, $i = 1, \dots, n$, lowers it by 2. Thus, if we let

$$\langle \beta \rangle = \sum_1^v \beta_i + 2 \sum_1^n \beta_{v+j} \, , \tag{20}$$

we see that $\partial_x^\alpha \partial_\xi^\beta \sigma$ is homogeneous of degree $q - \langle\beta\rangle$. It follows that, for x in a compact subset of $\mathbf{R}^m$ and α, β fixed,

$$\left|\partial_x^\alpha \partial_\xi^\beta \, \sigma(x,\xi)\right| \leq C_{\alpha,\beta} \left(1 + \|\xi\|'\right)^{q - \langle\beta\rangle} \, . \tag{21}$$

To employ the usual classes of ΨDO one needs to relate the right hand side to the expression

$$\left(1 + \|\xi\|\right)^{q - \frac{1}{2}|\beta|} \, .$$

With $a = \|\xi_v\|$, $b = \|\xi_n\|$, one has

$$\|\xi\|^2 = a^2 + b^2 \, , \quad \|\xi\|'^4 = a^4 + b^2 \, ,$$

so that

$$\|\xi\|' \leq \left(1+\|\xi\|^2\right)^{1/2} \leq 1+\|\xi\|\,, \quad 1+\left(\|\xi\|'\right)^4 \geq \|\xi\|^2\,,$$
$$\|\xi\|^{1/2} \leq 1+\|\xi\|' \leq 2+\|\xi\|\,. \tag{22}$$

It thus follows that, if $q \geq 0$, a homogeneous symbol σ of degree q is of class $S^q_{0,1/2}$, i.e.

$$\left|\partial_x^\alpha\, \partial_\xi^\beta\, \sigma(x,\xi)\right| \leq C_{\alpha,\beta}\left(1+\|\xi\|\right)^{q-\frac{1}{2}|\beta|}\,, \tag{23}$$

while for $q<0$ it is of class $S^{q/2}_{0,1/2}$.

This implies in particular that for $q \leq 0$ the operator P_σ is bounded in L^2 (cf. [BG]).

We can now introduce the relevant class of symbols for the proof of Theorem I.1. We consider symbols σ such that:

$$\begin{gathered}\text{there exists a sequence } (\sigma_q) \text{ of homogeneous symbols,}\\ \sigma_q \text{ of degree } q,\ \text{with } \sigma \sim \sum_{q\leq q_0} \sigma_q\end{gathered} \tag{24}$$

where $\sim$ means that for any N the difference

$$\sigma_N = \sigma - \sum_{N<q\leq q_0} \sigma_q$$

satisfies the inequalities (21) for $q=N$.

Definition I.2. *We shall say that an operator P is a $\Psi DO'$ if it is of the form P_σ, with σ as in (24).*

Let us now describe the composition $P_\sigma \circ P_{\sigma'}$ of two $\Psi DO'$. Denoting as usual

$$D_x^\alpha = (-i)^{|\alpha|}\partial_x^\alpha\,, \tag{25}$$

the action of P_σ can be formally expanded as

$$P_\sigma = \sum \frac{1}{\alpha!}\,\partial_\xi^\alpha\,\sigma(x,\xi) D_x^\alpha\,.$$

Let us consider the expansion:

$$\sum \frac{1}{\alpha!}\,\partial_\xi^\alpha\,\sigma_{q_1}(x,\xi) D_x^\alpha\,\sigma'_{q_2}(x,\xi)\,. \tag{26}$$

The degree of homogeneity of each of the terms is

$$q_1+q_2-\langle\alpha\rangle\,,$$

where $\langle\alpha\rangle$ is defined in (20).

This shows that (26) makes sense as an asymptotic expansion and corresponds to the product $P_\sigma \circ P_{\sigma'}$.

Note that every differential operator P is a $\Psi DO'$; the only difference is in the notion of degree, since while $\partial/\partial x_j = \partial_j$ has degree 1 for $j=1,\dots,v$

it has degree 2 for $j = v+1, \ldots, v+n$. It is not true however that an ordinary ΨDO is a $\Psi DO'$, even in the order 0 case.

We need to define the notions of principal symbol and ellipticity for $\Psi DO'$. To this end we shall first consider what happens under a change of foliation chart.

Let φ be such a change of charts. It defines a local $\mathbb{R}^n$ diffeomorphism φ_n, in such a way that, with the notations $x = (x_v, x_n)$, $y = (y_v, y_n)$, one has

$$\varphi(x) = \big(\varphi_v(x_v, x_n), \varphi_n(x_n)\big) \; . \tag{27}$$

Similarly, $\psi = \varphi^{-1}$ is of the same form, $\psi = (\psi_v, \psi_n)$. Given a covector $\xi \in T_x^*$ the corresponding covector at $y = \varphi(x)$ is

$$\eta = \psi_*^t(\xi) \tag{28}$$

i.e. $\langle \eta, Y\rangle = \langle \xi, \psi_*(Y)\rangle$, $\forall Y \in T_y$.

With $\sigma = \sigma(y,\eta)$ a homogeneous symbol, let us consider the composition

$$\widetilde{\sigma}(x,\xi) = \sigma\big(\varphi(x), \psi^t_{*,\varphi(x)}(\xi)\big) \; . \tag{29}$$

To compare it with $\widetilde{\sigma}(x, \lambda\cdot\xi)$ we just need to understand the linear map ψ_*^t at $\varphi(x) = y$. This map, L, preserves the natural subspace $N^* \subset T^*$ of covectors orthogonal to the leaves. On the subspace $N^* = \{(0,\xi_n)\}$ one has $\lambda\cdot\xi = \lambda^2\xi$. Thus

$$L = \begin{bmatrix} L_{vv} & 0 \\ L_{nv} & L_{nn} \end{bmatrix}$$

$$L(\lambda\cdot\xi) = \big(\lambda L_{vv}(\xi_v), \lambda^2 L_{nn}(\xi_n) + \lambda L_{nv}(\xi_v)\big) \; .$$

With obvious notation, we have

$$\begin{aligned} \sigma\big(L(\xi)\big) &= \sigma\big(L_{vv}(\xi_v), L_{nn}(\xi_n) + L_{nv}(\xi_v)\big) \\ &= \sum \frac{1}{\alpha!}\partial^\alpha_{\eta_v}\sigma\big(L_{vv}(\xi_v), L_{nn}(\xi_n)\big)\big(L_{n,v}(\xi_v)\big)^\alpha \; , \end{aligned}$$

which gives the desired expansion of $\sigma\circ L$ as a sum of homogeneous symbols. This shows that formula (29) defines a transformation of symbols.

To obtain the symbol of the operator P_σ in the new coordinates one writes its kernel as

$$k(x,x') = (2\pi)^{-m}\int e^{i\langle\varphi(x)-\varphi(x'),\eta\rangle}\sigma\big(\varphi(x),\eta\big)d^m\eta \; . \tag{30}$$

One first changes variables from η to ξ using (28) and the fact that k is a 1-density, so that no Jacobian enters in the change of variables, due to the invariance of the symplectic Liouville measure. Thus,

$$k(x,x') = (2\pi)^{-m}\int e^{i(\langle x-x',\xi\rangle + \alpha(x,x',\xi))}\widetilde{\sigma}(x,\xi)d^m\xi \; , \tag{31}$$

where $\alpha(x,x',\xi)$ is the non-linearity at x of the map φ:

$$\langle \varphi(x)-\varphi(x'),\eta\rangle = \langle x-x',\xi\rangle + \alpha(x,x',\xi)\ . \tag{32}$$

The variable ξ appears linearly in α and the coefficient of ξ_n only invokes $\varphi_n(x_n)-\varphi_n(x'_n)$ and not x_v or x'_v. By construction, α and its first derivatives in x' vanish at $x'=x$. This implies that the Taylor expansion at $x'=x$ of $e^{i\alpha(x,x',\xi)}$ is of the form

$$e^{i\alpha(x,x',\xi)} = \sum P_\beta(x,\xi)(x-x')^\beta\ , \tag{33}$$

where $P_\beta(x,\xi)$ is a polynomial in ξ whose degree, in the sense that ξ^α has degree $\langle\alpha\rangle$, is smaller than $\frac{1}{2}\langle\beta\rangle$. Thus, using $\partial_\xi^\beta e^{i\langle x-x',\xi\rangle} = i^{|\beta|}(x-x')^\beta e^{i\langle x-x',\xi\rangle}$ and integrating by parts, one gets the full symbol for P_σ:

$$\sigma(x,\xi) = \sum i^{-|\beta|}\partial_\xi^\beta\, P_\beta(x,\xi)\widetilde{\sigma}(x,\xi)\ . \tag{34}$$

In particular, at a given x the value of $\sigma(x,\xi)$ only involves $\widetilde{\sigma}$ restricted to T_x^* and not its restriction to any T_y^*, $y\neq x$. To this restriction $\widetilde{\sigma}$ one applies a differential operator with polynomial coefficients.

Let us now turn to the *principal symbol*. Let σ be a symbol of order q; then its principal part is

$$\lim_{\lambda\to\infty} \lambda^{-q}\sigma(x,\lambda\cdot\xi)\ . \tag{35}$$

We see that it gives a well defined function on the bundle

$$V^*\oplus N^*\ ,$$

the direct sum of the subbundle $N^*\subset T^*$ and of $V^* = T^*/N^*$. The above limit exists and under a change of coordinates it behaves, in view of the above formulae, as a function on $V^*\oplus N^*$.

As an example, let us compute the principal symbol of

$$[|D|,P_\sigma]$$

where σ is of order 0 and where the principal symbol of $|D|$ is

$$\sigma_1(x,\xi) = \|\xi\|' \qquad \text{(cf. 18)}\ . \tag{36}$$

We can use formula (26) in local coordinates. It gives

$$\sum_{\langle\alpha\rangle=1} \left(\partial_\xi^\alpha\sigma_1(x,\xi)D_x^\alpha\sigma(x,\xi) - \partial_\xi^\alpha\,\sigma(x,\xi)\,D_x^\alpha\,\sigma_1(x,\xi)\right)\ . \tag{37}$$

Note that since $\langle\alpha\rangle=1$, this formula only involves the longitudinal differentiation $D_{x_v}^\alpha$.

4. Proof of Theorem I.1 The operator D is defined by equation (14), where Q is a selfadjoint differential operator by construction. We shall first show that D is a $\Psi DO'$ of order 1.

As noted before, any differential operator is $\Psi DO'$ in the above sense, but the notions of degree and of principal symbol differ from the usual ones. In particular, Q is *elliptic* of degree 2 and its principal symbol is, for $\xi_v \in V^*$, $\xi_n \in N^*$, the endomorphism of $\wedge V^* \otimes \wedge N^*$

$$\sigma_2(\xi_v, \xi_n) = (e_{\xi_v} i_{\xi_v} - i_{\xi_v} e_{\xi_v}) \otimes (-1)^{\partial} + 1 \otimes \left(ie_{\xi_n} + (ie_{\xi_n})^*\right) .$$

The two sides anticommute, so that when we square σ_2 we get

$$\sigma_2^2(\xi_v, \xi_n) = \left(\|\xi_v\|^4 + \|\xi_n\|^2\right) \cdot 1 ,$$

which shows that Q^2 is a $\Psi DO'$ of order 4, elliptic and with principal symbol a multiple of the identity.

As in the ordinary pseudodifferential calculus, this is enough (cf. [G, p. 52]) to construct an asymptotic $\Psi DO'$, $R(\mu)$, which is an asymptotic resolvent for Q^2.

We shall use for D and $|D|$ the following formulae:

$$|D| = \frac{\sqrt{2}}{2\pi} \int_0^\infty \frac{Q^2}{Q^2 + \mu} \, \mu^{-3/4} \, d\mu \tag{38}$$

$$D = \frac{\sqrt{2}}{2\pi} \int_0^\infty \frac{Q}{Q^2 + \mu} \, \mu^{-1/4} \, d\mu . \tag{39}$$

Let us replace, in these two formulae, $(Q^2 + \mu)^{-1}$ by the asymptotic resolvent $R(\mu)$. Then $R(\mu)(Q^2 + \mu) - 1$ is smoothing of any order, with the corresponding norms controlled by $(1 + \mu)^{-k}$. Thus, when we replace $Q^2/(Q^2 + \mu)$ by $Q^2 R(\mu)$, we use

$$\frac{Q^2}{Q^2 + \mu}(1 - (Q^2 + \mu)R(\mu)) = \frac{Q^2}{Q^2 + \mu} - Q^2 R(\mu) ,$$

which is therefore also smoothing of any order if $\mu \geq 0$. Using the boundedness of

$$\frac{\mu^{1/2} Q^2}{Q^2 + \mu} \quad \text{for} \quad \mu \geq 0 ,$$

a similar statement holds for

$$\mu^{1/2} \left(\frac{Q}{Q^2 + \mu} - QR(\mu) \right) .$$

Since the asymptotic symbol of $(Q^2 + \mu)^{-1}$ is, at the principal level,

$$\sigma_{-4} = \left(\|\xi_v\|^4 + \|\xi_n\|^2 + \mu\right)^{-1}$$

one obtains the principal symbols of the operators

$$|D|_r = \frac{\sqrt{2}}{2\pi}\int_0^\infty Q^2 R(\mu)\mu^{-3/4} d\mu \ , \tag{40}$$

$$D_r = \frac{\sqrt{2}}{2\pi}\int_0^\infty QR(\mu)\mu^{-1/4}\, d\mu \tag{41}$$

as the integrals

$$\frac{\sqrt{2}}{2\pi}\int_0^\infty \frac{\|\xi\|'^4}{\|\xi\|'^4+\mu}\,\mu^{-3/4}\, d\mu = \|\xi\|' \ , \tag{42}$$

$$\frac{\sqrt{2}}{2\pi}\int_0^\infty \sigma_2(Q)\big(\|\xi\|'^4+\mu\big)^{-1}\mu^{-1/4} d\mu = \frac{\sigma_2(Q)(\xi)}{\|\xi\|'} \ . \tag{43}$$

This is enough to show that both $|D|$ and D are $\Psi DO'$ of order 1 and to give in local coordinates the asymptotic expansion of their symbol.

Since the principal symbol of $|D|$ is $\|\xi\|' \cdot 1$, a multiple of the identity matrix, it commutes with the symbol of any $\Psi DO'$ of order 0. This shows that, with δ denoting the derivation

$$\delta(T) = \big[|D|, T\big] \ ,$$

one has:

LEMMA I.2. *Any $\Psi DO'$ of order 0 belongs to $\bigcap_{n\geq 1} \operatorname{Dom} \delta^n$.*

This applies to the multiplication operator f as well as to $[D, f]$ and proves Theorem I.1 1).

For assertion 2) of the theorem, we shall prove the more precise result

$$|D|^{-1} \in \mathcal{L}^{(v+2n,\infty)} \ , \tag{44}$$

which in turn will follow from:

LEMMA I.3. *Let P be a $\Psi DO'$ of order $-(v+2n)$. Then*

$$\mu_k(P) = O(k^{-1}) \ ,$$

where $\mu_k(P)$ is the kth characteristic value of P.

Proof: It is enough to check this locally, so that one can assume without loss of generality that

$$M = \mathbf{T}^v \times \mathbf{T}^n \quad \text{and} \quad P = (\Delta_v^2 \otimes 1 + 1 \otimes \Delta_n + 1)^{-\frac{v+2n}{4}} \ .$$

One just has to bound the number of eigenvalues $\lambda > \varepsilon$ of P by C/ε for some $C < \infty$. Equivalently, one has to bound the number of eigenvalues $\lambda < \varepsilon^{-\frac{4}{v+2n}}$ of $\Delta_v^2 \otimes 1 + 1 \otimes \Delta_n + 1$ by C/ε. But this number is less than $N_v(E)\ N_n(E)$, where $E = \varepsilon^{-\frac{4}{v+2n}}$, $N_v(E)$ is the number of eigenvalues of Δ_v^2 less than E, while $N_n(E)$ stands for the number of eigenvalues of Δ_n less than E.

One has $N_v(E) \leq C_v\ E^{v/4}$, $N_n(E) \leq C_n\ E^{n/2}$ and we get the required bound. □

For Theorem I.1 3), we just note that the choice of H does not affect the principal symbols of either D or $|D|$, while any $\Psi DO'$ of order 0 is bounded.

It remains to prove assertion 4). The operators $U_\varphi D U_\varphi^{-1}$, $U_\varphi |D| U_\varphi^{-1}$ are $\Psi DO'$ of order 1, with the same principal symbols as D and $|D|$ respectively. This shows that the following are $\Psi DO'$ of order 0:

$$[D, U_\varphi]\, U_\varphi^{-1}\ , \quad [|D|, U_\varphi] U_\varphi^{-1}\ .$$

In particular, they are bounded and belong to $\bigcap_{n\geq 1} \operatorname{Dom} \delta^n$. Thus, $U_\varphi \in \operatorname{Dom}\delta$, $\delta(U_\varphi)U_\varphi^{-1} \in \operatorname{Dom}\delta$, hence $U_\varphi \in \operatorname{Dom}\delta^2$. By induction, using $\delta(U_\varphi)\cdot U_\varphi^{-1} \in \operatorname{Dom}\delta^n$, one gets $U_\varphi \in \bigcap_{n\geq 1}\operatorname{Dom}\delta^n$ and thus $[D, U_\varphi] = ([D, U_\varphi]\, U_\varphi^{-1})\cdot U_\varphi \in \cap_{n\geq 1} \operatorname{Dom}\delta^n$. □

5. The Dixmier trace of $\Psi DO'$ of order $-(v+2n)$. We shall now describe the kernels $k(x,y)$ for operators of order $-(v+2n)$ and compute their Dixmier trace. As we shall see, the relevant question is that of extending a homogeneous symbol $\sigma(\xi)$, $\xi \in \mathbb{R}^{v+n}\backslash\{0\}$,

$$\sigma(\lambda\cdot\xi) = \lambda^{-(v+2n)}\sigma(\xi) \qquad \forall\,\lambda \in \mathbb{R}_+^* \tag{45}$$

to a homogeneous distribution.

The degree of homogeneity $q = -(v+2n)$ considered in (45) is the limit case for integrability both near 0 and near ∞. Indeed, the Jacobian of $\xi \to \lambda\cdot\xi$ is λ^{v+2n} and on each orbit of the flow F,

$$F_s(\xi) = e^s\cdot\xi\ ,$$

the measure $\sigma(\xi)d^{v+n}\xi$ is proportional to $\frac{d\lambda}{\lambda}$. It thus has a logarithmic divergency both at 0 and at ∞. They turn out to be intimately related and we shall investigate the divergency at 0, following closely ([BG]).

We consider the linear form on the space

$$\{f \in \mathcal{S}(\mathbb{R}^{v+n})\ ;\ f(0) = 0\} = \mathcal{S}_0$$

given by

$$L(f) = \int f(\xi)\sigma(\xi)d^{v+n}\xi\ .$$

It makes sense because σ is bounded at ∞ (polynomial growth would be enough) and $f(0) = 0$ takes care of the non-integrability at 0.

With $f_\lambda(\xi) = f(\lambda^{-1}\cdot\xi)$ the homogeneity of σ means:

$$L(f_\lambda) = L(f) \qquad \forall\, f \in \mathcal{S}_0\ . \tag{46}$$

By the Hahn-Banach theorem the linear form L extends from the hyperplane $\mathcal{S}_0$ to all of $\mathcal{S}$ as a continuous linear form and we get a one dimensional affine subspace of $\mathcal{S}'$:

$$E = \{\tau \in \mathcal{S}' \; ; \; \tau|\mathcal{S}_0 = L\} \; ; \tag{47}$$

the corresponding linear space is the space of multiples of δ_0 the Dirac mass at 0.

The dilations θ_λ,

$$\langle \theta_\lambda(\tau), f \rangle = \langle \tau, f_\lambda \rangle$$

act on E; since they act trivially on the associated linear space, their action on E is given, for some constant c, by

$$\theta_\lambda \tau = \tau + c \log \lambda \delta_0 \,, \qquad \forall \tau \in E \,, \; \lambda \in \mathbb{R}_+^* \,. \tag{48}$$

To determine c, let $\psi \in C_c^\infty([0,\infty[)$ be identically 1 near 0, and let τ be given by

$$\tau(f) := L(f - f(0)\psi(\| \, \|')) = \int (f(\xi) - f(0)\psi(\|\xi\|'))\sigma(\xi) d\xi \,. \tag{49}$$

One has,

$$\begin{aligned} \tau(f_\lambda) - \tau(f) &= \int (f(\lambda^{-1} \cdot \xi) - f(0)\psi(\|\xi\|'))\sigma(\xi)\, d\xi \\ &\quad - \int (f(\xi) - f(0)\psi(\|\xi\|'))\sigma(\xi)\, d\xi \\ &= f(0) \int (\psi(\|\xi\|') - \psi(\lambda\|\xi\|'))\sigma(\xi)\, d\xi \\ &= f(0) c_\sigma \int_0^\infty (\psi(\mu) - \psi(\lambda\mu)) \frac{d\mu}{\mu} \,, \end{aligned}$$

where c_σ is obtained as the pairing between any transversal cycle $\|\xi\|' =$ *constant* to the foliation of $\mathbb{R}^{v+n} \backslash \{0\}$ by the orbits of the flow F and the closed de Rham current obtained as the contraction $i_e(\sigma\, d\xi)$ of the differential form $\sigma\, d\xi$ by the vector field $e = \lambda^{-1}\, d/d\lambda$ generating the flow.

Letting $\mu = e^u$, $\lambda = e^s$ one gets

$$\int_0^\infty (\psi(\mu) - \psi(\lambda\mu)) \frac{d\mu}{\mu} = \int_{-\infty}^\infty (\psi(e^u) - \psi(e^{u+s}))\, du = s = \log \lambda \,.$$

Thus we have shown that $c = c_\sigma$ is exactly the obstruction to extending σ as a homogeneous distribution on $\mathbb{R}^{v+n}$ (cf. [BG]).

We can write, in a formal way,

$$c_\sigma = \int_{\|\xi\|'=1} i_e \sigma(\xi) d\xi \,. \tag{50}$$

Let us now relate this obstruction to the behavior of the inverse Fourier transform of σ

$$\check{\sigma}(y) = (2\pi)^{-(v+n)} \int e^{i\langle y, \xi \rangle} \sigma(\xi) d\xi \,. \tag{51}$$

We first need to relate the oscillatory integral definition (51) to the Fourier transform for tempered distributions.

For $y \neq 0$, the oscillatory integral is defined as the value of the convergent integral

$$(2\pi)^{-v+n} \int e^{i\langle y,\xi\rangle} \big(P^k \sigma(\xi)\big) d\xi \qquad k \geq 1 \tag{52}$$

where the symbol $\sigma(\xi)$ has been smoothed for ξ small, and P_y is a differential operator of degree 1 in $\frac{\partial^\alpha}{\partial \xi^\alpha}$ such that

$$P_y^t(e^{i\langle y,\xi\rangle}) = e^{i\langle y,\xi\rangle} .$$

The smoothing of σ near $\xi = 0$ introduces an ambiguity of the addition of arbitrary elements of $(C_c^\infty)^\vee \subset \mathcal{S}$. But this does not affect the behavior at $y = 0$, which we are after.

Having chosen an extension τ of σ as a distribution, we need to check that the inverse Fourier transform $\check{\tau}$, a tempered distribution, is represented by a locally integrable tempered function h whose behavior at $y = 0$ is the same as for the oscillatory integral.

Since the Fourier transform of a distribution with compact support is represented by a nice smooth function, we just need to know that if σ_1 is a symbol, smooth on all of $\mathbb{R}^{v+n}$, then its Fourier transform as a tempered distribution is given by the oscillatory integral expression (52).

To check this, let $P\big(y, \frac{\partial}{\partial \xi}\big) = \frac{1}{i} \sum y^j \frac{\partial}{\partial \xi^j}$; then

$$P e^{i\langle y,\xi\rangle} = \|y\|^2 e^{i\langle y,\xi\rangle} .$$

With $f \in \mathcal{S}(\mathbb{R}_{n+v})_0$, we can write:

$$\begin{aligned}
\int \check{f}(\xi)\sigma_1(\xi) d\xi &= (2\pi)^{-(n+v)} \int\int f(y) e^{i\langle y\cdot\xi\rangle} dy\, \sigma_1(\xi) d\xi \\
&= (2\pi)^{-(n+v)} \int\int P(e^{i\langle y\cdot\xi\rangle}) \|y\|^{-2} f(y) \sigma_1(\xi) dy\, d\xi \\
&= (2\pi)^{-(n+v)} \int\int e^{i\langle y,\xi\rangle} \|y\|^{-2} f(y) (P\sigma_1)(\xi) dy\, d\xi \\
&= (2\pi)^{-(n+v)} \int\int e^{i\langle y,\xi\rangle} \|y\|^{-2} P\sigma_1(\xi) d\xi\, f(y)\, dy .
\end{aligned}$$

Thus, we know that the distribution $\check{\tau}$ is represented outside 0 by a smooth function with tempered growth. This function is then unique and the homogeneity property (48) implies, using $\langle \lambda \cdot \xi, y\rangle = \langle \xi, \lambda \cdot y\rangle$,

$$\check{\tau}(\lambda^{-1} \cdot y) = (\theta_\lambda \tau)^\vee(y) = \check{\tau}(y) + c' \log \lambda . \tag{53}$$

Thus,

$$\check{\tau}(y) = \check{\tau}(y/\|y\|') - c' \log(\|y\|') , \qquad c' = (2\pi)^{-(n+v)} c .$$

We are now ready to deal with the Dixmier trace of $\Psi DO'$ of order $-(v+2n)$.
We let (M, V) be as above, with M compact.

PROPOSITION I.2. *Let T be a $\Psi DO'$ of order $-(v+2n)$. Then*

1) *T is measurable, $T \in \mathcal{L}^{(1,\infty)}$, with its Dixmier trace $\mathrm{Tr}_\omega(T)$ independent of ω and given by*

$$\mathrm{Tr}_\omega = (2\pi)^{-(n+v)} \frac{1}{v+2n} \int_{\|\xi\|'=1} \sigma(x,\xi) i_e(dx\, d\xi) \ ,$$

where e is the generator of the flow

$$F_s(\xi_v, \xi_n) = (e^s \xi_v, e^{2s} \xi_n)$$

and the choice of transversal $\|\xi\|' = 1$ is irrelevant. Also $dx\, d\xi$ corresponds to the symplectic volume form.

2) *The kernel $k(x,y)$ for T has the following behavior near the diagonal*

$$k(x,y) = c(x) \log \left(\|x-y\|' \right) + 0(1) \ ,$$

where the 1-density $c(x)$ is given by the formula

$$c(x) = (2\pi)^{-(n+v)} \int_{\|\xi\|'=1} \sigma(x,\xi) i_e\, d\xi \ .$$

Remark I.1: Before beginning the proof, let us note that 1) reduces to the usual formula (cf. [C1]) for the Dixmier trace of ordinary ΨDO when either $n = 0$ or $v = 0$; in the latter case the exponent $2s$ accounts for the $2n$. Note also that in 2) the choice of the local distance function $\|x - y\|'$ has no effect on the value of $c(x)$. The statement 2) is a special case of [BG].

Proof: 1) By Lemma I.2, one has $T \in \mathcal{L}^{(1,\infty)}$ so that $\mathrm{Tr}_\omega(T)$ is well defined. Since operators of lower order are (by the argument of Lemma I.2) of trace class, it follows that $\mathrm{Tr}_\omega(T)$ only depends upon the principal symbol $\sigma(T)$. The map

$$\sigma \to \mathrm{Tr}_\omega(T_\sigma) \tag{54}$$

is then a positive linear form on the space $\mathcal{F}$ of homogeneous symbols of order $-(v + 2n)$. It is therefore given by a positive measure on the (non-canonical) unit sphere

$$S^* \{(x,\xi) \in V^* \oplus N^* \ ; \ \|\xi\|' = 1\} \ . \tag{55}$$

The unitary invariance of Tr_ω together with the use of translations in a foliation chart, show that this measure is absolutely continuous with respect to the smooth measure dx on M. The diffeomorphism invariance of the $\Psi DO'$-calculus shows that the conditional measures on the fibers of

$$p : S^* \to M \tag{56}$$

must be, in the appropriate sense, invariant under all maps

$$(\xi_v, \xi_n) \to (L_v \xi_v, L_n \xi_n)$$

with both L_v and L_n invertible.

Such maps do not act transitively on $V_x^* \oplus N_x^*$ but the two invariant subspaces $V_x^* \oplus 0$, $0 \oplus N_x^*$ do not carry any measure with the correct homogeneity.

This implies that there exists a constant $a(\omega)$ such that

$$\mathrm{Tr}_\omega(T) = a(\omega) \int_{\|\xi\|'=1} \sigma(x,\xi) i_e \, dx \, d\xi \ . \tag{57}$$

To show that this constant $a(\omega)$ does not depend on ω and to determine it, one just needs to compute $\mathrm{Tr}_\omega(T)$ for one specific example for each value of v and n. With the notations of Lemma I.2, we take:

$$T = (\Delta_v^2 \otimes 1 + 1 \otimes \Delta_n + 1)^{-(\frac{v+2n}{4})} \ . \tag{58}$$

In order to compute $\mathrm{Tr}_\omega(T)$, we just use the following general fact (cf. Appendix A).

(59) *Let Δ be a positive (unbounded) operator such that $\Delta^{-1} \in \mathcal{L}^{(p,\infty)}$ for some $p \geq 1$, and*

$$t^p \,\mathrm{Trace}\,(e^{-t\Delta}) \underset{t\to 0}{\longrightarrow} L \ .$$

Then $\mathrm{Tr}_\omega(\Delta^{-p})$ is independent of ω and given by

$$\mathrm{Tr}_\omega(\Delta^{-p}) = \frac{L}{\Gamma(p+1)} \ .$$

We take $\Delta = \Delta_v^2 \otimes 1 + 1 \otimes \Delta_n + 1$, $p = \frac{v+2n}{4}$. To compute L it is enough to determine

$$\lim_{t\to 0} t^{p_v} \,\mathrm{Trace}\,(e^{-t\Delta_v^2}) = L_v \ , \qquad p_v = v/4 \ ,$$

$$\lim_{t\to 0} t^{p_n} \,\mathrm{Trace}\,(e^{-t\Delta_n}) = L_n \ , \qquad p_n = n/2$$

which then gives $L = L_v \, L_n$.

By (59) one has $L_v = \Gamma(p_v+1)\mathrm{Tr}_\omega(\Delta_v^{-v/2})$ and $L_n = \Gamma(p_n+1)\mathrm{Tr}_\omega(\Delta_n^{-n/2})$. Choosing the standard metric $\sum d\theta_j^2$ on both $\mathbb{T}^v$ and $\mathbb{T}^n$ gives, with $|S^k|$ the volume of the k-dimensional sphere,

$$L_v = \Gamma\left(\frac{v}{4}+1\right)\frac{1}{v}|S^{v-1}| \ , \quad L_n = \Gamma\left(\frac{n}{2}+1\right)\frac{1}{n}|S^{n-1}| \tag{60}$$

and

$$\mathrm{Tr}_\omega(\Delta^{-p}) = \Gamma\left(\frac{v+2n}{4}+1\right)^{-1} L_v L_n \ .$$

The principal symbol of Δ^{-p} is

$$\sigma(x,\xi) = \left(\|\xi\|'\right)^{-(v+2n)} . \tag{61}$$

If we let

$$|S|' = \int_{\|\xi\|'=1} i_e(d\xi) ,$$

we have the equalities

$$\int_{\|\xi\|'=1} \sigma(x,\xi) i_e(dx\,d\xi) = (2\pi)^{n+v}|S|' \tag{62}$$

and

$$\int f(\|\xi\|')d\xi = |S|' \int_0^\infty f(\rho)\rho^{v+2n}\frac{d\rho}{\rho} , \tag{63}$$

whenever both sides make sense.

Using $f(\rho) = e^{-\lambda\rho^4}$ yields

$$|S|' = \frac{1}{2}\Gamma\left(\frac{v+2n}{4}\right)^{-1} \Gamma\left(\frac{v}{4}\right)|S^{v-1}|\Gamma\left(\frac{n}{2}\right)|S^{n-1}| . \tag{64}$$

Together with (60) and (62), this gives the normalization:

$$\mathrm{Tr}_\omega(\Delta^{-p}) = (2\pi)^{-(n+v)}(v+2n)^{-1} \int_{\|\xi\|'=1} \sigma(x,\xi) i_e(dx\,d\xi) . \tag{65}$$

2) The kernel $k(x,y)$ of T is given by

$$k(x,y) = (2\pi)^{-(v+n)} \int e^{i\langle x-y,\xi\rangle}\sigma(x,\xi)d\xi ,$$

so that, for fixed x, it is, as a function of $y-x$, the Fourier transform of $\sigma(x,\xi)$. Thus, using (53) we get the required answer. For a more detailed proof, the reader is referred to [BG]. □

Proposition I.2 extends immediately to non-scalar operators, with $\sigma(x,\xi)$ replaced by its trace $\mathrm{tr}(\sigma(x,\xi))$ taken in the fiber over x. We can therefore apply it to the operator $|D|^{-(v+2n)}$ of Theorem I.1, whose symbol is the identity matrix (for $\|\xi\|'=1$) on a space of dimension $2^{(v+n)}$. We get:

$$\mathrm{Tr}_\omega\left(f|D|^{-(v+2n)}\right) = \pi^{-(n+v)}(v+2n)^{-1}|S|' \int f(x)\,dx \tag{66}$$

where $|S|'$ is given by (64) and dx is the volume form on M corresponding to the given Euclidean structures on both V and N.

6. The analogue of the Wodzicki residue for $\Psi DO'$ operators. Let us now go back to the obstruction c_σ (cf. (50)), and exploit its definition involving the behavior of T near 0 rather than (the ultraviolet behavior) near ∞.

LEMMA I.4 ([BG]). *Let $\sigma \in C^\infty(\mathbb{R}^{n+v}\backslash\{0\})$ be homogeneous of order q, i.e. $\sigma(\lambda \cdot \xi) = \lambda^q\, \sigma(\xi)$, $\forall \xi \neq 0$, $\forall \lambda \in \mathbb{R}_+^*$, with $(q \in \mathbb{C})$.*

a) *If $q \notin \{-(v+2n) - k; k \in \mathbb{N}\}$ then σ extends to a homogeneous distribution on $\mathbb{R}^{n+v}$.*

b) *If $q = -(v+2n) - k$, then the obstruction to homogeneous extension is given by the $c_{\xi^\alpha \sigma}$, $|\alpha| = k$.*

Proof: Let ψ be as in (49) and k the integral part of

$$-\mathrm{Re}(q) - (v+2n) = a\ .$$

The size of $\sigma(\xi)$ for ξ small is comparable to $(\|\xi\|')^{\mathrm{Re}\, q} = \|\xi\|'^{-(v+2n+a)}$. Thus $\sigma(\xi)\xi^\alpha$ is locally integrable if $|\alpha| > k$ and the following is an extension of σ:

$$\tau(f) = \int \Big(f(\xi) - \sum_{|\alpha| \leq k} \frac{\xi^\alpha}{\alpha!} f^{(\alpha)}(0)\psi(\xi) \Big) \sigma(\xi) d\xi\ . \tag{67}$$

One gets then,

$$\begin{aligned} \tau(f_\lambda) - \lambda^{q+(v+2n)}\tau(f) &= \int \Big(f(\lambda^{-1}\cdot\xi) - \sum_{|\alpha|\leq k} \frac{\lambda^{-|\alpha|}\xi^\alpha}{\alpha!} f^{(\alpha)}(0)\psi(\xi) \Big)\sigma(\xi)d\xi \\ &\quad - \lambda^{q+(v+2n)} \int \Big(f(\xi) - \sum_{|\alpha|\leq k} \frac{\xi^\alpha}{\alpha!} f^{(\alpha)}(0)\psi(\xi) \Big)\sigma(\xi)d\xi \\ &= \lambda^{q+(n+v)} \sum_{|\alpha|\leq k} \frac{f^\alpha(0)}{\alpha!} \Big(\int_{\|\xi\|'=1} \xi^\alpha \sigma(\xi) i_e\, d\xi \Big) \rho_\alpha\ , \end{aligned}$$

$$\rho_\alpha = \int_0^\infty (\psi(\mu) - \psi(\lambda\mu))\, \mu^{q+(v+2n)+|\alpha|}\, d\mu/\mu\ .$$

To prove a), it is enough to choose ψ so that the ρ_α all vanish. With $\psi(\mu) = h(\log \mu)$, we thus look for $h' \in C_c^\infty(\mathbb{R})$, $\int h'(s)ds = -1$, such that

$$\int_{-\infty}^{\infty} h'(s)e^{bs}\, ds = 0 \qquad \forall b = q + (v+2n) + |\alpha|\ ,\ 0 \leq |\alpha| \leq k\ . \tag{68}$$

One just lets $h' = \prod(d/ds + b)f$, where the product is taken over all values of b in (68) and $f \in C_c^\infty(\mathbb{R})$, with $\int f(s)ds = (\prod b)^{-1}$. This is possible since $b \neq 0$ by hypothesis.

Assertion b) can be proved in a similar fashion, since for $q = -(v+2n) - k$ one can get $\rho_\alpha = 0$ for any α, $|\alpha| < k$. □

The ambiguity in the extension of σ is, a priori, of the form $\sum_{|\alpha|\leq k} a_\alpha \partial^\alpha \delta_0$. By a), if $q \notin \{-(v+2n) - \mathbb{N}\}$, the homogeneous extension exists and is unique, except when $q \in \mathbb{N}$.

COROLLARY I.1 ([BG]). *Let $\sigma \in C^\infty(\mathbf{R}^{n+v})$ have an asymptotic expansion $\sigma \sim \sum_{k\leq q} \sigma_k$ in homogeneous symbols (for $\lambda\cdot$). Then the Fourier transform $\check{\sigma}$ (of σ smooth at 0) has a behavior at $y = 0$ of the form:*

$$\check{\sigma}(y) = \sum_{-(q+(v+2n))}^{0} a_j(y) + c \log \|y\|' + 0(1) ,$$

where a_j is homogeneous of degree j.

Proof: We can neglect the difference $\sigma - \sum_{-(v+2n)\leq k\leq q} \sigma_k$ since it is integrable at ∞ and yields a $O(1)$ contribution for small y. By Lemma I.4, each σ_k, $k > -(v+2n)$ extends to a homogeneous distribution, so that the Fourier transform $\check{\sigma}_k$ of σ_k (smoothed at 0) has the indicated divergency at 0. The case $k = v + 2n$ follows from (53). ◻

We see that

(69) *the coefficient of* $\log \|y\|'$ *only depends on* $\sigma_{-(v+2n)}$ *and is equal to*

$$(2\pi)^{-(n+v)} \int_{\|\xi\|'=1} \sigma_{-(v+2n)}(\xi) i_e \, d\xi \ .$$

One can check directly, using (29) and (44), that the density

$$(2\pi)^{-(n+v)} \int_{\|\xi\|'=1} \sigma_{-(v+2n)}(x,\xi) i_e \, d\xi \tag{70}$$

is invariantly defined under the action of diffeomorphisms on the $\Psi DO'$ calculus. The obtained density is the coefficient of $\log \|x - y\|'$, in the expansion near the diagonal, of the kernel $k(x, y)$ for $P = P_\sigma$.

PROPOSITION I.3. *The Dixmier trace Tr_ω has a canonical extension to a trace on the algebra of $\Psi DO'$ operators of arbitrary order. It is given globally by the equality*

$$\mathrm{Tr}_\omega(T) = \frac{1}{v+2n} \int_M c(x) \ ,$$

where $c(x)$ is the 1-density occurring in the expansion of the kernel k of T near the diagonal, as a coefficient of $\log \|x - y\|'$. In local coordinates it is given by $(v + 2n)^{-1}$ times the expression (70).

The fact that the asserted extension is a trace will be proved in greater generality in section II (cf. Prop. II.1).

We proceed to show that there is a natural extension, similar to that of [KV], of the ordinary trace of operators to $\Psi DO'$ of complex order. To this end, we go back to the space of symbols, where we need to define what is meant by a holomorphic map $z \to \sigma_z$ with values in the space of symbols. We want the order $f(z)$ to be holomorphic in z, the bounds in the asymptotic expansion $\sigma(z) \sim \sum \sigma_{f(z)-p}$ to be uniform and the pointwise values of $\sigma(z)(\xi)$, $\sigma(z) \in C^\infty(\mathbf{R}^{v+n})$, to be holomorphic in the variable z. Then each $\sigma_{f(z)-p}(\xi)$ is also holomorphic in z. The functional

$$L(\sigma) = \check{\sigma}(0) = (2\pi)^{-(v+n)} \int \sigma(\xi)\, d\xi \tag{71}$$

is well defined on symbols of order $< -(v+2n)$ (for the real part of the order) and is holomorphic, inasmuch as $L(\sigma_z)$ is a holomorphic function of z for any holomorphic map $z \to \sigma_z$.

LEMMA I.5. *The functional L has a unique holomorphic extension $\widetilde{L}$ to the space of symbols of non-integral order ($z \notin \mathbb{Z}$). The value of $\widetilde{L}$ on $\sigma \sim \sum \sigma_{z-p}$ is given by*

$$\widetilde{L}(\sigma) = (2\pi)^{-(v+n)} \int \left(\sigma - \sum_0^N \tau_{z-p}\right)(\xi) d\xi\,, \quad N \geq \mathrm{Re}(z) + (v+2n)$$

where τ_{z-p} is the unique homogeneous extension of σ_{z-p} (given by Lemma I.4).

Proof: First the value of N used in Lemma I.5 is irrelevant, since the Fourier transform of a homogeneous distribution such as τ_{z-p} vanishes at 0 if $p > \mathrm{Re}\, z + (v+2n)$. Also the uniqueness is clear since any σ of order z can be connected to integrable order by a holomorhic path. It remains to show that if $z \to \sigma(z)$ is holomorphic, with order $f(z) \notin \mathbb{Z}$ then $L(\sigma(z))$ is holomorphic.

For large ξ the pointwise value of $\sigma(z) - \sum_0^N \sigma_{f(z)-p}$ is holomorphic in z and has uniformly integrable behavior at ∞; thus it is enough to control the behavior in z of

$$\int \left(\sigma - \sum_0^N \tau_{f(z)-p}\right) \varphi(\xi) d\xi\,,$$

where φ has compact support. In fact we can consider separately the term $\int \tau_{f(z)-p} \varphi(\xi) d\xi$, which is holomorphic in z by the very construction of τ (cf. (67) and (68)). □

Remark I.2: To a symbol of order z and to any given $p \in \mathbb{N}$ one can assign the number

$$\int_{\|\xi\|'=1} \sigma_{z-p}(\xi) i_e\, d\xi$$

and the functional thus obtained is holomorphic, but it does not vanish on σ with integrable order and cannot be added to $\widetilde{L}$ to yield another extension.

Let us now return to the (compact) manifold (M,V) as above, and consider the product $\Psi^0\times\mathbb{C}$ of the space of $\Psi DO'$ of order 0 by $\mathbb{C}$, endowed with the product structure of complex manifold. We shall adapt the method of [KV] to our context to obtain:

PROPOSITION I.4. *The function* $(P,z)\to \mathrm{Trace}(P|D|^{-z})$ *is holomorphic on* $\Psi^0\times\{z;\mathrm{Re}\,z>(v+2n)\}$ *and extends uniquely to a holomorphic function* TR *on* $\Psi^0\times(\mathbb{C}\backslash\mathbb{Z})$.

Proof: In a local chart, the trace of a $\Psi DO'$ $P=P_\sigma$ of order $<-(v+2n)$ is given by

$$\mathrm{Trace}(P_\sigma)=(2\pi)^{-(n+v)}\int\sigma(x,\xi)dx\,d\xi\ , \tag{72}$$

where the total symbol σ is smooth.

Thus, in a local chart, the following formula provides the required extension of Trace to $\Psi DO'$ of arbitrary order $z\notin\mathbb{Z}$

$$\mathrm{TR}(P_\sigma)=\int\widetilde{L}(\sigma(x,\cdot))dx\ . \tag{73}$$

The ambiguity of smoothing operators does not alter the existence of this extension globally on all $\Psi DO'$ of order $z\notin\mathbb{Z}$ and yields the required extension of *Trace*, provided one knows that $|D|^{-z}$ is a $\Psi DO'$ of order $-z$ and a holomorphic function of z. This follows from the proof of Theorem I.1 (see (38)), using the asymptotic resolvent of Q^2 (cf. [S]). □

Let us now relate the value of $\mathrm{Tr}_\omega(P)$, $P\in\Psi DO'$ to the residue at $z=0$ of the function:

$$\zeta(z)=\mathrm{TR}(P|D|^{-z})\ . \tag{74}$$

We can work first at the level of symbols and consider a fixed symbol σ of integral order q. We let σ_z, $\sigma_z(\xi)=\sigma(\xi)(\|\xi\|')^{-z}$ and investigate the behavior of $\widetilde{L}(\sigma_z)$ at $z=0$. Let $N\geq q+(v+2n)$, then

$$\widetilde{L}(\sigma_z)=(2\pi)^{-(v+n)}\int\Big(\sigma-\sum_0^N\sigma_{q-k}\Big)(\xi)\big(\|\xi\|'\big)^{-z}d\xi$$

where $\sigma_{q-k}(\xi)\big(\|\xi\|'\big)^{-z}$ is replaced near 0 by its unique extension as a homogeneous distribution.

The singularity at $z=0$ comes from ξ in the neighborhood of 0. When $q-k>-(v+2n)$, $\sigma_{q-k}(\xi)d\xi$ is integrable at 0 and the unique extension of $\sigma_{q-k}(\xi)\big(\|\xi\|'\big)^{-z}d\xi$ is holomorphic in z at $z=0$. Thus none of these terms contribute to the singularity of $\widetilde{L}(\sigma_z)$ at $z=0$.

We can choose $N=q+(v+2n)$ since, by Lemma I.5, any larger value gives the same answer. We thus need to understand the behavior at $z=0$ of

$$\int_{\|\xi\|'\leq a} \sigma_{-(v+2n)}(\xi)\left(\|\xi\|'\right)^{-z} d\xi \tag{75}$$

where $\sigma_{-(v+2n)}(\xi)(\|\xi\|')^{-z}d\xi$ is extended uniquely as a homogeneous distribution at $\xi = 0$. But for $\operatorname{Re} z < 0$ one has integrability near 0 so that this unique extension is the obvious one and one can write (75) as

$$\left(\int_{\|\xi\|'=1} \sigma_{-(v+2n)}(\xi) i_e\, d\xi\right)\left(\int_0^a \mu^{-z}\frac{d\mu}{\mu}\right). \tag{76}$$

As $\int_0^a \mu^{-z}\frac{d\mu}{\mu} = -\left[\frac{\mu^{-z}}{z}\right]_0^a = -a^{-z}/z$, one gets that the singularity of $\widetilde{L}(\sigma_z)$ at $z = 0$ is a simple pole with residue:

$$(2\pi)^{-(v+2n)} \int_{\|\xi\|'=1} \sigma_{-(v+2n)}(\xi) i_e\, d\xi\ . \tag{77}$$

Next, when we investigate (74) the situation is more complicated since in a local (foliation) chart we have an intricate expression for the total symbol of $|D|^{-z}$.

Let us first remark that the above discussion of the behavior of $\widetilde{L}(\sigma_z)$ continues to hold when σ_z is of the form

$$\sigma_z = \sigma(\xi, z)\left(\|\xi\|'\right)^{-z}, \tag{78}$$

where $z \to \sigma(\cdot, z)$ is a holomorphic map to symbols of fixed order q. Moreover, the residue at $z = 0$ is given by (77), with $\sigma(\xi, 0)$. Using this, we should be able to replace $|D|$ by the operator $|D_1|$ which in the given (foliation) chart involves the flat metric

$$\|\xi\|'^4 = \|\xi_v\|^4 + \|\xi_n\|^2 , \tag{79}$$

independently of $x = (x_v, x_n)$.

Since the total symbol of D_1^4, assumed to be given by (79), does not involve x, the corresponding (differential) operator is translation invariant. So are the complex powers $|D_1|^z$ and their total symbol is given by (a smoothed version of)

$$\sigma_z(x, \xi) = \left(\|\xi\|'\right)^z . \tag{80}$$

The computation of the total symbol of $P|D_1|^{-z}$ for any $\Psi D0'$, is then obvious, by (26), and the above discussion shows that the function $\zeta_1(z) = \mathrm{TR}(P|D_1|^{-z})$ has a simple pole at $z = 0$, which is given by (77). This continues to hold for any holomorphic map $z \to P_z$ to symbols of fixed order q.

We can now write

$$P|D|^{-z} = PU(z)|D_1|^{-z}\ , \quad U(z) = |D|^{-z}\ |D_1|^z \tag{81}$$

and it just remains to show that $U(z)$ is a holomorphic map to operators of order 0, with $U(0) = 1$.

It is not a polynomial in z because the principal symbols σ of D^4 and σ_1 of D_1^4 are different. Thus, the principal symbol of $U(z)$ will be $\sigma(\xi)^{-z/4}\sigma_1(\xi)^{z/4}$. By construction, $U(z)$ is the $\Psi DO'$ product of two holomorphic maps and hence is holomorphic. We thus proved the following result:

THEOREM I.2. *Let P be a $\Psi DO'$ of integral order q. Then the function $z \to \mathrm{Trace}(P|D|^{-z})$ is holomorphic for $\mathrm{Re}\, z > q + (v + 2n)$ and admits a (unique) analytic continuation to $\mathbb{C}\backslash\mathbb{Z}$ with at most simple poles at integers $k \leq q + (v + 2n)$. Its residue at $z = 0$ is given by*

$$\mathrm{Re}\, s_{z=0}\mathrm{Trace}\big(P|D|^{-z}\big) = (v + 2n)\,\mathrm{Tr}_\omega(P) ,$$

where Tr_ω is defined in Proposition I.3.

II. The Local Index Formula for Spectral Triples

1. Dimension spectrum. In this section, we shall describe a general local index formula in terms of the Dixmier trace, extended to operators of arbitrary order, for our spectral triples:

$$(\mathcal{A}, \mathcal{H}, D) . \tag{1}$$

Contrary to the standard practice, we shall focus on the odd case, the point being that in the even case there is a natural obstruction to express the (cyclic cocycle) character (cf. [C1]) of the triple (1) in terms of a residue or Dixmier trace. Indeed, the latter vanishes on any finite rank operator and thus will give the result 0 whenever $\mathcal{H}$ is finite-dimensional. Since it is easy to construct finite-dimensional (i.e. $\dim \mathcal{H} < \infty$) even triples with $\mathrm{Ind}(D) \neq 0$, one cannot expect to cover this case as well. However, one can convert any even triple into an odd one by crossing it with S^1, i.e. with the triple

$$\left(C^\infty(S^1)\ ,\ L^2(S^1)\ ,\ D = \frac{1}{i}\,\frac{\partial}{\partial\theta}\right) . \tag{2}$$

Thus, there is no real loss of generality in treating the odd case only.

The next point is that the usual notion of dimension (cf. [C1]) for spectral triples, provided by the degree of summability

$$D^{-1} \in \mathcal{L}^{(p,\infty)} , \tag{3}$$

gives only an upper bound on dimension and cannot detect the individual dimensions of the various pieces of a space which is a union of pieces of different dimensions $(\mathcal{A}_k, \mathcal{H}_k, D_k)$, $k = 1, ..., N$,

$$\mathcal{A} = \oplus\mathcal{A}_k\ ,\quad \mathcal{H} = \oplus\mathcal{H}_k\ ,\quad D = \oplus D_k\ . \tag{4}$$

In [C1] we gave a formula for the p-dimensional Hochschild cohomology class of the character, namely:

$$\tau(a^0,\dots,a^p)=\mathrm{Tr}_\omega\big(a^0[D,a^1]\dots[D,a^p]|D|^{-p}\big)\ . \tag{5}$$

Clearly, this Hochschild cocycle cannot account for lower-dimensional pieces in a union such as (4).

As it turns out, the correct notion of dimension is given not by a single real number p but by a subset

$$Sd\subset\mathbb{C} \tag{6}$$

which shall be called the *dimension spectrum* of the given triple. We shall assume that Sd is a discrete subset of $\mathbb{C}$, condition which will be incorporated in the following definition:

DEFINITION II.1. *A spectral triple* (1) *has discrete dimension spectrum Sd, if $Sd\subset\mathbb{C}$ is discrete and for any element of the algebra $\mathcal{B}$ generated by the $\delta^n(a)$, $a\in\mathcal{A}$, the function*

$$\zeta_b(z)=\mathrm{Trace}\big(b|D|^{-z}\big)$$

extends holomorphically to $\mathbb{C}\backslash Sd$.

Here δ denotes the derivation $\delta(T)=[|D|,T]$ and we assume that $\mathcal{A}\subset\bigcap_{n>0}\mathrm{Dom}\,\delta^n$ (see also Appendix B). The operator $b|D|^{-z}$ of Definition II.1 is then of trace class for $\mathrm{Re}\,z>p$, with p as in (3). On the technical side, we shall assume that *the analytic continuation of ζ_b is such that $\Gamma(z)\zeta_b(z)$ is of rapid decay on vertical lines $z=s+it$, for any s with* $\mathrm{Re}\,s>0$.

It is not difficult to check that Sd has the correct behavior with respect to the operations of sum and product for spectral triples:

$$Sd\ (\text{Sum of two spaces})=\cup Sd(\text{Spaces}) \tag{7}$$

$$Sd\ (\text{Product of two spaces})=Sd(\text{Space}_1)+Sd(\text{Space}_2)\ . \tag{8}$$

According to Theorem I.2 of section I, the dimension spectrum of the hypoelliptic spectral triple constructed there is contained in

$$\{q\in\mathbb{Z}\ ;\ q\le v+2n\}\ . \tag{9}$$

It is easy to give many examples of spectral triples with discrete dimension spectrum, but we shall now concentrate on the general theory of such spaces.

Our first task will be to extend the Wodzicki residue to this general framework, or equivalently, to extend the Dixmier trace to operators $P|D|^{-z}$ of arbitrary order, where P is an element of $\mathcal{B}$. In fact it is more convenient (cf. Appendix B) to introduce the algebra $\Psi^*(\mathcal{A})$ of operators which have an expansion:

$$P\simeq b_q|D|^q+b_{q-1}|D|^{q-1}+\cdots\ ,\quad b_q\in\mathcal{B}\ , \tag{10}$$

where the equality with $\sum_{-N<n\le q}b_n|D|^n$ holds modulo OP^{-N}.

To see that it is an algebra one uses Theorem B.1 of Appendix B, which gives an identity of the form:

$$|D|^\alpha b \simeq \sum_0^\infty c_{\alpha,k}\delta^k(b)|D|^{\alpha-k} , \tag{11}$$

where $c_{\alpha,k}$ is the coefficient of ε^k in the expansion of

$$(1+\varepsilon)^\alpha = \sum_0^\infty \frac{\alpha(\alpha-1)\cdots(\alpha-k+1)}{k!}\varepsilon^k , \tag{12}$$

with $\varepsilon(b) = \delta(b)|D|^{-1}$.

We shall say that the dimension spectrum Sd is *simple*, when the singularities of the functions $\zeta_b(z)$ of Definition II.1 at $z \in Sd$ are at most simple poles. Similarly, we say that Sd has finite multiplicity k when ζ_b has at most a pole of order k. We shall assume for simplicity that *Sd has finite multiplicity* throughout this section.

PROPOSITION II.1. *Let $p < \infty$ be the degree of summability of D.*

a) *For $P \in \Psi^*(\mathcal{A})$ the function $h(z) = \mathrm{Trace}(P|D|^{-2z})$ is holomorphic for $\mathrm{Re}\, z > \frac{1}{2}(\mathrm{Order}\, P + p)$ and extends to a holomorphic function on the complement of a discrete subset of $\mathbb{C}$.*

b) *Let $\tau_k(P)$ be the residue at 0 of $z^k\, h(z)$, $k \geq 0$; then*

$$\tau_k(P_1P_2 - P_2P_1) = \sum_{n>0} \frac{(-1)^{n-1}}{n!}\tau_{k+n}\big(P_1L^n(P_2)\big) ,$$

where L is the derivation $L = 2\log(1+\varepsilon)$.

Proof: a) The statement follows immediately from Definition II.1, for any finite sum of operators $b_n\ |D|^n$. Furthermore, if P is of order less than $-N$ then $h(z)$ is holomorphic if $\mathrm{Re}\, z > \frac{1}{2}(p-N)$, and for any given z this is achieved for N large enough.

b) First, the derivation $L = 2\log(1+\varepsilon)$ makes sense as a power series in ε and can be viewed at the formal level as implemented by $\log|D|^2$.

One has, for any P, an expansion near 0

$$\mathrm{Trace}\big(P|D|^{-2z}\big) = \sum_{k\geq 0} \tau_k(P) z^{-(k+1)} + 0(1) . \tag{13}$$

We can then write

$$\mathrm{Trace}\big(P_2P_1|D|^{-2z}\big) = \mathrm{Trace}\big(P_1(1+\varepsilon)^{-2z}(P_2)|D|^{-2z}\big) \tag{14}$$

and, since

$$(1+\varepsilon)^{-2z} = \exp(-zL) , \tag{15}$$

we get

$$\mathrm{Trace}\big(P_2P_1|D|^{-2z}\big) = \sum \frac{(-z)^n}{n!}\mathrm{Trace}\big(P_1\ L^n(P_2)\ |D|^{-2z}\big)\ . \tag{16}$$

By (13) we can expand:

$$\mathrm{Trace}\big(P_1L^n(P_2)|D|^{-2z}\big) = \sum \tau_q\big(P_1\ L^n(P_2)\big)z^{-(q+1)} + 0(1)$$

and, when multiplied by z^n, we see that we get the exponent $z^{-(k+1)}$ for $n-q=-k$. Thus, the coefficient of $z^{-(k+1)}$ in the expansion (16) is

$$\sum_{n=q-k} \frac{(-1)^n}{n!}\tau_q\big(P_1\ L^n(P_2)\big) = \sum \frac{(-1)^n}{n!}\tau_{n+k}\big(P_1\ L^n(P_2)\big)\ .$$

Therefore, we obtain:

$$\tau_k(P_2P_1) - \tau_k(P_1P_2) = \sum_{n>0} \frac{(-1)^n}{n!}\tau_{n+k}\big(P_1\ L^n(P_2)\big)\ . \tag{17}$$

□

It follows, of course, that if q is the multiplicity of Sd, i.e. the highest order of poles, then τ_q is a trace.

By Appendix A, in the case of a simple spectrum the trace $\tau=\tau_0$ is an extension of the Dixmier trace, the latter being defined only when the operator $P \in \Psi^*(\mathcal{A})$ belongs to OP^{-p}.

2. Local formula of the Chern character. Before giving the general local formula for the Chern character of a triple $(\mathcal{A},\mathcal{H},D)$ with discrete dimension spectrum, we need to recall a few basic definitions from [C1].

First, the cyclic cohomology $HC^n(\mathcal{A})$ is defined as the cohomology of the complex of cyclic cochains, i.e. those satisfying

$$\psi(a^1,\dots,a^n,a^0) = (-1)^n\psi(a^0,\dots,a^n)\ , \qquad \forall a^j \in \mathcal{A}\ , \tag{18}$$

under the coboundary operation b given by

$$(b\psi)(a^0,\dots,a^{n+1}) = \tag{19}$$

$$\sum_0^n(-1)^j\psi(a^0,\dots,a^ja^{j+1},\dots,a^{n+1}) + (-1)^{n+1}\psi(a^{n+1}a^0,\dots,a^n)\ ,\ \forall a^j \in \mathcal{A}.$$

Equivalently, $HC^n(\mathcal{A})$ can be described in terms of the second filtration of the (b,B) bicomplex of arbitrary (non-cyclic) cochains on $\mathcal{A}$, where $B: C^m \to C^{m-1}$ is given by

$$(B_0\varphi)(a^0,\dots,a^{m-1}) = \varphi(1,a^0,\dots,a^{m-1}) - (-1)^m\varphi(a^0,\dots,a^{m-1},1)\ , \tag{20}$$

$$B = AB_0\ ,\quad (A\psi)(a^0,\dots,a^{m-1}) = \sum(-1)^{(m-1)j}\psi(a^j,\dots,a^{j-1})\ .$$

To an n-dimensional cyclic cocycle ψ one associates the (b,B) cocycle $\varphi \in Z^p(F^q\ C)$, $n=p-2q$ given by

$$(-1)^{[n/2]}(n!)^{-1}\psi = \varphi_{p,q} \tag{21}$$

where $\varphi_{p,q}$ is the only non-zero component of φ.

Given a spectral triple $(\mathcal{A}, \mathcal{H}, D)$, with $D^{-1} \in \mathcal{L}^{(p,\infty)}$, its Chern character in cyclic cohomology is obtained from the following cyclic cocycle (τ_n), $n \geq p$, n odd,

$$\tau_n(a^0, \ldots, a^n) = \lambda_n \mathrm{Tr}'(a^0[F, a^1] \ldots [F, a^n]) , \qquad \forall a^j \in \mathcal{A} , \tag{22}$$

where $F = \mathrm{Sign} D$, $\lambda_n = \sqrt{2i}(-1)^{\frac{n(n-1)}{2}} \Gamma(\frac{n}{2} + 1)$ and

$$\mathrm{Tr}'(T) = \frac{1}{2} \mathrm{Trace}\big(F(FT + TF)\big) . \tag{23}$$

Theorem IV.2.8 of [C1] gives the following general formula for the Hochschild cohomology class of τ_n in terms of the Dixmier trace:

$$\varphi_n(a^0, \ldots, a^n) = \lambda_n \mathrm{Tr}_\omega(a^0[D, a^1] \ldots [D, a^n]|D|^{-n}) , \quad \forall a^j \in \mathcal{A} . \tag{24}$$

Our local formula for the *cyclic cohomology* Chern character, i.e. for a cyclic cocycle cohomologous to (22), will be expressed in terms of the (b, B) bicomplex. Bearing this in mind we see that if we want to regard the cochain φ_n of (24) as a cochain of the (b, B) bicomplex, we should use, instead of λ_n, the normalization constant

$$\mu_n = (-1)^{[n/2]}(n!)^{-1}\lambda_n = \sqrt{2i}\frac{\Gamma\left(\frac{n}{2} + 1\right)}{n!} \qquad (\text{for } n \text{ odd}) . \tag{25}$$

Let us now state the result. We let $(\mathcal{A}, \mathcal{H}, D)$ be a spectral triple with discrete dimension spectrum and $D^{-1} \in \mathcal{L}^{(p,\infty)}$. We shall use the following notations:

$$da = [D, a] , \quad \forall a \text{ operator in } \mathcal{H} , \tag{26}$$

$$\nabla(a) = [D^2, a] ; \quad a^{(k)} = \nabla^k(a) , \quad \forall a \text{ operator in } \mathcal{H} . \tag{27}$$

THEOREM II.1. a) *The following formula defines an odd cocycle in the (b, B) bicomplex of $\mathcal{A}$:*

$$\varphi_n(a^0, \ldots, a^n) = \sqrt{2i} \sum_{q \geq 0, k_j \geq 0} c_{n,k,q}\ \tau_q\big(a^0 (da^1)^{(k_1)} \ldots (da^n)^{(k_n)} |D|^{-(n+2\Sigma k_j)}\big) ,$$

where

$$c_{n,k,q} = (-1)^{k_1 + \cdots + k_n}(k_1! \ldots k_n!)^{-1}\Gamma^{(q)}\left(k_1 + \cdots + k_n + \frac{n}{2}\right) \times$$
$$\frac{1}{q!}\big((k_1 + 1)(k_1 + k_2 + 2) \ldots (k_1 + \cdots + k_n + n)\big)^{-1},$$

with $\Gamma^{(q)}$ the qth derivative of the Γ function.

b) *The cohomology class of the cocycle (φ_n) in $HC^{odd}(\mathcal{A})$ coincides with the cyclic cohomology Chern character $ch_*(\mathcal{A}, \mathcal{H}, D)$.*

Before starting the proof, a few comments are in order. First, let us note that the term $\tau_q(T_{n,k})$ with coefficient $c_{n,k,q}$ in the above sum vanishes when

$$n + \sum k_j > p \,, \tag{28}$$

since the operator $T_{n,k}$ is in $\mathcal{L}_0^{(1,\infty)}$ when (28) holds. This implies that for fixed n the sum involved contains only finitely many terms. (We assume that Sd has finite multiplicity so that only finitely many q's are involved.) It also implies that

$$\varphi_n = 0 \,, \qquad \text{if} \qquad n > p \,. \tag{29}$$

Next, we note that all the operators $T_{n,k}$ involved in the above formula are *homogeneous of degree* 0 in D, i.e. they are unaffected by the scaling

$$D \to \lambda D \,, \qquad \lambda \in \mathbb{R}_+^* \,. \tag{30}$$

Lastly, let us remark that assertion a), i.e. the equality

$$b\varphi_n + B\varphi_{n+2} = 0 \,, \qquad \forall n \,, \tag{31}$$

is actually a consequence of our proof of b). Nevertheless, it will be instructive to check it directly. We shall do so by making use of the following properties (with $\tau = \tau_0$):

$$D\, da + da\, D = \nabla(a) \,, \qquad \forall\, a \text{ operator in } \mathcal{H} \tag{32}$$

$$\tau\big((da)^{(k)} |D|^{-q}\big) = 0 \,, \qquad \forall\, a \in \mathcal{A} \,,\ \forall\, k \geq 0 \,,\ \forall\, q \tag{33}$$

$$\tau\big(\nabla(T)\, |D|^{-q}\big) = 0 \,, \qquad \forall\, T \,,\ \forall\, q \tag{34}$$

$$D_{(k)} b = \sum_0^\infty \frac{(-1)^\ell}{\ell!} b^{(\ell)} D_{(k+\ell)} \,, \qquad \forall\, b \in B \,, \tag{35}$$

where by definition $D_{(k)} = \Gamma(k)|D|^{-2k}$.

The meaning of (32) is that, if we view the graded commutator with D as a graded derivation in the appropriate way, then $d^2 = \nabla$. The meaning of (33) and (34) is that integration by parts is possible, since both d and ∇ are derivations. Finally, (35) follows from Theorem B.1 of Appendix B and will be used only under the trace τ so that only finitely many non-zero terms of the sum of the right hand side will appear.

Proof: a) We shall perform the computations under the simplifying assumption of simple spectrum. Only minor modifications will be required to treat the general case.

Let us first show that $B\varphi_1 = 0$. One has (up to the overall factor $\sqrt{2i}$ which we shall ignore) :

$$(B_0\varphi_1)(a) = \sum c_{1,k} \tau\big((da)^{(k)} |D|^{-1-2k}\big) \,, \tag{36}$$

hence, using (33), one gets

$$B_0\varphi_1 = 0 \quad \text{and} \quad B\varphi_1 = 0] \ .$$

We shall now compare $b\varphi_1$ and $-B\varphi_3$. The Leibnitz rule gives, in general,

$$\frac{1}{q!}(b_1 \dots b_n)^{(q)} = \sum_{\substack{q_j \geq 0 \\ \Sigma q_j = q}} \frac{b_1^{(q_1)}}{q_1!} \cdots \frac{b_n^{(q_n)}}{q_n!} \ . \tag{37}$$

One has

$$\varphi_1(a^0, a^1) = \sum_k \frac{(-1)^k}{(k+1)!} \tau\big(a^0 (da^1)^{(k)} D_{(k+\frac{1}{2})}\big) \ , \qquad \forall\, a^j \in \mathcal{A} \ . \tag{38}$$

To compute $b\varphi_1$ we need to apply τ, for each k, to

$$\big(a^0 a^1 (da^2)^{(k)} - a^0 d(a^1 a^2)^{(k)}\big) D_{(k+\frac{1}{2})} + a^0 (da^1)^{(k)} D_{(k+\frac{1}{2}} a^2 \ .$$

Using $d(a^1 a^2) = a^1 da^2 + (da^1) a^2$, (37) and (35) we thus get

$$-\sum_{\substack{k_1+k_2=k \\ k_1 \neq 0}} k! a^0 \frac{(a^1)^{(k_1)}}{k_1!} \frac{(da^2)^{(k_2)}}{k_2!} D_{(k+\frac{1}{2})} - \sum_{k_1+k_2=k} k! a^0 \frac{(da^1)^{(k_1)}}{k_1!} \frac{(a^2)^{(k_2)}}{k_2!} D_{(k+\frac{1}{2})}$$

$$+ \sum_{\ell \geq 0} (-1)^\ell a^0 (da^1)^{(k)} \frac{(a^2)^{(\ell)}}{\ell!} D_{(k+\ell+\frac{1}{2})} \ .$$

Thus, if we introduce the cochains

$$\tau(1, k_1, k_2)(a^0, a^1, a^2) = (-1)^{k_1+k_2} \tau\big(a^0 (a^1)^{(k_1)} (da^2)^{(k_2)} D_{(k_1+k_2+\frac{1}{2})}\big) \ , \tag{39}$$

$$\tau(2, k_1, k_2)(a^0, a^1, a^2) = (-1)^{k_1+k_2} \tau\big(a^0 (da^1)^{(k_1)} (a^2)^{(k_2)} D_{(k_1+k_2+\frac{1}{2})}\big) \ ,$$

we can express $b\varphi_1$ as follows:

$$b\varphi_1 = -\sum_{\substack{k_1 \neq 0 \\ k_j \geq 0}} \frac{(k_1+k_2+1)^{-1}}{k_1! k_2!} \tau(1, k_1, k_2) \tag{40}$$

$$+ \sum_{k_j \geq 0} \frac{\big((k_1+1)^{-1} - (k_1+k_2+1)^{-1}\big)}{k_1! k_2!} \tau(2, k_1, k_2) \ .$$

We shall now express $B\varphi_3$ in a similar manner, as a linear combination of the cochains (39). We have

$$B_0\varphi_3(a^0, a^1, a^2) = \sum c_{3,k}\, \tau\big((da^0)^{(k_1)} (da^1)^{(k_2)} (da^2)^{(k_3)} |D|^{-(3+2|k|)}\big) \ , \tag{41}$$

where $|k| = k_1 + k_2 + k_3$.

The cochain $B\varphi_3$ is the sum of the three cochains obtained from $B_0\varphi_3$ by cyclic permutations,

$$B\varphi_3 = B_0\varphi_3 + (B_0\varphi_3)' + (B_0\varphi_3)'' \ , \tag{42}$$

where

$$(B_0\varphi_3)'(a^0,a^1,a^2) = (B_0\varphi_3)(a^1,a^2,a^0)$$

and

$$(B_0\varphi_3)''(a^0,a^1,a^2) = (B_0\varphi_3)(a^2,a^0,a^1).$$

Using integration by parts (i.e. (33) and (34)), we can express $B_0\varphi_3$ as

$$\sum_{k_j\geq 0,\ell\geq 0}(-1)^{k_1+1}\alpha_k\frac{1}{\ell!}\frac{1}{(k_1-\ell)!}\frac{1}{k_2!}\frac{1}{k_3!}\tau(1,k_2+1+\ell,k_3+k_1-\ell) \quad (43)$$
$$+\sum(-1)^{k_1}\alpha_k\frac{1}{\ell!}\frac{1}{(k_1-\ell)!}\frac{1}{k_2!}\frac{1}{k_3!}\tau(2,k_2+\ell,k_3+1+k_1-\ell)\ ,$$

where

$$\alpha_k = (k_1+1)^{-1}(k_1+k_2+2)^{-1}(k_1+k_2+k_3+3)^{-1}\ .$$

Let us now compute $(B_0\varphi_3)'$:

$$(B_0\varphi_3)'(a^0,a^1,a^2) = \quad (44)$$
$$\sum(-1)^{|k|}\alpha_k\frac{1}{k_1!}\frac{1}{k_2!}\frac{1}{k_3!}\tau\big((da^1)^{(k_1)}(da^2)^{(k_2)}(da^0)^{(k_3)}D_{(3/2+|k|)}\big)\ .$$

In order to obtain terms of the form (39) we must move $(da^0)^{(k_3)}$ in front, using (35), and then integrate by parts. We use the trace property of τ and apply (35), for each term, with

$$b = (da^1)^{(k_1)}(da^2)^{(k_2)}\ . \quad (45)$$

Thus, we get

$$(B_0\varphi_3)'(a^0,a^1,a^2) = \quad (46)$$
$$\sum(-1)^{|k|+\ell}\alpha_k\frac{1}{k_1!}\frac{1}{k_2!}\frac{1}{k_3!}\frac{1}{\ell!}\tau\big((da^0)^{(k_3)}((da^1)^{(k_1)}(da^2)^{(k_2)})^{(\ell)}D_{(\frac{3}{2}+|k|+\ell)}\big)\ .$$

Integration by parts gives:

$$\begin{aligned}
&\tau\big((da^0)^{(k_3)}((da^1)^{(k_1)}(da^2)^{(k_2)})^{(\ell)}D_{(3/2+|k|+\ell)}\big)\\
&= (-1)^{k_3}\tau\big(da^0((da^1)^{(k_1)}(da^2)^{(k_2)})^{(\ell+k_3)}D_{(3/2+|k|+\ell)}\big)\\
&= (-1)^{k_3+1}\tau\big(a^0((a^1)^{(k_1+1)}(da^2)^{(k_2)})^{(\ell+k_3)}D_{(3/2+|k|+\ell)}\big)\\
&\quad + (-1)^{k_3}\tau\big(a^0((da^1)^{(k_1)}(a^2)^{(k_2+1)})^{(\ell+k_3)}D_{(3/2+|k|+\ell)}\big)\ .
\end{aligned}$$

We then use (37) to get

$$\left((a^1)^{(k_1+1)}(da^2)^{(k_2)}\right)^{(\ell+k_3)}$$
$$= \sum \frac{(\ell+k_3)!}{m!(\ell+k_3-m)!}(a^1)^{(k_1+1+m)}(da^2)^{(k_2+\ell+k_3-m)}$$
$$\left((da^1)^{(k_1)}\,(a^2)^{(k_2+1)}\right)^{(\ell+k_3)}$$
$$= \sum \frac{(\ell+k_3)!}{m!(\ell+k_3-m)!}(da^1)^{(k_1+m)}(a^2)^{(k_2+1+\ell+k_3-m)}\ .$$

This gives the following formula for $(B_0\,\varphi_3)'$:

$$(B_0\varphi_3)' = \sum_{k_j\geq 0,\ell,m\geq 0} (-1)^{k_3+1}\alpha_k \frac{1}{k_1!}\frac{1}{k_2!}\frac{1}{k_3!}\frac{1}{\ell!}\frac{1}{m!} \tag{47}$$
$$\frac{(\ell+k_3)!}{(\ell+k_3-m)!}\tau(1,k_1+1+m,k_2+\ell+k_3-m)+$$
$$\sum_{k_j\geq 0,\ell,m\geq 0} (-1)^{k_3}\alpha_k\ \frac{1}{k_1!}\frac{1}{k_2!}\frac{1}{k_3!}\frac{1}{\ell!}\frac{1}{m!}\frac{(\ell+k_3)!}{(\ell+k_3-m)!}$$
$$\tau(2,k_1+m,k_2+1+\ell+k_3-m)\ .$$

The computation of $(B_0\varphi_3)''$ is completely similar and gives:

$$(B_0\varphi_3)'' = \sum_{k_j\geq 0,\ell,m\geq 0} (-1)^{k_2+1}\alpha_k \frac{1}{k_1!}\frac{1}{k_3!}\frac{1}{\ell!}\frac{1}{m!}\frac{1}{(k_2-m)!} \tag{48}$$
$$\tau(1,k_3+1+m,k_1+\ell+k_2-m)+$$

$$\sum (-1)^{k_2}\alpha_k \frac{1}{k_1!}\frac{1}{k_3!}\frac{1}{\ell!}\frac{1}{m!}\frac{1}{(k_2-m)!}\tau(2,k_3+m,k_1+1+\ell+k_2-m)\ .$$

In order to compare $b\varphi_1$ with $B_0\varphi_3 + (B_0\varphi_3)' + (B_0\varphi_3)'' = B\varphi_3$ we just need to compare the coefficients of $\tau(j,k_1,k_2)$ in the formulae (40) and (43) + (47) + (48). The most convenient way to proceed is to introduce generating functions $f_j(x,y)$, $j=1,2$, in which we replace $\tau(j,k_1,k_2)$ by $x^{k_1}y^{k_2}$.

Let us first compute $f_j(x,y)$ for the expression (40); we get

$$f_1^b(x,y) = -\sum_{k_1\neq 0}(k_1+k_2+1)^{-1}\frac{x^{k_1}}{k_1!}\frac{y^{k_2}}{k_2!}\ , \tag{49}$$
$$f_2^b(x,y) = \sum \left((k_1+1)^{-1}-(k_1+k_2+1)^{-1}\right)\frac{x^{k_1}}{k_1!}\frac{y^{k_2}}{k_2!}\ .$$

Thus, $f_1^b(x,y) = -\int_0^1 (e^{u(x+y)}-e^{uy})du$, hence

$$f_1^b(x,y) = \frac{1-e^{x+y}}{x+y} + \frac{e^y - 1}{y} \ . \tag{50}$$

Similarly,

$$f_2^b(x,y) = \frac{1-e^{x+y}}{x+y} + e^y \left(\frac{e^x - 1}{x}\right) \ . \tag{51}$$

Let us next compute f_1 and f_2 for the expression (43); we get

$$\begin{aligned} f_1(x,y) &= \sum (-1)^{k_1+1} \alpha_k \frac{1}{\ell!} \frac{1}{(k_1-\ell)!} \frac{1}{k_2!} \frac{1}{k_3!} x^{k_2+1+\ell} y^{k_3+k_1-\ell} \ , \\ f_2(x,y) &= \sum (-1)^{k_1} \alpha_k \frac{1}{\ell!} \frac{1}{(k_1-\ell)!} \frac{1}{k_2!} \frac{1}{k_3!} x^{k_2+\ell} y^{k_3+k_1+1-\ell} \ . \end{aligned} \tag{52}$$

In order to compute them, we introduce the function

$$f(x,y,z) = \sum \alpha_k \frac{x^{k_1}}{k_1!} \frac{y^{k_2}}{k_2!} \frac{z^{k_3}}{k_3!} \ . \tag{53}$$

With this notation,

$$\begin{aligned} f_1(x,y) &= -f(-(x+y),x,y)x \ , \\ f_2(x,y) &= f(-(x+y),x,y)y \ . \end{aligned} \tag{54}$$

With the selfexplanatory notation f_1', f_2' for (47), we have

$$\begin{aligned} f_1'(x,y) &= \sum (-1)^{k_3+1} \alpha_k \frac{1}{k_1!} \frac{1}{k_2!} \frac{1}{k_3!} \frac{1}{\ell!} \frac{1}{m!} \frac{(\ell+k_3)!}{(\ell+k_3-m)!} x^{k_1+1+m} y^{k_2+\ell+k_3-m} \\ f_2'(x,y) &= \sum (-1)^{k_3} \alpha_k \frac{1}{k_1!} \frac{1}{k_2!} \frac{1}{k_3!} \frac{1}{\ell!} \frac{1}{m!} \frac{(\ell+k_3)!}{(\ell+k_3-m)!} x^{k_1+m} y^{k_2+1+\ell+k_3-m} \ . \end{aligned} \tag{55}$$

It follows that

$$\begin{aligned} f_1'(x,y) &= -e^{(x+y)} f\big(x,y,-(x+y)\big)x \ , \\ f_2'(x,y) &= e^{x+y} f\big(x,y,-(x+y)\big)y \ . \end{aligned} \tag{56}$$

Similarly, for f_1'', f_2'', we have

$$\begin{aligned} f_1''(x,y) &= \sum (-1)^{k_2+1} \alpha_k \frac{1}{k_1!} \frac{1}{k_3!} \frac{1}{\ell!} \frac{1}{m!} \frac{1}{(k_2-m)!} x^{k_3+1+m} y^{k_1+\ell+k_2-m} \\ f_2''(x,y) &= \sum (-1)^{k_2} \alpha_k \frac{1}{k_1!} \frac{1}{k_3!} \frac{1}{\ell!} \frac{1}{m!} \frac{1}{(k_2-m)!} x^{k_3+m} y^{k_1+1+\ell+k_2-m} \end{aligned} \tag{57}$$

which gives

$$\begin{aligned} f_1''(x,y) &= -e^y f\big(y,-(x+y),x\big)x \ , \\ f_2''(x,y) &= e^y f\big(y,-(x+y),x\big)y \ . \end{aligned} \tag{58}$$

We can thus express the generating functions for $B\varphi_3$ as follows:

$$\begin{aligned} f_1^B(x,y) &= -\big(f(-(x+y),x,y) + e^{x+y} f(x,y,-(x+y)) \\ &\qquad + e^y f(y,-(x+y),x)\big)x \\ f_2^B(x,y) &= \big(f(-(x+y),x,y) + e^{x+y} f(x,y,-(x+y)) \\ &\qquad + e^y f(y,-(x+y),x)\big)y \,. \end{aligned} \tag{59}$$

Let us compute $f(x,y,z)$; one has

$$f(x,y,z) = \int_{0\le u_1\le u_2\le u_3\le 1} e^{(u_1x+u_2y+u_3z)} du_1\, du_2\, du_3 \ ,$$

since

$$\int_{0\le u_1\le u_2\le u_3\le 1} u_1^{k_1}\, u_2^{k_2}\, u_3^{k_3}\, du_1\, du_2\, du_3 = \alpha_k \ .$$

We then obtain:

$$f(x,y,z) = \frac{e^{x+y+z}-1}{x(x+y)(x+y+z)} - \frac{e^z-1}{x(x+y)z} - \frac{(e^{y+z}-1)}{xy(y+z)} + \frac{(e^z-1)}{xyz} \,. \tag{60}$$

Since only the restriction of f to the hyperplane $x+y+z=0$ is involved, we can use the equality $\frac{e^a-1}{a} = 1$ for $a=0$, to handle the first term. We get

$$f(-(x+y),x,y) = \tag{61}$$

$$\frac{1}{-(x+y)(-y)} + \frac{(e^{x+y}-1)}{(x+y)x(x+y)} + (e^y-1)\left(\frac{1}{-(x+y)xy} - \frac{1}{(x+y)y^2}\right) ,$$

$$e^{x+y} f(x,y,-(x+y)) = \frac{e^{x+y}}{x(x+y)} + \frac{e^y - e^{x+y}}{x^2y} + \frac{e^{x+y}-1}{(x+y)^2y} \,, \tag{62}$$

$$e^y f(y,-(x+y),x) = -\frac{e^y}{xy} + \frac{e^y-1}{y^2(x+y)} + \frac{(e^x-1)e^y}{x^2(x+y)} \,. \tag{63}$$

Let us first compute the coefficient of e^{x+y} in the sum (61) + (62) + (63). We have

$$\begin{aligned} &\frac{1}{x(x+y)^2} + \frac{1}{x(x+y)} - \frac{1}{x^2y} + \frac{1}{(x+y)^2y} + \frac{1}{x^2(x+y)} \\ &= \frac{1}{x(x+y)} + \left(x^2y(x+y)^2\right)^{-1}\left(xy - (x+y)^2 + x^2 + y(x+y)\right) \\ &= \frac{1}{x(x+y)} \,. \end{aligned}$$

The coefficient of e^y is given by

$$-\frac{1}{(x+y)y^2}-\frac{1}{xy(x+y)}+\frac{1}{x^2y}-\frac{1}{xy}+\frac{1}{y^2(x+y)}-\frac{1}{x^2(x+y)}$$
$$=-\frac{1}{xy}+\left(x^2y^2(x+y)\right)^{-1}\left(-x^2-xy+y(x+y)+x^2-y^2\right)$$
$$=-\frac{1}{xy}\,.$$

The rational term which remains is then

$$\frac{1}{(x+y)y}-\frac{1}{x(x+y)^2}+\frac{1}{xy(x+y)}+\frac{1}{y^2(x+y)}-\frac{1}{y(x+y)^2}-\frac{1}{y^2(x+y)}$$
$$=\frac{1}{(x+y)y}+\left(xy^2(x+y)^2\right)^{-1}\left(-y^2+y(x+y)+x(x+y)-xy-x(x+y)\right)$$
$$=\frac{1}{(x+y)y}\,.$$

Thus, the sum (61) + (62) + (63) is equal to

$$\frac{e^{x+y}}{x(x+y)}-\frac{e^y}{xy}+\frac{1}{y(x+y)}\,. \tag{64}$$

When we multiply (64) by $-x$ we get

$$\frac{1-e^{x+y}}{x+y}-\left(\frac{1-e^y}{y}\right)\,. \tag{65}$$

When we multiply (64) by y we get:

$$-\frac{e^y}{x}+\frac{1}{x+y}+\frac{e^{x+y}y}{x(x+y)}\,. \tag{66}$$

Comparing these expressions with (50) and (51) we then obtain

$$f_1^b+f_1^B=0\quad,\quad f_2^b+f_2^B=0\,. \tag{67}$$

This shows that in the expression for $b\varphi_1+B\varphi_3$ the coefficient of any of the $\tau(j;k_1,k_2)$ vanishes for $j=1,2$ and any k_1,k_2.

We have thus shown that $B\varphi_1=0$ and that $b\varphi_1+B\varphi_3=0$. The proof for the general identity

$$b\varphi_n+B\varphi_{n+2}=0$$

is based on a similar computation.

The above discussion only covers the case of simple poles. The general case follows in exactly the same way, by introducing the expression

$$D_{(z)}=\Gamma(z)|D|^{-2z}\,,$$

for complex values of z, and performing the above manipulations with τ and $D_{(k)}$ replaced respectively by the usual trace *Trace* and by $D_{(k+\epsilon)}$, where ϵ is a small complex number. Taking the residue at $\epsilon=0$ gives the desired identities.

b) With $a^0,\dots,a^n \in \mathcal{A}$ fixed, we let

$$\zeta(z^0,\dots,z^n)=\mathrm{Trace}\big(a^0|D|^{-2z_0}da^1|D|^{-2z_1}da^2\dots|D|^{-2z_{n-1}}da^n|D|^{-2z_n}\big)\,. \tag{68}$$

This expression makes sense if $\sum \mathrm{Re}\, z_i > \frac{p}{2}$ and we shall first express it in terms of the following functions of a single complex variable:

$$h_k(z) = \mathrm{Trace}\big(a^0(da^1)^{(k_1)}(da^2)^{(k_2)}\dots(da^n)^{(k_n)}D^{-2\Sigma k_j-2z}\big)\ , \tag{69}$$

where $k=(k_1,\dots,k_n)$ is a multi-index.

As in (35), one has

$$|D|^{-2z}da = \sum \frac{(-1)^k}{k!} z^{(k)}(da)^{(k)}|D|^{-2z-2k}\ , \tag{70}$$

where $z^{(k)} = z(z+1)\dots(z+k-1)$.

We can then write the expansion

$$\zeta(z_0,\dots,z_n) = \sum P_k(z_0,\dots,z_{n-1})\ h_k\Big(\sum_0^n z_j\Big)\ , \tag{71}$$

where the polynomial P_k is given by

$$P_k(z_0,\dots,z_{n-1}) = \frac{(-1)^{k_1+\dots+k_n}}{k_1!\dots k_n!}\ z_0^{(k_1)}\ (z_0+k_1+z_1)^{(k_2)} \tag{72}$$

$$(z_0+k_1+z_1+k_2+z_2)^{(k_3)}\dots(z_0+k_1+z_1+k_2+z_2+k_3+\dots+z_{n-1})^{(k_n)}\,.$$

In the expansion (71), if we sum the terms for which $|k|>p$, they contribute by a function of $(z_0,\dots,z_n)$ which is holomorphic and bounded for $\sum \mathrm{Re}\, z_i > 0$. This follows from the Hölder inequality and the control (Theorem B.1 of Appendix B) of the remainder in the Taylor expansion (70).

Let $\lambda \in \mathbb{R}_+$, $\lambda > \frac{p}{2(n+1)}$; then using the equality

$$e^{-uD^2} = \frac{1}{2\pi}\int_{-\infty}^{\infty} \Gamma(\lambda+is)|D|^{-2(\lambda+is)}u^{-(\lambda+is)}ds\ , \qquad \forall u>0\ , \tag{73}$$

we obtain:

$$\mathrm{Trace}\big(a^0e^{-u_0D^2}da^1e^{-u_1D^2}da^2\dots e^{-u_{n-1}D^2}da^ne^{-u_nD^2}\big) \tag{74}$$

$$= (2\pi i)^{-(n+1)}\int_{C_\lambda^{n+1}} \Gamma(z_0)\dots\Gamma(z_n)u^{-z}\zeta(z_0,\dots,z_n)dz_0\dots dz_n\ ,$$

where $u_j>0$, $u^{-z} = u_0^{-z_0}\dots u_n^{-z_n}$, and $C_\lambda = \{\lambda+is;\, s\in\mathbb{R}\}$.

We let $\theta(u_0,u_1,\dots,u_n)$ be the function defined by (74) and we want to compute the coefficient of $\epsilon^{-n/2}$ in the expansion of

$$\theta(\epsilon v_0,\epsilon v_1,\dots,\epsilon v_n)\ ,\ \sum v_i = 1\ ,\ \epsilon\to 0\ . \tag{75}$$

From (71) and the hypothesis on the dimension spectrum we see, using the boundedness of $\Gamma(z_0)\ldots\Gamma(z_n)\Gamma(z_0+\cdots+z_n)^{-1}$ on C_λ^{n+1}, that except for finitely many values of λ the function

$$\Gamma(z_0)\ldots\Gamma(z_n)\zeta(z_0,\ldots,z_n)$$

is integrable on C_λ^{n+1} (for $\lambda > 0$ say). It follows then that the right hand side of (74), when evaluated for such a λ, at (ϵv_i) is a $O(\epsilon^{-(n+1)\lambda})$. It is not equal to (75) because of the contribution of the residues of the differential form

$$\omega = \Gamma(z_0)\ldots\Gamma(z_n)u^{-z}\zeta(z_0,\ldots,z_n)dz_0\ldots dz_n\ . \tag{76}$$

Using the expansion (71), we let

$$c(k,q) = \text{coeff of } \epsilon^{-q} \text{ in } h_k\left(\frac{n}{2}+\epsilon\right) \text{ at } \quad \epsilon = 0 \tag{77}$$

and concentrate on the contribution of the residue of the differential form

$$\omega = \Gamma(z_0)\ldots\Gamma(z_n)P_k(z_0,\ldots,z_{n-1})u^{-z}\left(z_0+\cdots+z_n-\frac{n}{2}\right)^{-q}\prod_0^n dz_i \tag{78}$$

(which then needs to be multiplied by $c(k,q)$).

If we denote by X the differential operator

$$X = \frac{1}{n+1}\sum_0^n \frac{\partial}{\partial z_j}\ , \tag{79}$$

the contribution of the residue is given by

$$\int_{\Sigma z_i=\frac{n}{2}} \frac{X^{q-1}}{(q-1)!}\left(\Gamma(z_0)\ldots\Gamma(z_n)P_k(z_0,\ldots,z_{n-1})u^{-z}\right)\prod_1^n dz_i\ , \tag{80}$$

where $\operatorname{Re} z_i = \lambda = \frac{n}{2(n+1)}$.

Thus, with $u_j = \epsilon v_j$, $\sum_0^n v_j = 1$, the coefficient of $\epsilon^{-n/2}$ is given by (80) with v instead of u, since the derivatives of $\epsilon^{-\Sigma z_i}$ contribute by terms involving $\epsilon^{-n/2}(\log\epsilon)^k$.

Introducing an additional variable $t \in \mathbf{R}$ we can rewrite the result as

$$\frac{1}{2\pi}\int \frac{t^{q-1}}{(q-1)!}\Gamma(z_0)\ldots\Gamma(z_n)P_k(z_0,\ldots,z_{n-1})v^{-z}e^{-t(\Sigma z_i-\frac{n}{2})}\prod_0^n dz_i\,dt\ , \tag{81}$$

where $\operatorname{Re} z_i = \lambda = \frac{n}{2(n+1)}$ and one integrates in the $n+2$ remaining variables.

Before taking care of the polynomial P_k, we can already compute, for fixed t,

$$\int_{\operatorname{Re} z_i=\lambda} \Gamma(z_0)\ldots\Gamma(z_n)v^{-z}e^{-t\Sigma z_i}\prod_0^n dz_i = (2\pi i)^{(n+1)}\prod_0^n e^{-v_j e^t}\ , \tag{82}$$

which holds for any value of $v_j > 0$.

Next, we have

$$\left(\frac{\partial}{\partial v_0}\right)^k v_0^{-z_0} = (-1)^k z_0^{(k)} v_0^{-(z_0+k)} , \tag{83}$$

or better

$$\left(\frac{\partial}{\partial u}\right)^k (uv_0)^{-z_0} = (-1)^k z_0^{(k)} u^{-(z_0+k)} v_0^{-z_0} . \tag{84}$$

This means that the effect of $P_k(z_0, \ldots, z_{n-1})$ in the above integral is obtained as follows, for the term

$$(-1)^{\Sigma k_j} z_0^{(k_1)} (z_0 + k_1 + z_1)^{(k_2)} \cdots (z_0 + k_1 + z_1 + \cdots + z_{n-1})^{(k_n)} .$$

One starts with the integral (82) written as $f(v_0, \ldots, v_n)$, then one applies

$$\left(\frac{\partial}{\partial u}\right)^{k_1} f(uv_0, v_1, \ldots, v_n) = f_1(u, v_0, v_1, \ldots, v_n) \tag{85}$$

and one continues with

$$\left(\frac{\partial}{\partial u}\right)^{k_2} f_1(u, v_0, uv_1, v_2, \ldots, v_n) = f_2(u, v_0, \ldots, v_n) , \tag{86}$$

$$\left(\frac{\partial}{\partial u}\right)^{k_3} f_2(u, v_0, v_1, uv_2, \ldots, v_n) = f_3(u, v_0, \ldots, v_n) ,$$

$$\ldots$$

$$\left(\frac{\partial}{\partial u}\right)^{k_n} f_{n-1}(u, v_0, \ldots, uv_{n-1}, v_n) = f_n(u, v_0, \ldots, v_n) ,$$

which is finally evaluated at $u = 1$.

Using (82), we are just applying this rule to

$$f(v_0, \ldots, v_n) = e^{-(\Sigma v_j)e^t} .$$

We get

$$\begin{aligned}
f_1(u, v_0, \ldots, v_n) &= (-v_0 e^t)^{k_1} f(uv_0, v_1, \ldots, v_n) , \\
f_2(u, v_0, \ldots, v_n) &= (-v_0 e^t)^{k_1} \big(- (v_0 + v_1)e^t \big)^{k_2} f(uv_0, uv_1, v_2, \ldots, v_n) , \\
f_3(u, v_0, \ldots, v_n) &= (-v_0\, e^t)^{k_1} \big(- (v_0 + v_1)e^t \big)^{k_2} \big(- (v_0 + v_1 + v_2)e^t \big)^{k_3} \\
&\qquad f(uv_0, uv_1, uv_2, v_3, \ldots) \\
f_n(1, v_0, \ldots, v_n) &= (-1)^{\Sigma k_j} e^{t \Sigma k_j} v_0^{k_1} (v_0 + v_1)^{k_2} \cdots (v_0 + \cdots + v_{n-1})^{k_n} \\
&\qquad f(v_0, \ldots, v_n) .
\end{aligned}$$

We can thus write (81) as

$$(2\pi)^{-1}(2\pi i)^{n+1}\frac{(-1)^{\Sigma k_j}}{k_1!\dots k_n!}\int\frac{t^{q-1}}{(q-1)!}e^{t(\Sigma k_j+\frac{n}{2})} \tag{87}$$
$$e^{-(\Sigma v_j)e^t}dt\, v_0^{k_1}(v_0+v_1)^{k_2}\dots(v_0+\dots+v_{n-1})^{k_n}\ .$$

We then have to integrate the result on the simplex $\sum v_j = 1$. The first task is thus to compute

$$\int_{-\infty}^{\infty}\frac{t^{q-1}}{(q-1)!}e^{t\alpha}e^{-e^t}dt\ ,\quad \alpha=\sum k_j+\frac{n}{2}\ . \tag{88}$$

It is obtained from the Taylor expansion at α of

$$\int_{-\infty}^{\infty}e^{t\alpha}e^{-e^t}dt=\Gamma(\alpha)$$

and is given by

$$\frac{\Gamma^{(q-1)}(\alpha)}{(q-1)!}\ . \tag{89}$$

The remaining part of the integral is

$$\int_{\sum_0^n v_j=1}v_0^{k_1}(v_0+v_1)^{k_2}\dots(v_0+\dots+v_{n-1})^{k_n}\prod dv_i$$
$$=(k_1+1)^{-1}(k_1+k_2+2)^{-1}\dots(k_1+\dots+k_n+n)^{-1}\ .$$

To complete the proof of part b) of the statement, we shall use the Chern character of $(\mathcal{A},\mathcal{H},D)$ in entire cyclic cohomology (cf. [C1]), given in the most efficient manner by the JLO formula, which defines the components of an entire cocycle in the (b,B) bicomplex:

$$\psi_n(a^0,\dots,a^n)=\sqrt{2i}\int_{\sum_0^n v_i=1,v_i\geq 0} \tag{90}$$
$$\mathrm{Trace}\big(a^0e^{-v_0D^2}[D,a^1]e^{-v_1D^2}\dots e^{-v_{n-1}D^2}[D,a^n]e^{-v_nD^2}\big)\ ,\qquad \forall a^j\in\mathcal{A}$$

where n is odd.

We introduce a parameter ϵ by replacing D^2 by ϵD^2, which yields a cocycle ψ_n^ϵ which is cohomologous to ψ_n. One has moreover

$$\psi_n^\epsilon(a^0,\dots,a^n)=\sqrt{2i}\Big(\int_{\sum_0^n v_i=1}\theta(\epsilon v_0,\dots,\epsilon v_n)\prod dv_i\Big)\epsilon^{n/2}\ , \tag{91}$$

where θ was defined by (74). The factor $\epsilon^{n/2}$ comes from the n terms $[D,a^j]$ in formula (90). Since n is odd, $n/2\neq 0$. The above computation of the behavior of $\epsilon^{n/2}\theta(\epsilon v_0,\dots,\epsilon v_n)$ from the residues of the differential form ω (76) gives an expansion of the form

$$\theta(\epsilon v_0,\dots,\epsilon v_n)=\sum\alpha_{m,\ell}\epsilon^{-p_m}(\log\epsilon)^\ell+O(\epsilon^{-n/2})\ , \tag{92}$$

where the exponents p_m correspond to the poles of h_k whose real part is larger than $\frac{n}{2}$. Moreover, we have already computed above the coefficient of $\epsilon^{-n/2}$ in this expansion, and after integration of the result in v_i we get:

$$\psi_n^\epsilon(a^0,\ldots,a^n) = \sum \beta_{m,\ell}\epsilon^{n/2-p_m}(\log\epsilon)^\ell + O(1)\,, \tag{93}$$

$$\beta_{0,0} = \sqrt{2i} \sum_{k_1,\ldots,k_n,q} (-1)^{\Sigma k_j} \frac{1}{k_1!\cdots k_n!}(k_1+1)^{-1}(k_1+k_2+2)^{-1}\cdots$$
$$(k_1+\ldots+k_n+n)^{-1}\frac{\Gamma^{(q)}\left(k_1+\ldots+k_n+\frac{n}{2}\right)}{q!}c(k,q)\,. \tag{94}$$

When we pair the (b,B) cocycle ψ^ϵ with a cyclic cycle $c=(c_n)$ in entire cyclic homology (cf. [C1]), the pairing gives a scalar independent of ϵ and written as a sum of terms of the form (93). The total contribution of the terms ψ_n^ϵ, $n>p$ converges to 0, by the argument of [CM1, §4]. Thus, we can assert that

$$\sum_{n\leq p}\langle\psi_n^\epsilon, c_n\rangle \underset{\epsilon\to 0}{\longrightarrow} \langle ch_*(\mathcal{H},D),c\rangle\,. \tag{95}$$

In view of (93) this is possible only if the asymptotic expansion of the left hand side in terms of $\epsilon^{-p_m}(\log\epsilon)^\ell$, $\mathrm{Re}\,p_m\leq 0$, only contains a constant term, and by (94) we know the value of this constant term: it is given by the pairing $\langle\varphi,c\rangle$ of the cyclic cocycle of Theorem II.1 a) with c.

Finally, to prove that the class of (φ_n) actually coincides with $ch_*(\mathcal{A},\mathcal{H},D)$, it suffices to recall that there is a canonical transgressed cochain $(\widetilde{\psi}_k)$ such that

$$\frac{d}{d\epsilon}\psi_n^\epsilon = b\widetilde{\psi}_{n-1}^\epsilon + B\widetilde{\psi}_{n+1}^\epsilon\,, \tag{96}$$

and then to note that $(\widetilde{\psi}_k)$ also has an asymptotic expansion of the form (93). It then follows from [CM2, §4] that (ψ_n) remains cohomologous with the *finite part* of (ψ_n^ϵ), which in turn, from the above discussion, is seen to be precisely the cocycle (φ_n). □

3. Renormalization. There is one unpleasant feature of the formula II.1 a) for the cyclic cocycle φ, namely the occurence of the transcendental numbers which enter in the Taylor expansion of the Γ function at the points $\Gamma\left(\frac{1}{2}+q\right)$, $q\in\mathbb{N}$. Also, the sum

$$\sum\frac{\Gamma\left(|k|+\frac{n}{2}\right)^{(q)}}{q!}\operatorname*{Res}_{s=0}\left(s^q\zeta(s)\right) \tag{97}$$

is an infinite sum when ζ is not meromorphic at $s=0$. We can of course rewrite it as

$$\operatorname*{Res}_{s=0} \Gamma\left(|k| + \frac{n}{2} + s\right) \zeta(s) . \tag{98}$$

We shall however proceed to show how to obtain a modified cyclic cocycle φ', giving the same result Thm. II.1 b) as φ, but involving a *finite* linear combination *with rational coefficients* of the terms

$$\sqrt{2i}\Gamma\left(\tfrac{1}{2}\right) \tau_q\left(a^0(da^1)^{(k_1)} \dots (da^n)^{(k_n)} |D|^{-2\Sigma k_j - n}\right) . \tag{99}$$

To achieve this, we shall exploit the freedom of replacing the operator D by $\mu^{-1}D$, $\mu \in \mathbf{R}_+^*$ without affecting Thm. II.1 b). The effect of this transformation on the functionals τ_q is as follows:

$$\tau_q^\mu = \sum \frac{(\log \mu)^m}{m!} \tau_{q+m} . \tag{100}$$

This implies that for any integer $m \geq 1$ the following formula defines the components of a cyclic cocycle which pairs trivially with cyclic homology:

$$\varphi_n^{(m)}(a^0, \dots, a^n) = \sum c_{n,k,q} \tau_{q+m}\left(a^0(da^1)^{(k_1)} \dots (da^n)^{(k_n)} |D|^{-(n+2|k|)}\right) . \tag{101}$$

What we shall do now is add a suitable linear combination of counterterms $\varphi^{(m)}$ in order to cancel all the transcendental coefficients occurring in the Taylor expansion of $\Gamma\left(\frac{1}{2}\right)^{-1} \Gamma(s)$ at half integers. Even though we could right away write down the list of the coefficients β_m needed in front of $\varphi^{(m)}$, we shall instead explain carefully how they are obtained.

To begin with, there is no problem at all if Sd is *simple*, i.e. if one has at most a simple pole. In that case one simply writes

$$\Gamma\left(\frac{1}{2} + q\right) = \frac{1}{2} \frac{3}{2} \cdots \left(\frac{1}{2} + q - 1\right) \Gamma\left(\frac{1}{2}\right) \tag{102}$$

and since all τ_q's with $q \geq 1$ vanish one gets the desired answer.

Let us see what happens when Sd has multiplicity two, i.e. when we have at most a double pole. In that case by Proposition II.1 we know that τ_1 is a trace, while $\tau_q = 0$ for $q \geq 2$. This means that the formula for φ_n involves the combination

$$\Gamma\left(|k| + \frac{n}{2}\right) \tau_0(A) + \Gamma'\left(|k| + \frac{n}{2}\right) \tau_1(A) , \tag{103}$$

where A is some operator. Now since the Hochschild coboundary $b\tau_0$ is given rationally in terms of τ_1 (cf. Proposition II.1) we do not expect to need the transcendental coefficient

$$\frac{\Gamma'\left(\frac{1}{2} + m\right)}{\Gamma\left(\frac{1}{2} + m\right)} \tag{104}$$

in order to compensate for the lack of trace property of τ_0. If we replace the term $\Gamma'\left(|k|+\frac{n}{2}\right)\tau_1(A)$ by $\Gamma\left(|k|+\frac{n}{2}\right)\tau_1(A)$, then we get exactly the components of $\varphi_n^{(1)}$ which we can subtract from φ without affecting Thm. II.1 b). Thus, we shall look for a coefficient λ such that

$$\Gamma'\left(\frac{1}{2}+m\right)=\lambda\Gamma\left(\frac{1}{2}+m\right)+c_m\Gamma\left(\frac{1}{2}+m\right)\,, \quad m\in\mathbb{N}\,, \tag{105}$$

where the c_m are rational numbers.

To obtain (105) one just uses the equality

$$\frac{\Gamma'\left(\frac{1}{2}+m+\epsilon\right)}{\Gamma\left(\frac{1}{2}+m+\epsilon\right)}=\sum_{a=0}^{m-1}\frac{1}{\frac{1}{2}+a+\epsilon}+\frac{\Gamma'\left(\frac{1}{2}+\epsilon\right)}{\Gamma\left(\frac{1}{2}+\epsilon\right)}\,, \tag{106}$$

which we write with ϵ for later use.

Thus, the constant λ is

$$\frac{\Gamma'\left(\frac{1}{2}\right)}{\Gamma\left(\frac{1}{2}\right)}=-(\gamma_E+2\log 2)\,,$$

where γ_E is Euler's constant.

If we replace φ by $\varphi-\lambda\varphi^{(1)}$, then using (105) we find that in the formula giving φ the terms $\Gamma'\left(|k|+\frac{n}{2}\right)\tau_1(A)$ should be replaced simply by $c_{|k|+\frac{n-1}{2}}\Gamma\left(|k|+\frac{n}{2}\right)\tau_1(A)$, where

$$c_\ell=\sum_{a=0}^{\ell-1}\frac{1}{\frac{1}{2}+a}\,. \tag{107}$$

Let us consider the next case, when Sd has multiplicity 3, i.e. when we have triple poles. This time we shall get the combination

$$\Gamma\left(|k|+\frac{n}{2}\right)\tau_0(A)+\Gamma'\left(|k|+\frac{n}{2}\right)\tau_1(A)+\frac{\Gamma''}{2!}\left(|k|+\frac{n}{2}\right)\tau_2(A)\,. \tag{108}$$

We want to use a further subtraction, say of $\lambda_2\varphi^{(2)}$, to get coefficients for $\tau_1(A)$ and $\tau_2(A)$ of the form $\Gamma\left(|k|+\frac{n}{2}\right)\times\mathbb{Q}$. From (106) we get the formula

$$\Gamma'\left(\tfrac{1}{2}+m+\epsilon\right)=R_m(\epsilon)\Gamma\left(\tfrac{1}{2}+m+\epsilon\right)+f(\epsilon)\Gamma\left(\tfrac{1}{2}+m+\epsilon\right)\,, \tag{109}$$

where R_m is the rational fraction

$$R_m(\epsilon)=\sum_{a=0}^{m-1}\frac{1}{\frac{1}{2}+a+\epsilon} \tag{110}$$

and where the function f is given by

$$f(\epsilon)=\frac{\Gamma'\left(\frac{1}{2}+\epsilon\right)}{\Gamma\left(\frac{1}{2}+\epsilon\right)}\,. \tag{111}$$

If we differentiate (109) we get:

$$\Gamma''\left(\tfrac{1}{2}+m+\epsilon\right) = R_m'(\epsilon)\,\Gamma\left(\tfrac{1}{2}+m+\epsilon\right) + R_m(\epsilon)\Gamma'\left(\tfrac{1}{2}+m+\epsilon\right) \tag{112}$$
$$+f'(\epsilon)\Gamma\left(\tfrac{1}{2}+m+\epsilon\right) + f(\epsilon)\Gamma'\left(\tfrac{1}{2}+m+\epsilon\right) .$$

We have to transform the term $R_m(\epsilon)\Gamma'\left(\frac{1}{2}+m+\epsilon\right)$ because it involves at the same time a function of m, $R_m(\epsilon)$ and a derivative of Γ. To do this we replace $\Gamma'\left(\frac{1}{2}+m+\epsilon\right)$ by its value (109) which yields:

$$R_m(\epsilon)^2\Gamma\left(\tfrac{1}{2}+m+\epsilon\right) + R_m(\epsilon)f(\epsilon)\Gamma\left(\tfrac{1}{2}+m+\epsilon\right) \tag{113}$$

and we use again (109) to replace the second term of the formula by

$$f(\epsilon)\Gamma'\left(\tfrac{1}{2}+m+\epsilon\right) - f(\epsilon)^2\Gamma\left(\tfrac{1}{2}+m+\epsilon\right) . \tag{114}$$

Coming back to all the terms of (112) we thus proved:

$$\Gamma''\left(\tfrac{1}{2}+m+\epsilon\right) = \left(R_m'(\epsilon)+R_m(\epsilon)^2\right)\Gamma\left(\tfrac{1}{2}+m+\epsilon\right) \tag{115}$$
$$+\left(f'(\epsilon)-f(\epsilon)^2\right)\Gamma\left(\tfrac{1}{2}+m+\epsilon\right) + 2f(\epsilon)\Gamma'\left(\tfrac{1}{2}+m+\epsilon\right) .$$

This shows that if we replace φ by $\varphi-\lambda_1\varphi^{(1)}-\lambda_2\varphi^{(2)}$, where $\lambda_1=\lambda=f(0)$ as above and

$$\lambda_2 = \frac{1}{2!}f_2(0) \quad , \quad f_2(\epsilon) = f'(\epsilon) - f(\epsilon)^2 , \tag{116}$$

then the combination (108) gets replaced by

$$\Gamma\left(|k|+\frac{n}{2}\right)\tau_0(A) + c_{|k|+\frac{n-1}{2}}\Gamma\left(|k|+\frac{n}{2}\right)\tau_1(A) + c'_{|k|+\frac{n-1}{2}}\Gamma\left(|k|+\frac{n}{2}\right)\tau_2(A) , \tag{117}$$

where the rational number c_m' is $\frac{1}{2!}R_m^{(1)}(0)$, with

$$R_m^{(1)}(\epsilon) = R_m'(\epsilon) + R_m(\epsilon)^2 . \tag{118}$$

We can now proceed by induction to the general case. One proves by induction on ℓ the following formula on the ℓth derivative $\Gamma^{(\ell)}$ of the Γ function

$$\Gamma^{(\ell)}\left(\tfrac{1}{2}+m+\epsilon\right) = R_m^{(\ell-1)}(\epsilon)\Gamma\left(\tfrac{1}{2}+m+\epsilon\right) + \sum_{j=1}^{\ell} C_\ell^j f_j(\epsilon)\Gamma^{(\ell-j)}\left(\tfrac{1}{2}+m+\epsilon\right) \tag{119}$$

where $R_m^{(\ell)}$ and f_j are defined inductively by

$$R_m^{(\ell+1)}(\epsilon) = R_m^{(\ell)}(\epsilon)' + R_m(\epsilon)\,R_m^{(\ell)}(\epsilon) , \tag{120}$$
$$f_{j+1}(\epsilon) = f_j'(\epsilon) - f(\epsilon)\,f_j(\epsilon) . \tag{121}$$

The proof uses the same patern as above and is straightforward. It shows that if we replace φ by

$$\varphi' = \varphi - \lambda_1 \varphi^{(1)} - \lambda_2 \varphi^{(2)} - \cdots - \lambda_\ell \varphi^{(\ell)} \cdots \tag{122}$$

where $\lambda_\ell = \frac{1}{\ell!} f_\ell(0)$, then the combination of terms

$$\sum \frac{1}{q!} \Gamma\left(|k| + \frac{n}{2}\right)^{(q)} \tau_q(A) \tag{123}$$

in the expression of the cocycle φ, gets replaced by

$$\Gamma\left(|k| + \frac{n}{2}\right) \sum \frac{1}{q!} R_m^{(q-1)}(0) \tau_q(A) \qquad m = |k| + \frac{n-1}{2} . \tag{124}$$

Now the functions $R_m^{(\ell)}(\epsilon)$ are easy to compute, since

$$R_m(\epsilon) = \sum_{a=0}^{m-1} T_a(\epsilon) \quad \text{with} \quad T_a'(\epsilon) = -T_a(\epsilon)^2 \ , \quad T_a(\epsilon) = \frac{1}{\frac{1}{2} + a + \epsilon} .$$

One obtains that $R_m^{(\ell)}(\epsilon)$ is the $(\ell+1)$th symmetric function of the m terms $\frac{1}{\frac{1}{2}+a+\epsilon}$:

$$\prod_{a=0}^{m-1} \left(1 + \frac{z}{\frac{1}{2} + a + \epsilon}\right) = \sum R_m^\ell(\epsilon)\, z^{\ell+1} . \tag{125}$$

We can then easily compute the product $\Gamma\left(\frac{1}{2} + m\right) R_m^{(q-1)}(0)$, which appears in (124), and get

$$\Gamma\left(\tfrac{1}{2} + m\right) R_m^{(q-1)}(0) = \Gamma\left(\tfrac{1}{2}\right) \sigma_{m-q}(m) \ , \tag{126}$$

where $\sigma_j(m)$ is the jth symmetric function of the first m odd 1/2 integers:

$$\prod_{\ell=0}^{m-1} \left(z + \frac{(2\ell+1)}{2}\right) = \sum z^j \sigma_{(m-j)}(m) . \tag{127}$$

What is remarkable now is that these coefficients vanish if q is larger than m so that, not only have we transformed them to numbers in $\Gamma\left(\frac{1}{2}\right)\mathbb{Q}$, but we have also eliminated all except finitely many.

We note that the function $f_j(\epsilon)$ is not difficult to compute, and by induction we get the formula

$$f_j(\epsilon) = -\Gamma\left(\frac{1}{2} + \epsilon\right) \left(\frac{1}{\Gamma\left(\frac{1}{2} + \epsilon\right)}\right)^{(j)} , \tag{128}$$

as can be seen by interpreting the transformation

$$T(h) = h' - fh \tag{129}$$

as $T(h) = \Gamma\,(h/\Gamma)'$, and using (121).

It follows that

$$\lambda_\ell = -\Gamma\left(\frac{1}{2}\right)\frac{1}{\ell!}\left(\frac{1}{\Gamma\left(\frac{1}{2}+\epsilon\right)}\right)^{(\ell)}\Bigg|_{\epsilon=0} \tag{130}$$

and also that the above operation of subtraction has the following simple interpretation.

In the proof of Theorem II.1 a), we were applying the linear form $\operatorname*{Res}_{s=0}$ to meromorphic functions of the form

$$\Gamma\left(|k|+\frac{n}{2}+s\right)\mathrm{Trace}\left(A|D|^{-2s}\right)=\zeta(s)\ , \tag{131}$$

where A is an operator. But any other linear form such as

$$\zeta \to \operatorname*{Res}_{s=0} g(s)\zeta(s)\ , \tag{132}$$

with g holomorphic at 0, would have worked equally well. What the subtraction of $\sum \lambda_\ell\, \varphi^{(\ell)}$ is doing is exactly to take

$$g(s)=\frac{\Gamma(1/2)}{\Gamma(1/2+s)}\ .$$

If we combine this with

$$\frac{\Gamma\left(\frac{1}{2}+m+s\right)}{\Gamma\left(\frac{1}{2}+s\right)}=\prod_0^{m-1}\left(\frac{1}{2}+a+s\right)\ , \tag{133}$$

we can summarize the above discussion in the following variant of Theorem II.1.

THEOREM II.2. *The statements of Theorem* II.1 *are true for the cocycle* φ_n' *given, for n-odd, $n \le p$, by the formula*

$$\varphi_n'(a^0,\dots,a^n)=$$

$$\sqrt{2\pi i}\sum_{k,q}\frac{(-1)^{|k|}}{k_1!\cdots k_n!}\alpha_k\frac{1}{q!}\sigma_{m-q}(m)\tau_q\left(a^0(da^1)^{(k_1)}\cdots(da^n)^{(k_n)}|D|^{-(2|k|+n)}\right)\ ,$$

with $m=|k|+\frac{n-1}{2}$, $\alpha_k^{-1}=(k_1+1)(k_1+k_2+2)\cdots(k_1+\ldots+k_n+n)$ *and* σ *defined in* (127).

Not only are all coefficients rational multiples of the overall factor $\sqrt{2\pi i}$, but also the total number of terms in the formula is now finite and bounded in terms of p alone and not the order of the poles. Indeed,

$$q\le |k|+\frac{n-1}{2} \quad \text{and} \quad |k|+n\le p,$$

so that the formula does not involve more than p terms in the Laurent expansion.

Let us see what this formula looks like for small values of p.

Case $p = 1$. Then only φ'_1 is non-zero; we have $k = 0$ and $q = 0$, also

$$\varphi'_1(a^0, a^1) = \sqrt{2\pi i}\,\tau_0\big(a^0 da^1 |D|^{-1}\big)\ . \tag{134}$$

This shows that, even if we had poles of arbitrary order for the function $\zeta(s) = \mathrm{Trace}\big(a^0 da^1 |D|^{-1-2s}\big)$ at $s = 0$, they do not contribute to φ'_1 except for the residue of ζ at $s = 0$.

If we had used the formula of Theorem II.1 we would be taking the residue of $\Gamma\left(\frac{1}{2} + s\right)\zeta(s)$ at $s = 0$ which involves infinitely many of the functionals τ_k. Note also that here τ_0 is a trace.

Case $p = 2$. Again, only φ'_1 is non-zero, but now we can have $k_1 = 1$ and also $q = 1$ if $k_1 = 1$. Thus, we get three terms:

$$\begin{aligned}\varphi'_1(a_0, a_1) = \sqrt{2\pi i}\Big(&\tau_0\big(a^0 da^1 |D|^{-1}\big) - \tfrac{1}{2}\tau_0\big(a^0 (da^1)^{(1)} |D|^{-3}\big) \\ &-\tfrac{1}{2}\tau_1\big(a^0 (da^1)^{(1)} |D|^{-3}\big)\Big)\ .\end{aligned} \tag{135}$$

This time τ_0 is no longer a trace, as one can see using Proposition II.1, and the formula involves τ_1, i.e. the coefficient of s^{-2} in some ζ-function. However, no higher order coefficient is involved, unlike the formula for φ in Thm. II.1. a).

Case $p = 3$. Let us look at φ'_3 in this case. Here, we must have $k = 0$ but since $q \le |k| + \frac{n-1}{2}$, we can have $q = 1$. Thus, we get two terms for φ'_3:

$$\varphi'_3(a_0, a_1, a_2, a_3) = \sqrt{2\pi i}\big(\tau_0\big(a^0 da^1 da^2 da^3 |D|^{-3}\big) + \tau_1\big(a^0 da^1 da^2 da^3 |D|^{-3}\big)\big)\ . \tag{136}$$

This shows that, even for φ'_3, the coefficient of s^{-2} in the expansion of the ζ-function is playing a role, i.e. that τ_1 enters into play.

4. Local index formula. To get more insight into the content of Theorems II.1 and II.2, we shall now write down a corollary whose statement does not involve cyclic cohomology or noncommutative geometry but computes a Fredholm index (called *spectral flow*) as a sum of residues of ζ-functions attached to the problem.

To formulate the problem we just need a pair (D, U) of operators in Hilbert space, where D is selfadjoint with discrete spectrum, while U is unitary. The main assumption we need is that $[D, U]$ is bounded, which implies immediately that the compression PUP of U of the *positive* part of D, $\big(P = \frac{1+F}{2},\ F = \mathrm{Sign} D\big)$ is a Fredholm operator. The index

$$\mathrm{Index}\, PUP = \dim \mathrm{Ker}\, PUP - \dim \mathrm{Ker}\, PU^*P \tag{137}$$

can be interpreted as spectral flow, i.e. as the net number of eigenvalues which cross the origin in the linear homotopy between D and $UDU^* =$

$D+U[D,U^*]$. In any case, it is an integer, and we shall compute it as a sum of residues.

We also make the following additional hypotheses.

(138) a) If S is the spectrum of D (with multiplicity), then

$$\sum_{\lambda\in S}|\lambda|^{-s}<\infty \quad \text{for some finite} \quad s\,.$$

(We call p the lower bound of such s).

b) The operators U and $[D,U]$ are in the domain of δ^k, $\delta=[|D|,\cdot]$ for $1\le k\le N$, $N\gg 0$.

c) The following functions, holomorphic for $\mathrm{Re}\, s\gg 0$, are meromorphic with finitely many poles for $\mathrm{Re}\, s>-\epsilon$:

$$\zeta_{(k,n)}(s)=\mathrm{Trace}\big(U^{-1}\,[D,U]^{(k_1)}[D,U^{-1}]^{(k_2)}\dots[D,U]^{(k_n)}|D|^{-2|k|-n-s}\big)\,,$$

where we use the notation $X^{(k)}=\nabla^k(X)$, $\nabla(X)=[D^2,X]$.

In c) only *finitely many* functions are involved because of the inequality $|k|+n\le p$. At the technical level, we need to assume

d) $\Gamma(s)\zeta(s)$ restricted to vertical lines is of rapid decay.

COROLLARY II.1. *Let D and U be as above. Then*

$$\mathrm{Index}\,PUP=\sum_{n\le p}(-1)^{\frac{n-1}{2}}\left(\frac{n-1}{2}\right)!\sum_{k,q}\frac{(-1)^{|k|}}{k_1!\dots k_n!}\alpha_k\frac{1}{q!}\sigma_{m-q}(m)\operatorname*{Res}_{s=0}s^q\zeta_{(k,n)}(s)\,,$$

with $m=|k|+\frac{n-1}{2}$.

Proof: We just apply Theorem II.2 to the special case when $\mathcal{A}=C^\infty(S^1)$, acting on $\mathcal{H}$ by the unique representation which sends the function $f(e^{i\theta})$ to $f(U)$. We use the formula for the pairing between K^1-theory and odd cyclic cohomology, together with the index formula (cf. [C1]),

$$\langle\varphi',U\rangle=\frac{1}{\sqrt{2\pi i}}\sum_{n\ \mathrm{odd}}(-1)^{\frac{n-1}{2}}\left(\frac{n-1}{2}\right)!\ \varphi'_n(U^{-1},U,\dots,U^{-1},U)\,. \qquad (139)$$

The proof of Theorem II.1 b) shows that the hypothesis (138) is sufficient to conclude. □

At this point we should stress the considerable freedom that one has in applying Corollary II.1. The data is a discrete subset (perhaps with multiplicity) of $\mathbb{R}$,

$$S=\mathrm{Spectrum}\,D\ , \qquad (140)$$

together with a unitary matrix, $u(\lambda,\lambda')_{\lambda,\lambda'\in S}$ which signifies a "unitary correspondence" on the list S. The main hypothesis is that when D is shifted by this correspondence (i.e. UDU^* is considered), it stays at bounded distance from D. Then, one writes down a finite number of ζ-functions, the $\zeta_{(k,n)}$ above, which can be expressed as Dirichlet series of the form

$$\sum a_n^{\pm} |\lambda_n^{\pm}|^{-s} \tag{141}$$

when one computes the trace in the basis of eigenvectors

$$De_n^{\pm} = \lambda_n^{\pm} e_n^{\pm} \tag{142}$$

for the operator D.

The statement is that a certain rational combination of residues of these functions gives the index of PUP or spectral flow. In particular, one has:

COROLLARY II.2. *If* Index $PUP \neq 0$, *at least one of the functions* $\zeta_{(k,n)}(s)$ *has a non-trivial pole at* $s = 0$.

5. Concluding remarks. We shall now briefly discuss the analogues of Theorems II.1 and II.2 in the even case, i.e. when we have a $\mathbb{Z}/2$-grading γ of the Hilbert space $\mathcal{H}$ such that:

$$\gamma a = a\gamma \qquad \forall\, a \in \mathcal{A}\,, \qquad D\gamma = -\gamma D\,. \tag{143}$$

Let us go back to Theorem II.1 and consider for n *even*, $n > 0$, the cochain given by the similar formula, i.e.

$$\varphi_n(a^0, \ldots, a^n) = \sum_{q \geq 0, k_j \geq 0} c_{n,k,q} \tau_q\big(\gamma a^0 (da^1)^{(k_1)} \ldots (da^n)^{(k_n)} |D|^{-(n+2\Sigma k_j)}\big)\,, \tag{144}$$

$\forall\, a^j \in \mathcal{A}$, where the $c_{n,k,q}$ are given by

$$c_{n,k,q} = (-1)^{k_1 + \ldots + k_n} \tag{145}$$

$$(k_1! \cdots k_n!)^{-1} \Gamma^{(q)}\left(k_1 + \ldots + k_n + \frac{n}{2}\right) \frac{1}{q!} \big((k_1+1)(k_1+k_2+2)\ldots\big)^{-1}\,.$$

One can compute $B\varphi_2$ exactly as we did in the proof of Thm. II.1 a). One obtains:

$$B\varphi_2(a^0, a^1) = -\sum_{\substack{k,q \\ k \geq 1}} (-1)^k \frac{\Gamma(k)^{(q)}}{k!q!} \tau_q\big(\gamma a^0 (da^1)^{(k)} |D|^{-1-2k}\big)\,. \tag{146}$$

One checks, as in the proof of a), that $b\varphi_n + B\varphi_{n+2} = 0$ for all $n > 0$ (n even), but to get the correct expression for φ_0 we need to go back to the proof of b) for the case $n = 0$.

In the case at hand we get

$$\psi_0^{\epsilon}(a) = \operatorname{Trace}(\gamma a e^{-\epsilon D^2}) \qquad \forall\, a \in \mathcal{A} \tag{147}$$

and this time we look for the *constant* terms in the asymptotic expansion at 0 of this function of ϵ. It is of course related to the analogue of (68) for $n = 0$,

$$\zeta(z) = \text{Trace}(\gamma a|D|^{-2z}) . \tag{148}$$

However, this time it is no longer the residue of (148) which matters, but rather its value at 0, which governs the residue of the product $\Gamma(z)\zeta(z)$, i.e.

$$\varphi_0(a) = \operatorname*{Res}_{s=0} \left(\Gamma(s)\text{Trace}(\gamma a|D|^{-2s})\right) . \tag{149}$$

The right hand side is equal to

$$\sum_{q\geq 0} \frac{\Gamma(1)^{(q)}}{q!} \tau_{q-1}(\gamma a) ,$$

where we need to define τ_{-1}, following Proposition II.1, by

$$\tau_{-1}(b) = \operatorname*{Res}_{s=0} s^{-1}\text{Trace}(b|D|^{-2s}) . \tag{150}$$

We can now simplify the cocycle φ as we did in Theorem II.2. The expression (144) for φ_n involves

$$\operatorname*{Res}_{s=0} \Gamma\left(|k| + \frac{n}{2} + s\right) \text{Trace}(A|D|^{-2s}) , \tag{151}$$

for some operator A. As in the proof of Theorem II.2, we replace this by

$$\operatorname*{Res}_{s=0} \Gamma(1+s)^{-1}\Gamma\left(|k| + \frac{n}{2} + s\right) \text{Trace}(A|D|^{-2s}) \tag{152}$$

and we use the equality

$$\Gamma(m+s)\Gamma(1+s)^{-1} = \sum_{0}^{m-1} \sigma_j(m)s^j , \tag{153}$$

where the σ_j are the elementary symmetric function of the numbers $1, 2, \ldots, m-1$.

We thus obtain the graded analogue of Theorem II.2:

THEOREM II.3. a) *The following formula defines an even cocycle in the (b, B) bicomplex of $\mathcal{A}$:*

$$\varphi_n(a^0, \ldots, a^n) =$$

$$\sum_{k,q} \frac{(-1)^{|k|}}{k_1! \ldots k_n!} \alpha_k \sigma_q \left(|k| + \frac{n}{2}\right) \tau_q\left(\gamma a^0 (da^1)^{(k_1)} \ldots (da^n)^{(k_n)} |D|^{-(2|k|+n)}\right)$$

for $n \neq 0$ even, while

$$\varphi_0(a^0) = \tau_{-1}(\gamma a^0) .$$

b) *The cohomology class of the cocycle (φ_n) in $HC^{ev}(\mathcal{A})$ coincides with the cyclic cohomology Chern character $ch_*(\mathcal{A}, \mathcal{H}, D)$.*

Remark II.1: To see where the classical index theorem for manifolds fits in this picture, let us consider the spectral triple $(\mathcal{A}, \mathcal{H}, D)$ consisting of the Dirac operator D acting on the Hilbert space $\mathcal{H}$ of L^2-spinors over a closed Riemannian Spin manifold M of dimension p, and with $\mathcal{A} = C^\infty(M)$.

Then:

a) *the dimension spectrum of the triple $(\mathcal{A}, \mathcal{H}, D)$ is simple and contained in the set*

$$\{n \in \mathbf{Z}\ ;\ n \leq p\}\ ;$$

b) *with $\tau = \tau_0$ and otherwise using the above notation, one has*

$$\tau\big(a^0(da^1)^{(k_1)} \dots (da^n)^{(k_n)}|D|^{-(n+2|k|)}\big) = 0\ , \quad \text{for} \quad |k| \neq 0\ ;$$

c) *for $k_1 = k_2 = \dots = k_n = 0$, one has*

$$\tau\big(a^0[D, a^1] \dots [D, a^n]|D|^{-n}\big) = \nu_n \int_M \hat{A}(R) \wedge a^0 da^1 \wedge \dots \wedge da^n\ ,$$

where ν_n is a numerical factor and $\hat{A}(R)$ stands for the $\hat{A}$-form of the Riemannian curvature of M.

Assertion a) follows from standard pseudodifferential calculus, while b) and c) can be checked, for example, by means of the symbol calculus for asymptotic operators, as in [CM1, §3].

A much more interesting example for the general local index formalism developed in this section is provided by the spectral triples of section 1, associated to triangular structures on manifolds. This application will be discussed in a forthcoming paper.

Appendix A. Inequalities for eigenvalues and the Dixmier trace

Let $\mathcal{H}$ be a (separable) Hilbert space. We let $\mathcal{K} \subset \mathcal{L}(\mathcal{H})$ be the ideal of compact operators and $\mathcal{L}^1 \subset \mathcal{K}$ the ideal of trace class operators:

$$T \in \mathcal{L}^1 \Leftrightarrow \sum_0^\infty \mu_n(T) < \infty \tag{1}$$

where $\mu_n(T) = \mathrm{Inf}\{\|T|E^\perp\|; \dim E = n\}$ is the nth characteristic value of T. Note that

$$\mu_n(T) = \mathrm{dist}(T, \mathcal{R}_n)\ , \qquad \mathcal{R}_n = \{\text{operators of rank } \leq n\} \tag{2}$$

$$\mu_n(T) = n + 1\text{'th eigenvalue of } |T|\ , \tag{3}$$

the last equality being the minimax principle.

One defines $\mathrm{Trace}(T)$ for $T \in \mathcal{L}^1$ as

$$\mathrm{Trace}(T) = \sum \langle T\xi_n, \xi_n\rangle\ , \qquad (\xi_n) \text{ orthonormal basis}\ . \tag{4}$$

This converges because T is an ℓ^1 sum of rank one operator $|\eta\rangle\langle\eta|$ and is independent of the basis, as can be seen for such an operator.

LEMMA A.1. $$\sum_0^\infty \mu_n(T) = \text{Sup}\left|\text{Trace}(TX)\right| \, ; \qquad \|X\| \leq 1 \, .$$

Proof: First note that $|\text{Trace}(T)| \leq \sum_0^\infty \mu_n(T)$, using the polar decomposition $T = U|T|$, $|T| = \sum \mu_n |\eta_n\rangle\langle\eta_n|$ and $|\text{Trace}|U\eta_n\rangle\langle\eta_n|| \leq 1$.

Then $|\text{Trace}(TX)| \leq \sum_0^\infty \mu_n(TX) \leq \sum_0^\infty \mu_n(T)$, if $\|X\| \leq 1$. For the converse use $U^* = X$. □

One lets $\|T\|_1 = \text{Trace}(|T|)$; it is a norm (the *trace norm*).

DEFINITION A.1. *For each integer $N \geq 1$, let, for $T \in \mathcal{K}$*

$$\sigma_N(T) = \text{Sup}\left\{\|TE\|_1 \, ; \; E \textit{ subspace} \, , \; \dim E = N\right\} .$$

By construction σ_N is a norm:

$$\sigma_N(T_1 + T_2) \leq \sigma_N(T_1) + \sigma_N(T_2) \qquad \forall T_j \in \mathcal{K} \, . \tag{5}$$

PROPOSITION A.1. $$\sigma_N(T) = \sum_0^{N-1} \mu_n(T) \, .$$

Proof: We can assume $T \geq 0$. Then, taking $E = E_N(T)$ the spectral projection on the spectral subspace corresponding to the first N eigenvalues, one gets the inequality $\geq$. Conversely, if $\dim E = N$ then $\|TE\|_1 = \sum_0^\infty \mu_n(TE) = \sum_0^{N-1} \mu_n(TE) \leq \sum_0^{N-1} \mu_n(T)$ (since $\text{dist}(TE, \mathcal{R}_N) = 0$). □

We shall now extend the definition of σ_N to a function σ_λ of the positive "scale" parameter λ as follows:

LEMMA A.2. *Let $T \in \mathcal{K}$. The following function of $\lambda \in \mathbb{R}_+^*$ agrees with $\sigma_N(T)$ at integer values:*

$$\sigma_\lambda(T) = \text{Inf}\left\{\|x\|_1 + \lambda\|y\|_\infty \, ; \; x + y = T\right\} .$$

Proof: One can assume $T \geq 0$. With $N \in \mathbb{N}^*$, we compare $\sigma_N(T)$ of Definition A.1 with the r.h.s. of (4). Assume $T = x + y$, with $\|x\|_1 + N\|y\|_\infty \leq 1$. Then

$$\sigma_N(T) \leq \sigma_N(x) + N\|T - x\|_\infty \leq \|x\|_1 + N\|y\|_\infty \leq 1 \, .$$

Conversely, write $T = (T - \mu_N(T)1)E_N + (\mu_N(T)E_N + T(1 - E_N)) = x + y$. One has $\|x\|_1 = \sigma_N(T) - N\mu_N(T)$ and $\|y\|_\infty = \mu_N(T)$. Thus,

$$\|x\|_1 + N\|y\|_\infty = \sigma_N(T).$$ □

By construction, the function $\sigma_\lambda(T)$ is increasing as a function of λ. Also, the unit ball for the norm σ_λ is the convex hull of the unit ball of $\mathcal{L}_1$ and λ^{-1} times the unit ball of $\mathcal{K}$. The slope of σ_λ, such as $\sigma_{N+1}-\sigma_N=\mu_N$, is decreasing with λ, thus σ_λ is a concave function of λ.

Between 0 and 1 one has $\sigma_\lambda(T)=\lambda\|T\|$ as follows from:

$$\|x\|_\infty \leq \|x\|_1 .$$

One can check as in the proof of Lemma A.2 that σ_λ is affine between N and $N+1$ for any N. In particular (5) holds for all real values $\lambda>0$.

For a positive operator T we view $\sigma_\lambda(T)$ as the trace of T *cutoff* at the (inverse) scale λ.

We shall now investigate the additivity of σ_λ for $T\geq 0$.

LEMMA A.3. *For $T_1,T_2\geq 0$ and $\lambda_1,\lambda_2\in\mathbf{R}_+^*$*

$$\sigma_{\lambda_1+\lambda_2}(T_1+T_2)\geq\sigma_{\lambda_1}(T_1)+\sigma_{\lambda_2}(T_2) .$$

Proof: Let us assume that N_1,N_2 are integers. First, for $T\geq 0$ one has:

$$\sigma_N(T)=\operatorname{Sup}\left\{\operatorname{Trace}(TE)\ ;\ \dim E=N\right\} . \tag{6}$$

(The r.h.s. is smaller than $\|TE\|_1$ and hence than $\sigma_N(T)$, the other inequality is clear.)

Then $\operatorname{Trace}(T_1E_1)+\operatorname{Trace}(T_2E_2)\leq\operatorname{Trace}(T_1E)+\operatorname{Trace}(T_2E)$ where $E=E_1\vee E_2$ has dimension $\leq N_1+N_2$. One can then deduce the result for arbitrary λ_1,λ_2 by piecewise linearity. □

We shall now concentrate on the $\log\lambda$ divergence of $\sigma_\lambda(T)$ and average the coefficient of $\log\lambda$ over various scales by considering the following function:

$$\tau_\lambda(T)=\frac{1}{\log\lambda}\int_a^\lambda\frac{\sigma_u(T)}{\log(u)}\frac{du}{u}\qquad(\text{we fix } a>e) . \tag{7}$$

By construction τ_λ is subadditive (cf. (5)). Let us now evaluate a lower bound for $\tau_\lambda(T_1+T_2)$, $T_j\geq 0$.

LEMMA A.4. *Let T_1,T_2 be such that $\sigma_\lambda(T_j)\leq C_j\log\lambda$, $\forall\,\lambda\geq a$. If $T_j\geq 0$, then:*

$$\left|\tau_\lambda(T_1+T_2)-\tau_\lambda(T_1)-\tau_\lambda(T_2)\right|\leq(C_1+C_2)\big((\log\log\lambda+2)\log 2\big)/\log\lambda .$$

Proof: One has $\sigma_{2u}(T_1+T_2)\geq\sigma_u(T_1)+\sigma_u(T_2)$. Thus:

$$\tau_\lambda(T_1)+\tau_\lambda(T_2)\leq\frac{1}{\log\lambda}\int_a^\lambda\frac{\sigma_{2u}(T_1+T_2)}{\log u}\frac{du}{u}=\frac{1}{\log\lambda}\int_{2a}^{2\lambda}\frac{\sigma_u(T_1+T_2)}{\log(u/2)}\frac{du}{u} .$$

On $[a,+\infty[$ one has $\sigma_u(T_1+T_2)\le (C_1+C_2)\log u$, thus:

$$\left|\frac{1}{\log\lambda}\int_a^\lambda \frac{\sigma_u(T_1+T_2)}{\log u}\frac{du}{u} - \frac{1}{\log\lambda}\int_{2a}^{2\lambda}\frac{\sigma_u(T_1+T_2)}{\log u}\frac{du}{u}\right| \le (C_1+C_2)\frac{2\log 2}{\log\lambda}\,.$$

Next,

$$\frac{1}{\log\lambda}\int_{2a}^{2\lambda}\left|\frac{\sigma_u(T_1+T_2)}{\log(u/2)} - \frac{\sigma_u(T_1+T_2)}{\log u}\right|\frac{du}{u}$$
$$\le (C_1+C_2)\frac{1}{\log\lambda}\int_{2a}^{2\lambda}\frac{\log u-\log u/2}{\log u/2}\frac{du}{u}$$

$$\int_{2a}^{2\lambda}\frac{\log 2}{\log(u/2)}\frac{du}{u} = \log 2\int_a^\lambda \frac{du}{u\log u} = \log 2(\log\log\lambda - \log\log a)\ . \square$$

Let $\mathcal{A}$ be the space of functions $h(\lambda)$; $\lambda\in[a,\infty[$, which are bounded and are taken modulo those of order $O(\log\log\lambda/\log\lambda)$. The latter form an ideal for the obvious pointwise product and thus $\mathcal{A}$ is an algebra.

By Lemma A.4 we have a well defined additive map,

$$\tau : \mathcal{L}^{(1,\infty)} \to \mathcal{A}\,, \tag{8}$$

defined by $\tau(T)=$ class of $(\tau_\lambda(T))_{\lambda\ge a}$ for any $T\ge 0$. Here we used:

DEFINITION A.2. *$\mathcal{L}^{(1,\infty)}$ is the normed ideal with norm:*

$$\mathrm{Sup}_{u\ge a}\frac{\sigma_u(T)}{\log u} = \|T\|_{1,\infty}\,.$$

Note that the image by τ of the ideal $\mathcal{L}^1$ of trace class operators is $\{0\}$.

The above definition of $\tau(T)$ for $T\ge 0$ has been extended to any T using linearity. For instance if we write $T=T^*$ in two ways as $T_1-T_2=T_1'-T_2'$ (all $T_i, T_j'\ge 0$) we have

$$\tau(T_1)-\tau(T_2)=\tau(T_1')-\tau(T_2')\,, \tag{9}$$

since the equality $T_1+T_2'=T_2+T_1'$ can be combined with Lemma A.4.

PROPOSITION A.2. *τ is a linear positive map from $\mathcal{L}^{(1,\infty)}$ to $\mathcal{A}$ such that for any bounded operator S in $\mathcal{H}$:*

$$\tau(ST)=\tau(TS) \qquad \forall T\in\mathcal{L}^{(1,\infty)}\,.$$

Proof: One has, for any unitary U and $T\ge 0$ in $\mathcal{L}^{(1,\infty)}$,

$$\tau(UTU^*)=\tau(T)\,.$$

This equality extends by linearity to arbitrary T's. Using TU instead of T, one gets $\tau(UT)=\tau(TU)$ and the conclusion follows by linearity. $\square$

COROLLARY A.1. *For $T \in \mathcal{L}^{(1,\infty)}$ the class $\tau(T)$ only depends upon the locally convex topology of $\mathcal{H}$, not on the inner product.*

Proof: $\tau(STS^{-1}) = \tau(T)$ for any invertible S. □

Let us now consider states ω on the C^*-algebra

$$A = C_b([a,\infty[)/C_0([a,\infty[) \ . \tag{10}$$

LEMMA A.5. *If $f \in C_b([a,\infty[)$ then $f(\lambda)$ has a limit for $\lambda \to \infty$ iff $\omega(f)$ is independent of ω.*

Proof: If $f \to L$ then $f - L \in C_0$ on which any ω vanishes. Conversely, if f has two distinct limit points, one gets two states ω_1, ω_2 whose values on f are different. □

Remark A.1: For any separable C^*-subalgebra of $C_b([a,\infty[)/C_0$ the construction of states can be effectively performed without using the axiom of uncountable choice. So, whenever we apply Tr_ω (cf. infra) to any separable subspace of $\mathcal{L}^{(1,\infty)}$, we may assume that it is effectively constructed.

DEFINITION A.3. *For any ω we let*

$$\mathrm{Tr}_\omega(T) = \omega(\tau(T)) \qquad \forall T \in \mathcal{L}^{(1,\infty)} \ .$$

One has, using Proposition A.2:

PROPOSITION A.3. a) Tr_ω *is a positive linear form on* $\mathcal{L}^{(1,\infty)}$ *and* $\mathrm{Tr}_\omega(ST) = \mathrm{Tr}_\omega(TS)$ $\forall S \in \mathcal{L}(\mathcal{H})$.

b) $\mathrm{Tr}_\omega(T)$ *is independent of ω iff $\tau(T)$ converges for $\lambda \to \infty$ (and the limit is then equal to* $\mathrm{Tr}_\omega(T)$).

By construction, Tr_ω is continuous for the norm $\| \ \|_{1,\infty}$ and vanishes on the closure $\mathcal{L}_0^{(1,\infty)}$ of finite rank operators in this norm.

One has

$$T \in \mathcal{L}_0^{(1,\infty)} \Leftrightarrow \sigma_\lambda(T) = o(\log \lambda) \ . \tag{11}$$

In particular,

$$\mu_n(T) = o\left(\frac{1}{n}\right) \Rightarrow T \in \mathcal{L}_0^{(1,\infty)} \ . \tag{12}$$

The following is an easy case where Tr_ω is independent of ω and can be computed.

PROPOSITION A.4. *Let $T \geq 0$, $\mu_n(T) = 0(1/n)$ and $\zeta(s) = \mathrm{Trace}(T^s)$. Then the following are equivalent:*

a) $$(s-1)\zeta(s) \to L \quad as \quad s \to 1+ \quad ,$$

b) $$(\log N)^{-1} \sum_0^{N-1} \mu_n(T) \to L \quad as \quad N \to \infty \ .$$

When this holds, one has $\mathrm{Tr}_\omega(T) = L$ independently of ω.

In particular take $T = \Delta^{-p}$, $p > 0$, where one knows the asymptotic behavior of $\mathrm{Trace}(e^{-t\Delta}) \sim Lt^{-p}$ for $t \to 0$. Assume $\Delta \geq c > 0$ and write:

$$\Delta^{-s} = \frac{1}{\Gamma(s)} \int_0^\infty e^{-t\Delta} t^s \frac{dt}{t} \,. \tag{13}$$

With $s = p(1+\varepsilon)$ and applying the trace on both sides we get:

$$\zeta(1+\varepsilon) = \frac{1}{\Gamma(p+\varepsilon p)} \int_0^\infty \mathrm{Trace}(e^{-t\Delta}) t^{p+\varepsilon p} \frac{dt}{t} \;;$$

for $\varepsilon \to 0$ it is equivalent to

$$\frac{1}{\Gamma(p)} L \int_0^1 t^{\varepsilon p} \frac{dt}{t} = \frac{1}{\varepsilon} \frac{1}{\Gamma(p+1)} L \,.$$

Thus, we get in that case:

$$\mathrm{Tr}_\omega(\Delta^{-p}) = \frac{1}{\Gamma(p+1)} \lim_{t\to 0} t^p \, \mathrm{Trace}(e^{-t\Delta}) \,. \tag{14}$$

For Δ the Laplacian on the n-torus $\mathbf{T}^n$, where the length of $\mathbf{T}$ is 2π, one can use

$$\sum_{\mathbf{Z}} e^{-tk^2} \sim \frac{\pi^{1/2}}{\sqrt{t}} \qquad t \to 0 \,,$$

which gives $\mathrm{Tr}_\omega(\Delta^{-n/2}) = \frac{\pi^{n/2}}{\Gamma(\frac{n}{2}+1)}$.

More generally, for ordinary pseudodifferential operators (ΨDO) on a manifold, the Dixmier trace is given by the Wodzicki residue,

PROPOSITION A.5. *Let V be an n-dimensional manifold and $P \in OP^{-n}(V)$ a ΨDO of order $-n$ then*

$$1) \;\; P \in \mathcal{L}^{(1,\infty)} \;; \qquad 2) \;\; \mathrm{Tr}_\omega(P) = \frac{1}{n}(2\pi)^{-n} \int_{S^*} a_{-n}(x,\xi) d^n x \, d^{n-1}\xi \,.$$

Recall that, in local coordinates x, a ΨDO is written as

$$(P\eta)(x) = (2\pi)^{-n} \int e^{i(x-y)\cdot\xi} \alpha(x,\xi) \eta(y) \, dy \, d\xi$$

where $\alpha \sim a_q + a_{q-1} + \cdots$ and $a_\ell(x,\xi)$ is homogeneous of degree ℓ in ξ.

The principal symbol $a_q(x,\xi)$ is invariantly defined as a function on the cotangent bundle T^*V by:

$$\big(P(e^{i\tau\varphi}\eta)\big)(x) \sim \tau^q a_q(x, d\varphi)\eta(x) \qquad \forall \eta \,.$$

Next, let us consider on the complement of the 0-section V in T^*V the measure $dx\,d\xi$ associated to the symplectic structure. For any homogeneous function of degree $-n$, $a_{-n}(x,\xi)dx\,d\xi$ is now invariant under $\xi \to \lambda\xi$.

If we let $E = r\frac{\partial}{\partial r}$ be the vector field generating this one parameter group, we have $\partial_E \mu = 0$, $d i_E\,\mu = 0$, where μ is viewed as a form of top degree. Thus $i_E\mu$ is a form of degree $2n-1$ and its integral on any two homologous cycles of dimension $2n-1$ gives the same result.

In particular we can choose a Riemannian metric on V and take as cycle its unit sphere bundle S^*. Then we get

$$\int_{S^*} a_{-n}(x,\xi)d^n x\, d^{n-1}\xi$$

where $d^{n-1}\xi$ is the volume form on the $n-1$ sphere with its induced metric. If we compute it for the constant 1 we get $|S^{n-1}|\int_V d^n x$, where

$$|S^{n-1}| = \frac{2\pi^{n/2}}{\Gamma\left(\frac{n}{2}\right)}\ .$$

Let us check the equality 2) for the above torus $\mathbf{T}^n$ and the Laplacian. The l.h.s. is $\frac{\pi^{n/2}}{\Gamma\left(\frac{n}{2}+1\right)}$ and the r.h.s. is

$$\frac{1}{n}(2\pi)^{-n}(2\pi)^n|S^{n-1}| = \frac{2}{n}\frac{\pi^{n/2}}{\Gamma\left(\frac{n}{2}\right)}\ .$$

The general conclusion follows by positivity and invariance of Tr_ω. □

Appendix B. Spectral Triple and Pseudodifferential Calculus

Let $(\mathcal{A}, \mathcal{H}, D)$ be a spectral triple. For each $s \in \mathbf{R}$ we let $\mathcal{H}^s = \text{Domain}\,(|D|^s)$ and

$$\mathcal{H}^\infty = \bigcap_{s\geq 0} \mathcal{H}^s \quad , \quad \mathcal{H}^{-\infty} = \text{dual of } \mathcal{H}^\infty\ .$$

We obtain in this way a scale of Hilbert spaces, and for each r we define op^r to be the linear space of operators in $\mathcal{H}^\infty$ which are continuous for every s, thus

$$op^r : \mathcal{H}^s \to \mathcal{H}^{s-r}\ .$$

We shall use the following smoothness condition on $\mathcal{A}$: $\forall\, a \in \mathcal{A}$, both a and $[D,a]$ are in the domain of all powers of the derivation $\delta = [|D|, \cdot]$.

LEMMA B.1. *Under the above assumption, a, $[D,a]$ are in op^0 and*

$$b - |D|b|D|^{-1} \in op^{-1} \qquad (b = a \text{ or } [D,a])\ .$$

Proof: Let us first check that $|D|^n b|D|^{-n}$ is bounded for $n \geq 0$. With $\sigma(\cdot) = |D| \cdot |D|^{-1}$, one has:

$$\sigma = id + \varepsilon \ , \qquad \varepsilon(b) = \delta(b)|D|^{-1} \ .$$

Since $\varepsilon^k(b)$ is bounded, equal to $\delta^k(b)|D|^{-k}$, we get the result using $\sigma^n = \sum C_n^k \varepsilon^k$.

Moreover, $\sigma^{-1}(b) = |D|^{-1}b|D| = b - |D|^{-1}\delta(b)$ and the same argument shows that $\sigma^n(b)$ is bounded for $n < 0$. Then one uses interpolation.

For the second part one applies the above argument to $\delta(b)$; thus,

$$\delta(b) \in op^0 \ , \qquad \delta(b)|D|^{-1} \in op^{-1} \ . \qquad \square$$

It is important to note that the above smoothness hypothesis can be replaced by:

$$a \text{ and } [D,a] \in \cap \mathrm{Dom}\, L^k R^q \ , \ L(b) = |D|^{-1}[D^2,b] \ , \ R(b) = [D^2,b]|D|^{-1} \ .$$

Indeed, assuming the above, one has

$$L(b) = |D|^{-1}\left(|D|\delta(b) + \delta(b)|D|\right) \in op^0 \ , \qquad R(b) \in op^0$$

and the same applies to $L^k R^q(b)$.

COROLLARY B.1. *Under the above hypothesis one has*

$$\Big[D^2, [D^2, \ldots [D^2, b]\underbrace{] \cdots}_{n}\Big] \in op^n \qquad \forall\, b \in \mathcal{A} \ , \ \text{or } [D,\mathcal{A}] \ .$$

Let us now show that if $b \in \cap \mathrm{Dom}\, L^k R^q$ then $b \in \mathrm{Dom}\, \delta$. The proof is more subtle than one would first expect, because the obvious argument, using

$$|D| = \pi^{-1} \int_0^\infty \frac{D^2}{D^2+\mu} \mu^{-1/2} d\mu \ ,$$

requires some care. Indeed, one gets from the above

$$[|D|, b] = \pi^{-1} \int_0^\infty (D^2+\mu)^{-1}[D^2,b](D^2+\mu)^{-1}\mu^{1/2} d\mu \ .$$

We can replace $[D^2,b]$ by $|D|$, which has the same size, and get

$$\int_0^\infty (D^2+\mu)^{-2}|D|\mu^{1/2}d\mu = \int_0^\infty (1+t)^{-2}t^{1/2}dt \ \cdot \mathrm{Id} \ .$$

For this to work, we need to move $[D^2,b]$ in front of the above integral, i.e. use the finiteness of the norm of

$$\int_0^\infty \underbrace{[(D^2+\mu)^{-1}, [D^2,b]]}_{-(D^2+\mu)^{-1}[D^2,[D^2,b]](D^2+\mu)^{-1}} \quad (D^2+\mu)^{-1}\mu^{1/2} d\mu \ .$$

This finiteness follows from:

1) $(D^2+\mu)^{-1}[D^2,[D^2,b]]$ bounded (as $L^2(b)$) ;

2) $\int_0^\infty \|(D^2+\mu)^{-2}\|\mu^{1/2}\, d\mu \leq C\int_0^1 \mu^{1/2}d\mu + \int_1^\infty \mu^{-3/2}d\mu < \infty$.

Once $[D^2,b]$ is moved in front, the above calculation applies.

It follows that $b\in \operatorname{Dom}\delta$ and iterating the argument one gets $b\in \cap \operatorname{Dom}\delta^k$. We thus obtain:

LEMMA B.2. $$\bigcap_{k,q} \operatorname{Dom} L^k R^q = \bigcap_n \operatorname{Dom}\delta^n\ .$$

We shall define the order of an operator by means of the following filtration:

$$P\in OP^\alpha \quad \text{iff} \quad |D|^{-\alpha}P\in \cap \operatorname{Dom}\delta^n\ .$$

Thus, $OP^0 = \cap\operatorname{Dom}\delta^n$ and we have:

$$OP^\alpha\subset op^\alpha \qquad \forall\alpha\ .$$

Let us now describe the general pseudodifferential calculus.

We let ∇ be the derivation: $\nabla(T)=[D^2,T]$, and consider the algebra generated by the $\nabla^n(T)$, $T\in\mathcal{A}$ or $[D,\mathcal{A}]$. We view this algebra $\mathcal{D}$ as an analogue of the algebra of differential operators. In fact, by Corollary B.1 we have a natural filtration of $\mathcal{D}$ by the total power of ∇ applied, and moreover

$$\mathcal{D}^n\subset OP^n\ . \tag{1}$$

We want to develop a calculus for operators of the form

$$A|D|^z \qquad z\in\mathbb{C}\ ,\quad A\in\mathcal{D}\ . \tag{2}$$

We shall use the notation $\Delta = D^2$ and begin by understanding the action of $\mathbb{C}$ given by

$$\sigma^{2z}=\Delta^z\cdot\Delta^{-z}. \tag{3}$$

By construction, $\mathcal{D}$ is stable under the derivation ∇ and

$$\nabla(\mathcal{D}^n)\subset\mathcal{D}^{n+1}\ . \tag{4}$$

Also, for $A\in\mathcal{D}^n$ and $z\in\mathbb{C}$, one has

$$A|D|^z\in OP^{n+\mathrm{Re}(z)}\ . \tag{5}$$

We shall use the group σ^{2z} to understand how to multiply operators of complex order modulo OP^{-k} for any k. One has

$$\sigma^2 = 1+\mathcal{E},\ \mathcal{E}(T)=\nabla(T)\Delta^{-1}\ . \tag{6}$$

LEMMA B.3. *Let $T\in\mathcal{D}^q$ then $\mathcal{E}^k(T)\in OP^{q-k}$ $\forall k\geq 0$.*

Proof: $$\mathcal{E}^k(T)=\nabla^k(T)\Delta^{-k}\in OP^{q+k}\Delta^{-k}\subset OP^{q-k}\ .$$ □

We now wish to justify the formal expansion

$$\sigma^{2z}(T) = \left(1 + z\mathcal{E} + \frac{z(z-1)}{2!}\mathcal{E}^2 + \cdots\right)(T) .$$

It should give a control of $\sigma^{2z}(T)$ modulo OP^{q-k-1} if we stop at $\mathcal{E}^k(T)$.

To this end, we need to control the remainder in the Taylor formula:

$$(1+\mathcal{E})^{n+1-\alpha} = 1 + (n+1-\alpha)\mathcal{E} + \frac{(n+1-\alpha)(n-\alpha)}{2!}\mathcal{E}^2 + \cdots + \tag{7}$$

$$(n+1-\alpha)\cdots(n+1-k-\alpha)\frac{\mathcal{E}^{k+1}}{(k+1)!} + \cdots + (n+1-\alpha)\cdots(2-\alpha)\frac{\mathcal{E}^n}{n!} +$$

$$\mathcal{E}^{n+1}\int_0^1 (n+1-\alpha)\cdots(1-\alpha)(1+t\mathcal{E})^{-\alpha}\frac{(1-t)^n}{n!}dt .$$

The main lemma is the following:

LEMMA B.4. *Let $\alpha \in \mathbb{C}$, $0 < \operatorname{Re}\alpha < 1$ and $\beta > 0$, $\beta < a = \operatorname{Re}\alpha$. Then the following operator preserves the space OP^α, for any α:*

$$\Psi = \sigma^{2\beta}\int_0^1 (1+t\mathcal{E})^{-\alpha}(1-t)^n dt .$$

Proof: This will be done by expressing Ψ as an integral of the form

$$\Psi = \int \sigma^{2is} d\mu(s) \qquad \|\mu\| < \infty . \tag{8}$$

One writes

$$(1+t\mathcal{E})^{-\alpha} = \frac{\sin\pi\alpha}{\pi}\int_0^\infty \frac{1}{1+t\mathcal{E}+\mu}\mu^{-\alpha}d\mu \tag{9}$$

using the standard formula

$$x^{-\alpha} = \frac{\sin\pi\alpha}{\pi}\int_0^\infty \frac{1}{x+\mu}\mu^{-\alpha}d\mu . \tag{10}$$

Let us then consider the resolvent of $-\sigma^2$, namely

$$R(\lambda) = (\lambda + \sigma^2)^{-1} .$$

One has, with $\beta \in \,]0,1[$ as above,

$$R(\lambda) = \frac{1}{2}\int_{-\infty}^\infty \sigma^{-2(\beta+is)}\lambda^{\beta+is-1}\frac{ds}{\sin\pi(\beta+is)} , \tag{11}$$

which follows from

$$\frac{1}{1+y} = \frac{1}{2}\int_{-\infty}^\infty y^{-(\beta+is)}\frac{ds}{\sin\pi(\beta+is)} . \tag{12}$$

(With $y = e^u$, this means that $\frac{e^{\beta u}}{1+e^u}$ is the Fourier transform of $\frac{1}{\sin \pi(\beta+is)}$, which also follows from (10) written as $\frac{\pi}{\sin(\pi(1-\beta-is))} = \int_{-\infty}^{\infty} \frac{e^{(1-\alpha)u}}{1+e^u} du$, $\alpha = 1-\beta-is$.)

Thus, from (11) we get,

$$\sigma^{2\beta} R(\lambda) = \frac{1}{2} \int_{-\infty}^{\infty} \sigma^{-2is} \lambda^{\beta-1} \frac{\lambda^{is} ds}{\sin \pi(\beta+is)} , \tag{13}$$

where the measure $\frac{\lambda^{is} ds}{\sin \pi(\beta+is)}$ is well controled by $e^{-|s|}ds$. By (9), we have

$$(1+t\mathcal{E})^{-\alpha} = \frac{\sin \pi\alpha}{\pi} \int_0^\infty \frac{1}{t} R\left(\frac{\mu+1}{t} - 1\right) \mu^{-\alpha} d\mu ,$$

$$\sigma^{2\beta}(1+t\mathcal{E})^{-\alpha} = \frac{\sin \pi\alpha}{\pi} \frac{1}{2} \int_0^\infty \frac{1}{t} \int_{-\infty}^{\infty} \sigma^{-2is} \left(\frac{\mu+1}{t} - 1\right)^{\beta-1} \mu^{-\alpha} \frac{\lambda^{is} ds}{\sin \pi(\beta+is)} d\mu ,$$

with $\lambda = \left(\frac{\mu+1}{t} - 1\right)$.

For fixed s, we are thus dealing with the size

$$\frac{1}{t} \int_0^\infty \left(\frac{\mu+1}{t} - 1\right)^{\beta-1} \mu^{-\alpha} d\mu = I .$$

One has $0 < t < 1$, so that the behavior at $\mu = 0$ is fine. Also, the integral converges for $\mu \to \infty$ as $\mu^{(\beta-\alpha)-1}$, since $\beta < \alpha$. Thus,

$$\begin{aligned} I &= \frac{1}{t} \int_0^\infty \left(u + \frac{1}{t} - 1\right)^{\beta-1} t^{-\alpha} u^{-\alpha} t \, du \\ &= \frac{1}{t} \int_0^\infty \left(\frac{1}{t} - 1\right)^{\beta-1} (v+1)^{\beta-1} t^{-\alpha} \left(\frac{1}{t} - 1\right)^{-\alpha} v^{-\alpha} t \left(\frac{1}{t} - 1\right) dv \\ &= t^{-\alpha} \left(\frac{1}{t} - 1\right)^{\beta-\alpha} \int_0^\infty (v+1)^{\beta-1} v^{-\alpha} dv \\ &= (1-t)^{\beta-\alpha} t^{-\beta} c(\alpha,\beta) . \end{aligned}$$

Finally, we get the equality

$$\int_0^1 \sigma^{2\beta}(1+t\mathcal{E})^{-\alpha} \frac{(1-t)^n}{n!} dt = \int_{-\infty}^{\infty} \sigma^{2is} d\nu(s) ,$$

where the total mass of the measure ν is finite.

Alternatively, this can be checked by looking directly at the L^1-norm of the Fourier transform of the function

$$u \to \int_0^1 e^{\beta u} (1 + t(e^u - 1))^{-\alpha} \frac{(1-t)^n}{n!} dt .$$

Indeed, it is enough to check that the following function of u is in the Schwartz space $\mathcal{S}(\mathbf{R})$:

$$\varphi_n(u) = (e^u - 1)^{-(n+1)} e^{\beta u}\Big(e^{(n+1-\alpha)u} - 1 - (n+1-\alpha)(e^u-1) -$$
$$\frac{(n+1-\alpha)(n-\alpha)}{2!}(e^u-1)^2 - \cdots - \frac{(n+1-\alpha)(n-\alpha)\cdots(2-\alpha)}{n!}(e^u-1)^n\Big).$$

First, for $u \to \infty$ the size is $\sim e^{-(n+1)u} e^{\beta u} e^{(n+1-\alpha)u} = e^{(\beta-\alpha)u} \to 0$. For $u \to -\infty$ it behaves like $e^{\beta u} \to 0$. We need to know that it is smooth at $u = 0$ but this follows from the Taylor expansion. The same argument applies to all derivatives. This gives another proof of the lemma. □

We are now ready to prove:

THEOREM B.1. *Let $T \in \mathcal{D}^q$ and $n \in \mathbf{N}$. Then for any $z \in \mathbf{C}$*

$$\sigma^{2z}(T)$$
$$- \Big(T + z\mathcal{E}(T) + \frac{z(z-1)}{2!}\mathcal{E}^2(T) + \cdots + \frac{z(z-1)\cdots(z-n+1)}{n!}\mathcal{E}^n(T)\Big)$$

$\in OP^{q-(n+1)}$.

Proof: First, for any $z \in \mathbf{C}$ and $k \in \mathbf{N}$ one has

$$\mathcal{E}^k(\sigma^{2z}(T)) \in OP^{q-k}\ . \tag{14}$$

Indeed, by (8) we know that σ^{2z} leaves any OP^n invariant; as $\mathcal{E}^k \circ \sigma^{2z} = \sigma^{2z} \circ \mathcal{E}^k$, we just use Lemma B.3.

For $0 < \operatorname{Re}\alpha < 1$, β as above and $z = (n+1) - \alpha$, one has

$$\sigma^{2(\beta+(n+1)-\alpha)}(T)$$
$$- \Big(\sigma^{2\beta}(T) + z\sigma^{2\beta}\mathcal{E}(T) + \cdots + \frac{z(z-1)\cdots(z-n+1)}{n!}\sigma^{2\beta}\mathcal{E}^n(T)\Big)$$
$$= \text{const} \cdot \Psi(\mathcal{E}^{n+1}(T))\ . \tag{15}$$

If we apply this equality to $\sigma^{2s}(T)$ and use (14) and Lemma B.4 we see that for any s there exists a polynomial $P_s(\alpha)$ of degree n in α such that:

$$\sigma^{2(s-\alpha)}(T) - P_s(\alpha) \in OP^{q-(n+1)} \qquad \beta < \operatorname{Re}\alpha < 1\ .$$

The polynomials $P_s(\alpha + s)$ have to agree modulo $OP^{q-(n+1)}$ on the overlap of the bands $\beta < \operatorname{Re}\alpha < 1$ and thus the difference between any two of them will belong to $OP^{q-(n+1)}$ for all z. It follows that there is $P(z)$ which works for all z. To obtain its coefficients, it suffices to take $z = 0, 1, \ldots, n$, which yields the formula of Theorem B.1. □

References

[BG] R. BEALS, P. GREINER, Calculus on Heisenberg manifolds, Annals of Math. Studies 119, Princeton Univ. Press, Princeton, N. J., 1988 .

[C1] A. CONNES, Noncommutative Geometry, Academic Press, Inc. 1994.

[C2] A. CONNES, Cyclic cohomology and the transverse fundamental class of a foliation, Geometric methods in operator algebras, (Kyoto, 1983), pp. 52-144, Pitman Res. Notes in Math. 123 Longman, Harlow (1986).

[C3] A. CONNES, Noncommutative Geometry and Physics, Les Houches Conference, Preprint IHES M/93/32, 1993.

[CL] A. CONNES, J. LOTT, Particle models and noncommutative geometry, Nuclear Physics B 18B (1990), suppl. 29-47 (1991).

[CM1] A. CONNES, H. MOSCOVICI, Cyclic cohomology, the Novikov conjecture and hyperbolic groups, Topology 29 (1990), 345-388.

[CM2] A. CONNES, H. MOSCOVICI, Transgression and the Chern character of finite-dimensional K-cycles, Commun. Math. Phys. 155 (1993), 103-122.

[G] P. GILKEY, Invariance theory, the heat equation and the Atiyah-Singer index theorem, Math. Lecture Ser. 11, Publish or Perish, Wilmington, Del., 1984.

[HSk] M. HILSUM, G. SKANDALIS, Morphismes K-orientés d'espaces de feuilles et fonctorialité en théorie de Kasparov, Ann. Sci. École Norm. Sup. (4) 20 (1987), 325-390.

[KV] M. KONTSEVICH, S. VISHIK, Determinants of elliptic pseudodifferential operators, Preprint, 1994.

[S] R.T. SEELEY, Complex powers of elliptic operators, Proc. Symp. Pure Math. 10 (1967), 288-307.

[W] M. WODZICKI, Noncommutative residue, Part I. Fundamentals, K-theory, arithmetic and geometry (Moscow, 1984-86), 320-399; Springer Lecture Notes Math. 1289, Berlin, (1987).

Alain Connes
Collège de France, Paris and
Institut des Hautes Etudes Scientifiques
91440 Bures-sur-Yvette,
France

Henri Moscovici
Department of Mathematics
Ohio State University
Columbus, OH 43210
USA

Submitted: January 1995

Geometric And Functional Analysis
Vol. 5, No. 2 (1995)
1016-443X/95/0200244-26$1.50+0.20/0

LAGRANGIAN INTERSECTIONS IN CONTACT GEOMETRY

Y. ELIASHBERG, H. HOFER AND D. SALAMON

1. Introduction

It is well-known that all problems of Contact geometry can be reformulated as problems of Symplectic geometry. This can be done via *symplectization* (see 2.1 below). In particular, the problem of Lagrangian intersections naturally arises in connection with several contact geometric questions (see 2.5 example, and below). However, there is one major difficulty when one tries to realize this approach: *the symplectizations of contact manifolds are non-compact* and, what is even worse, *non-convex* (see [EGr1]). This leads to the loss of compactness for the spaces of holomorphic curves and thus creates serious difficulties for the traditional Floer homology approach. The goal of this paper is to show that this problem can be successfully overcome by using an idea from [H].

We begin with an exposition of the main notions of contact geometry and their symplectic analogues. We develop then an analogue of Floer homology theory for the Lagrangian intersection problem in symplectizations of contact manifolds and give applications of this theory to contact geometry.

There exist other methods for handling similar problems in contact geometry. Let us mention here Givental's approach through the, so-called, non-linear Maslov index (see [G]), as well as the approach based on the theory of generating functions and hypersurfaces described in [EGr2]. Kaoru Ono ([On]) independently proved a result similar to our Theorem 2.5.4. All these methods, and the method considered in this paper, have common as well as complementary areas of application.

A part of this paper was written while the first and third authors visited IHES. They thank the institute for its hospitality.

2. Contact Geometry

2.1 Contact manifolds and their symplectizations. We recall in this section some basic definitions of contact geometry and their symplectic counterparts (see also [AG]). Let ζ be a *contact structure* on a $(2n+1)$-dimensional manifold M, i.e. ζ is a completely non-integrable tangent plane distribution of codimension 1. Thus, at least locally, ζ can be defined by the equation $\{\gamma = 0\}$ where the 1-form γ satisfies the condition $\gamma \wedge (d\gamma)^n \neq 0$. The global existence of such a form γ is equivalent to the coorientability of ζ.

Only cooriented contact structures are considered in this paper. The general case requires a $\mathbf{Z}/2$-equivariant analogue of the theory described here.

Let $S_\zeta(M)$ be the (trivial) subbundle of the cotangent bundle $T^*(M)$ whose fiber over a given point $q \in M$ consists of all non-zero linear forms from $T_q^*(M)$ which annihilate the hyperplane $\zeta_q \subset T_q(M)$ and define its given coorientation. The bundle $S_\zeta(M)$ is a principal $\mathbf{R}$-bundle where the action of $\mathbf{R}$ is defined by

$$\lambda * v = e^\lambda \cdot v\,, \qquad \lambda \in \mathbf{R}\,,\ v \in S_\zeta(M)\,.$$

Let us denote by α_ζ the restriction $pdq|_{S_\zeta(M)}$ of the canonical form pdq on T^*M to $S_\zeta(M) \subset T^*M$. Then the 2-form $\omega_\zeta = d\alpha_\zeta$ is a symplectic structure on $S_\zeta(M)$. The symplectic manifold $(S_\zeta(M), \omega_\zeta)$ is called the *symplectization* of the contact manifold (M, ζ).

Let us denote by X_ζ the vector field on $S_\zeta(M)$ which is ω_ζ-dual to α_ζ, i.e. $X_\zeta \rfloor \omega_\zeta = \alpha_\zeta$. The field X_ζ generates the $\mathbf{R}$-action described above:

$$(X_\zeta)^\lambda(v) = e^\lambda v\,, \qquad \lambda \in \mathbf{R}\,,\ v \in S_\zeta(M)\,.$$

The sections of the bundle $S_\zeta(M) \to M$ are called *contact forms*. The space of all contact forms will be denoted by $\mathrm{Cont}(\zeta)$.

A choice of a contact form $\gamma \in \mathrm{Cont}(\zeta)$ defines a splitting $H_\gamma : S_\zeta(M) \to M \times \mathbf{R}$. In terms of this splitting we have

$$\alpha_\zeta = e^\theta \gamma\,, \quad \omega_\zeta = d(e^\theta \gamma)\,, \quad X_\zeta = \frac{\partial}{\partial\theta}\,,$$

where $\theta \in \mathbf{R}$ and we identify γ defined on M with its pullback on $M \times \mathbf{R}$.

It is useful to observe the following

PROPOSITION 2.1.1. *A fiberwise splitting $H : S_\zeta(M) \to M \times \mathbf{R}$ has the form H_γ for a contact form $\gamma \in \mathrm{Cont}(\zeta)$ if and only if H commutes with the $\mathbf{R}$-actions on $S_\zeta(M)$ and $M \times \mathbf{R}$.*

A diffeomorphism $f : M \to M$ lifts canonically to a symplectomorphism $F : T^*M \to T^*M$. Moreover, F preserves the 1-form pdq as well. If f is a contactomorphism of the contact manifold (M, ζ) then F leaves the subbundle $S_\zeta(M)$ invariant and thus induces an $\mathbf{R}$-equivariant symplectomorphism $S_\zeta(M) \to S_\zeta(M)$. We will denote this symplectomorphism by $\hat{f}$ and call it the *symplectization* of the contactomorphism f.

The converse is also true: *any* **R**-*equivariant symplectomorphism of* $S_\zeta M$ *has the form* $\hat{f}$ *for a uniquely defined suitable contactomorphism* $f : (M, \zeta) \to (M, \zeta)$.

The vector field X on (M, ζ) is called *contact* if the flow generated by X consists of contactomorphism $(M, \zeta) \to (M, \zeta)$. Equivalently, pick a 1-form $\gamma \in \mathrm{Cont}(\zeta)$. Then the vector field X is contact iff the Lie derivative $L_X\gamma$ is proportional to γ.

Each contact vector field X on (M, ζ) admits a lift to an **R**-invariant Hamiltonian vector field $\hat{X}$ on $(S_\zeta(M), \omega_\zeta)$. Conversely, each **R**-invariant Hamiltonian vector field Y on $(S_\zeta(M), \omega_\zeta)$ projects to a contact vector field on (M, ζ). An important example of a contact vector field is provided by the *Reeb vector field.* Notice that the choice of a contact form $\gamma \in \mathrm{Cont}(\zeta)$ defines on M a Hamiltonian flow which is transversal to the contact structure ζ. Indeed, there exists a unique vector field Y tangent to M such that $Y \rfloor d\gamma = 0$ and $\gamma(Y) = 1$. The vector field Y is called the Reeb vector field generated by the contact form γ. The field Y is contact. Indeed, we have $L_Y\gamma = d(\gamma(Y)) - Y \rfloor d\gamma = 0$.

2.2 Legendrian, Lagrangian, pre-Lagrangian. An n-dimensional submanifold $\Lambda \subset (M, \zeta)$ is called *Legendrian* if it is tangent to the distribution ζ. If γ is a contact form from $\mathrm{Cont}(\zeta)$ then Λ is Legendrian iff $\alpha|_\Lambda = 0$. The preimage $\hat{\Lambda} = \pi^{-1}(\Lambda) \subset S_\zeta M$ under the canonical projection $S_\zeta M \to M$ is an **R**-invariant Lagrangian cone. We call $\hat{\Lambda}$ the *symplectization* of Λ. Conversely, any Lagrangian cone in the symplectization projects onto a Legendrian submanifold in (M, ζ).

The following notion was suggested to us by D. Bennequin.

An $(n+1)$-dimensional submanifold L of the $(2n+1)$-dimensional contact manifold (M, ζ) is called pre-*Lagrangian* if it satisfies the following two conditions:

— L is transverse to ζ;

— The distribution $\zeta \cap T(L)$ is integrable and can be defined by a closed 1-form.

Remark 2.2.1: It is useful for applications to extend the definition of a pre-Lagrangian submanifold allowing certain types of tangency of L and ζ instead of their transversality. It will be done in one of our subsequent papers.

The motivation for the term pre-Lagrangian is provided by the following

PROPOSITION 2.2.2. *For any pre-Langrangian submanifold $L \subset M$ there exists a Lagrangian submanifold $\hat{L} \subset S_\zeta M$ such that $\pi(\hat{L}) = L$. The cohomology class $\lambda \in H^1(L; \mathbf{R})$, such that $\pi^*\lambda = [\alpha_\zeta|_{\hat{L}}]$, is defined uniquely up* *to multiplication by a non-zero constant. Conversely, if $L \subset M$ is the (embedded) image of a Lagrangian submanifold $\hat{L} \subset S_\zeta M$ under the projection $S_\zeta M \to M$ then L is pre-Lagrangian.*

Proof: By the definition of a pre-Lagrangian submanifold there exists a contact form $\beta \in \mathrm{Cont}_\zeta(M)$ whose restriction to L is closed. The required lift $\hat{L}$ of L is the graph of the form $\beta|_L$. Suppose that $\beta' \in \mathrm{Cont}(\zeta)$ is another form whose restriction to L is closed. Then $\beta'|_L = f\beta|_L$ for a non-vanishing function f, and we have $df \wedge \beta|L = 0$. Thus the function f must be constant on leaves of the foliation $\beta = 0$ on L. If the cohomology class $\lambda = [\beta|_L]$ is proportional to the integral class from $H^1(L;\mathbf{Z})$ then we can think that λ itself is integral and, therefore, $\beta|_L = h^*(d\theta)$, where h is a map $L \to S^1$ and the cohomology class of the closed form $d\theta$ generates $H^1(S^1)$. Thus the function f is constant on the fibers $h^{-1}(\theta)$, $\theta \in S^1$, i.e. f can be written as $\varphi \circ h$ for a function $\varphi : S^1 \to \mathbf{R}$. Set $C = \int_{S^1} \varphi d\theta$. Then there exists a diffeomorphism $g : S^1 \to S^1$ such that $g^*(d\theta) = (\varphi/C)d\theta$. Thus

$$\frac{1}{C}\beta'|_L = \frac{f}{C}\beta|_L = h^*\left(g^*(d\theta)\right) ,$$

and, therefore, the cohomology class $[\beta'|_L]$ coincides with $C\lambda$. If λ is not proportional to an integral class then the foliation defined by the form β on L has everywhere dense leaves. This implies that the function f has to be constant on all L. □

Thus with any pre-Lagrangian submanifold $L \subset M$ one can canonically associate a projective class of the form λ. A curve $\Gamma \subset L$ is called a *vanishing cycle* of L if its homology class annihilates λ. Examples of vanishing cycles are provided by curves which are contained in a Legendrian submanifold of L.

Let us recall that if $\delta : S^1 \to L$ is a loop in a Lagrangian, possibly immersed submanifold Lag of a symplectic manifold V then given a symplectic trivialization of the bundle $f^*T(V)$ one can define the Maslov index $\mu(\delta)$ (see, for instance, [RS1]). Of course, the index $\mu(\delta)$ depends on the trivialization. However, if $\Delta : S^1 \times [0,1] \to V$ is a homotopy between the loops $\delta_0 = \Delta|_{S^1\times 0} : S^1 \to Lag$, and $\delta_1 = \Delta|_{S^1\times 1} : S^1 \to Lag$ then the difference $\mu(\delta_0, \delta_1) = \mu(\delta_0) - \mu(\delta_1)$ can be invariantly defined. To do this one just needs to trivialize the bundle $\Delta^*T(V)$ over $S^1 \times [0,1]$.

The procedure of symplectization allows us to define the relative Maslov index $\mu(\delta_0, \delta_1)$ for a pair of homotopic loops in a contact manifold provided they are contained in its Legendrian or pre-Lagrangian submanifolds.

2.3 Contactization of symplectic manifolds. If a symplectic manifold (N,ω) is exact, i.e. $\omega = d\alpha$, then it can be contactized. The *contactization*

$C(N,\omega)$ is the manifold $M = N \times S^1$ (or $N \times \mathbf{R}$) endowed with the contact form $dz - \alpha$. Here we denote by z the projection to the second factor and still denote by α its pull-back under the projection $M \to N$.

However, the contactization can be defined sometimes, even when ω is not exact. Suppose that there exists an $\hbar > 0$ such that the form $\omega/\hbar$ represents an integral cohomology class $[\omega/\hbar] \in H^2(N)$. The *contactization* $C(N,\omega)$, or as it is also called, *pre-quantization* of the symplectic manifold (N,ω) can be constructed in this case as follows (see [W]). Let $M \to N$ be a principal circle bundle with the Euler class equal to $[\omega/\hbar]$. This bundle admits a connection whose curvature form equals $\omega/\hbar$. This connection can be viewed as a S^1-invariant 1-form α on M. The non-degeneracy of ω implies that α is a contact form and, therefore $\zeta = \{\alpha = 0\}$ is a contact structure on M. The contact manifold (M,ζ) is, by the definition, the contactization $C(N,\omega)$ of the symplectic manifold (N,ω). A change of the connection α leads to a contactomorphic manifold. However, a change of $\hbar$ (for instance, $\hbar \to \hbar/2$) affects not only the contact structure ζ but also the topology of the manifold M itself.

2.4 Examples. We give here examples of pre-Lagrangian and Legendrian submanifolds.

2.4.1 SYMPLECTIZATION OF THE SPACE OF CONTACT ELEMENTS. Let $M = P^+T^*N$ be the projectivized cotangent bundle of a n-manifold N, or the *space of cooriented contact elements* of N. Thus a point of M is a cooriented tangent hyperplane $T \subset T(V)$. The manifold M carries a canonical contact structure ζ (see [AG]) which is uniquely defined by the following property:

The symplectization $S_\zeta(M)$ coincides with $T^*N \setminus N$, the symplectic form ω_ζ is the restriction of the canonical symplectic form $d(pdq)$, and the $\mathbf{R}$-action is given by the multiplication by e^θ.

If we fix a Riemannian metric on N then the space P^+T^*N can be identified with the unit cotangent bundle. The restriction of the canonical 1-form pdq is a contact form for ζ. Thus the flow generated by the Reeb vector field for this contact form coincides with the geodesic flow.

Suppose now that α is a non-vanishing closed 1-form on N. Then it corresponds to a Lagrangian section $\hat{L}_\alpha \subset T^*N \setminus N = S_\zeta M$. The image $L_\alpha \subset M$ of $\hat{L}_\alpha$ under the canonical projection $S_\zeta M \to M$ is a pre-Lagrangian submanifold. The form α defines on L_α a foliation with Legendrian leaves. If a multiple $C\alpha$ for a constant $C > 0$ represents an integer cohomology class in $H^1(N;\mathbf{R})$ then all leaves of the foliation are closed Legendrian submanifolds of M.

Equivalently, the above example can be rephrased as follows. Suppose that a closed manifold N can be fibered over the circle S^1. Let $\pi : N \to S^1$ be the projection. Then $d\pi$ is a non-vanishing closed 1-form on N and its graph $\hat{L}$ is a Lagrangian submanifold in $T^*N \setminus N$. Then the image $L = L_{d\pi}$ of $\hat{L}_{d\pi}$ under the projection $T^*N \setminus N \to PT^*N$ is a pre-Lagrangian submanifold in the space of co-oriented contact elements of N. Notice that L is foliated by Legendrian lifts of hypersurfaces $\pi^{-1}(T) \subset N$, $T \in S^1$.

For instance, if N is the torus T^n, then the contact manifold $M = P^*T^*N$ admits a splitting $M = T^n \times S^{n-1}$ such that each torus $T^n \times p$, $p = (p_1, \dots, p_n) \in S^{n-1}$, is a pre-Lagrangian torus of the form L_α for the non-vanishing closed 1-form $\alpha = \sum_1^n p_i dq_i$, $q_i \in S^1$. An everywhere dense set of these tori can be further split as products $T^{n-1} \times S^1$ where all tori $T^{n-1} \times q$, $q \in S^1$, are Legendrian.

2.4.2 PRE-LAGRANGIAN SURFACES IN 3-MANIFOLDS. Let (M, ζ) be a three-dimensional contact manifold and $T \subset M$ be an embedded 2-torus, transversal to ζ. The line bundle $T(T) \cap \zeta$ integrates to a 1-dimensional, so-called *characteristic* foliation $\mathcal{F}_\zeta$. The torus T is pre-Lagrangian if and only if the foliation $\mathcal{F}_\zeta$ is diffeomorphic to a linear foliation of the torus $T \cong \mathbf{R}^2/\mathbf{Z}^2$.

Remark 2.4.1: The above example indicates that the class of smoothness of the Lagrangian lift can be of crucial importance even in the case of a C^∞-smooth pre-Lagrangian manifold.

2.4.3 SYMPLECTIZATION OF CONTACTIZATION. Let (N, ω) be a symplectic manifold with the symplectic form $\omega/\hbar$ representing an integral cohomology class $[\omega/\hbar]$. Let (M, ζ) be the contactization $C(N, \omega)$ of the manifold (V, ω) and α be the connection on V as described in §2.3 above.

If $L \subset N$ is a Lagrangian submanifold then the connection α over it is flat. The pull-back $\pi^{-1}(L) \subset M$ under the projection $\pi : M \to N$ is a pre-Lagrangian submanifold L_0 foliated by Legendrian leaves obtained by integrating the flat connection over L. If this foliation is a fibration, i.e. when the holonomy defined by the connection α is integral over L then the pre-Lagrangian submanifold L is foliated by closed Legendrian manifolds. In particular, this is the case when the connection form is exact over L, i.e. the connection over L is trivial. If this condition is satisfied then L is called a *Bohr-Sommerfeld orbit*. In this case the pre-Lagrangian submanifold L_0 is foliated by closed Legendrian lifts of L. These lifts are called sometimes, *Planckian* submanifolds (see [W] and [So]). The integrality of the holonomy is independent of the choice of the connection α but the Bohr-Sommerfeld condition depends on this choice, unless the image

$$\mathrm{Im}(H_1(L; \mathbf{R}) \to H_1(N; \mathbf{R}))$$

is trivial.

2.5 Lagrangian intersections in contact manifolds. In this section, we formulate theorems which give lower bounds for the number of transversal intersection points of Legendrian and pre-Lagrangian submanifolds of a contact manifold. These estimates will be proven in §3.8 below as an application of Floer homology theory which we are going to build in the next sections.

2.5.1 INTERSECTIONS IN THE SPACE OF CONTACT ELEMENTS. Suppose that a closed manifold N admits a Riemannian metric without contractible closed geodesics (e.g. a metric of non-positive sectional curvature). Let $M = P^+T^*N$ be the space of co-oriented contact elements. Suppose that there exists a non-vanishing closed 1-form α which represents an integral class $[\alpha] \in H^1(N)$. Let L_α be the pre-Lagrangian submanifold constructed in §2.4.1. In other words, L_α is the image of the graph $\hat{L}_\alpha \subset T^*N$ of the form α under the projection $T^*N \setminus N \to M = P^+T^*N$. As explained in 2.4.1, L_α carries a foliation by closed Legendrian leaves. Let Λ be one of the leaves.

THEOREM 2.5.1. *Let $\varphi_t : M \to M$, $t \in [0,1]$, $\varphi_0 = \mathrm{Id}$, be a contact isotopy of M such that $\varphi_1(\Lambda)$ is transversal to L_α. Then*

$$\#\varphi_1(\Lambda) \cap L_\alpha \geq \mathrm{rank}(\mathrm{H}_*(\Lambda; \mathbf{Z}/2)) .$$

In particular, suppose $M = T^n$ is the n-torus. Then we have the splitting $P^+T^*T^n = T^n \times S^{n-1}$ and all tori $T^n \times a$, $a \in S^{n-1}$, are pre-Lagrangian. For an everywhere dense subset $A \subset S^{n-1}$, the tori $T^n \times a$, $a \in A$, are foliated by Legendrian $(n-1)$-dimensional tori. Let L be one of these pre-Lagrangian tori $T^n \times a$ and Λ, $\Lambda \subset L$, be one of its Legendrian subtori. Let $\varphi_t : P^+T^*T^n \to P^+T^*T^n$, $t \in [0,1]$, be a contact isotopy with $\varphi_0 = \mathrm{Id}$ such that $\varphi_1(\Lambda)$ is transversal to L.

Then we have

COROLLARY 2.5.2. $\#\varphi_1(\Lambda) \cap L \geq 2^{n-1}$.

Remark 2.5.3: A Legendrian submanifold $\Lambda \subset M$ has a neighborhood U contactomorphic to the 1-jet space $J^1(\Lambda)$. The pre-Lagrangian submanifold $L \cap U$ can be identified under the contactomorphism with the "0-wall" $W = \Lambda \times \mathbf{R} \subset J^1(\Lambda)$, i.e. the set of 1-jets of functions with zero differential. Thus, Theorem 2.5.1 can be considered as a global version of the well-known fact that Λ cannot be disjoined with W via a contact isotopy (Chekanov's theorem).

2.5.1 INTERSECTIONS IN THE SPACE OF PRE-QUANTIZATION. Let us now turn to the situation described in section 2.4.3. Let (N, ω) be a symplectic manifold such that the symplectic form $\omega/\hbar$ represents an integral cohomology class $[\omega/\hbar] \in H^2(N)$. Let $(M,\zeta) = C(N,\omega)$ be the contactization of (N,ω) (see 2.4.3 above) and $L \subset N$ be a Lagrangian submanifold which satisfies the Bohr-Sommerfeld condition. Let Λ_1, $\Lambda_1 \subset M$ be a Legendrian lift of L and $\Lambda_0 = \pi^{-1}(L)$ be the pre-Lagrangian pull-back of L under the projection $\pi : M \to N$.

Let $\varphi_t : M \to M$, $t \in [0,1]$, be a contact isotopy with $\varphi_0 = \mathrm{Id}$ such that $\varphi_1(\Lambda_1)$ is transversal to Λ_0.

THEOREM 2.5.4. *Suppose that $\pi_2(M,\Lambda_0) = 0$. Then*

$$\#\varphi_1(\Lambda_1) \cap \Lambda_0 \geq \ \textit{rank}\ H_*(\Lambda_1; \mathbb{Z}/2)\ .$$

For instance, let N be a surface of positive genus, ω an area form with $\int_N \omega = 1$ and $(M,\zeta) = C(N,\omega)$ be the contactization with $\hbar = 1/n$. Let $L \subset N$ be a non-contractible Bohr-Sommerfeld orbit, $\Lambda_1 \subset M$ its Legendrian lift and $\Lambda_0 = \pi^{-1}(L) \subset M$ its pre-Lagrangian pull-back. Then, we have

COROLLARY 2.5.5. *For the contact isotopy $\varphi_t : M \to M, t \in [0,1], \varphi_0 = \mathrm{Id}$, such that $\varphi_1(\Lambda_1)$ is transversal to Λ_0 we have*

$$\#\varphi_1(\Lambda_1) \cap \Lambda_0 \geq 2\ .$$

3. Floer Homology

3.1 Admissible Legendrian and pre-Lagrangian submanifolds. Let Λ_0 and Λ_1 be a pre-Lagrangian and a Legendrian submanifold, respectively, of a contact manifold (M,ζ). We will always assume in what follows that the submanifold Λ_1 is connected.

Let us denote by $\mathcal{P}(\Lambda_0,\Lambda_1)$ the space of paths $\delta : [0,1] \to M$ with $\delta(0) \in \Lambda_0$ and $\delta(1) \in \Lambda_1$. A component $\mathcal{P}_0$ of the space $\mathcal{P}(\Lambda_0,\Lambda_1)$ is called *admissible* if it satisfies the following two conditions $\mathcal{P}_1$ and $\mathcal{P}_2$.

$\mathcal{P}_1$ For any map $\Delta : S^1 \times [0,1] \to M$ such that $\Delta(u,0) \in \Lambda_0$, $\Delta(u,1) \in \Lambda_1$, and $\Delta|_{u\times[0,1]} \in \mathcal{P}_0$ for $u \in S^1$, the curve $\Delta|_{S^1\times 0}$ is a vanishing cycle on Λ_0 (see 2.2).

$\mathcal{P}_2$ For any map $\Delta : S^1 \times [0,1] \to M$, as in $\mathcal{P}_1$, the relative Maslov class $\mu(\Delta|_{S^1\times 0}, \Delta|_{S^1\times 1})$ vanishes (see 2.2).

LEMMA 3.1.1. *The condition $\mathcal{P}_1$ implies that for any map $F : (D^2, \partial D^2) \to (M,\Lambda_0)$ the curve $F|_{\partial D^2} : \partial D^2 \to \Lambda_0$ is a vanishing cycle in Λ_0.*

Proof: Any such map can be deformed, keeping the boundary fixed, to a map $\widetilde{F}$ such that there exists a map Δ as in $\mathcal{P}_1$, which can factored as

$\Delta = \widetilde{F} \circ p$ where $p : S^1 \times [0,1] \to D^2$ is the projection which collapses the circle $S^1 \times 1$ to the center of the disc D^2. □

Of course, existence of an admissible component of the space $\mathcal{P}(\Lambda_0, \Lambda_1)$ is a very restrictive condition on manifolds M, Λ_0 and Λ_1. However, there is an important case when it does exist.

LEMMA 3.1.2. *Suppose that $\Lambda_1 \subset \Lambda_0$ and the boundary homomorphism*

$$\pi_2(M, \Lambda_0) \to \pi_1(\Lambda_0)$$

is trivial. Then the component $\mathcal{P}_0 \subset \mathcal{P}(\Lambda_0, \Lambda_1)$ which contains constant paths from Λ_1 is admissible.

Proof: The proof follows immediately from the observation that every loop in $\mathcal{P}_0$ is homotopic to a loop of constant paths. □

In order to develop a Floer homology theory for the intersection problem of Λ_0 and Λ_1 we fix a path component $\mathcal{P}_0 \subset \mathcal{P}(\Lambda_0, \Lambda_1)$ and impose two severe restrictions, including the admissibility of $\mathcal{P}_0$.

O_1 The path component $\mathcal{P}_0 \subset \mathcal{P}(\Lambda_0, \Lambda_1)$ is admissible.

O_2 There exists a contact form $\delta \in \mathrm{Cont}(\zeta)$ such that the flow defined by its Reeb vector field Y has no contractible periodic orbits and each orbit with two ends on Λ_1 represents a non-trivial class from $\pi_1(M, \Lambda_1)$.

The set of contact forms $\delta \in \mathrm{Cont}(\zeta)$ which satisfy the condition O_2 will be denoted by $\mathrm{Adm}(\zeta, \Lambda_0, \Lambda_1)$.

Our goal is to define Floer homology groups of the triple $\Lambda_0, \Lambda_1, \mathcal{P}_0$. To understand the relevance of the component $\mathcal{P}_0$ note that every intersection point $x \in \Lambda_0 \cap \Lambda_1$ determines a constant path $\delta(t) \equiv x$ and these constant paths may lie in different path components for different intersection points. The Floer homology groups $HF_*(\Lambda_0, \Lambda_1, \mathcal{P}_0)$ will arise from a chain complex which is generated by all those intersection points which correspond to constant paths in $\mathcal{P}_0$. In most of our applications there is only one relevant path component which corresponds to all the fixed points and the Floer homology groups of all other path components are zero. Hence we shall sometimes neglect the dependence on $\mathcal{P}_0$ in our notation when the choice of the path component is clear from the context.

3.2 Examples of admissible submanifolds. We will verify in this section that the conditions O_1 and O_2 hold in all theorems from section 2.5.

LEGENDRIAN AND PRE-LAGRANGIAN SUBMANIFOLDS IN P^+T^*N. Let $M = P^+T^*N$, $\Lambda_1 = \Lambda$ and $\Lambda_0 = L_\alpha$ be as in Theorem 2.5.1. Fix a point $\tilde{q} \in \Lambda_0$. Let us denote by $p : M \to N$ the canonical projection and set $q = p(\tilde{q})$,

$S = p^{-1}(q)$. Let us verify that the boundary homomorphism $\pi_2(M, \Lambda_0) \to \pi_1(\Lambda_0, \tilde{q})$ is trivial. Indeed, let f be a map $(D^2, \partial D^2) \to (M, \Lambda_0)$ and $g_t : D^2 \to N$, $t \in [0,1]$, be a homotopy of the projection $g_0 = p \circ f$ to a constant map g_1 to the point $q \in Q$. This homotopy can be lifted, using the covering homotopy property, to a homotopy $f_t : (D^2, \partial D^2) \to (M, \Lambda_0)$. In particular, f_1 maps D^2 to the sphere $S = p^{-1}(q)$ and $f_1(\partial D^2)$ is the point $\tilde{q} = \Lambda_0 \cap p^{-1}(q)$. Thus the conditions of Lemma 3.1.2 are satisfied, and therefore the component $\mathcal{P}_0$ is admissible.

To verify the condition O_2 take a metric on N without contractible closed geodesics. Identifying $M = P^+T^*N$ with the unit cotangent bundle with respect to this metric we get a contact form β on M whose Reeb flow is the geodesic flow for the chosen metric. Thus the Reeb flow for the form β has no contractible periodic orbits. Let $q_\alpha : N \to S^1$ be the map corresponding to the form α, i.e. $\alpha = q_\alpha^*(d\theta)$. Notice that the projection $p : P^+T^*N \to N$ maps Λ onto one of the fibers $N_1 = q_\alpha^{-1}(\text{point})$. Let Γ be a piece of trajectory of the Reeb flow with two ends on Λ. Then the projection

$$(P^+T^*N, \Lambda) \xrightarrow{p} (N, N_1) \xrightarrow{q_\alpha} (S^1, \{\text{point}\})$$

projects Γ onto a non-trivial element of $\pi_1(S^1)$. Thus Γ represents a non-trivial element of $\pi_1(M, \Lambda)$ which verifies the condition O_2.

LEGENDRIAN AND PRE-LAGRANGIAN SUBMANIFOLDS IN THE SPACE OF CONTACTIZATION. Under the assumptions of Theorem 2.5.4 we have $\pi_2(M, \Lambda_0) = 0$ and thus, according to Lemma 3.1.2, the component $\mathcal{P}_0 \subset \mathcal{P}(\Lambda_0, \Lambda_1)$, which contains constant paths from Λ_1, is admissible.

Let us check the condition O_2. Let us recall that the contact structure ζ on the space M of pre-quantization can be defined by an S^1-invariant contact form α on the principal S^1-bundle $M \to N$. The trajectories of the Reeb flow for the form α are fibers of the fibration. Thus all trajectories are closed and all simple closed trajectories are homotopic. Let Γ be one of the trajectories which is contained in Λ_0. Then $\int_\Gamma \alpha \neq 0$. Suppose that Γ bounds a disc $D \subset M$. Then $\int_D d\alpha \neq 0$ and, therefore, D represents a non-trivial element from $\pi_2(M, \Lambda_0)$. This contradicts the assumption of Theorem 2.5.4, and, therefore, Γ, and all its multiples, are non-contractible.

Notice that a trajectory of the Reeb flow with both ends on Λ_0 has to coincide with the periodic orbit Γ considered above. If Γ represents a trivial element of $\pi_1(M, \Lambda_1)$ then it bounds, together with a curve $\Gamma' \subset \Lambda_1$, a disc $D \subset M$, i.e. $\partial D = \Gamma \cup \Gamma'$. Then

$$\int_D d\alpha = \int_\Gamma \alpha + \int_{\Gamma'} \alpha \,.$$

But $\gamma|_{\Lambda_1} = 0$ and therefore the second integral equals 0. Thus, as in the case of the closed orbit, we have $\int_D d\alpha \neq 0$ and hence D represents a non-trivial element of $\pi_2(M, \Lambda_0)$ which again contradicts to assumption of Theorem 2.5.4.

3.3 Almost complex structures on the symplectization. Suppose that the contact manifold (M, ζ), its pre-Lagrangian submanifold Λ_0 and Legendrian submanifold Λ_1 satisfy the conditions O_1 and O_2. Let $(V = S_\zeta(M), \omega = \omega_\zeta)$ be the symplectization of (M, ζ).

Let us recall that an almost complex structure J is called *compatible* with ω, if the bilinear form $\langle v, w\rangle = \omega(v, Jw)$ is a metric, invariant under J.

A fiberwise splitting $H : V \to M \times \mathbf{R}$ is called *admissible* if it coincides at infinity with H_γ for an admissible form $\gamma \in \mathrm{Adm}(\zeta, \Lambda_0, \Lambda_1)$.

Notice that the push-forward $(H^{-1})^*\alpha_\zeta$ of the canonical form α_ζ on $S_\zeta(M)$ can be written as $\exp\theta\gamma_\theta$ where $\gamma_\theta \in \mathrm{Cont}(\zeta)$, $\theta \in \mathbf{R}$, and γ_θ coincides with γ when $|\theta|$ is sufficiently large. In other words, the pre-image $H^{-1}(M \times \theta) \subset S_\zeta(M)$, $\theta \in \mathbf{R}$, is the graph of the 1-form $\exp\theta\gamma_\theta$.

We also have $\omega = H^*(d(\exp\theta\gamma_\theta))$ and $dH(X_\zeta) = h\frac{\partial}{\partial\theta}$ for a positive function $h : M \times \mathbf{R} \to \mathbf{R}$ which is equal to 1 at infinity.

Having fixed an admissible splitting $H : V \to M \times \mathbf{R}$ we will not distinguish between an almost complex structure J on V and its push-forward $H_*(J)$ on $M \times \mathbf{R}$.

An almost complex structure J compatible with ω is called *admissible* for $(M, \zeta), \Lambda_0$ and Λ_1 if there exists an admissible splitting $H : S_\zeta(M) \to M \times \mathbf{R}$ of the space of symplectization such that

- for each $a \in \mathbf{R}$ the contact structure $\zeta = \{\gamma_a = 0\}$ on $M_a = M \times a$ is invariant under J;
- the vector field $J \cdot \frac{\partial}{\partial\theta}|_{M_a}$ belongs to the kernel of the form $(H^{-1})^*\omega|_{M_a} = d(\exp\theta\gamma_\theta)|_{M_a}$, $a \in \mathbf{R}$;
- J is invariant under the $\mathbf{R}$-action at infinity.

Notice that the above conditions imply that all the levels M_a, $a \in \mathbf{R}$, are J-convex being cooriented by the vector field $\frac{\partial}{\partial\theta}$.

Suppose we are given two admissible structures J and J'. Viewing them as defined on $M \times \mathbf{R}$ we say that a sequence of admissible almost complex structures J_n, $n = 1, \ldots$, *interpolates* between J' and J if there exists a constant $N > 0$ and an increasing sequence $d_n \to \infty$ such that $J_n = J$ on $M \times [-d_n, d_n]$, $J_n = J'$ outside of $M \times [-(d_n + N), d_n + N]$, and the restrictions $J_n|_{M\times[-(d_n+N),-d_n]}$ coincide up to translations for all $n = 1, \ldots$.

3.4 Action functional. Suppose that $(M, \zeta), \Lambda_0, \Lambda_1$ and the path component $\mathcal{P}_0 \subset \mathcal{P}(\Lambda_0, \Lambda_1)$ satisfy the condition O_1. Let (V, ω) be the symplectization of (M, ζ), L_1 the symplectization of Λ_1, and L_0 a Lagrangian lift of Λ_0. Denote by $\mathcal{P}(L_0, L_1)$ the space of paths $\delta : [0,1] \to V$ with $\delta(0) \in L_0$ and $\delta(1) \in L_1$. Note that every path $\delta \in \mathcal{P}(\Lambda_0, \Lambda_1)$ lifts to a path $\hat{\delta} : [0,1] \to V$ with $\hat{\delta}(0) \in L_0$ and $\hat{\delta}(1) \in L_1$ and that the homotopy class of the lift is uniquely determined by δ. Hence the path component $\mathcal{P}_0 \subset \mathcal{P}(\Lambda_0, \Lambda_1)$ determines a unique path component in $\mathcal{P}(L_0, L_1)$ which we shall also denote by $\mathcal{P}_0$. This slight abuse of notation should not create any confusion.

Fix a path $\delta_0 \in \mathcal{P}_0 \subset \mathcal{P}(L_0, L_1)$ and for any other path $\delta \in \mathcal{P}_0$ choose a homotopy $\delta_u \in \mathcal{P}_0$, $u \in [0,1]$, which connects δ_0 with $\delta_1 = \delta$. Set $\Delta(u,t) = \delta_u(t)$ for $(u,t) \in [0,1] \times [0,1]$. Define now the action

$$\mathcal{A}_{\delta_0}(\delta) = \int_{[0,1]\times[0,1]} \Delta^* \omega \ .$$

We will omit δ_0 in the notation for the action when the choice of the base path is clear or irrelevant.

The property O_1 ensures that $\mathcal{A}_{\delta_0}(\delta)$ does not depend on the choice of the homotopy Δ (but it does depend on the choice of the path δ_0).

Critical points of the functional $\mathcal{A}_{\delta_0}$ are constant paths corresponding to the intersection points of L_0 and L_1. In order to count their number we need to define (and compute) Floer homology groups for the action functional $\mathcal{A}_{\delta_0}$.

3.5 Gradient flow. Choose an admissible almost complex structure J on V. This choice allows us to define a quasi-Kählerian metric on V:

$$g(v,w) = \omega(v, Jw) \ , \qquad v, w \in T_x(V) \ , \ x \in V \ .$$

Given a family J^t, $t \in [0,1]$, of admissible almost complex structures, we can define a metric on the path space $\mathcal{P}(L_0, L_1)$ by the formula

$$||v||^2 = \int_0^1 \omega(v, J^t v) dt \ , \qquad v \in T_\delta \mathcal{P}(L_0, L_1) \ , \ \delta \in \mathcal{P}(L_0, L_1) \ .$$

The gradient of the symplectic action $\mathcal{A}_{\delta_0}$ with respect to this metric on $\mathcal{P}(L_0, L_1)$ is given by

$$\operatorname{grad} \mathcal{A}_{\delta_0}(\delta) = -J^t \dot{\delta} \ .$$

Thus a gradient flow line is a smooth map $u : \mathbf{R} \times [0,1] \to V$ which satisfies the partial differential equation

$$\frac{\partial u}{\partial s} + J^t(u) \cdot \frac{\partial u}{\partial t} = 0 \tag{1}$$

with boundary conditions

$$u(s,0) \in L_0 \ , \quad u(s,1) \in L_1 \text{ for } s \in \mathbf{R} \ . \tag{2}$$

When $J^t \equiv J$ then this is just the usual Cauchy-Riemann equations and, therefore, the gradient line u is a J-holomorphic curve in V with boundary in $L_0 \cup L_1$.

In the general case the gradient trajectories of the action functional can also be interpreted as holomorphic curves but in an auxiliary manifold, and not in the manifold V itself (comp., for instance, [Gr], [F3] and [SZ]).

3.6 Energy. Given a solution $u : \mathbf{R} \times [0,1] \to V$ of (1) and (2), the symplectic area $\int_B u^*\omega$ will be denoted by $E(u)$ and called the *energy* of the solution u. When $J^t \equiv J$ then the energy $E(u)$ coincides with the area of the J-holomorphic curve u computed in terms of the almost Kählerian metric

$$g(u,v) = \omega(u, Jv) .$$

The following proposition is standard in Floer homology theory (cf. [F1]) and me omit its proof here.

THEOREM 3.6.1. *Suppose that L_0 and L_1 intersect transversally and J^t is a family of admissible almost complex structures. Let u be a solution of (1) and (2) with $E(u) < \infty$. Then there exist the limits*

$$\lim_{t \to \pm\infty} u(s,t) = x^\pm , \qquad x^\pm \in L_0 \cap L_1 .$$

We will call such a u a *connecting orbit* between the two critical points x^+ and x^- of the action functional $\mathcal{A}_{\delta_0}$. The definition of the action functional implies that

$$E(u) = \mathcal{A}_{\delta_0}(x^+) - \mathcal{A}_{\delta_0}(x^-) .$$

If $J^t \equiv J$ then a solution u of (1) and (2) of finite energy can be viewed as a J-holomorphic disc with boundary in $L_0 \cup L_1$ passing through two points $x^\pm \in L_0 \cap L_1$.

A Floer complex can be defined now as usual by counting the connecting orbits when the relative Morse index is 1.

The crucial point for the construction of the theory is the following compactness theorem for the solutions of (1) and (2) with bounded energy. The proof will be given in §3.9.

THEOREM 3.6.2. *Assume that the contact manifold (M,ζ), its pre-Lagrangian submanifold Λ_0 and a Legendrian submanifold Λ_1 satisfy the hypotheses O_1 and O_2. Let J^t, $t \in [0,1]$, be a family of admissible almost complex structures on the symplectization V. Then for every $c > 0$ the space*

$$\mathcal{M}^c = \mathcal{M}^c(L_0, L_1; J^t)$$

of all smooth solutions u of the boundary value problem (1) and (2) which satisfy the energy bound

$$E(u) \leq c$$

is compact (with respect to the topology of uniform convergence with all derivatives on compact sets).

We will need also a slightly stronger Theorem 3.6.3.

THEOREM 3.6.3. *Suppose that a sequence J_n^t, $t \in [0,1]$, $n = 1, \ldots,$ of families of admissible almost complex structures on V interpolates between two families of admissible structures $(J')^t$ and J^t. Then given a sequence $u_n \in \mathcal{M}^c(L_0, L_1, J_n^t)$, $n = 1, \ldots,$ one can find a subsequence which converges, uniformly on compact sets, to a solution $u \in \mathcal{M}^c(L_0, L_1, J^t)$.*

Remark 3.6.4: Theorem 3.6.3 holds even in a stronger form: it is sufficient to require that the sequence J_n^t converges to J^t uniformly on compact sets. However we will not need this stronger version in this paper.

Notice that the condition O_1 prohibits bubbling-off of the solutions at boundary points while the bubbling-off at interior points is impossible because the symplectic manifold (V, ω) is exact.

Thus, if we knew à priori that all the solutions of (1) and (2) would take values in a compact subset of V then the above theorem would follow directly from the usual compactness theory for Gromov's pseudoholomorphic curves (cf. [Gr] or [MS]). Hence our goal is to prove this bound for solutions from $\mathcal{M}^c$. The main ingredient to the proof is a rescaling trick which was first applied by Hofer in [H].

3.7 Floer homology. Suppose that (M, ζ), Λ_0 and Λ_1 satisfy the conditions $O_1 - O_2$. Let $(V, \omega), L_0, L_1$ be their symplectic counterparts and J^t a family of admissible almost complex structures. Pick an admissible component $\mathcal{P}_0 \subset \mathcal{P}(\Lambda_0, \Lambda_1)$ and a path $\delta_0 \in \mathcal{P}_0$. Let $\hat{\delta}_0$ be a lift of δ_0 to the space $\mathcal{P}(L_0, L_1)$. The component of $\hat{\delta}_0$ in $\mathcal{P}(L_0, L_1)$ will be still denoted by $\mathcal{P}_0$.

The Floer homology groups

$$HF_*(\Lambda_0, \Lambda_1; J^t) = HF_*(L_0, L_1; J^t) = HF_*(L_0, L_1, \mathcal{P}_0, J^t)$$

can roughly be described as the *middle dimensional homology groups* of the path space $\mathcal{P}_0 \subset \mathcal{P}(L_0, L_1)$ (compare [Wi]). They are obtained from the gradient flow of the symplectic action

$$\mathcal{A} : \mathcal{P}_0 \to \mathbf{R}$$

as in Floer's original work on Lagrangian intersections in compact symplectic manifolds [F1-3]. See also [O]. We summarize the main points of Floer's construction.

Assume that Λ_0 and Λ_1 (and hence L_0 and L_1) intersect transversally. Then all the critical points of $\mathcal{A}$ are nondegenerate. Given two intersection points $x^\pm \in L_0 \cap L_1$ denote by

$$\mathcal{M}(x^-, x^+) = \mathcal{M}(x^-, x^+, J^t)$$

the space of all solutions $u : B \to V$ of (1) and (2) with limits (3.6.1). Linearizing the differential equation (1) gives rise to an operator

$$D_u : W_L^{1,2}(u^*(TV)) \to L^2(u^*(TV)) .$$

Here $W_L^{1,2}(u^*(TV))$ denotes the Sobolev space of all vector fields $Y(s,t) \in T_{u(s,t)}V$ along u which satisfy the boundary condition

$$Y(0,t) \in T_{u(0,t)}L_0 , \quad Y(1,t) \in T_{u(1,t)}L_1 .$$

The space $L^2(u^*(TV))$ is defined similarly and D_u is a Cauchy-Riemann operator. This operator is Fredholm whenever L_0 and L_1 intersect transversally. Its index is a relative Maslov class and can be defined as follows. Given $u \in \mathcal{M}(x^-, x^+)$ choose a symplectic trivialization

$$\Phi(s,t) : \mathbf{R}^{2n+2} \to T_{u(s,t)}V$$

of $u^*(TV)$ such that

$$\Phi(s,t)^*\omega = \sum_{j=0}^{n} dx_j \wedge dy_j$$

and there exist limits

$$\lim_{t\to\pm\infty} \Phi(s,t) = \Phi^\pm : \mathbf{R}^{2n+2} \to T_{x^\pm}V .$$

This gives rise to two Lagrangian paths in $\mathbf{R}^{2n+2}$:

$$\lambda_0(t) = \Phi(0,t)^{-1} T_{u(0,t)}(L_0)$$

and

$$\lambda_1(t) = \Phi(1,t)^{-1} (T_{u(1,t)}(L_1)) .$$

These paths are transverse at $t = \pm\infty$ and therefore have a relative Maslov index $\mu(\lambda_0, \lambda_1)$ (cf. [F1] and [RS1]). This index is independent of the choice of the trivialization. The Fredholm index of D_u agrees with this Maslov index

$$\text{INDEX } D_u = \mu(u) = \mu(\lambda_0, \lambda_1)$$

whenever u satisfies the boundary condition (2) and the limit condition (3.6.1) (cf. [F1] and [RS2]).

Now recall that not all the intersection points from $L_0 \cap L_1$, viewed as constant paths, belong to the component $\mathcal{P}_0$. We denote by $(L_0 \cap L_1)_0$ the subset of $L_0 \cap L_1$ which consists of those intersection points which belong to $\mathcal{P}_0$. The condition $\mathcal{P}_2$ implies:

LEMMA 3.7.1. *If* $x^- = x^+ \in (L_0 \cap L_1)_0$ *then* $\mu(u) = 0$.

The previous lemma shows that there exists a map $\mu : (L_0 \cap L_1)_0 \to \mathbf{Z}$ such that

$$\text{INDEX } D_u = \mu(x^-) - \mu(x^+)$$

whenever u and θ satisfy (2) and (3.6.1). Now everything is as usual. A family of admissible almost complex structures J^t, $t \in [0,1]$, is called *regular* if the operator D_u is onto for every $u \in \mathcal{M}(x^-, x^+)$ and every pair of intersection points $x^\pm \in L_0 \cap L_1$. By the Sard-Smale theorem the set

$$\mathcal{REG} = \mathcal{REG}(L_0, L_1)$$

of regular J^t is dense in the set of all admissible families.

The argument is as in [F2] or [SZ]. See also [FSH] for a detailed discussion of transversality properties.

Remark 3.7.2: We need to consider families J^t rather than individual J just to ensure this genericity condition.

Now for every $J^t \in \mathcal{REG}$ the space $\mathcal{M}(x^-, x^+)$ is a finite dimensional manifold with

$$\dim \mathcal{M}(x^-, x^+) = \mu(x^-) - \mu(x^+) .$$

If $\mu(x^-) - \mu(x^+) = 1$ then, by Theorem 3.6.2, the quotient space $\mathcal{M}(x^-, x^+)/\mathbf{R}$ consists of finitely many orbits and the numbers

$$n_2(x^-, x^+) = \#\mathcal{M}(x^-, x^+)/\mathbf{R} \pmod 2$$

determine the Floer chain complex as follows. The chain groups are defined by

$$CF_k = CF_k(L_0, L_1, \mathcal{P}_0) = \sum_{\substack{x \in (L_0 \cap L_1)_0 \\ \mu(x) = k}} \mathbf{Z}_2 \langle x \rangle .$$

and the boundary operator $\partial : CF_k \to CF_{k-1}$ is given by

$$\partial \langle x \rangle = \sum_{\mu(y) = k-1} n_2(x, y) \langle y \rangle$$

for $x \in (L_0 \cap L_1)_0$ with $\mu(x) = k$. As in Floer's original proof one uses gluing techniques to prove that $\partial \circ \partial = 0$ (cf. [F3] and [SZ]).

The *Floer homology groups* are now defined as the homology of this chain complex

$$HF_*(L_0, L_1; J^t) = HF_*(L_0, L_1, \mathcal{P}_0; J^t) := H_*(CF, \partial) .$$

The Floer homology groups depend on the path component $\mathcal{P}_0$ but when the choice of the path component is clear from the context we shall drop $\mathcal{P}_0$ from the notation.

THEOREM 3.7.3. (i) *For any two admissible families of almost complex structures* J^t, $(J')^t \in \mathcal{REG}$ *there is a natural isomorphism*

$$HF_*(L_0, L_1, \mathcal{P}_0; J^t) \to HF_*(L_0, L_1, \mathcal{P}_0; (J')^t) .$$

(ii) *For any* $J^t \in \mathcal{REG}$ *and any compactly supported Hamiltonian isotopy* ψ_t, $t \in [0,1]$, *there exists a natural isomorphism*

$$HF_*(L_0, L_1, \mathcal{P}_0; J^t) \to HF_*({\psi_0}^{-1}(L_0), {\psi_1}^{-1}(L_1), \psi^*\mathcal{P}_0; J^t)$$

where $\psi^*\mathcal{P}_0 \subset \mathcal{P}({\psi_0}^{-1}(L_0), {\psi_1}^{-1}(L_1))$ *denotes the component of the path* $t \mapsto {\psi_t}^{-1}(\delta(t))$ *for* $\delta \in \mathcal{P}_0 \subset \mathcal{P}(L_0, L_1)$.

(iii) *For any symplectomorphism* $f : V \to V$, *which commutes at infinity with the* $\mathbf{R}$*-action, there exists a natural isomorphism*

$$HF_*(L_0, L_1, \mathcal{P}_0; J^t) \to HF_*(f(L_0), f(L_1), f_*\mathcal{P}_0; f_*J^t) .$$

Proof: Statement (iii) is obvious. The invariance under compactly supported change of the regular family J^t is standard in Floer's theory. To prove the invariance under Hamiltonian isotopies of the Lagrangian submanifolds L_0 and L_1 it is convenient to introduce a Hamiltonian term in the action functional $\mathcal{A}$. Hence let $H^t = H^{t+1} : V \to \mathbf{R}$ be a smooth family of compactly supported Hamiltonian functions with corresponding Hamiltonian vector fields X^t. Then the critical points of the perturbed action functional are solutions $x : [0,1] \to V$ of $\dot{x}(t) = X^t(x(t))$ with $x(0) \in L_0$ and $x(t) \in L_1$ and the gradient flow lines are solutions $u : \mathbf{R} \times [0,1] \to V$ of

$$\partial_s u + J^t(u)(\partial_t u - X^t(u)) = 0 \tag{3}$$

with the same boundary conditions $u(s,0) \in L_0$ and $u(s,1) \in L_1$ (compare with equations (1) and (2)). This gives rise to Floer homology groups $HF_*(L_0, L_1, \mathcal{P}_0; J^t, H^t)$ and as in the usual Floer theory one proves that these groups are independent of J and H up to natural isomorphisms. Now let $\psi_t : V \to V$ be a Hamiltonian isotopy generated by X^t via $\frac{d}{dt}\psi_t = X^t \circ \psi_t$ and define $v(s,t) = \psi_t^{-1}(u(s,t))$ where u is a solution of (3). Then, by a simple calculation, we find that

$$\partial_s v + \psi_t^* J^t(v) \partial_t v = 0$$

and $v(s,0) \in \psi_0^{-1}(L_0)$, $v(s,1) \in \psi_1^{-1}(L_1)$. This shows that there is a natural isomorphism

$$HF_*(L_0, L_1, \mathcal{P}_0; J^t, H^t) \to HF_*(\psi_0^{-1}(L_0), \psi_1^{-1}(L_1), \psi^*\mathcal{P}_0; \psi_t^* J^t, 0)$$

Thus we have proved (ii) as well as (i) for compactly supported variations of the almost complex structure. The only additional thing we need to check is that the groups $HF_*(L_0, L_1; J^t)$ and $HF_*(L_0, L_1; (J')^t)$ are isomorphic even when J^t and $(J')^t$ differ at infinity.

There exists a sequence J_n^t, $n = 1, \dots,$ of admissible almost complex structures which interpolates between $(J')^t$ and J^t. In view of Theorem 3.6.3 one can find a compact set K such that all connecting orbits for all J_n^t, as well as for J^t, are contained in K. If n is sufficiently large then J_n^t coincides with J^t on K. Thus J^t and J_n^t have the same set of connecting orbits, and therefore the Floer homology groups $HF_*(L_0, L_1; J^t)$ and $HF_*(L_0, L_1; J_n^t)$ coincide. On the other hand, J_n^t coincides with $(J')^t$ at infinity. Thus we have a canonical isomorphism between the groups $HF_*(L_0, L_1; J^t)$ and $HF_*(L_0, L_1; J_n^t)$ in view of the argument above while the groups $HF_*(L_0, L_1; J_n^t)$ and $HF_*(L_0, L_1; (J')^t)$ are isomorphic according to the conventional Floer theory. □

Theorem 3.7.3 shows, in particular, that we can drop J^t from the notation of Floer homology groups and that the groups $HF_*(L_0, L_1, \mathcal{P}_0)$, also denoted by $HF_*(\Lambda_0, \Lambda_1, \mathcal{P}_0)$, are well defined even when Λ_0 and Λ_1 are not transversal. It should be noted, however, that these groups do depend on the choice of the admissible path component $\mathcal{P}_0$.

3.8 Contact manifolds. Let us return now to the contact environment. Theorem 3.7.3 implies

THEOREM 3.8.1. *Suppose that the contact manifold (M, ξ), the pre-Lagrangian submanifold Λ_0, the Legendrian submanifold Λ_1, and the path component $\mathcal{P}_0 \subset \mathcal{P}(\Lambda_0, \Lambda_1)$ satisfy the conditions O_1 and O_2. Then the groups*

$$HF_*(\Lambda_0, \Lambda_1, \mathcal{P}_0)$$

are well defined and invariant under Legendrian isotopy of the submanifold Λ_1 as well as under a contactomorphism $f : M \to M$, i.e.

$$HF_*\big(f(\Lambda_0), f(\Lambda_1), f_*\mathcal{P}_0\big) = HF_*(\Lambda_0, \Lambda_1, \mathcal{P}_0)\ .$$

Theorem 3.8.1 has the following standard application for counting the number of intersection points $\#\Lambda_0 \cap \Lambda_1 = \#L_0 \cap L_1$.

THEOREM 3.8.2. *Let Λ_0, Λ_1, and $\mathcal{P}_0$ be as in Theorem 3.8.1. Suppose that Λ_0 and Λ_1 intersect transversally. Then*

$$\#\Lambda_0 \cap \Lambda_1 \geq \#(\Lambda_0 \cap \Lambda_1)_0 \geq \operatorname{rank} HF_*(\Lambda_0, \Lambda_1, \mathcal{P}_0)\ .$$

In particular, if all path components are admissible, then

$$\#\Lambda_0 \cap \Lambda_1 \geq \sum_{\mathcal{P}_0} \operatorname{rank} HF_*(\Lambda_0, \Lambda_1, \mathcal{P}_0)\ .$$

We have to impose an additional restriction on Λ_1 and Λ_0 in order to be able to compute Floer homology groups $HF_*(\Lambda_0, \Lambda_1)$.

THEOREM 3.8.3. *Suppose that in addition to the assumptions of Theorem 3.8.2 we have $\Lambda_1 \subset \Lambda_0$. Then there is a natural isomorphism*

$$HF_*(\Lambda_0, \Lambda_1, \mathcal{P}_0) \to H_*(\Lambda_1; \mathbf{Z}/2)$$

where $\mathcal{P}_0$ denotes the component of the space of constant paths. In particular,

$$\#\Lambda_0 \cap \Lambda_1' \geq \operatorname{rank}\bigl(H_*(\Lambda_1; \mathbf{Z}/2)\bigr)$$

for any Legendrian submanifold Λ_1' which is Legendrian isotopic to Λ_1 and transverse to Λ_0.

Proof: As has already been mentioned (see 2.5.3), a neighborhood U of the Legendrian submanifold Λ_1 in M is contactomorphic to a neighborhood of the 0-section in the 1-jet space $J^1(\Lambda_1)$. This contactomorphism moves $\Lambda_0 \cap U$ onto the 0-wall W, i.e. the space of 1-jets of functions with 0 differential. Thus a Legendrian submanifold Λ_1', which is C^1-close to Λ_1 and transverse to W, corresponds to a Morse function $\varphi : \Lambda_1 \to \mathbf{R}$ so that the intersection points of Λ_0 and Λ_1' are in one-to-one correspondence with the critical points of the function φ. One can explicitly choose a metric on Λ_1 and an admissible almost complex structure J on the symplectization of M in such a way that the connecting orbits of the action functional would be in one-to-one correspondence with the gradient trajectories of the function φ connecting the corresponding critical points of this function. This identifies the Floer chain complex $CF_*(\Lambda_0, \Lambda_1')$ with the Morse chain complex for the function φ (cf. [Sc]) and thus defines a canonical isomorphism between the groups $HF_*(\Lambda_0, \Lambda_1)$ and $H_*(\Lambda_1; \mathbf{Z}/2)$. See [P] for a detailed proof (in the general case of clean Lagrangian intersections). □

Proof of Theorems 2.5.1 and 2.5.4: We already verified in 3.2 the conditions O_1 and O_2 in the situation of 2.5.1 and 2.5.4. Thus both statement follow from Theorem 3.8.3. □

3.9 Compactness. To clarify the main ideas of the proof we will assume in this section that all considered families of almost complex structures are constant. Thus the solutions of (1) and (2) can be treated as holomorphic curves for the corresponding almost complex structures. The general case, when the almost complex structures may depend on the parameter t, is similar, but less geometrically transparent.

As it was mentioned in Section 3.6 a solution $u : B = \mathbf{R} \times [0,1] \to V$ from $\mathcal{M}^c(L_0, L_1, J)$ can be equivalently viewed as a J-holomorphic disc in V with boundary in $L_0 \cup L_1$. We will employ both points of view.

The Theorem 3.6.3 is an immediate corollary of the following

THEOREM 3.9.1. *Suppose that a contact manifold (M, ζ), a pre-Lagrangian*

submanifold $\Lambda_0 \subset M$ and a Legendrian submanifold $\Lambda_1 \subset M$ satisfy the conditions O_1 and O_2. Let $(V,\omega), L_1$ and L_0 be the symplectization of $(M,\zeta), \Lambda_1$ and a Lagrangian lift of Λ_0, respectively. Let J_n, $n = 1,\ldots,$ be a sequence of admissible almost complex structures on V which interpolates between two admissible almost complex structures J' and J. Let $u_n : B = \mathbf{R} \times [0,1] \to V, n = 1,\ldots,$ be a sequence of J_n-holomorphic curves from $\mathcal{M}^c(L_0, L_1, J_n)$. Then all discs $\Delta_n = u_n(B)$ are contained in a common compact set $K \subset V$.

Proof: Set $J_0 = J'$, $J_\infty = J$. As in 3.3 we will consider the almost complex structures J, J' and J_n, $n = 0,\ldots,\infty$, as defined on the product $M \times \mathbf{R}$ so that the following conditions are satisfied:

— there exists an integer $d > 0$ such that J, J' are invariant under the $\mathbf{R}$-action (by translations) outside of $M \times [-d, d]$;

— there exists a constant $N > 0$ and an increasing sequence $d_n \to \infty$ such that $d_1 = d$ and for all $n < \infty$ we have $J_n = J$ on $M \times [-d_n, d_n]$, $J_n = J'$ outside of $M \times [-(d_n+N), d_n+N]$, and the restrictions $J_n|_{[-(d_n+N),-d_n]}$ coincide up to translations for all $n = 1,\ldots$;

— for each $n = 0,\ldots,\infty$ the almost complex structure J_n is compatible with the symplectic form $\omega_n = d(\exp\theta\gamma_{n,\theta})$, $\gamma_{n,\theta} \in \mathrm{Cont}(\zeta)$; $\gamma_{n,\theta} = \gamma_\infty$ for $|\theta| \leq d_n$, $\gamma_{n,\theta} = \gamma_0$ for $|\theta| \geq d_n + N$, and $\gamma_{n,\pm\theta\pm d_n} = \gamma_{m,\pm\theta\pm d_m}$ for all $m, n \geq 1$ and $\theta \geq 0$;

— for each $n = 0,\ldots,\infty$ and $a \in \mathbf{R}$ the contact structure $\zeta = \{\gamma_{n,a} = 0\}$ on the level $M_a = M \times a$ is invariant under J_n, and $J_n \cdot \frac{\partial}{\partial\theta}|_{M_a}$ belongs to the kernel of the form $\omega_n|_{M_a}$.

The last condition implies, in particular, that all levels M_a, being cooriented by the vector field $\frac{\partial}{\partial\theta}$, are (pseudo)convex for each of the almost complex structures J_n.

Without loss of generality we can also assume that $L_0 \subset M \times (-d, d)$, $L_1 = \Lambda_1 \times \mathbf{R}$. According to Sard's theorem there exists a constant a, arbitrarily close to 1 such that u_n are transversal to M_{ka} for all integers k and all $n \geq 1$. To simplify the notation we will assume that $a = 1$.

Set $\Omega_{a,b} = M \times [a, b]$.

First we observe

LEMMA 3.9.2. *All discs Δ_n are contained in $\Omega_{-\infty,d}$.*

Proof: Suppose that a disc Δ_n intersects $\Omega_{d,\infty}$. Then we have $\sup\theta\circ u_n \geq d$. The maximum of the function $\theta|_{\Delta_n}$ is achieved in a point $p \in \Delta_n$ because u_n converges to $x^\pm$ at infinity, and, on the other hand, $\theta(x^\pm) < d$. Thus $a = \theta(p) \geq d$. The point p cannot be an interior point of Δ_n because this would contradict the pseudoconvexity of M_a (maximum principle). Suppose that $p \in \partial\Delta_n$. Let τ be a vector tangent to $\partial\Delta_n$ at the point p. Then τ is tangent to L_1 and, therefore, $\tau \in \zeta_p \subset T_p(M_a)$. By the assumption, ζ is J_n-invariant and hence we have $J_n \cdot \tau \in \zeta_p \subset T_p(M_a)$. Therefore, the disc Δ_n is tangent to M_a at the boundary point p. But this is again impossible in view of pseudoconvexity of M_a (strong maximum principle). □

Set $\bar{\omega}_n = d_M(\gamma_{n,\theta})$, $n = 0, \dots, \infty$. Here d_M denotes the differential with respect to the variable $x \in M$. Thus for a point $p = (x, a) \in M \times \mathbf{R}$ we have

$$\bar{\omega}_n|_{T_p(M\times\mathbf{R})} = \exp(-a)(d\pi)^*(\omega|_{T_p(M_a)})$$

where π is the projection $M \times \mathbf{R} \to M$.

Denote by κ_n the plane field formed by kernels of the form $\bar{\omega}_n$. It is generated by the vector field $X = \frac{\partial}{\partial\theta}$ and the vector field $Y_n = J_n \cdot X$. Notice that Y_n is tangent to the level-sets M_a and $Y_n|_{M_a}$ is proportional to the Reeb vector field of the form $\gamma_{n,a}$.

LEMMA 3.9.3. *For any J_n-holomorphic curve $v : C \to M \times \mathbf{R}$ we have*

$$v^*\bar{\omega}_n = h\exp(-\theta \circ v)v^*\omega_n|_C \quad \text{for a function} \quad h : C \to [0,1] \,.$$

The function h vanishes only at singular points of v and the points of tangency of the curve $h(C)$ and the vector field $X = \frac{\partial}{\partial\theta}$.

Proof: Outside the branching points of v, the function h is the determinant of the projection of $v(C)$ to the contact distribution ζ along the plane field κ_n. According to the choice of the J_n this is an orthogonal projection which is a pointwise complex linear map. Hence, $0 \le h \le 1$ and h vanishes only at the points where the vector field X is tangent to $v(C)$. □

Observe also

LEMMA 3.9.4. *For each $n = 1, \dots$ and $i \ge d$ the domain $C_n^i = u_n^{-1}(\Omega_{-\infty,-i})$ is a union of discs and the following inequality*

$$0 \le \int_{C_n^i} u_n^*\bar{\omega}_n \le \exp(i)\int_B u_n^*\omega_n < c\exp(i)$$

holds.

Proof: The first statement of the lemma follows from J_n-convexity of the levels M_a, similarly to the proof of Lemma 3.9.2. Set $P_n^i = u_n^{-1}(M_{-i})$ and $R_n^i = u_n^{-1}(L_1 \cap \Omega_{-\infty,-i})$. Thus $\partial C_n^i = P_n^i \cup R_n^i$. Taking into the account that $\gamma_{n,\theta}|_{L_1} = 0$ we get

$$\begin{aligned}\int_{C_n^i} u_n^*\bar{\omega}_n &= \int_{\partial C_n^i} u_n^*\gamma_{n,\theta} = \int_{P_n^i} u_n^*\gamma_{n,-i} \\ &= \exp(i)\int_{C_n^i} u_n^*\omega_n \le \exp(i)\int_B u_n^*\omega_n < c\exp(i)\,.\end{aligned}$$

□

Let u be a map $B \to V$. A subdomain $U \subset B$ is called a *special domain of level k for u* if
— U is either a disc or annulus;
— $u|_U$ is transversal to $M_{-k} \cup M_{-k-1}$;
— $u(\partial U) \subset M_{-k} \cup M_{-k-1} \cup L_1$, $f(\partial U \cap \partial B) \subset L_1$;
— $u(\partial U) \cap M_{-j} \neq \emptyset$ for $j = k, k+1$;
— $u(U) \subset \Omega_{-\infty,-d}$.

LEMMA 3.9.5. *Let U be a special domain of level k for a J_n-holomorphic map $u : B \to V$. Then*

$$\int_U u^*\omega_n \leq 2\exp(d-k)\int_B u^*\omega_n < 2c\exp(d-k) = C_1\exp(-k)\ .$$

Proof: Similarly to the proof of 3.9.4 set

$$P_+ = u^{-1}(M_{-k})\ ,\quad P_- = u^{-1}(M_{-k-1})\ ,\quad R = \partial U \setminus (P_+ \cup P_-)\ .$$

Notice that $f(R) \subset L_1$ and thus $(u^*\gamma_{n,\theta})|_R = 0$. Thus, properly orienting $P_\pm$ we get

$$\begin{aligned} 0 < \int_U u^*\omega_n &= \int_{\partial U} \exp(-\theta \circ f)u^*\gamma = \\ &= \exp(-k)\int_{P_+} u^*\gamma_{n,-k} + \exp(-k-1)\int_{P_-} u * \gamma_{n,-k-1} \leq \\ &\leq 2\exp(-k)\int_{C_n^k} u^*\bar{\omega}_n \leq \\ &\leq 2\exp(-k)\int_{C_n^d} u^*\bar{\omega}_n \leq \\ &\leq 2\exp(d-k)\int_B u^*\omega_n \leq 2c\exp(d-k)\ . \end{aligned}$$

□

The following combinatorial lemma plays the crucial role in the proof of Theorem 3.9.1.

LEMMA 3.9.6. *Suppose that the sequence of J_n-holomorphic discs $u_n : B \to V$ is not contained in any compact set. Then there exists a subsequence u_{n_k}, $k = 1, \ldots,$ and a sequence G_k, $G_k \subset B$, such that*
- *G_k is special for u_{n_k};*
- *$\int_{G_k} u^*_{n_k}\bar{\omega}_{n_k} \underset{k\to\infty}{\longrightarrow} 0.$*
- *G_k is either*
  - a) *on the level j, $d \leq j < d_{n_k}$ or $j \geq d_{n_k} + N$, and is contained in $\Omega_{-(j+2),-j}$* or
  - b) *on the level d_{n_k}, and is contained in $\Omega_{-(d_{n_k}+N+1),-d_{n_k}+1}$.*

Proof: According to the assumption, the holomorphic discs Δ_n are not contained in any compact set. In view of Lemma 3.9.2 one can choose a subsequence u_{n_k}, $k = 1, \ldots$, such that $d_{n_k} \geq d + k + 1$ and $\Delta_{n_k} \cap M_{-k-1-d} \neq \emptyset$.

Let $d \leq i \leq d + k$. Set $\varphi_k = -\theta \circ u_{n_k}$ and $B_k^i = C_{n_k}^i \setminus \mathrm{Int} C_{n_k}^{i+1} = \{i \leq \varphi_k \leq i+1\}$. Let B be a component of B_k^i which has non-empty intersections with $\varphi_k^{-1}(i)$ and $\varphi_k^{-1}(i+1)$. Then B is a disc, possibly with several holes. One gets a *saturation* $\widehat{B}$ of the domain B by filling either all of these holes, or all but one in such a way that both intersections $\partial \widehat{B} \cap \varphi_k^{-1}(i)$ and $\partial \widehat{B} \cap \varphi_k^{-1}(i+1)$ are still non-empty. Notice that $\widetilde{B}$ is a *special domain of level* i for the map u_{n_k}.

For each $k \geq 1$ we can find a sequence of these special domains $\widetilde{B}_k^j$, $j = d, \ldots, d+k$, such that $\widetilde{B}_k^j$ is on the level j and $\mathrm{Int} \widetilde{B}_k^j \cap \mathrm{Int} \widetilde{B}_k^i = \emptyset$ for $i \neq j$. Notice that $\bigcup_{j=d}^{d+k} \widetilde{B}_k^j \subset C_{n_k}^d$. Thus according to 3.9.3 and 3.9.4 we have

$$\sum_{j=d}^{k+d} \int_{\widetilde{B}_k^j} u_{n_k}^* \bar{\omega}_{n_k} \leq \int_{C_{n_k}^d} u_{n_k}^* \bar{\omega}_{n_k} \leq c \exp(d) = C_1 \ ,$$

where all terms of the sum are positive. Hence, at least for some of the domains $\widetilde{B}_k^j$ we have $\int_{\widetilde{B}_k^j} u_{n_k}^* \bar{\omega}_{n_k} \leq C_1/k$.

Now choose a special domain G_k for u_{n_k} which has the smallest value of $\int_{G_k} u_{n_k}^* \bar{\omega}_{n_k}$ among all special domains on levels $j \in [d, d_{n_k} - 1] \cup [d_{n_k} + N, \infty)$. Then we have $\int_{G_k} u_{n_k}^* \bar{\omega}_{n_k} \leq C_1/k$. Let $j = j(k)$ be the level of G_k. In all cases we have $G_k \cap M_{-j+1} = \emptyset$ in view of 3.9.4. If $j(k) < d_{n_k}$ or $j(k) \geq d_{n_k} + N$ then $u_{n_k}(G_k)$ does not intersect M_{-j-2} because otherwise we could choose a smaller special domain. By the same reason if $j(k) = d_{n_k}$ then $u_{n_k}(G_k)$ does not intersect $M_{-d_{n_k}-N-1}$ and thus $u_{n_k}(G_k) \subset \Omega_{-(d_{n_k}+N+1), -d_{n_k}}$. □

Now we apply the trick from [H]. Passing, if necessary, to a subsequence, we can think that all domains G_k were chosen either on the level

(∗) $j < d_k$, or

(∗∗) $j \geq d_k + N$ or

(∗∗∗) $j(k) = d_k$.

Let us denote by J'', ω'' and $\bar{\omega}''$ the almost complex structure $J_n|_{\Omega_{-d_n-N-1,-d_n}}$ and the forms $\omega_n|_{\Omega_{-d_n-N-1,-d_n}}$, $\bar{\omega}_n|_{\Omega_{-d_n-N-1,-d_n}}$, respectively, translated by the **R**-action to the domain $\Omega = \Omega_{-d-N-1,-d}$. Set $\mu = \omega_\infty$, $\bar{\mu} = \bar{\omega}_\infty$ in the case (∗), $\mu = \omega_0$, $\bar{\mu} = \bar{\omega}_0$ in the case (∗∗) and $\mu = \bar{\omega}''$, $\bar{\mu} = \bar{\omega}''$ in the case (∗∗∗). Set also $\widetilde{J} = J$ in the case (∗), $\widetilde{J} = J'$ in the case (∗∗) and $\widetilde{J} = J''$ in the case (∗∗∗). Notice that J'', ω'' and $\bar{\omega}''$ coincide on $\Omega_{-d-1,-d}$ with $J = J_\infty, \omega_\infty$ and $\bar{\omega}_\infty$, respectively. Let us

translate now holomorphic maps $u_{n_k} : G_k \to V$ to the same common level d. Thus we get a sequence of maps $\tilde{u}_{n_k} : G_k \to \Omega$ such that
— each $\tilde{u}_{n_k}$ is holomorphic with respect to the almost complex structure $\widetilde{J}$;
— $\int_{G_k} \tilde{u}^*_{n_k}\bar{\mu} \underset{k\to\infty}{\to} = 0.$

We also have

$$\int_{G_k} \tilde{u}^*_{n_k}\mu = \exp\left(j(k)\right) \int_{G_k} u^*_{n_k}\omega_{n_k}$$

and in combination with Lemma 3.9.5 we get

$$\int_{G_k} \tilde{u}^*_{n_k}\mu < 2C_1 \ .$$

Let us consider all maps $\tilde{u}_{n_k}$ as being parametrized by the same unit disc Δ or a fixed annulus A (with a variable conformal structure). The sequence viewed this way will still be denoted by $\tilde{u}_{n_k}$.

We are now in a position to apply Gromov's compactness theorem (see [Gr]).

LEMMA 3.9.7. *There exists a subsequence of the sequence $\tilde{u}_{n_k}$ which converges uniformly on compact sets to a non-constant $\widetilde{J}$-holomorphic curve $\tilde{u}_\infty$. The set of boundary values of the map $\tilde{u}_\infty$ is contained in $L_1 \cup M_{-d} \cup M_{-d-1}$ and it is smooth at the boundary points which are maped into L_1.*

This lemma is a standard application of Gromov's theory (see [L] for the statement about the set of boundary values) for the case when the sequence $\tilde{u}_{n_k}$ is defined on the disc Δ, and would be for the case when it is defined on the annuli if we knew à priori that conformal moduli of the annuli were bounded. This is actually the case in our situation (see [La] for the proof). However, even without this knowledge Gromov's theory assures the convergence to a holomorphic *cusp*-curve. In our case the cusp degeneration would imply the existence of non-constant J-holomorphic discs with boundary values in $M_{-d-1} \cup M_{-d}$. The next lemma shows, in particular, that this is impossible.

LEMMA 3.9.8. *Let B be either a disc or an annulus and $u_\infty : \mathrm{Int}B \to \Omega$ be a non-constant $\widetilde{J}$-holomorphic curve with (possibly empty) boundary such that its boundary values are contained $L_1 \cap M_{-d} \cup M_{-d-1}$. Suppose that $\int_B u^*_\infty \bar{\mu} = 0$. Then $u_\infty(B)$ is a cylinder over an integral curve $P \subset M$ of the Reeb vector field of the contact form γ_0 in the case (**) and of the contact form γ_∞ in the cases (*) and (* * *). In other words,*

$$u_\infty(B) = P \times (-d-1, -d) \subset \mathrm{Int}\, \Omega_{-d-1,-d} \ .$$

The curve P is either a closed orbit or an arc connecting two points from Λ_1.

Proof: According to Lemma 3.9.3 we have $u_\infty^*\bar{\mu} = hu_\infty^*\mu$, where the function h takes values in $[0,1]$ and vanishes at the points where the vector field X is tangent to $u_\infty(B)$. Therefore the condition $\int_B u_\infty^*\bar{\mu} = 0$ implies that $h \equiv 0$ which means that $u_\infty(B)$ is a cylinder $P\times(-d-1,-d) \subset \mathrm{Int}\Omega_{-d-1,-d}$. The form $\bar{\mu}$ on $\Omega_{-d-1,-d}$ equals $d\gamma_0$ in the case (**) and $d\gamma_\infty$ in the cases (*) and (***). Thus the vector field $\widetilde{J}\cdot\frac{\partial}{\partial\theta}$ is proportional to the Reeb vector field for the contact forms γ_0 or γ_∞, respectively. P is a closed orbit if B is an annulus and P is an arc connecting two points of Λ_1 if B is a disc. □

Although Lemma 3.9.7 by itself does not provide any information about the boundary smoothness, or even continuity of the map u_∞ away from L_1, we can conclude from 3.8.9 that the curve B_∞ is smooth up to the boundary and transversal to M_{-d} and M_{-d-1}. This implies that the (subsequence of the) sequence $\tilde{u}_{n_k}$ converges to u_∞ on the closed domain B. In particular, the curve $P\times(-d)$ is a C^∞-limit of contractible loops in M_{-d} or arcs representing the trivial element of $\pi_1(M_{-d},\Lambda_1\times(-d))$. Summarizing we get that $P \subset M$ is a trajectory of the Reeb vector field of one of the forms γ_0 or γ_∞. P is either a closed contractible trajectory or an arc with ends on Λ_1 which represents the trivial class from $\pi_1(M,\Lambda_1)$. In both cases we get a contradiction with the admissibility of the almost complex structures $J_\infty = J$ or $J_0 = J'$.

This concludes the proof of Theorem 3.9.1.

References

[AG] V.I. Arnold, A.B. Givental, Symplectic geometry, in "Dynamical Systems–IV, Encyclopedia of Math. Sciences, Springer, 1990, 1–136.

[EGr1] Y. Eliashberg, M. Gromov, Convex symplectic manifolds, Proc. of Symp. in Pure Math. 52 (1991), part II, 135–162.

[EGr2] Y. Eliashberg, M. Gromov, in preparation.

[F1] A. Floer, A relative Morse index for the symplectic action, Comm. Pure Appl. Math. 41 (1988), 393–407.

[F2] A. Floer, The unregularized gradient flow of the symplectic action, Comm. Pure Appl. Math. 41 (1988), 775–813.

[F3] A. Floer, Morse theory for the symplectic action, J. Diff. Geom. 28 (1988), 513–547.

[F4] A. Floer, Symplectic fixed points and holomorphic spheres, Comm. Math. Phys. 120 (1989), 575–611.

[FHS] A. Floer, H. Hofer, D. Salamon, Transversality in elliptic Morse theory for the symplectic action, to appear in Duke Math. Journal.

[G] A. Givental, The non-linear Maslov index, Lect. Notes (London Math. Soc.), Cambridge Univ. Press 151 (1990), 35–44.

[Gr] M. Gromov, Pseudoholomorphic curves in symplectic manifolds, Inv. Math. 82 (1985), 307–347.

[H] H. Hofer, Pseudoholomorphic curves in symplectizations with applications to the Weinstein conjecture in dimension three, Preprint, Ruhr-Universität Bochum, 1993.

[HS] H. HOFER, D.A. SALAMON, Floer homology and Novikov rings, to appear in Floer memorial volume.

[K] S. KOBAYASHI, Prinicipal fibre bundles with the 1-dimensional toroidal group, Tôhoku Math. Journal 8 (1956), 29–45.

[L] F. LABOURIE, Examples of courbes pseudo-holomorphes en géométrie Riemannienne, in "Holomorphic Curves in Symplectic Geometrie", Birkhäuser Verlag, 1994.

[La] F. LAUDENBACH, Orbites periodiques et courbes pseudo-holomorphes, application à la conjecture de Weinstein en dinension 3 (d'après H. Hofer at al.), Seminaire Bourbaki, exposé 786, Juin 1994.

[MS] D. MCDUFF, D.A. SALAMON, J-holomorphic Curves and Quantum Cohomology, AMS University Lectures Series 6 (1994).

[O] Y.-G. OH, Floer cohomology of Lagrangian intersections and pseudo-holomorphic discs, Comm. Pure and Appl. Math. 46 (1993), 949–993.

[On] K. ONO, Legendrian intersections in pre-quantization bundles, Talk at the Symplectic Geometry Workshop at the Isaak Newton Institute, Cambridge, October 1994.

[P] M. POZNIAK, Floer homology, Novikov rings, and clean intersections, PhD thesis, University of Warwick, 1994.

[RS1] J.W. ROBBIN, D.A. SALAMON, The Maslov index for paths, Topology, 32 (1993), 827–844.

[RS2] J.W. ROBBIN, D.A. SALAMON, The Spectral flow and the Maslov index, to appear in Bulletin L.M.S.

[SZ] D.A. SALAMON, E. ZEHNDER, Morse theory for periodic orbits of Hamiltonian systems and the Maslov index, Comm. Pure Appl. Math. 45 (1992), 1303–1360.

[Sc] M. SCHWARZ, Morse Homology, Progress in Mathematics 111 (1994), Birkhäuser.

[So] J.M. SOURIAU, Groupes Différentiels, Dunod, Paris, 1970.

[W] A. WEINSTEIN, Connections of Berry and Hannay type for moving Lagrangian submanifolds, Advances in Math. 82 (1990), 133–159.

[Wi] E. WITTEN, Morse theory and supersymmetry, J. Diff. Equations 17 (1982), 661–692.

Y. Eliashberg
Dept. of Math.
Stanford University
Stanford, CA 94305
USA

H. Hofer
Mathematik
ETH-Zentrum
CH-8092 Zürich
Switzerland

D. Salamon
Mathematics Institute
University of Warwick
Coventry CV4 7AL
Great Britain

Submitted: October 1994

Geometric And Functional Analysis
Vol. 5, No. 2 (1995)
1016-443X/95/0200270-59$1.50+0.20/0

PROPERTIES OF PSEUDO-HOLOMORPHIC CURVES IN SYMPLECTISATIONS II: EMBEDDING CONTROLS AND ALGEBRAIC INVARIANTS

H. HOFER, K. WYSOCKI, E. ZEHNDER

Dedicated to M. Gromov on the occasion of his 50th birthday

1. Introduction

In the following we look for conditions on a finite energy plane $\tilde{u} := (a, u) : \mathbb{C} \to \mathbb{R} \times M$, which allow us to conclude that the projection into the manifold M, $u : \mathbb{C} \to M$, is an embedding. For this purpose we shall introduce several algebraic invariants. Finite energy planes have been introduced in [H] for the solution of A. Weinstein's conjecture about closed characteristics on three dimensional contact manifolds. In order to recall the concept, we first start with some definitions from contact geometry.

We denote by (M, λ) a three-dimensional compact manifold equipped with a contact form λ which, by definition, is a one-form on M such that $\lambda \wedge d\lambda$ is a volume form. Associated to the pair (M, λ) we have the contact structure $\xi = \operatorname{kern}(\lambda)$. The contact structure is a 2-dimensional subbundle of TM and $d\lambda|\xi \oplus \xi$ defines on every fibre a symplectic form. Hence $(\xi, d\lambda) \to M$ is a symplectic vector bundle. Further, we find a unique vector field X, called the Reeb vector field, defined by the conditions $i_X \lambda = 1$ and $i_X d\lambda = 0$. The vector field X spans a line bundle ℓ with the preferred section X. Summing up we see that a contact form λ on M defines a natural splitting

$$TM = (\ell, X) \oplus (\xi, d\lambda)$$

of the tangent bundle into a line bundle with a preferred section and a symplectic vector bundle.

A compatible complex multiplication J for the contact structure $\xi \to M$ is a smooth fibre preserving, fibrewise linear map $J : \xi \to \xi$ satisfying $J^2 = -1$ so that, in addition,

$$g_J(x)(h, k) = d\lambda(x)\big(h, J(x)k\big)$$

defines a smooth fibrewise metric for ξ. It is well known and not difficult to show that the space of all such J's equipped with the C^∞-topology is contractible, see [AH],[Gr],[HZ],[MS].

Given a J as above there is an associated almost complex structure $\tilde{J}$ on $\mathbf{R} \times M$ defined by

$$\tilde{J}(a,u)(h,k) = \big(-\lambda(u)(k), J(u)(\pi k) + hX(u)\big) .$$

Here $\pi : TM \to \xi$ is the projection along X.

Let (S,i) be a compact Riemannian surface and $\Gamma \subset S$ be a finite collection of points. We are interested in maps $\tilde{u} = (a,u) : S \setminus \Gamma \to \mathbf{R} \times M$ satisfying the first order elliptic system

$$\begin{aligned} \pi \circ Tu \circ i &= J(u) \circ \pi \circ Tu \\ (u^*\lambda) \circ i &= da . \end{aligned} \tag{1}$$

We also impose an energy condition. To do so, we denote by Σ the set of all smooth maps $\phi : \mathbf{R} \to [0,1]$ satisfying $\phi' \geqslant 0$. For $\phi \in \Sigma$ we define a 1-form λ_ϕ on $\mathbf{R} \times M$ by $\lambda_\phi(a,u)(h,k) = \phi(a)\lambda(u)(k)$. Then $E(\tilde{u})$ is defined by

$$E(\tilde{u}) = \sup \left\{ \int_{S\setminus\Gamma} \tilde{u}^* d\lambda_\phi \mid \phi \in \Sigma \right\} \tag{2}$$

We require the solution of (1) to satisfy $0 < E(\tilde{u}) < \infty$. It is important to note that the integrand in (2) is nonnegative. Indeed, in local holomorphic coordinates $s + it$ one computes, for a solution $\tilde{u} = (a,u)$,

$$\begin{aligned} 2\tilde{u}^* d\lambda_\phi = \phi'(a)[a_s^2 + a_t^2 + \lambda(u_s)^2 + \lambda(u_t)^2] ds \wedge dt \\ + \phi(a)[|\pi u_s|_J^2 + |\pi u_t|_J^2] ds \wedge dt , \end{aligned} \tag{3}$$

where we have used the norm $|h|_J^2 = g_J(h,h)$.

A nontrivial solution of (1) having a finite energy (2) will be called in the following a finite energy surface. In the special case $\mathbf{C} = S^2 \setminus \{\text{point}\}$ we call the associated solutions $\tilde{u} = (a,u)$ finite energy planes.

The equation (1) can be written in the form

$$\tilde{J} \circ T\tilde{u} = T\tilde{u} \circ i , \tag{4}$$

so that $\tilde{u} : S \setminus \Gamma \to \mathbf{R} \times M$ is a pseudo holomorphic curve. In order to describe the behavior near a puncture we first observe that a neighborhood of a puncture in S looks biholomorphically like the complement of the closed unit ball D in $\mathbf{C}$. Hence a solution of (1) and (2) gives, for every puncture, a map $\tilde{v} = (b,v) : \mathbf{C} \setminus D \to \mathbf{R} \times M$ satisfying, with the usual coordinates on $\mathbf{C}$ being $s + it$,

$$\begin{aligned} \pi v_s + J(v) \pi v_t &= 0 \\ v^* \lambda \circ i &= db \\ 0 < E(\tilde{v}) &< \infty . \end{aligned}$$

The last statement follows from the following observation: if the integral in (2) over an open set $U \subset S \setminus \Gamma$ vanishes, then $\tilde{u}_{|U}$ is constant, in view of (3), and hence, by the continuation theorem (similarity principle), the solution

$\tilde{u}$ of (4) must be a constant, in contradiction to $E(\tilde{u}) > 0$. We distinguish two different types of punctures in Γ. Namely a puncture is a

removable singularity: if the $\mathbf{R}$-component of $\tilde{v}$ is bounded near the puncture.

non-removable singularity: if the $\mathbf{R}$-component of $\tilde{v}$ is unbounded near the puncture.

It follows from Gromov's removable singularity theorem, that in the first case $\tilde{v}$ can be smoothly extended over the puncture. Indeed, the image of the neighbourhood of the puncture is mapped into a compact region $K \subset \mathbf{R}\times M$. On K the almost complex structure $\tilde{J}$ is calibrated by a symplectic structure of the form $d(\phi\lambda)$. Therefore the removable singularity theorem applies, see [Gr] and [Au],[MS]. We shall assume in the following that all punctures in Γ are non-removable. The concept "singularity" will refer to "non-removable singularity". We should note that $\Gamma \neq \emptyset$ if there exists a non-constant finite energy surface $\tilde{u} : S \setminus \Gamma \to \mathbf{R} \times M$. Indeed, if $\Gamma = \emptyset$ then $\tilde{u} : S \to \mathbf{R} \times M$, and since S has no boundary we conclude by Stokes theorem and by (3) that $\tilde{u}^* d\lambda_\phi = 0$, so that $\tilde{u}$ is a constant map, in contradiction to $E(\tilde{u}) > 0$. It follows from the analysis in [H],[HWyZ3], that for a solution $\tilde{u}$ the set Γ can be split into two distinguished subsets. Namely a puncture in Γ is called a

negative (non-removable) singularity if the $\mathbf{R}$-component is unbounded, but bounded from above.

positive (non-removable) singularity if the $\mathbf{R}$-component is unbounded, but bounded from below.

In particular, there are no punctures for which the $\mathbf{R}$-component of a solution $\tilde{u} = (a, u)$ is unbounded both from above and below. We shall denote in the following by Γ^+ the set of positive singularities and by Γ^- the set of negative singularities, so that $\Gamma = \Gamma^- \cup \Gamma^+$. Let us first recall the argument which leads to this classification of punctures. Near a puncture in Γ we consider the solution $\tilde{v}$ (4) in holomorphic cylindrical coordinates and set $\tilde{u}(s,t) = \tilde{v}(e^{2\pi(s+it)})$, $s \geqslant 0$ and $t \in \mathbf{S}^1$, and write $\tilde{u} = (a, u)$. From $E(\tilde{u}) < \infty$ we conclude by Stokes theorem that the limit

$$T := \lim_{s\to\infty} \int_{\mathbf{S}^1} u(s)^* \lambda \in \mathbf{R} \tag{5}$$

exists. Now, let us assume at first that $T \neq 0$. Then we claim that

$$\frac{\partial}{\partial s} a(s,t) \to T \ , \tag{6}$$

as $s \to \infty$, so that in this case the sign of T determines the type of singularity. In order to prove (6) we argue by contradiction, and assume that for a sequence (s_k, t_k)

$$\left|\frac{\partial}{\partial s}a(s_k,t_k)-T\right| \geqslant \epsilon\,, \qquad s_k \to \infty\,, \tag{7}$$

for some $\epsilon > 0$. Define a sequence of pseudo holomorphic maps $\tilde{u_k}$: $[-s_k, \infty) \times \mathbb{S}^1 \to \mathbb{R} \times M$ by

$$\tilde{u_k} = (a_k, u_k) = \big(a(s+s_k,t) - a(s_k,t_k), u(s+s_k,t)\big)\ .$$

Then by the uniform gradient estimates for solutions of (1) proved in [H] we conclude that $\tilde{u_k} \to \tilde{v}$ in C^∞_{loc} for a smooth pseudo holomorphic curve $\tilde{v} : \mathbb{R} \times \mathbb{S}^1 \to \mathbb{R} \times M$, defined on the whole cylinder $\mathbb{R} \times M$ and satisfying

$$\tilde{v_s} + \tilde{J}(\tilde{v})\tilde{v_t} = 0$$

$$\int_{\mathbb{R}\times\mathbb{S}^1} \tilde{v}^* d\lambda = 0$$

$$E(\tilde{v}) < \infty\ .$$

It follows from the classification in the appendix that $\tilde{v}(s,t) = (Ts + d, x(Tt))$ for some constant $d \in \mathbb{R}$ and some T-periodic solution $x(t)$ of the Reeb vector field $\dot{x} = X(x)$ on M. Consequently,

$$\frac{\partial}{\partial s}a_k(0,t) = \frac{\partial}{\partial s}a(s_k, s_k) \to T$$

in contradiction to (7). The case $T = 0$ requires a more subtle argument which shows that in this case the function a is bounded, so that the puncture is a removable singularity. We, again, proceed by contradiction and assume e.g. that a is not bounded from above. Then given any $R > 0$ sufficiently large and any $b > 0$ we find $r_1 \geqslant R$ and $r_2 \geqslant r_1 + b$ which are regular values of the function $a : [0,\infty) \times \mathbb{S}^1 \to \mathbb{R}$. Then $\Omega = a^{-1}([r_1, r_2])$ is a compact surface with boundary, the two boundary components will be denoted by ∂^- and ∂^+. Now $\tilde{u}_{|\Omega} : \Omega \to \mathbb{R} \times M$ is a pseudo holomorphic curve with $\tilde{u}(\partial^-) \subset \{r_1\} \times M$ and $\tilde{u}(\partial^+) \subset \{r_2\} \times M$. Moreover, from $E(\tilde{u}) < \infty$ one concludes using Stokes theorem, that

$$E_\Omega(\tilde{u}) := \int_\Omega \tilde{u}^* d(\phi\lambda) = \int_{\partial^+ - \partial^-} \tilde{u}^*(\phi\lambda) \to 0 \tag{8}$$

as $r_1 \to \infty$, independently of b, for every bounded function $\phi : \mathbb{R} \times M \to \mathbb{R}$. In view of the compactness of M and the translation invariance of $\tilde{u}$ one can show that (8) contradicts Gromov's isoperimetric inequality for pseudo holomorphic curves.

The behavior of a pseudo holomorphic curve $\tilde{v}$ near a puncture, which is assumed to be a non-removable singularity, has been studied in [H],[HWyZ3]. It is intimately related to the dynamics of the Reeb vector field X on (M, λ). If $\tilde{v}$ satisfies (5), then there exists a sequence $R_k \to \infty$, a number $T > 0$ and a T-periodic solution $\dot{x} = X(x)$ on M such that, as $k \to \infty$,

$$v(R_k e^{2\pi i t}) \to x(Tt)$$

in $C^\infty(\mathbb{S}^1, M)$, see [H]. Moreover, if the T-periodic solution $x(t)$ is non-degenerate, then the limit exists for $R \to \infty$ and the convergence is of an exponential nature, see [HWyZ3]. Therefore, if the periodic solutions of X on M are non-degenerate, a solution $\tilde{u}$ of (1) and (2) gives a smooth map $u : S \setminus \Gamma \to M$ which converges, near the singular points, exponentially to periodic solutions of the Reeb vector field on M, if one looks at the images of small concentric circles around the punctures.

As we shall see, the geometry of a finite energy surface is quite intricate and some of its properties can be encoded into algebraic invariants. One of the key technical steps in understanding the geometry of finite energy surfaces is the study of the behavior near a puncture as given in [HWyZ3]. Let $\dot{S} = S \setminus \Gamma$. For example given a $\tilde{u} : \dot{S} \to \mathbf{R} \times M$ we may view $\pi \circ Tu$ as a section of the bundle $\mathrm{Hom}_{\mathbf{C}}(T\dot{S}, u^*\xi) \to \dot{S}$. It will turn out that if $\pi \circ Tu$ does not vanish at one point, then this section has only isolated zeros, which all have a strictly positive index. From the results in [HWyZ3] we know that in the above case $\pi \circ Tu(z)$ is always nonzero for z sufficiently close to the punctures. Making use of the results in [HWyZ3], we shall define (under some weak topological hypothesis) an asymptotic winding number $\mathrm{wind}_\infty(u, z)$, where z is a puncture. This number describes the way in which the image of a small concentric circle around z winds around the asymptotic limit $P_u = \{x(t); t \in \mathbf{R}\}$. We can also define another winding number $\mathrm{wind}_\pi(u, z)$. It measures how the section $\tau \to \pi \circ Tu(\tau)$ winds in $\mathrm{Hom}_{\mathbf{C}}(T\dot{S}, u^*\xi)$ as τ varies on a small concentric circle around z. One of our goals is to establish some inequalities for the different winding numbers associated to the punctures. We shall also relate these algebraic and geometric invariants to the dynamical invariant $\mu(x, u) \in \mathbf{Z}$, associated to the non-degenerate periodic orbit $x(t)$ of the Reeb vector field, which appears in the limit. The Maslov-type index $\mu(x, u)$ is a winding in $Sp(1)$ and is defined by means of the linearized Reeb vector field X along the periodic orbit, see section 3.

In order to describe some of the results in detail we consider the special case $\tilde{u} = (a, u) : \mathbf{C} = \mathbb{S}^2 \setminus \{\text{point}\} \to \mathbf{R} \times M$ of a finite energy plane. It turns out that in this case the puncture belongs to Γ^+. We shall assume that the limit orbit $x(t)$ is non-degenerate. We then call the finite energy plane non-degenerate and abbreviate the set $P_u = \{x(t); t \in \mathbf{R}\} \subset M$. In section 2 we shall prove:

THEOREM 1.1. *Let $\tilde{u} = (a, u) : \mathbf{C} \to \mathbf{R} \times M$ be a non-degenerate finite energy plane. If $u(\mathbf{C}) \cap P_u = \emptyset$ and if $\tilde{u}$ is somewhere injective, then the map $u : \mathbf{C} \to M$ is injective.*

The proof makes use of the precise asymptotic behavior of $\tilde{u}$ as $|z| \to \infty$ studied in [HWyZ3], and of D. McDuff's intersection theory of pseudo holomorphic curves in [M]. In the section 3 and 4 we introduce some algebraic and dynamical invariants and use them to prove

THEOREM 1.2. *Let $\tilde{u} = (a, u) : \mathbb{C} \to \mathbb{R} \times M$ be a non-degenerate finite energy plane. Then its asymptotic limit $x(t)$ is a contractible loop in M having an index $\mu(x, u) \geqslant 2$. Moreover, if $\mu(x, u) \leqslant 3$ and if $\tilde{u}$ is somewhere injective and $u(\mathbb{C}) \cap P_u = \emptyset$, then the map $u : \mathbb{C} \to M \setminus P_u$ is an embedding.*

The next statement does not require that the energy surface does not intersect with its limit P_u. It is more subtle and makes use of Giroux's elimination lemma. Moreover, we shall restrict the class of contact structures considered.

THEOREM 1.3. *Let M be a compact, connected and oriented 3-manifold equipped with a tight contact form λ. Assume $\tilde{u} = (a, u) : \mathbb{C} \to \mathbb{R} \times M$ is a finite energy plane with non-degenerate limit P_u having covering number $cov(u) = k$. Assume P_u is k-unknotted with respect to the homotopy class of u. If $\mu(x, u) \leqslant 3$, then $u(\mathbb{C}) \cap P_u = \emptyset$ and $u : \mathbb{C} \to M \setminus P_u$ is an embedding.*

We shall also study the intersections of two finite energy planes $\tilde{u}$ and $\tilde{v}$ having the same non-degenerate asymptotic limit x. We define a map $u \sharp v : \mathbb{S}^2 = \mathbb{C} \cup_{\mathbb{S}^1_\infty} \overline{\mathbb{C}} \to M$ by gluing the domain $\mathbb{C}$ of v with reversed orientation to the domain $\mathbb{C}$ of u along the circle x at infinity. Let $c(u, v) \in \mathbb{Z}$ be the Chern number of the complex bundle $(u \sharp v)^* \xi \to \mathbb{S}^2$ evaluated at the fundamental class of the two-sphere.

THEOREM 1.4. *Let $\tilde{u} = (a, u)$ and $\tilde{v} = (b, v) : \mathbb{C} \to \mathbb{R} \times M$ be two finite energy planes having the same non-degenerate asymptotic limit $x(t)$. Abbreviating $P = \{x(t); t \in \mathbb{R}\} \subset M$ we assume*

$$c(u, v) = 0$$
$$cov(u) = cov(v)$$
$$\mu(x, u) = \mu(x, v) \leqslant 3$$
$$u(\mathbb{C}) \cap P = \emptyset = v(\mathbb{C}) \cap P$$

Then $u, v : \mathbb{C} \to M \setminus P$ are embeddings, and

$$u(\mathbb{C}) = v(\mathbb{C}) \quad \text{or} \quad u(\mathbb{C}) \cap v(\mathbb{C}) = \emptyset \, .$$

For results on general finite energy surfaces we refer to section 5. The results will be useful in our applications of pseudo holomorphic curves to problems in low dimensional topology and Hamiltonian dynamics, see [CoZ],[E1-3]. In particular, we shall use these methods in order to construct open book decompositions in certain three-manifolds, as well as global surfaces of sections for Hamiltonian flows on 3-dimensional energy surfaces in $\mathbf{R}^4$, see [HWyZ1,2].

2. Injectivity of Non-degenerate Finite Energy Planes

We concentrate at first on Riemannian spheres with one puncture, that is, on $\mathbf{C}$. We study a map

$$\tilde{u} = (a, u) : \mathbf{C} \to \mathbf{R} \times M ,$$

which is non-constant and solves the partial differential equation:

$$\begin{aligned} \pi u_s + J(a)\pi u_t &= 0 \\ u^*\lambda \circ i &= da , \end{aligned}$$

where $\pi = \pi_u$ is the projection onto the contact plane in the tangent space at u. We require in addition,

$$0 < E(\tilde{u}) < \infty .$$

Then there exists a sequence $R_k \to \infty$ such that $u(R_k e^{2\pi i t}) \to x(Tt)$ uniformly in t, where $x(t)$ is a T periodic solution of the Reeb vector field $\dot{x} = X(x)$ on M, associated with the contact structure. The period T is equal to the energy

$$T = E(\tilde{u}) .$$

This is proved in [H]. If one assumes, in addition, that this T-periodic solution $x(t)$ is isolated in the set of all periodic solutions of $\dot{x} = X(x)$ having periods close to T, then the finite energy plane has $x(t)$ as asymptotic limit:

$$u(Re^{2\pi i t}) \to x(Tt) , \quad R \to \infty ,$$

uniformly in t. This is Theorem 1.2 in [HWyZ3]. Clearly, the T-periodic solution is isolated, if it is non-degenerate in the sense that it has only one Floquet multiplier equal to 1. In this case we call the finite energy plane $\tilde{u}$ non-degenerate.

The period T need not be the minimal period of $x(t)$. We shall denote by $T(u)$ the minimal period of the asymptotic periodic solution belonging to the finite energy plane $\tilde{u}$. Then $T = kT(u)$, $k \geqslant 1$ and $k \in \mathbf{N}$. This integer k we shall call the asymptotic covering number of u and denote it by $cov(u)$. We have

$$cov(u) = \frac{\int_{\mathbf{C}} u^* d\lambda}{\int_{P_u} \lambda} ,$$

where $P_u \equiv x(\mathbf{R}) \subset M$. This follows from $\lambda(X) = 1$, indeed abbreviating $\tau = T(u)$,

$$\int_{P_u} \lambda = \int_0^T x^*\lambda\, dt = \int_0^T \lambda(\dot{x})dt = \int_0^T \lambda(X)dt = \tau .$$

The periodic orbit $P_u \subset M$ gives a pseudo holomorphic cylinder $Z_u = \mathbf{R} \times P_u \subset \mathbf{R} \times M$ defined by the mapping $\tilde{v} = (a, v) : \mathbf{C} \to \mathbf{R} \times M : \tilde{v}(s,t) = (s, x(t))$, where $z = s + it \in \mathbf{C}$ and where $x(t+T) = x(t)$ for all $t \in \mathbf{R}$. One verifies readily that indeed $\tilde{v}_s + J(v)\tilde{v}_t = 0$. This special holomorphic map is trivial in the sense that $\int_{\mathbf{C}} v^* d\lambda = 0$. We introduce the following definition.

DEFINITION 2.1. *A finite energy plane $\tilde{u}$ is called somewhere injective if there exists a point z_0 satisfying*

$$\tilde{u}^{-1}\big(\tilde{u}(z_0)\big) = \{z_0\}$$

and

$$T\tilde{u}(z_0) \neq 0 .$$

Similarly we define a somewhere injective finite energy surface.

Remark 2.2: The notion of "somewhere injective" is well known from the study of pseudo holomorphic spheres. Namely, a pseudo holomorphic sphere is either multiply covered or somewhere injective, see for example [MS].

The aim of this section is to prove the following

THEOREM 2.3. *Let $\tilde{u} := (a, u) : \mathbf{C} \to \mathbf{R} \times M$ be a non-degenerate finite energy plane. Assume that $\tilde{u}(\mathbf{C}) \cap Z_u = \emptyset$ (or equivalently $u(\mathbf{C}) \cap P_u = \emptyset$) and that $\tilde{u}$ is somewhere injective. Then the map $u : \mathbf{C} \to M$ is injective.*

The proof of this theorem is based on an asymptotic formula for $\tilde{u}(z)$ as $|z| \to \infty$, which we recall from [HWyZ3]. It is assumed that $\tilde{u}$ is a non-degenerate finite energy plane, such that $u(Re^{2\pi it}) \to x(Tt)$ as $R \to \infty$. Let $\varphi : \mathbf{R} \times \mathbf{S}^1 \to \mathbf{C} \setminus \{0\}$ be the biholomorphic map $\varphi(s,t) = e^{2\pi(s+it)}$. In the following $\mathbf{S}^1$ is taken to be $\mathbf{S}^1 = \mathbf{R}/\mathbf{Z}$. Introducing $\tilde{v} = \tilde{u} \circ \varphi := (a, v) : \mathbf{R} \times \mathbf{S}^1 \to M$ the finite energy plane becomes a holomorphic cylinder satisfying $\tilde{v}_s + \tilde{J}(\tilde{v})\tilde{v}_t = 0$, and

$$v(s,t) \to x(Tt) \text{ as } s \to \infty ,$$

with convergence in $C^\infty(\mathbf{S}^1)$. In a tubular neighbourhood of the periodic orbit $x(t)$ in M there exist local coordinates $(\vartheta, z) \in \mathbf{R} \times \mathbf{R}^2$, where $\mathbf{R}$ is the covering space of $\mathbf{S}^1 = \mathbf{R}/\mathbf{Z}$, in which the periodic orbit $x(t)$ lies on $\{(\vartheta, 0)\} = \mathbf{R} \times \{0\}$. The map is represented by functions $(a, v) := (a(s,t), \vartheta(s,t), z(s,t)) \in \mathbf{R}^4$ smoothly defined on $[s_0, \infty) \times \mathbf{R}$ and satisfying

$$v(s,t) \to (kt, 0) \quad \text{as} \quad s \to \infty ,$$

uniformly in t. The functions a and z are 1-periodic in t while ϑ satisfies

$\vartheta(s,t+1)=\vartheta(s,t)+k$. As proved in [HWyZ3], there exist constants $c\in\mathbf{R}$ and $d>0$, such that

$$\left|\partial^\beta[a(s,t)-Ts-c]\right|\leqslant M\,e^{-ds}$$
$$\left|\partial^\beta[\vartheta(s,t)-kt]\right|\leqslant M\,e^{-ds}$$

for all multi-indices β, with constants $M=M_\beta$. Moreover, for $z(s,t)\in\mathbf{R}^2$ we have the asymptotic formula

$$z(s,t)=e^{\int_{s_0}^{s}\gamma(\sigma)d\sigma}\left[e(t)+r(s,t)\right] .$$

Here $\partial^\beta r(s,r)\to 0$ as $s\to\infty$ uniformly in t, for all derivatives. The function $\gamma:[s_0,\infty)\to\mathbf{R}$ is smooth and $\gamma(s)\to\lambda<0$ as $s\to\infty$. The negative number λ is an eigenvalue of a selfadjoint operator in $L^2(\mathbb{S}^1,\mathbf{R}^2)$ related to the linearized Reeb-vector field X along the periodic solution $x(t)$, in the local coordinates given by

$$A=-J_0\frac{d}{dt}-S_\infty(t)\ , \tag{11}$$

where $S_\infty(t)$ is a smooth, symmetric 1-periodic 2×2 matrix-function equivalent to $S_\infty(t)=-J_0\pi_m dX(m)\pi_m$, where $m=(kt,0)\in P_u$. The function $e(t)=e(t+1)\neq 0\in\mathbf{R}^2$ is an eigenvector of the operator A belonging to the eigenvalue λ. We also recall from [HWyZ3] that, if we identify $T_m\mathbf{R}^3$ with $\mathbf{R}^3$, the contact plane ξ_m is spanned by the basis $(0,1,0)$ and $(-x,0,1)$ at the point $m=(\vartheta,x,y)\in\mathbf{R}^3$. With respect to this basis we have for the map v at the point $v(s,t)$

$$\pi_v\circ Tv\left(\frac{\partial}{\partial s}\right)=\gamma(s)e^{\int_{s_0}^{s}\gamma(\sigma)d\sigma}\left[e(t)+\hat r(s,t)\right]\in\mathbf{R}^2\ ,$$

where $\partial^\beta\hat r(s,t)\to 0$ as $s\to\infty$, uniformly in t, for all derivatives β. Therefore, if $v=v(s,t)$,

$$\frac{\pi_v\circ Tv\left(\frac{\partial}{\partial s}\right)}{\left\|\pi_v\circ Tv\left(\frac{\partial}{\partial s}\right)\right\|_J}\to\rho(t)e(t)\in\xi_{x(t)}$$

as $s\to\infty$, for some smooth nonvanishing function $\rho(t)=\rho(t+1)$. Clearly

$$\pi\circ Tv(s,t)\neq 0$$

for all s sufficiently large. After these preliminaries we are ready for the proof of Theorem 2.3.

Proof: Define, for $\tau\in\mathbf{R}$, the holomorphic map $\tilde u_\tau(z)=(a(z)+\tau,u(z))$, so that $\tilde u_0=\tilde u$ if $\tau=0$. We first show that there exists $\tau_*>0$, such that

$$\tilde u_\tau(\mathbf{C})\cap\tilde u(\mathbf{C})=\emptyset\quad\text{if}\quad|\tau|\geqslant\tau_*\ . \tag{12}$$

Arguing indirectly we find sequences (z_k) and (z'_k) in $\mathbf{C}$ and a sequence $\tau_k \in \mathbf{R}$ satisfying $|\tau_k| \to \infty$ and $\tilde{u}_{\tau_k}(z'_k) = \tilde{u}(z_k)$. We may assume that $\tau_k \to \infty$. Hence

$$u(z'_k) = u(z_k)$$
$$a(z'_k) + \tau_k = a(z_k)$$

and $\tau_k \to \infty$. In view of its asymptotic behavior the function $a(z)$ is bounded from below and we conclude $a(z_k) \to \infty$. This implies that $|z_k| \to \infty$ and hence $u(z_k) \to P_u$. Consequently also $u(z'_k) \to P_u$. By assumption, $u(\mathbf{C}) \cap P_u = \emptyset$. Consequently, also $|z'_k| \to \infty$. Summarizing we have, so far,

$$\begin{aligned} u(z'_k) &= u(z_k)\ , \\ a(z'_k) + \tau_k &= a(z_k) \end{aligned} \tag{13}$$

where τ_k, $|z_k|$ and $|z'_k|$ converge to ∞. Since for k large we are in a neighbourhood of P_u we can use the local coordinates, near P_u and the asymptotic description in the cylinder coordinates $z_k = e^{2\pi(s_k+it_k)}$. In the local coordinates $v(s,t)$ represents $u(e^{2\pi(s+it)})$. We have $s_k \to \infty$ and $s'_k \to \infty$. From the asymptotic behavior of the function a we find that $\tau_k = T(s_k - s'_k) + \varepsilon_k$ where $\varepsilon_k \to 0$. Therefore,

$$s_k - s'_k \to \infty\ . \tag{14}$$

From $z(s_k, t_k) = z(s'_k, t'_k)$ we find

$$e^{-\int_{s'_k}^{s_k} \gamma(\sigma)d\sigma}\left[e(t'_k) + \varepsilon'_k\right] = \left[e(t_k) + \varepsilon_k\right] \tag{15}$$

with $\varepsilon_k \to 0$ and $\varepsilon'_k \to 0$. Since the functions are periodic in t we may assume that $t_k \to t_* \in [0,1)$ and $t'_k \to t'_* \in [0,1)$. From $\vartheta(s_k, t_k) = \vartheta(s'_k, t'_k) \bmod 1$ and the asymptotic behavior of the function ϑ we conclude

$$k(t_* - t'_*) = 0 \bmod 1\ . \tag{16}$$

Hence we have

$$t_* - t'_* = \frac{j}{k}\ , \tag{17}$$

where $j \in \{0, 1, \ldots, k-1\}$. If $e(t_*)$ is not a positive multiple of $e(t'_*)$ we obtain a contradiction from equation (15). Hence, we may assume that $e(t_*)$ is a positive multiple of $e(t'_*)$, say

$$e(t_*) = pe(t'_*)\ , \tag{18}$$

where $p > 0$. Then it follows from (17) and the fact that the asymptotic limit is k-fold covered, that $e(t_*) = e(t'_*)$. Indeed, define $f(t) = e(t) - pe(t - \frac{j}{k})$. In view of (17) and (18) we have $f(t_*) = 0$. Recalling the definition of the operator A from (11) and using that $S_\infty(t)$ is $\frac{1}{k}$-periodic we see that

$Af = \lambda f$. Since $f(t_*) = 0$, this implies $f = 0$. Taking the norm in equation (15) we find

$$\lim_{k\to\infty} \int_{s'_k}^{s_k} \gamma(\tau)d\tau = 0 .$$

Since $\gamma(s) \to \lambda < 0$ as $s \to \infty$ we conclude that $s_k - s'_k \to 0$ contradicting (14). We have proved the statement (12).

Next we claim that for every given $\varepsilon > 0$ there exists an $R = R(\epsilon) > 0$, such that if

$$\tilde{u}_\tau(z') = \tilde{u}(z) \quad \text{and} \quad |\tau| \geqslant \varepsilon , \tag{19}$$

then $|z|, |z'| \leqslant R$.

This follows again from the asymptotics. Indeed, arguing indirectly we find unbounded sequences (z_k) and (z'_k) in $\mathbb{C}$ and a sequence $\tau_k \in \mathbb{R}$ satisfying $|\tau_k| \geqslant \varepsilon$ such that $\tilde{u}_{\tau_k}(z'_k) = \tilde{u}(z_k)$. In view of (12) the sequence τ_k is bounded and we may assume that $\tau_k \to \tau \geqslant \varepsilon$. Then, as above, $s_k - s'_k \to \frac{\tau}{T} \neq 0$ and $s_k - s'_k \to 0$. This is absurd and hence the claim (19) is proved.

Using (12) and (19) we shall conclude that

$$\tilde{u}_\tau(\mathbb{C}) \cap \tilde{u}(\mathbb{C}) = \emptyset \quad \text{if} \quad \tau \neq 0 . \tag{20}$$

Indeed, the intersection number $\text{Int}(\tilde{u}_\tau, \tilde{u})$ of the two holomorphic curves is well defined for $\tau \neq 0$ and, in view of the homotopy invariance of the intersection number we conclude from (12) that $\text{Int}(\tilde{u}_\tau, \tilde{u}) = 0$. But, by the results due to D. McDuff [M], the intersections of different pseudo holomorphic curves are isolated and have a positive local intersection index. Consequently there are no intersections for $\tau \neq 0$ and (20) is proved. We conclude:

$$u(z) = u(z') \Longleftrightarrow \tilde{u}(z) = \tilde{u}(z') . \tag{21}$$

Hence in order to prove the theorem we have to exclude self-intersection points of our pseudo holomorphic curve $\tilde{u}$. We first recall (Theorem 1.5 in [HWyZ3]) that for a non-degenerate energy plane the set of points $\{z \in \mathbb{C} \mid T\tilde{u}(z) = 0\}$ is finite. Consider now the set of points

$$S = \{z \in \mathbb{C} \mid T\tilde{u}(z) \neq 0 \text{ and } (*) \text{ holds}\} ,$$

where property $(*)$ is given by:

There exists a point $z' \neq z$ and sequences (z_k) and (z'_k) satisfying $z_k \to z$, $z'_k \to z'$ and $\tilde{u}(z'_k) = \tilde{u}(z_k)$.

(The elements of the sequence are, of course, assumed to be different from the limit.) Clearly, the set S is closed in the set $S' = \{z \in \mathbb{C} \mid T\tilde{u}(z) \neq 0\}$, and we claim that it is also open in S'. Indeed choosing a Darboux chart, locally in a neighbourhood of $u(z') = u(z) \in M$ one shows, by means of the similarly principle proceeding as in the proof of Theorem 4 in [FHS], that there are open neighbourhoods of z and z' which have the same image under $\tilde{u}$. Consequently the set S is either empty or agrees with S'. By our assumption, there exists a point $z_0 \in \mathbf{C}$ satisfying $\tilde{u}^{-1}(\tilde{u}(z_0)) = \{z_0\}$ and $T\tilde{u}(z_0) \neq 0$. Therefore, $S = \emptyset$ and the intersection points must be isolated. By the results of D. McDuff [M] the local index of an isolated intersection of two pseudo holomorphic curves is positive. Let $z \neq z'$ be two points satisfying $\tilde{u}(z) = \tilde{u}(z')$. Since the intersection is isolated and has a positive index we find that for small $\tau \neq 0$ we necessarily have an intersection between $\tilde{u}(\mathbf{C})$ and $\tilde{u}_\tau(\mathbf{C})$. This however, contradicts (20) and proves that $\tilde{u}$ has no self-intersection points. Hence, in view of (21), the map $u : \mathbf{C} \to M \setminus P_u$ is injective, and the proof of Theorem 2.3 is complete.

3. The μ-index and Winding Numbers

The next aim is to associate with a non-degenerate periodic orbit $x(t)$ of the Reeb-vector field X on the three dimensional manifold M two integers which are homotopy invariants. The first integer is a Maslov-type index μ, sometimes called the Conley-Zehnder index; it is a winding number in the symplectic group Symp(1). The second index describes a geometrical winding in the contact planes along the orbit $x(t)$.

In order to briefly recall the μ-index (from [CoZ],[RS],[SZ]) we introduce in the 2-dimensional vector space $\mathbf{R}^2$ the symplectic form $\omega = \langle J\cdot, \cdot\rangle$ determined by the matrix

$$J = \begin{pmatrix} 0 & -1 \\ 1 & 0 \end{pmatrix} . \tag{22}$$

The standard $2n$-dimensional symplectic vector space is then $\mathbf{R}^{2n} = \mathbf{R}^2 \oplus \ldots \oplus \mathbf{R}^2$ with its symplectic form $\omega \oplus \ldots \oplus \omega$.

A continuous path $S(t)$, $0 \leqslant t \leqslant 1$ of symmetric matrices $S(t) \in \mathcal{L}(\mathbf{R}^{2n})$ generates an arc $\Phi(t)$, $0 \leqslant t \leqslant 1$ in the symplectic group Symp(n) as the solution of

$$\frac{d}{dt}\Phi(t) = JS(t)\Phi(t) , \qquad \Phi(0) = \mathbf{1} . \tag{23}$$

The arc starts at the identity in Symp(n). Conversely, every path $\Phi(t)$ in Symp(n) starting at the identity, defines the path $S(t)$ of symmetric matrices by $S(t) = -J\dot{\Phi}(t)\Phi(t)^{-1}$. We shall consider non-degenerate arcs and define Symp$(n)^*$ as the subset of symplectic matrices M which do not have 1 in their spectrum. We let $\Sigma(n)$ be the collection of all smooth arcs $\Phi\colon [0,1] \to$ Symp(n) starting at the identity, $\Phi(0) = \text{Id}$, and ending in

$\mathrm{Symp}(n)^*$, i.e. $\Phi(1) \in \mathrm{Symp}(n)^*$. Two such arcs will be called equivalent if they are homotopic in $\Sigma(n)$. The collection of equivalence classes will be denoted by $\tilde{\Sigma}(n)$. The μ-index is a map

$$\mu \colon \Sigma(n) \to \mathbf{Z} , \tag{24}$$

which is a homotopy invariant and induces a bijection $\mu \colon \tilde{\Sigma}(n) \to \mathbf{Z}$. It is defined as follows. Given $\Phi \in \Sigma(n)$, we extend this arc within $\mathrm{Symp}(n)^*$ to either M^+ or M^-, where

$$\begin{aligned} M^- &= -\,\mathrm{Id} \in \mathrm{Symp}(n)^* \\ M^+ &= \begin{pmatrix} \frac{1}{2} & 0 \\ 0 & 2 \end{pmatrix} \oplus \mathbf{1} \oplus \ldots \oplus \mathbf{1} . \end{aligned} \tag{25}$$

Now recall that every symplectic matrix has a unique polar-decomposition $M = PO$, with $P \in \mathrm{Symp}(n)$ positive definite and $O \in \mathrm{Symp}(n)$ orthogonal. Hence

$$O = \begin{pmatrix} u_1 & -u_2 \\ u_2 & u_1 \end{pmatrix} , \qquad u = u_1 + iu_2 \in U(n) , \tag{26}$$

where $U(n)$ is the unitary group in $\mathbf{C}^n$. To our arc $\Phi(t) = P(t)O(t)$ starting at the identity and ending in $\{M^-, M^+\}$, we choose a continuous winding number $\alpha(t)$ satisfying $\det u(t) = e^{i\alpha(t)}$ and define the integer

$$\mu(\Phi) = \frac{1}{\pi}\big(\alpha(1) - \alpha(0)\big) \in \mathbf{Z} . \tag{27}$$

It can be shown that $\mu(\Phi)$ does not depend on the continuation to $\{M^+, M^-\}$ of the arc. Moreover, it is a homotopy invariant in $\Sigma(n)$ and $\mu \colon \tilde{\Sigma}(n) \to \mathbf{Z}$ is a bijection onto $\mathbf{Z}$. For our purpose it will be more convenient to describe the map μ axiomatically.

Let $G(n)$ be the fundamental group of $\mathrm{Symp}(n)$ with base point Id. Then $G(n) \simeq \mathbf{Z}$, as is well known. An explicit isomorphism is given by the Maslov-index

$$m^n \colon G(n) \to \mathbf{Z} , \tag{28}$$

defined as follows. The loop

$$\alpha^1 \colon t \to e^{2\pi t J} = \begin{pmatrix} \cos(2\pi t) & -\sin(2\pi t) \\ \sin(2\pi t) & \cos(2\pi t) \end{pmatrix}$$

in $\mathrm{Symp}(1)$, $0 \leqslant t \leqslant 1$, generates $G(1)$ and defines the class $\tilde{\alpha}^1$. Denote by C the constant loop $t \mapsto \mathrm{Id}_{\mathbf{R}^2}$, and with $\alpha^n(t) = \alpha(t) \oplus C \oplus \ldots \oplus C$, where $(n-1)$ copies of C are taken. The associated class in $G(n)$ is denoted by $\tilde{\alpha}^n$. The Maslov isomorphism (28) is determined by the requirement $\tilde{\alpha}^n \mapsto 1$. Next we observe that $G(n)$ operates on $\tilde{\Sigma}(n)$ via

$$G(n) \times \tilde{\Sigma}(n) \to \tilde{\Sigma}(n) \colon \big([g], [\Phi]\big) \mapsto [g\Phi] , \tag{29}$$

where $g\Phi(t) = g(t)\Phi(t)$, for $0 \leqslant t \leqslant 1$. Recall that $g(0) = \mathrm{Id} = g(1)$ since it represents an element in $G(n)$. The operation $(g, g') \mapsto g \oplus g'$ defines a natural map $G(n) \oplus G(m) \to G(n+m)$, similarly we have a natural map $\Sigma(n) \times \Sigma(m) \to \Sigma(n+m)$. Observe that with $\Phi \in \Sigma(n)$, we also have $\Phi^{-1} \in \Sigma(n)$, where $\Phi^{-1}(t) = \Phi(t)^{-1}$. The μ-index is now axiomatically characterized as follows.

THEOREM 3.1. *There exists a unique family of maps $\mu^n : \Sigma(n) \to \mathbf{Z}$, $n = 1, 2, \ldots$ which are homotopy invariant and which has the following properties:*

1. $$\mu^{n+m}(\Phi \oplus \Psi) = \mu^n(\Phi) + \mu^n(\Psi) ,$$
   for all $\Phi \in \Sigma(n)$, $\Psi \in \Sigma(m)$.
2. $$\mu^n(g\Phi) = 2m^n(g) + \mu^n(\Phi) ,$$
   for all $g \in G(n)$ and $\Phi \in \Sigma(n)$.
3. $$\mu^n(\Phi) = -\mu^n(\Phi^{-1}) ,$$
   for all $\Phi \in \Sigma(n)$. Finally, the following normalization holds
4. $$\mu^1(\Phi_0) = 1 ,$$
   where $\Phi_0 \in \Sigma(1)$ is given by $\Phi_0(t) = \exp(\pi t J)$, $0 \leqslant t \leqslant 1$.

Proof: The proof follows from the construction of μ in [CoZ], see also [RS] and [SZ] and will only be sketched. By the homotopy invariance of μ, the proof can be reduced to the case $n = 1$. Note that the eigenvalues of a symplectic matrix $M \in \mathrm{Symp}(1)$ occur in pairs: either $(\lambda, \bar{\lambda})$ and $\lambda\bar{\lambda} = 1$, in which case M is called elliptic or $(\mu, 1/\mu)$ and $\mu \in \mathbf{R}$, in which case M is called hyperbolic.

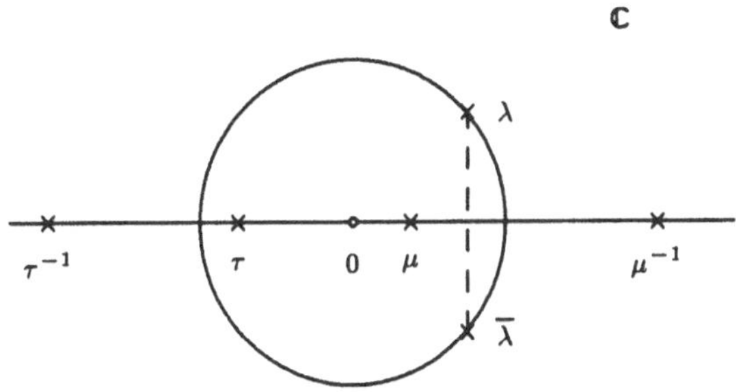

Assume the arc $\Phi \in \Sigma(1)$ ends at an elliptic point or at a hyperbolic point with negative eigenvalues. Then it can be extended in $\mathrm{Symp}(1)^*$ to $M^- = \mathbf{1}$ and we see that there is a unique $k \in \mathbf{Z}$, such that Φ is homotopic in $\Sigma(1)$ to the arc $e^{i(2k+1)\pi t}$, where we use complex notation. This arc is equal to $g \cdot \Phi_0$ with $g(t) = e^{i2\pi kt} \in G(1)$ and $\Phi_0(t) = e^{i\pi t} \in \Sigma(1)$. Consequently, by axiom 2 and 4, if we omit the index and write $\mu = \mu^1$:

$$\mu(\Phi) = 2k + 1 .$$

If, on the other hand, the arc $\Phi \in \Sigma(1)$ ends at a hyperbolic point with positive real eigenvalues, then it can be extended in $\mathrm{Symp}(1)^*$ to the point $M^+ = \mathrm{diag}(1/2, 2)$. Therefore there is a unique $k \in \mathbf{Z}$ such that Φ is homotopic in $\Sigma(1)$ to $g \cdot \Phi_0$ where $g(t) = e^{i2\pi kt} \in G(1)$, and $\Psi_0 \in \Sigma(1)$ is given by $\Psi_0(t) = \mathrm{diag}(2^{-t}, 2^t)$, $0 \leqslant t \leqslant 1$. Consequently, in view of axiom 2,

$$\mu(\Phi) = 2k + \mu(\Psi_0) .$$

Observe that Ψ_0 is homotopic in $\Sigma(1)$ to the arc Ψ_0^{-1}; the homotopy is given by $e^{\pi sJ}\Psi_0(t)e^{-\pi sJ}$, for $0 \leqslant s \leqslant 1$. Consequently, by the homotopy invariance and by axiom 3,

$$2\mu(\Psi_0) = \mu(\Psi_0) + \mu(\Psi_0) = \mu(\Psi_0^{-1}) + \mu(\Psi_0) = 0 ,$$

so that $\mu(\Psi) = 2k$. This finishes the sketch of the proof.

In the 2-dimensional case the μ-index is also characterized by the following axioms

THEOREM 3.2. *There exists a unique map $\mu: \Sigma(1) \to \mathbf{Z}$ which is a homotopy invariant, such that*

$$\mu(g \cdot \Psi) = 2m(g) + \mu(\Phi) \tag{30}$$

for $g \in G(1)$ and $\Phi \in \Sigma(1)$, and

$$\mu(\Phi^{-1}) = -\mu(\Phi) . \tag{31}$$

For a constant matrix S satisfying $\|S\| < 2\pi$, and $\Phi_0(t) = \exp tJS$, $0 \leqslant t \leqslant 1$, belonging to $\Sigma(1)$, we have

$$\phi(\Phi_0) = \mathrm{signature}(S) . \tag{32}$$

We omit the easy proof. Assume that x is a T-periodic solution of the Reeb-vector field X, $\dot{x} = X(x)$ so that $x(0) = x(T)$. Denote by ϕ_t the flow of X and abbreviate $x_0 = x(0)$. Then $T\phi_t(x_0): T_{x_0}M \to T_{x(t)}M$ maps the contact plane ξ_{x_0} onto $\xi_{x(t)}$ and this map is symplectic. This follows from $\phi_t^* \, d\lambda = d\lambda$. Assume $x: \mathbf{R}/T\mathbf{Z} \to M$ is a contractible loop and let $u: D \to M$ be a continuous map satisfying $x(t) = u(e^{2\pi it/T})$ for all t, with the disc $D = \{z \in \mathbf{C} \mid |z| \leqslant 1\}$. We take a symplectic trivialization of $u^*\xi$, denoted by $\beta: u^*\xi \to D \times \mathbf{R}^2$, and consider the arc $\Phi: [0, T] \to \mathrm{Symp}(1)$ determined by

$$\Phi(t) = \beta(e^{2\pi it/T})\big(T\phi_t(x_0) \mid \xi_{x_0}\big)\big(\beta(1)\big)^{-1} . \tag{33}$$

If x is non-degenerate, then $1 \notin \sigma(T\phi_T(x_0) \mid \xi_{x_0})$ and hence $\Phi \in \Sigma(1)$. Consequently, associated to the trivialization we have the μ-index, $\mu(x, u) \in \mathbf{Z}$.

It only depends on the homotopy class of the disc u with fixed boundary x. It is defined absolutely (i.e. without reference to a trivializing homotopy class of discs) if the topology of M is suitably restricted, namely if for every map $v: S^2 \to M$, the number $c_1(v^*\xi)([S^2])$ vanishes. Here c_1 denotes the first Chern class of the complex line bundle $v^*\xi$. The above assignment defines a homomorphism $c: \pi_2(M) \to \mathbb{Z}$, $[v] \mapsto c_1(v^*\xi)([S^2])$. If, for example, M is a homology three sphere, we have $c = 0$. If x is the asymptotic limit of the finite energy plane $\tilde{u}: \mathbb{C} \to \mathbb{R} \times M$, so that $u(Re^{2\pi it}) \to x(Tt)$ as $R \to \infty$, then we have a distinguished homotopy class of discs defined by the finite energy plane itself (see section 4), which prompts the following

DEFINITION 3.3. *Let $\tilde{u} = (a, u)$ be a finite energy plane with non-degenerate asymptotic orbit $x(t)$. Then we associate to $\tilde{u}$ the integer*

$$\mu(x, u) \in \mathbb{Z}\ . \tag{34}$$

In the special case Symp(1) we can relate the μ-index to winding numbers of vectors in $\mathbb{R}^2$. Consider a continuous arc $t \mapsto S(t) \in \mathcal{L}(\mathbb{R}^2)$, $0 \leqslant t \leqslant 1$ of symmetric matrices. Then we can associate with S the selfadjoint operator L_S in $L^2(\mathbb{S}^1, \mathbb{R}^2)$, which is defined, on its domain $W^{1,2}(\mathbb{S}^1, \mathbb{R}^2)$, where $\mathbb{S}^1 = \mathbb{R}/\mathbb{Z}$, by

$$(L_S x)(t) = -J\frac{d}{dt}x(t) - S(t)x(t)\ . \tag{35}$$

The operator is selfadjoint and has, since the embedding $W^{1,2}(\mathbb{S}^1, \mathbb{R}^2) \to L^2(\mathbb{S}^1, \mathbb{R}^2)$ is compact, a compact resolvent. The spectrum of L_S consists, therefore, of real eigenvalues of multiplicity at most 2 which only accumulate at $+\infty$ and at $-\infty$. Note that L_S is a bounded perturbation of the operator $-J\frac{d}{dt}$ whose spectrum is $2\pi\mathbb{Z}$.

Assume now that $x(t)$ is an eigenfunction of L_S with corresponding eigenvalue $\lambda \in \mathbb{R}$ and $x \neq 0$ in L^2. Then x solves the first order differential equation

$$-J\dot{x} - S(t)x(t) = \lambda x(t) \tag{36}$$

and the periodic boundary condition $x(t+1) = x(t)$. Hence $x(t) \neq 0$ for all t. Consequently x determines a map $\mathbb{S}^1 \to \mathbb{R}^2 \setminus \{0\}$ and has a winding number $w(x, \lambda) \in \mathbb{Z}$, defined by

$$w(x, \lambda) = \frac{1}{2\pi}\big(a(1) - a(0)\big) \in \mathbb{Z}\ , \tag{37}$$

where $a(t) = \arg(x(t))$, $0 \leqslant t \leqslant 1$, is any continuous argument.

LEMMA 3.4. *Assume x and y are (nonzero) eigenfunctions of L_S corresponding to the same eigenvalue λ. Then*

$$w(x, \lambda) = w(y, \lambda)\ .$$

Hence we can associate to the eigenspace belonging to the eigenvalue λ *of* L_S, *the integer:*

$$w(x,\lambda) = w(\lambda, S) \in \mathbf{Z} \; .$$

Proof: The conclusion of the lemma is obvious if both eigenvectors are linearly dependent. Let us assume that they are independent. It is convenient to identify $\mathbf{R}^2$ with $\mathbf{C}$ and to use complex notation. Define $z(t) = x(t) \cdot \overline{y(t)}$, $0 \leqslant t \leqslant 1$. Then $z \colon \mathbf{S}^1 \to \mathbf{C} \setminus \{0\}$ and the winding number of z is equal to $w(z) = w(x, \lambda) - w(y, \lambda)$. In order to prove that $w(z) = 0$ it is sufficient to show that $z(t) \notin \mathbf{R}$ for all t. Arguing by contradiction we assume $z(t_0) \in \mathbf{R}$ for some $t_0 \in \mathbf{R}$. Then $x(t_0) = \tau y(t_0)$ for some $\tau \in \mathbf{R} \setminus \{0\}$. Define $v(t) = x(t) - \tau y(t)$, then v solves the differential equation $L_S(v) = \lambda v$. Since $v(t_0) = 0$ we conclude that $v(t) \equiv 0$. This contradicts the assumption that x and y are linearly independent in L^2 and hence the lemma is proved. □

LEMMA 3.5. *Assume* x *and* y *are (nonzero) eigenfunctions of* L_S *corresponding to the eigenvalues* λ *and* μ. *If*

$$w(x, \lambda) = w(y, \mu) \quad \text{and} \quad \lambda \neq \mu \; ,$$

then the eigenvectors are pointwise linearly independent: for every $0 \leqslant t \leqslant 1$, $\mathrm{Span}\langle x(t), y(t) \rangle = \mathbf{R}^2$.

Proof: We may assume $\lambda > \mu$. We use complex notation, so that $x(t) \in \mathbf{C}$ and $y(t) \in \mathbf{C}$ and consider $z(t) = x(t) \cdot \overline{y(t)} \in \mathbf{C}$. Then the winding number of $z(t)$ vanishes since $w(z(t)) = w(x, \lambda) - w(y, \mu) = 0$, by assumption. Now $z(t_*) \in \mathbf{R}$ if and only if $x(t_*) = \tau y(t_*)$ for some $\tau \in \mathbf{R} \setminus \{0\}$ We claim that at such a point t_* the argument of $z(t_*)$ increases. Indeed, using $\dot{x} = i(S + \lambda)x$ and $\dot{y} = i(S + \mu)y$ one computes

$$\dot{z}(t) = i(\lambda - \mu) z(t) + H\big(x(t), y(t)\big)$$

where $H(x, y) = i\big[Sx \cdot \overline{y} - x \cdot \overline{Sy}\big]$. Therefore, $\overline{H(y, x)} = H(x, y)$. Consequently, if $z(t_*) \in \mathbf{R}$, then $x(t_*) = \tau y(t_*)$ for some $\tau \in \mathbf{R}$ and $H(x(t_*), y(t_*))$ is real. Since $\lambda > \mu$ and $z(t_*) \neq 0$ the argument increases as claimed. In polar coordinates, $z(t) = R(t) e^{i\phi(t)}$, we read off that $\dot{\phi}(t_*) = \lambda - \mu > 0$, whenever $\phi(t_*) = n\pi$ with $n \in \mathbf{Z}$. We deduce that if at some point t_*, the vectors $x(t_*)$ and $y(t_*)$ are linearly dependent, then necessarily $w(z(t)) \geqslant 1$, contradicting $w(z(t)) = 0$. The lemma is proved. □

LEMMA 3.6. *For every* $k \in \mathbf{Z}$ *there are exactly two eigenvalues (counting multiplicity)* μ *and* λ *of* L_S, *such that*

$$k = w(\mu, S) = w(\lambda, S) \; .$$

Proof: Define the family of selfadjoint operators L_τ by $L_\tau x = -J\dot{x} - \tau S(t)x$. The parameter varies in $[0,1]$. If $\tau = 0$, then the operator L_0 has the eigenvalues $2\pi k$, $k \in \mathbb{Z}$. The corresponding eigenspaces are spanned by $x(t) = e^{2\pi k t J}x(0)$ and have dimension 2. Hence the winding numbers are $w(2\pi k, 0) = k$. By Kato's perturbation theory of the spectral representation of selfadjoint operators (see [K, Chapter VII]) there exist continuous families $\mu_n(\tau)$, $\tau \in [0,1]$ of eigenvalues and continuous families of corresponding eigenfunctions $\phi_n(\tau)$, $\tau \in [0,1]$ such that for every given $\tau \in [0,1]$, the family $\phi_n(\tau)$, $n \in \mathbb{Z}$, is a Hilbert basis for L^2, and, moreover $\mu_n(0) = \hat{\mu}_n$. Here we define $\hat{\mu}_{2l+k}(0) = 2\pi l$ for $k \in \{0,1\}$, for the eigenvalues of L_0. It follows that the function $\tau \to w(\mu_n(\tau), \tau S)$ is constant. Consequently $w(\mu_{2l+k}(1), S) = l$ for $l \in \mathbb{Z}$ and $k \in \{0,1\}$. This completes the proof of the proposition. □

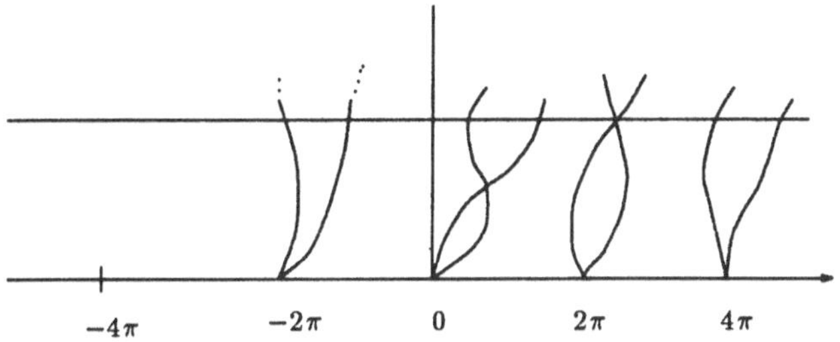

It follows from Lemma 3.4 and from the proof of Lemma 3.6, that two eigenvalue families whose associated winding numbers are different, never meet. Consequently we have the following monotonicity behavior:

LEMMA 3.7. *The map $\lambda \mapsto w(\lambda, S)$ from $\sigma(L_S)$ onto $\mathbb{Z}$ is monotonic:*

$$\lambda \leqslant \mu \quad \Rightarrow \quad w(\lambda, S) \leqslant w(\mu, S)\ .$$

As a consequence we can make the following construction. Given S we denote by $\lambda_k^i(S)$, where $i \in \{0,1\}$, the two eigenvalues of L_S having the same winding number k. Define

$$\lambda_k^- = \min\{\lambda_k^0(S), \lambda_k^1(S)\}$$
$$\lambda_k^+ = \max\{\lambda_k^0(S), \lambda_k^1(S)\}\ .$$

Hence we can associate with a winding number $k \in \mathbb{Z}$ an interval $I_k^S = [\lambda_k^-(S), \lambda_k^+(S)]$ whose endpoints are the eigenvalues of winding k. The interval degenerates to a point if the two eigenvalues agree and, therefore, have multiplicity 2. For $l < k$ the interval I_l^S lies to the left of I_k^S and the intervals are disjoint. We call these intervals the characteristic intervals associated with S. As in the proof of Lemma 3.6, one shows by Kato's perturbation theory:

LEMMA 3.8. *Let $\tau \mapsto S_\tau$ be a continuous map from $[0,1]$ into $C^\infty([0,1], \mathcal{L}_{\text{sym}}(\mathbf{R}^2))$. Then the map $\tau \mapsto I_k^{S_\tau}$ is continuous for every $k \in \mathbf{Z}$. Continuous means that the boundaries of the intervals depend continuously on the parameter τ.*

For a given arc $t \to S(t) \in \mathcal{L}(\mathbf{R}^2)$ of symmetric matrices we next define three integers $\alpha(S)$, $p(S)$ and $d(S)$. The integer $\alpha(S)$ is the maximal winding number occurring for negative eigenvalues of L_S:

$$\alpha(S) = \max\{w(\lambda, S) \mid \lambda \in \sigma(L_S) \cap (-\infty, 0)\} . \tag{38}$$

The number $p(S) \in \{0,1\}$, called parity in the following, is defined as follows

$$p(S) = \begin{cases} 0 & \text{if there exists } \lambda \in \sigma(L_S) \cap [0,\infty) \text{ with } \alpha(S) = w(\lambda, S) \\ 1 & \text{otherwise .} \end{cases} \tag{39}$$

By $d(S)$ we denote the dimension of the kernel of L_S, hence $0 \leqslant d(S) \leqslant 2$. Recall that to an arc S there corresponds in a canonical way an arc Φ: $[0,1] \to$ Symp(1) starting at the identity. We, therefore, can define the generalized μ-index:

DEFINITION 3.9. $\tilde{\mu}(\Phi) \equiv \tilde{\mu}(S) = 2\alpha(S) + p(S) \in \mathbf{Z}$.

Note that the definition does not require the arc Φ in Symp(1) to be non-degenerate, i.e. $\Phi \in \Sigma(1)$. Clearly $\Phi \in \Sigma(1)$ if and only if $d(S) = 0$. If $\tilde{\mu}(S) = 2k+p$, with $k \in \mathbf{Z}$ and $p \in \{0,1\}$, then it follows from the definitions that the special eigenvalue $\lambda_k^-(S)$ having winding number equal to k is < 0. If $p = 1$, then also $\lambda_k^+(S) < 0$, while if $p = 0$, then $\lambda_k^+(S) \geqslant 0$.

The main result of this section is the following relation between symplectical and geometrical winding.

THEOREM 3.10. *Let $S(t)$, $0 \leqslant t \leqslant 1$ be a continuous arc of symmetric matrices in $\mathcal{L}(\mathbf{R}^2)$ and denote by $\Phi(t)$ the associated arc in $Sp(1)$ starting at the identity. Assume, that $\ker L_S = \{0\}$, then $\Phi \in \Sigma(1)$, and*

$$\mu(\Phi) = \tilde{\mu}(\Phi) = 2\alpha(S) + p(S) .$$

Proof: Assume the two arcs S and S_1, give rise to the arcs Φ and Φ_1 in $\Sigma(1)$ which are homotopic in $\Sigma(1)$. During the homotopy we have $d(S) = 0$, so that no eigenvalue meets zero. Furthermore, eigenvalues belonging to different winding numbers do not meet during the homotopy. Consequently, in view of Lemma 3.8, $\tilde{\mu}(S) = \tilde{\mu}(S_1)$. Hence $\tilde{\mu}(\Phi)$ is a homotopy invariant of $\Sigma(1)$, and we have to show that $\tilde{\mu}$ satisfies the properties (30)–(32) of Theorem 3.2. We first prove (30). We identify $\mathbf{R}^2$ with $\mathbf{C}$ and use complex notation. Let $\Phi \in \Sigma(1)$ and $g \in G(1)$ with $m(g) = k \in \mathbf{N}$. By the homotopy invariance of $\tilde{\mu}$ we may assume that the loop g is given by $g(t) = e^{i2\pi kt}$,

$0 \leqslant t \leqslant 1$. Consider the arc $\Psi(t) = g(t)\Psi(t)$. It is generated by the arc $t \mapsto S_1(t)$ of symmetric matrices, where

$$S_1(t) = 2\pi k + g(t)S(t)g(t)^{-1} .$$

If v is an eigenfunction of L_S with eigenvalue λ, then $g(t)v(t)$ is an eigenfunction of L_{S_1} which belongs to the same eigenvalue λ. But its winding number is changed by k. Consequently $p(S) = p(S_1)$ and $\alpha(S_1) = \alpha(S) + k$, so that:

$$\tilde{\mu}(g\Phi) = 2\alpha(S_1) + p(S_1) = 2\alpha(S) + 2m(g) + p(S) = \tilde{\mu}(\Phi) + 2m(g) .$$

This proves (30). Using the homotopy invariance, the property (32) is verified by choosing suitable arcs. Taking the constant arcs $S_\pm : t \mapsto \pm\pi Id$ one computes that $\alpha(S_+) = 0$, $\alpha(S_-) = -1$, and $p(S_+) = p(S_-) = 1$ and consequently $\tilde{\mu}(S_\pm) = \pm 1$, as desired. Moreover, the constant arc $S_0(t) = \text{diag}(1, -1)$, $0 \leqslant t \leqslant 1$ leads to $\alpha(S_0) = p(S_0) = 0$ and hence $\tilde{\mu}(S_0) = 0$, so that property (32) is verified. Choosing suitable arcs, using the homotopy invariance and the properties (30) and (32) one shows that $\tilde{\mu}$ also meets the property (31). This completes the proof of Theorem 3.10. □

4. Algebraic Invariants and Embeddings

We consider a finite energy plane $\tilde{u} = (a, u) : \mathbf{C} \to \mathbf{R} \times M$ with a non-degenerate asymptotic limit $x(t)$ and abbreviate $P_u = x(\mathbf{R}) \subset M$. The second section was devoted to the injectivity of the map u. Now we look for conditions which allow us to conclude that u is an embedding. Observe that rank $Tu(z)$ is maximal (i.e. equal to 2) if $\pi \circ Tu(z) \neq 0$. This follows since, in view of $\pi \circ Tu \circ i = J \circ \pi \circ Tu$, the map $\pi \circ Tu(z)$ is complex linear:

$$\pi \circ Tu(z) \in \text{Hom}_{\mathbf{C}}(T_z\mathbf{C}, \xi_{u(z)}) .$$

Here J is an almost complex structure in ξ compatible with $d\lambda$. The map $z \to \pi \circ Tu(z)$ is a section of the complex line bundle $\text{Hom}_{\mathbf{C}}(\mathbf{C}, u^*\xi)$. We shall prove below that the set of points $z \in \mathbf{C}$ for which $\pi_{u(z)}Tu(z) = 0$ is a finite set. This allows the definition of the degree of $\pi \circ Tu$ using a trivialization of the bundle $u^*\xi$ denoted by

$$\varphi : \mathbf{C} \times \mathbf{C} \to u^*\xi .$$

The associated trivialization of $\text{Hom}_{\mathbf{C}}(\mathbf{C}, u^*\xi)$ is denoted by

$$\Phi : \mathbf{C} \times \text{Hom}_{\mathbf{C}}(\mathbf{C}, \mathbf{C}) \to \text{Hom}_{\mathbf{C}}(\mathbf{C}, u^*\xi) ,$$

where $\Phi(z, A)\lambda = \varphi(z, A\lambda)$ for $\lambda \in \mathbf{C}$. We identify $T_z\mathbf{C}$ with $\mathbf{C}$. In this trivialization the section $\pi \circ Tu$ is represented by $\gamma : \mathbf{C} \to \text{Hom}_{\mathbf{C}}(\mathbf{C}, \mathbf{C})$, given by $\gamma(z) = pr_2 \circ \Phi^{-1} \circ \pi \circ Tu(z)$. The section $\pi \circ Tu(z)\frac{\partial}{\partial x}$ of $u^*\xi$ becomes

$$U : \mathbb{C} \to \mathbb{C} , \quad U(z) = \gamma(z) \cdot 1 .$$

Here $\mathrm{Hom}_{\mathbb{C}}(\mathbb{C}, \mathbb{C})$ is identified with complex multiplication. We also used the notation $z = x + iy \in \mathbb{C}$. The zeros of the map U are the zeros of $\pi \circ Tu$. They are contained in a disc $D_R \subset \mathbb{C}$ of radius R centered at the origin, since there are only finitely many of them. Hence, the mapping degree of $U : \mathbb{C} \to \mathbb{C}$ is defined and we set:

$$\mathrm{wind}_\pi(u) := \deg(D_R, U, 0) \in \mathbb{Z} .$$

By the excision property of the mapping degree this integer is independent of R if $R \geqslant R_0$, for some large $R_0 > 0$. Since it is a homotopy invariant, it is also independent of the chosen trivialization. Moreover, it is the sum of the local degrees of the zeros of the section $\pi \circ Tu$:

$$\mathrm{wind}_\pi(u) = \sum_{\pi \circ Tu(z_j)=0} \deg(U, z_j) .$$

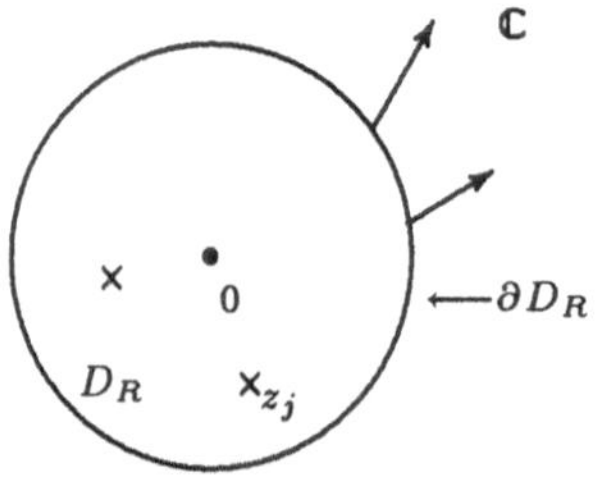

It is well known that the mapping degree agrees with the winding number of the vector field U along the boundary ∂D_R of the disc, denoted by $\mathrm{wind}_{\partial R}(U)$, hence

$$\mathrm{wind}_\pi(u) = \mathrm{wind}_{\partial D_R}(U) .$$

Recalling the definition we choose any continuous argument $a(t)=\arg U(Re^{it})$ and set $\mathrm{wind}_{\partial D_R}(U) = \frac{1}{2\pi}[a(2\pi) - a(0)]$.

Now we take a different section of $u^*\xi$ (away from 0) namely $z \to \pi \circ Tu(z)\frac{\partial}{\partial r}$. In the trivialization above it is represented by the map $V : \mathbb{C} \to \mathbb{C}$ given by $V(z) = \gamma(z) \cdot \frac{z}{|z|}$. We shall abbreviate its winding number

$$\mathrm{wind}_\infty(u) = \mathrm{wind}_{\partial D_R}(V) ,$$

and claim that

$$\mathrm{wind}_\infty(u) = \mathrm{wind}_\pi(u) + 1 .$$

Indeed, since in complex notation the winding number of a product is the sum of the winding numbers we have

$$\mathrm{wind}\left(\gamma(z) \cdot \frac{z}{|z|}\right) = \mathrm{wind}\left(\left[\gamma(z) \cdot 1\right] \cdot \frac{z}{|z|}\right) = \mathrm{wind}(\gamma(z) \cdot 1) + \mathrm{wind}\left(\frac{z}{|z|}\right) .$$

Clearly $\mathrm{wind}\left(\frac{z}{|z|}\right) = 1$ and the claim is proved.

PROPOSITION 4.1. *Assume $\tilde{u} = (a, u)$ is a non-constant finite energy plane with a non-degenerate limit $x(t)$. Then the set of zeros*

$$\{z \in \mathbf{C}\ ;\ \pi \circ Tu(z) = 0\}$$

is a finite set. Moreover, as a section of the vector bundle $u^*\xi \to \mathbf{C}$, the isolated zeros of $\pi \circ Tu\frac{\partial}{\partial x}$ have a positive degree.

Proof: It follows from the asymptotic description of a non-degenerate finite energy plane that there exists an $R_0 > 0$ such that

$$\pi \circ Tu(z) \neq 0 \quad \text{if} \quad |z| \geqslant R_0\ . \tag{40}$$

In order to prove that the zeros are isolated and have positive index, we work in local coordinates. Assume $\pi \circ Tu(z_0) = 0$, and put $m_0 = u(z_0) \in M$. By Darboux's theorem we find local coordinates in m_0, $\Phi : \mathcal{U} \subset \mathcal{M} \to \cap\mathcal{V}$, where $\mathcal{U}$ is an open neighbourhood of m_0 and $\mathcal{V} \subset \mathbf{R}^3$ an open neighbourhood of $\Phi(m_0) = 0$, such that

$$\Phi^*(d\vartheta + xdy) = \lambda\ .$$

Here we denote by $(\vartheta, x, y) \in \mathbf{R}^3$ the coordinates. In these coordinates the Reeb vector field X is constant and equal to $X = (1, 0, 0)$, and the contact plane ξ_m at $m = (\vartheta, x, y)$ is spanned by the vectors $e_1 = (-x, 0, 1)$, $e_2 = (0, 1, 0)$. The map $u : B_\varepsilon(z_0) \to \mathbf{R}^3$ satisfies

$$\pi u_s + J(u)\pi u_t = 0\ , \quad u^*(d\vartheta + x\, dy) = da \circ i\ .$$

Using $\pi(k) = k - \lambda_0(k)X$ for $k \in \mathbf{R}^3$ we can express the first equation in the basis e_1, e_2. With $u = (\vartheta, z)$ and $z(s,t) = (x(s,t), y(s,t))$, we find

$$z_s + J(s,t)z_t = 0\ .$$

The 2×2 matrix J satisfies $J^2 = -1$ and represents the complex structure $J(u(s,t))$ in the basis e_1, e_2. It is homotopic to the standard structure J_0. We may assume $z_0 = 0$. Differentiating the equation with respect to s, we find, for $\zeta = z_s$ the equation

$$\zeta_s + J(s,t)\zeta_t + A(s,t)\zeta = 0\ .$$

By assumption, $\zeta(0) = 0$. By the similarity principle, (see [FHS],[HZ]) we have, in a neighbourhood $D \subset \mathbf{C}$ of 0 the representation $\zeta(z) = G(z) \cdot \sigma(z)$, with $G(z) \in G\ell_{\mathbf{R}}(\mathbf{C})$ depending continuously on z, and satisfying $J(z)G(z) = G(z) \cdot i$. The function $\sigma : D \to \mathbf{C}$ is holomorphic. We conclude that either ζ vanishes identically on D or 0 is an isolated zero of ζ. In the first case we have $\pi \circ Tu(z) \equiv 0$ on D. This will lead to a contradiction. Indeed, define the set $S \subset \mathbf{C}$ by $S = \{z \in \mathbf{C} \mid \pi Tu(z) = 0 \text{ and satisfies property } (*)\}$, where $(*)$ requires z to be a cluster point of zeros of πTu. Then S is closed and open by the previous argument, hence equal to $\mathbf{C}$, which contradicts

(40). We have proved that the zeros of $\pi \circ Tu$ are isolated. Hence there are only finitely many of them in view of (40). Since $\zeta(z) = G(z)\sigma(z)$, the point $z = 0$ is an isolated zero of the holomorphic function σ and consequently has a positive local index. All the transformations carried out preserve the orientation. Therefore, we have proved that the zeros of $\pi \circ Tu$ have a positive index. This finishes the proof of the proposition. □

COROLLARY 4.2. *$\text{wind}_\pi(u) \geqslant 0$. Moreover, $\text{wind}_\pi(u) = 0$ if and only if $\pi \circ Tu \neq 0$ on $\mathbb{C}$.*

The winding number $\text{wind}_\pi(u) = \text{wind}_{\partial D_R}(V) \in \mathbb{Z}$ is constant for R large and, as we shall show next, related to the asymptotic orbit $x(t) = \lim u(s,t)$, as $s \to \infty$. Here we used the notation $u(s,t) = u(e^{2\pi(s+it)})$. To do so we extend $u : \mathbb{C} \to M \setminus P_u$ uniquely to a smooth map $\bar{u} : \overline{\mathbb{C}} = \mathbb{C} \cup S^1_\infty \to M$ such that $\bar{u}(e^{2\pi it}) = x(Tt)$, if $e^{2\pi it} \in S^1_\infty$.

Let D be the open unit disc $D = \{z \in \mathbb{C} \mid |z| < 1\}$ and define the diffeomorphism $\beta : D \to \mathbb{C}$ by $\beta(z) = \frac{z}{(1-|z|^2)^{1/2}}$. Define the differentiable structure on $\overline{\mathbb{C}} \equiv \mathbb{C} \cup S^1_\infty$ by means of the following chart:

$$\begin{array}{ccc} \mathbb{C} & \hookrightarrow & \mathbb{C} \cup S^1_\infty \equiv \bar{\mathbb{C}} \\ \beta \uparrow \cong & & \uparrow \bar{\beta} \\ D & \hookrightarrow & \bar{D} \end{array}$$

where $\bar{\beta} \mid D = \beta \mid D$ and $\bar{\beta}(e^{i\vartheta}) = e^{i\vartheta}$. In view of the exponential asymptotic behavior of $u(s,t) \to x(Tt)$ as $s \to \infty$, there exists a unique smooth extension $\overline{u \circ \beta} : \bar{D} \to M$ of $u \circ \beta : D \to M \setminus P$, such that $\overline{u \circ \beta}(e^{2\pi it}) = x(Tt)$. Therefore, we have the extension $\bar{u} : \bar{\mathbb{C}} \to M$ by $\bar{u} \circ \bar{\beta} = \overline{u \circ \beta}$, satisfying $\bar{u} \mid \mathbb{C} = u \mid \mathbb{C}$ and $\bar{u} \mid_{S^1_\infty} (e^{2\pi it}) = x(Tt)$. A trivialization of the bundle $\bar{u}^*\xi$ gives, in particular, a trivialization $\mathbb{S}^1 \times \mathbb{C} \to x^*\xi$. Recall now, that

$$\frac{\pi \circ Tu\left(\frac{\partial}{\partial s}\right)}{\left\|\pi \circ Tu\left(\frac{\partial}{\gamma s}\right)\right\|_J} \longrightarrow v(t) \quad \in \xi_{x(t)}$$

as $s \to \infty$, where $u(s,t) = u(e^{2\pi(s+it)})$, and where v is a distinguished section satisfying $v(t) \neq 0$. In the above trivialization it is represented by $\rho(t)e(t) \in \mathbb{R}^2 \cong \mathbb{C}$. Here $\rho(t) \neq 0$ is a non-vanishing function and $e(t)$ is an eigenvector of the selfadjoint operator A_∞ in $L^2(\mathbb{S}^1, \mathbb{R}^2)$:

$$A_\infty = -J_0 \frac{d}{dt} - S_\infty(t) \ ,$$

belonging to an eigenvalue $\lambda < 0$. The matrix function $S_\infty(t) = S_\infty(t+1)$ is symmetric. Since the winding number of a vector field does not change if the vector field is multiplied by a non-vanishing real function, we conclude for the winding numbers with respect to $\bar{u}$:

$$\text{wind}_{\partial\bar{D}}\big(e(t)\big) = \text{wind}_\infty(u) \ .$$

In view of the non-degeneracy condition we have $0 \notin \sigma(A_\infty)$, so that the indices $\mu(\bar{u}), \alpha(\bar{u})$ and $p(\bar{u})$ are defined (with respect to a trivialization of $\bar{u}^*\xi$). Changing the notation we recall that (Theorem 3.10)

$$\mu(u) = 2\alpha(u) + p(u) \ , \quad \text{with } p(u) \in \{0,1\} \ .$$

Since the eigenvalue associated to $e(t)$ is negative we have, by the definition of α, the estimate:

$$\text{wind}_{\partial\bar{D}}\big(e(t)\big) \leqslant \alpha(u) \ .$$

Consequently,

$$0 \leqslant \text{wind}_\pi(u) = \text{wind}_\infty(u) - 1 \leqslant \tfrac{1}{2}\mu(u) - 1 \ .$$

In particular, $\mu(u) \geqslant 2$, and we have proved:

THEOREM 4.3. *Assume $\tilde{u} = (a,u) : \mathbf{C} \to \mathbf{R} \times M$ is a non-constant finite energy plane with non-degenerate asymptotic limit $x(t)$. Then $x(t)$ is a contractible loop in M and has an index*

$$\mu(x,u) \geqslant 2 \ .$$

Next we prove:

THEOREM 4.4. *Assume $\tilde{u}$ is a non-constant finite energy plane with non-degenerate asymptotic limit $x(t)$ satisfying $\mu(x,u) \leqslant 3$. If $\tilde{u}$ is somewhere injective and $u(\mathbf{C}) \cap P_u = \emptyset$ then $u : \mathbf{C} \to M \setminus P_u$ is an embedding.*

Proof: From Theorem 2.3 we know that u is injective. From $\mu(x,u) \leqslant 3$ we conclude $0 \leqslant \text{wind}_\pi(u) \leqslant \frac{1}{2}$ and hence $\text{wind}_\pi(u) = 0$. Consequently $\pi \circ Tu(z) \neq 0$ for all $z \in \mathbf{C}$, by Corollary 4.2, so that u is an immersion. In view of the asymptotic behavior, u must be an embedding. The proof of the theorem is complete. □

So far we required that $u(\mathbf{C}) \cap P = \emptyset$, where $P = P_u$ is the limit periodic solution. A non-degenerate energy plane can intersect its limit P in at most finitely many points $z_j \in \mathbf{C}$:

$$u(z_j) \in P \ ,$$

see Theorem 5.1 in [HWyZ3]. This allows us to introduce an algebraic intersection index. We take a small embedded 2-disc $\mathcal{D}$ transversal to the periodic orbit $x(t) \in P$ at a point $m_0 \in P$ and tangent to ξ there: $T_{m_0}\mathcal{D} = \xi_{m_0}$. We orient the disc in such a way that its tangent space at m_0 has the same orientation as ξ_{m_0}. This introduces an orientation of the boundary, $\partial\mathcal{D}$. The disc induces a homology class, $[\mathcal{D}] \in H_2(M, M \setminus P)$ mapped into the homology class $[\partial\mathcal{D}] \in H_1(M \setminus P)$ under the connecting homomorphism. Let now $D \subset \mathbf{C}$ be a large disc centered at the origin, which contains all the intersection points z_j of u with P. If $\partial D = \mathbb{S}^1$ is the boundary, then $u(\partial D) \cap P = \emptyset$ and we study the mapping

$$u_* : H_1(\partial D) \longrightarrow H_1(M \setminus P) .$$

Clearly, the image in $H_1(M \setminus P)$ does not depend on the radius of the disc D chosen, as long as the intersection points are all in the interior.

LEMMA 4.5. *If α is the generator of $H_1(\partial D)$, then*

$$u_*(\alpha) = int(u)[\partial \mathcal{D}] \in H_1(M \setminus P) ,$$

where $int(u) \in \mathbb{Z}$ is called the intersection number of u with its limit P.

Proof: We pick a small tubular neighbourhood $\mathcal{U}$ of the periodic orbit $P \subset \mathcal{U}$. Since $u(z_j) \in P$ we can choose the open balls $B_j \subset \mathbb{C}$ around the intersection points z_j so small that $u(B_j) \subset \mathcal{U}$.

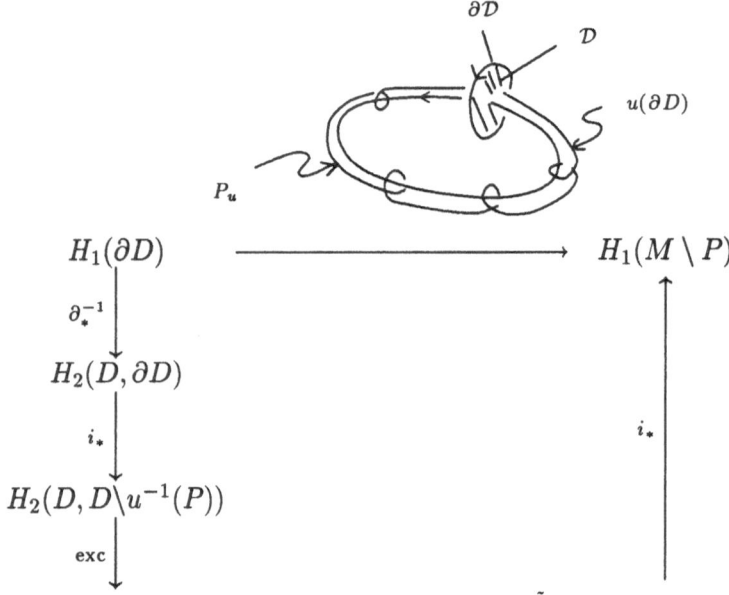

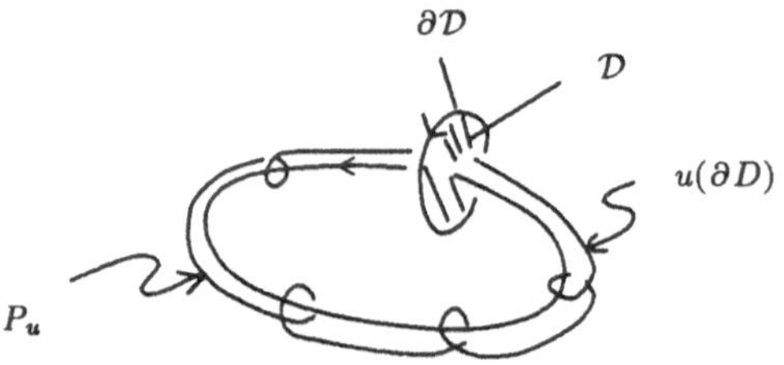

$$\begin{array}{ccccccc}
H_1(\partial D) & & \longrightarrow & & H_1(M \setminus P) & & \\
\downarrow {\scriptstyle \partial_*^{-1}} & & & & \uparrow {\scriptstyle i_*} & & \\
H_2(D, \partial D) & & & & & & \\
\downarrow {\scriptstyle i_*} & & & & & & \\
H_2(D, D \setminus u^{-1}(P)) & & & & & & \\
\downarrow {\scriptstyle \text{exc}} & & & & & & \\
\oplus H_2(B_j, B_j \setminus \{z_j\}) & \xrightarrow{u_*} & H_2(\mathcal{U}, \mathcal{U} \setminus P) & \xrightarrow{\tilde{\partial}_*} & H_1(\mathcal{U} \setminus P) & \xrightarrow{i_*} & H_1(\mathcal{U})
\end{array}$$

Here i_* stands for mappings induced by inclusions. By the long exact sequence for the pair $(\mathcal{U}, \mathcal{U} \setminus P)$ we conclude that the image of $\tilde{\partial}_*$ in $H_1(\mathcal{U} \setminus P)$ is equal to the kernel of i_*. Clearly $[\partial \mathcal{D}]$ is one of the generators (of the first homology) of the "2-torus" $\mathcal{U} \setminus P$. It generates the kernel of i_*. The diagram commutes, hence the statement follows. □

THEOREM 4.6. *Assume $\tilde{u} := (a, u) : \mathbb{C} \to \mathbb{R} \times M$ is a non-degenerate finite energy plane with asymptotic limit P. Then $int(u) \geqslant 0$ and $int(u) = 0$ if and only if $u(\mathbb{C}) \cap P = \emptyset$.*

Proof: We already know that there are at most finitely many points in $u(\mathbb{C}) \cap P$. Therefore, the oriented intersection number is the sum of the local intersection indices. Take a smooth approximation of u, to achieve transversality at an intersection. Then one concludes from the above diagram, in view of the piece $\to \oplus H_2(B_j, B_j \setminus \{z_j\}) \xrightarrow{u_*} H_2(\mathcal{U}, \mathcal{U} \setminus P)$ at the bottom, that indeed, by its definition, $int(u)$ agrees with the oriented intersection number. Moreover an intersection between u and P is (including the sign) an intersection of the two holomorphic curves $\tilde{u}$ and Z_u in $\mathbb{R} \times M$. Since the intersection points are isolated, the results of D. McDuff ([M]) apply and we deduce that every isolated intersection point has positive index. We remark that, since the holomorphic curve Z_u is actually an embedding, the positivity of the index can also be demonstrated using the similarity principle, we refer to C. Abbas and H. Hofer ([AH]). The proof is complete. □

In order to investigate the integer $int(u)$ we shall assume in the following that the covering number of the asymptotic limit x of u is equal to $k \geqslant 1$, and abbreviate $P = P_u = x_u(\mathbb{R}) \subset M$. We denote by $\Phi : \mathcal{U} \to M$ an embedding from an open neighbourhood $\mathcal{U}$ of the zero section of $\xi \mid P$. We require that $\Phi(0_p) = p$ and that the fibrewise derivative of Φ at 0_p, $p \in P$ is the inclusion of ξ_p into T_pM. Consider a non-vanishing section $v(t)$ of ξ along P which covers P precisely k-times and which is contained in $\mathcal{U}$. Define the loop $\beta(v)$ by $\beta(v)(t) = \Phi \circ v(t)$ for $0 \leqslant t \leqslant T$. It is contained in $M \setminus P$ and we denote by $[\beta(v)] \in H_1(M \setminus P)$ the homology class generated by this loop. Let $\text{wind}(v, u)$ be the winding number of the small section $v(t)$ with respect to the disc $\bar{u}$. Then

$$\text{wind}(v, u)[\partial\mathcal{D}] - [\beta(v)] \in H_1(M \setminus P) \tag{41}$$

is independent of v as described above. Indeed, two different sections covering P k times and having the same winding number with respect to $\bar{u}$ are homotopic. Therefore they induce the same homology class. Take any v and let us modify v in order to change the winding number without changing the fact that it covers P precisely k times. We can create additional winding in the fibre ξ_p over a given point $p \in P$. This however changes $\text{wind}(v, u)[\partial\mathcal{D}]$ and $[\beta(v)] \in H_1(M \setminus P)$ by the same amount. Moreover, since the asymptotic limit x_u is contractible, we conclude that the above homology class is a multiple of $[\partial\mathcal{D}]$, i.e. of the form $c(u)[\partial\mathcal{D}]$. Hence,

$$\left[\text{wind}(v, u) - c(u)\right][\partial\mathcal{D}] = \left[\beta(v)\right] . \tag{42}$$

If we choose, for example, the special section $v(t)$ representing $\Phi \circ v(t) = u(s^*, t)$ for some large $s^* \in \mathbb{R}$, then, by definition, $[\beta(v)] = int(u)[\partial\mathcal{D}]$, and we have the relation

$$\mathrm{wind}_\infty(u) - c(u) = int(u) . \tag{43}$$

In order to investigate $c(u)$, it is necessary to first recall some facts from contact geometry. We consider in the 3-dimensional contact manifold (M, λ) an embedded compact disc

$$F = \psi(D) , \tag{44}$$

where $D = \{z \in \mathbb{C}; |z| \leqslant 1\}$ and assume that the boundary ∂F is transversal to the contact planes ξ. The contact structure ξ is not integrable and hence induces the so called characteristic distribution on F defined by $T_m F \cap \xi_m$. This is a 1-dimensional subspace of $T_m F$ except at generically finitely many points $m_0 \in F$ where $T_{m_0} F = \xi_{m_0}$, called the singular points of the characteristic distribution. We shall assume that there are only finitely many singular points, they are, by assumption, in $F \setminus \partial F$. It is convenient to parameterize the distribution by a smooth vector field V on F in a canonical way.

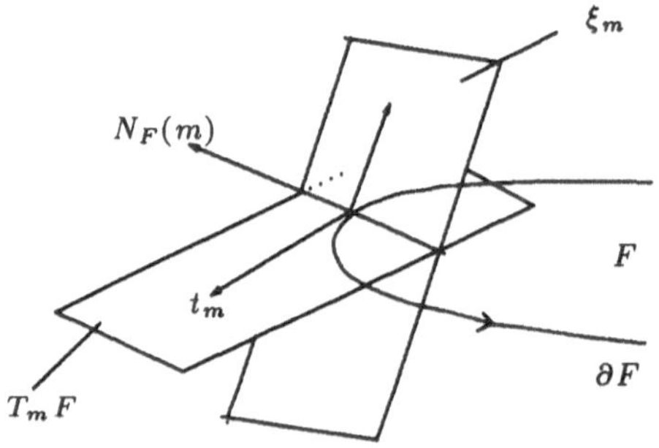

Recall that the orientation on M is defined by the volume form $\lambda \wedge d\lambda$. The orientation in ξ_m is defined by the non-degenerate 2 form $d\lambda | \xi_m$ so that the orientations of $T_m M = \mathbf{R} X(m) \oplus \xi_m$ coincide. Here $X(m)$ is the Reeb vector field, defined by $i_X d\lambda = 0$ and $\lambda(X) = 1$. The contact plane is defined by $\lambda(\xi_m) = 0$. We now orient the boundary ∂F of F such that the oriented tangent vector t_m points into the same direction as $X(m)$, i.e. $\lambda(t_m) > 0$, hence the orientations on

$$T_m M = T_m \partial F \oplus \xi_m , \quad m \in F \tag{45}$$

coincide. The surface F is now oriented the standard way, the "normal" vector $N_F(m)$ at $m \in \partial F$ pointing to the outside. We now take a vector field ν transversal to F and $\nu(m) \in \xi_m$ at the boundary $m \in \partial F$. We take it in such a way that the orientations of

$$T_m M = \big(\mathbf{R}\nu(m)\big) \oplus T_m F , \quad m \in F \tag{46}$$

coincide. At the boundary $m \in \partial F$ we then have $d\lambda(\nu(m), N_F(m)) > 0$. Now take a function $H : M \to \mathbf{R}$ defined in a neighbourhood of F which represents F as a regular surface: $F = \{H = 0\}$ satisfies $dH \neq 0$ on F, and $dH(\nu) > 0$. Consider the 1-form α near F defined by

$$\alpha = -dH + dH(X) \cdot \lambda \,. \tag{47}$$

Then $\alpha(X) = 0$. Hence at every point m, we have, for a tangent vector $Y(m) = aX(m) + b \in T_mM$, that $\alpha(Y(m)) = \alpha(b)$. There exists a unique vector $V(m) \in \xi_m$ satisfying $\alpha(b) = d\lambda(V(m), b)$ for all $b \in \xi_m$ and actually for all $b \in T_mM$. We see that there is a vector field V near F uniquely defined by

$$\alpha = i_V d\lambda \quad \text{and} \quad \lambda(V) = 0 \,. \tag{48}$$

It turns out that this vector field parameterizes the characteristic distribution on F and points outwards at the boundary ∂F, see [AH],[B]. We give the proof for the convenience of the reader.

LEMMA 4.7. *Let $m \in F$, then*

$$V(m) \in \xi_m \cap T_mF$$
$$V(m) = 0 \Longleftrightarrow T_mF = \xi_m$$
$$d\lambda(\nu(m), V(m)) > 0 \,, \quad m \in \partial F \,.$$

Proof: By definition, $\lambda(V) = 0$, so that $V(m) \in \xi_m$. Moreover, $-dH(V) = d\lambda(V, V) - dH(X)\lambda(V) = 0$, so that $V(m) \in T_mF$ proving the first claim.

In order to prove the second claim assume $V(m) = 0$. Then $d\lambda(V(m), \eta) = 0$ for all $\eta \in T_mM$ and hence, by (47) and (48), $\alpha(\eta) = 0$ and

$$dH(\eta) = dH(X)\lambda(\eta) \,, \quad \eta \in T_mM \,. \tag{49}$$

Observe that in this case $dH(X) \neq 0$ at m. Indeed if $\eta \in \xi_m$, then $dH(\eta) = 0$ by (49) and hence $dH(X) \neq 0$ since otherwise $dH = 0$ in contradiction to the assumption on H. Now it is immediate from (49) that $\eta \in T_mF$ if and only if $\eta \in \xi_m$. Conversely, if $T_mF = \xi_m$, then $\alpha(\eta) = -dH(\eta) + dH(X)\lambda(\eta) = 0$ for all $\eta \in \xi_m$, and since $\alpha(X) = 0$, we conclude that $\alpha = 0$ and hence $V(m) = 0$ as claimed. Finally, if $m \in \partial F$, then $d\lambda(\nu(m), V(m)) = -d\lambda(V, \nu(m)) = -[-dH(\nu) + dH(X)\lambda(\nu)] = dH(\nu) > 0$, since $\lambda(\nu) = 0$, proving the last claim of the lemma. □

Note that if H is replaced by another such H', then $V' = fV$ with a positive smooth function $f > 0$.

A singular point m_0 of the characteristic distribution is a singular point, $V(m_0) = 0$, of the smooth vector field V on F. It is called non-degenerate if the linearized map $V'(m_0) \in \mathcal{L}(T_{m_0}F)$ is an isomorphism. In the language of contact geometry it is called hyperbolic if $\lambda_1 \cdot \lambda_2 < 0$ and elliptic if $\lambda_1 \cdot \lambda_2 > 0$ with the eigenvalues λ_1 and $\lambda_2 \in \sigma(V'(m_0))$. We shall consider the generic case in which all the singular points are non-degenerate, see [AH],[G] for genericity results. Then a singular point m_0 is called positive (resp. negative) if the orientations of T_mF and ξ_{m_0} coincide (resp. are opposite). We shall denote by $e^{\pm}$ the number of ($\pm$)-elliptic points and by $h^{\pm}$ the number of ($\pm$)-hyperbolic points, and define

$$d^{\pm} = e^{\pm} - h^{\pm} \in \mathbf{Z} \,. \tag{50}$$

It is useful to interpret the singular points from the point of view of dynamical systems. We observe that the vector field V has, near a singularity, a non-vanishing divergence. Let $V(m_0) = 0$ then $T_{m_0}F = \xi_{m_0}$ and hence $j^*d\lambda$ is a non-degenerate form near $m_0 \in F$, where $j : F \to M$ is the inclusion mapping. One verifies readily, that

$$L_V\omega = dH(X)\omega \ , \quad \text{where } \omega = j^*d\lambda \tag{51}$$

at m_0. Since $dH(X) \neq 0$, we have $\operatorname{div}_\omega V(m_0) \neq 0$. If the orientations of $T_{m_0}F$ and ξ_{m_0} agree, then, by the definition of $\nu(m)$, the vectors X and ν point to the same side of $T_{m_0}F$. Therefore, since $dH(\nu) > 0$, we have $dH(X) > 0$ and hence $\operatorname{div}_\omega V(m_0) > 0$. We see that the (+)-elliptic point is, dynamically, a source (or repeller) of the vector field V. Similarly, a negative elliptic point is dynamically a sink (or attractor). Hyperbolic singular points are always saddle points of V : schematically:

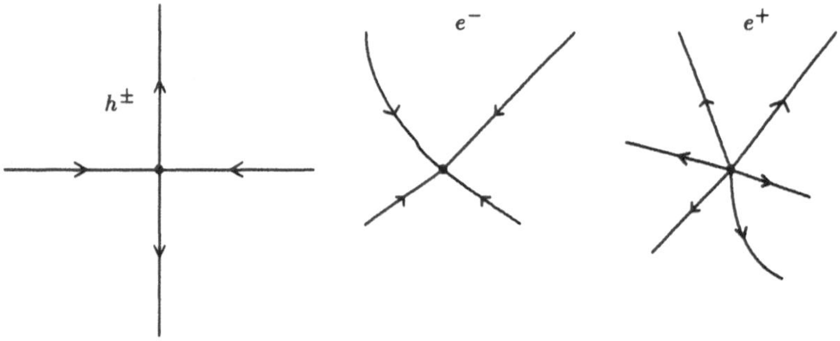

By definition, the vector field V on F is a section of two different bundles, namely of the tangent bundle $TF \to F$ and also of $\xi \mid F \to F$. Correspondingly we can associate with the singularities of the characteristic distribution two mapping degrees. If we trivialize the tangent bundle TF we find, by degree theory,

$$d^+ + d^- = 1 \ , \tag{52}$$

where, on the right hand side, the integer 1 is the winding number of the vector field $V(m)$, $m \in \partial F$, which points to the outside of the disc F.

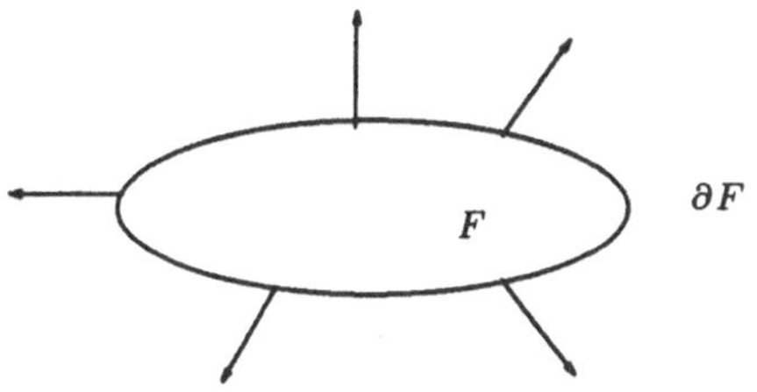

The left hand side is the number of the local degrees of the zeros the vector field V on F, in the trivialization. If we consider V as a section of $\xi \mid F$ we find in a trivialization of $\xi \mid F$, by the mapping degree, the formula

$$d^+ - d^- = \tau(\xi \mid F) , \tag{53}$$

where $\tau(\xi \mid F)$ is the winding number of $V(m)$ along the oriented boundary ∂F. This time, however, the zeros of the negative singular points contribute with a sign opposite to the previous case. Hence $\tau(\xi|F)$ is the algebraic number of zeros a section of $\xi|F \to F$ must have if it is identical to $V(m)$ at ∂F. Similar formulas for closed surfaces can be found in [E1-3], see also [AH].

Now, we restrict the class of contact structures under consideration and recall the

DEFINITION 4.8. *A contact structure ξ on M is called overtwisted, if there exists an embedded closed disc $D \subset M$ such that*
(i) $T_m\partial D \subset \xi_m$ *for all* $m \in \partial D$.
(ii) ∂D *does not contain any point* m*, at which* $T_mD = \xi_m$.
If such a ("overtwisted") disc does not exist, then the contact structure is called tight.

It should be recalled that the contact structure on $\mathbf{R}^3$, defined by the contact form $\lambda = dz + x\,dy$, is tight. This is a fundamental result due to D. Bennequin ([B]). It is not known, whether every orientable compact 3-manifold without boundary admits a tight contact structure.

We assume now that our contact form λ is tight and consider the above closed embedded disc F having the oriented boundary transversal to the contact structure. In order to compute the invariant $\tau(\xi \mid F)$ we shall deform F keeping the boundary ∂F fixed. By a C^∞-small perturbation of F, keeping the boundary fixed, we can achieve that the canonically induced vector field V on the perturbed disc F is a Morse-Smale vector field , see [AH],[G]. This means, by definition (see J. Palis and W. de Melo [PdM]), that V possesses only finitely many singular points and periodic orbits, which are all hyperbolic in the sense of dynamical systems and such that, in addition, there are no saddle-saddle-connections. Since λ is tight, there are no periodic orbits. Indeed, a periodic orbit would bound, in view of the classical Schoenfliess theorem (see in Rolfson [Ro]), an embedded disc which is overtwisted as in the above definition, in contradiction to the tightness assumption. Precisely at this point we used the assumption of tightness for the contact form λ, and we would like to point out that the assumption is not needed if the Reeb vector field X is transversal to the interior of F. In this case a limit cycle can be excluded by the following familiar argument. If X is transversal, then $d\lambda$ is an area form on $F \setminus \partial F$, since $d\lambda_{|T_m F} = 0$ if and only if $X(m) \in T_m F$. Therefore, if $\gamma = \partial\Omega$ is a limit cycle bounding the disc $\Omega \subset F$ we find, by using Stokes theorem, and by using $\lambda(V) = 0$,

$$0 < \int_\Omega d\lambda = \int_\gamma \lambda = 0 \, .$$

This contradiction shows that there are no periodic orbits for V on F. Assume now that there exists a connecting orbit between two singular points of the same sign as defined above. By Giroux's elimination theorem, see [G], and [AH],[E1-3],[H], we find a C^0-small smooth isotopy supported near the connecting orbit in such a way that the two critical points including their connecting orbit vanish, without creating new singular points. After this, we can achieve, by a C^∞-small perturbation which does not create new singular points, that the flow is Morse-Smale again. Proceeding by induction, we end up with a new disc of the same boundary which has, however, the property, that there are no connecting orbits between singular points of the same sign, is Morse-Smale and has no periodic orbits. All the perturbations we have carried out can be assumed to take place away from the boundary ∂F. Moreover, if the Reeb vector field X is transversal at the beginning, this property holds true for all deformations constructed. Let us prove now, that there are no positive hyperbolic and no negative elliptic points left. Otherwise we would be able to derive a contradiction as follows. Let h^+ be a positive hyperbolic point. It is a saddle point. Consider its stable manifold. Since the vector field V points outwards at the boundary ∂F, the stable manifold has a backward (in time) limit point in F. Since there is no saddle-saddle connection, by the Morse-Smale property, the backward limit point is necessarily a repeller, hence a positive elliptic point e^+, in the language of contact geometry:

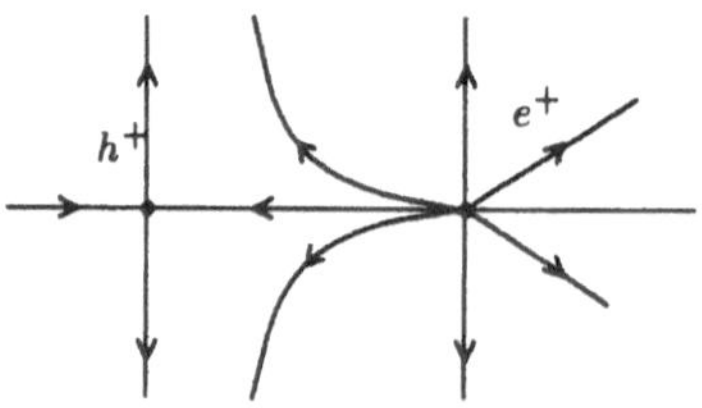

Hence we have a connection between two critical points of the same sign, which contradicts our assumption that these are already killed. Similarly one shows that there are no sinks. Indeed the basin of attraction, $B(e^-)$, of a sink in reverse time is a relatively compact set of the interior of F. Its boundary is connected, compact and invariant under the flow of V. It cannot be a periodic solution and consists of more than finitely many points. Therefore, there is a point on the boundary which is not a singular point and hence has a forward and a backward limit point which are critical points. Both of them must be repellers which, however, is absurd. Examples of possible situations we come across are depicted in the following picture:

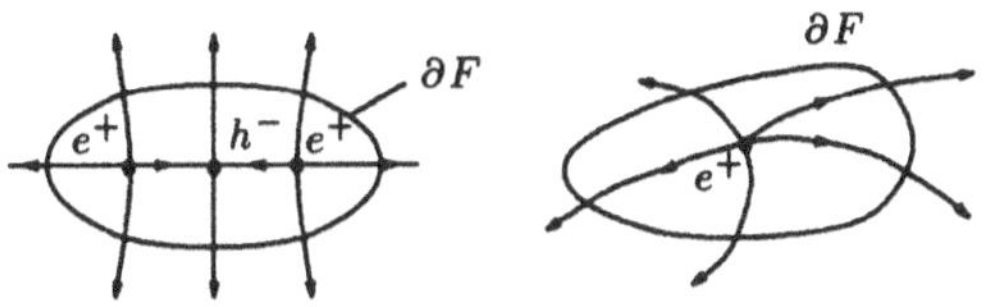

Since there are only positive elliptic and negative hyperbolic points left on F, we have

$$d^+ = e^+ \quad \text{and} \quad d^- = -h^- \ . \tag{54}$$

The index formulae (52) and (53), therefore, become

$$\begin{aligned} 1 &= e^+ - h^- \\ \tau(\xi \mid F) &= 2e^+ - 1 = 2h^- + 1 \ . \end{aligned} \tag{55}$$

In particular, $e^+ \geqslant 1$, $\tau(\xi \mid F) \geqslant 1$ and, in addition, $\tau(\xi \mid F)$ is an odd number. After these preliminaries in contact geometry, we are ready to prove the main result of this section.

DEFINITION 4.9. *We call an embedded loop P in a three-dimensional manifold M, k-unknotted, $k = 1, 2, \dots$ if there exists an immersion of a disc: $\varphi : D \to M$ such that φ : interior $(D) \to M \setminus P$ is an embedding and $\varphi : \partial D \to P$ is a k-fold covering.*

THEOREM 4.10. *Let M be a compact, connected and oriented 3-manifold equipped with a tight contact form λ. Assume $\tilde{u} : \mathbf{C} \to \mathbf{R} \times M$ is a non-constant finite energy plane with a non-degenerate asymptotic limit $x = x_u$ having covering number $cov(u) = k$. Assume further that $P_u = x_u(\mathbf{R})$ is k-unknotted with respect to the homotopy class of u. If $\mu = \mu(x_u, u)$ satifies*

$$\mu \leqslant 3 \ ,$$

then $u(\mathbf{C}) \cap P_u = \emptyset$ and $u : \mathbf{C} \to M \setminus P_u$ is an embedding.

Proof: By the discussion prior to the theorem we may assume that the disc $F = \varphi(D)$ with $\partial F = P$ satisfies $\tau(\xi \mid F) = 2e^+ - 1$ and $e^+ \geqslant 1$. Let V

be the canonical vector field associated to the characteristic distribution on F. By assumption, a loop $[t \mapsto \varphi(re^{2\pi it})]$ where $0 \leqslant t \leqslant 1$ and $0 < r < 1$ is in the interior of F and hence homologous to zero in $M \setminus P$. Moreover, if $0 < r < 1$ is close to 1, it is homotopic in $M \setminus P$ to the loop

$$\beta_v : \left[t \longmapsto \Phi(x(t), v(t))\right] , \quad 0 \leqslant t \leqslant T ,$$

where $v(t) = -a(t)V(x(t))$ with $a(t) > 0$, is a small non-vanishing section of $\xi \mid P$. Hence the winding number of $v(t)$ with respect to $\varphi(D) = F$ is equal to $\tau(\xi \mid F)$. Moreover, $[\beta_v] = 0$. Since φ and $\bar{u}$ are, by assumption, homotopic as maps with the boundary (circle at infinity) mapped to P as k-fold coverings, we see that the winding number of β_v with respect to $\bar{u}$ is also equal to $\tau(\xi \mid F)$. Hence $\text{wind}(v, u) = 2e^+ - 1$, and we find, in view of (42), that

$$c(u) = 2e^+ - 1 .$$

Consequently, in view of (43),

$$int(u) = \text{wind}_\infty(u) - (2e^+ - 1) .$$

On the other hand, we can use the estimate $0 \leqslant \text{wind}_\pi(u) = \text{wind}_\infty(u) - 1 \leqslant \frac{1}{2}\mu(u) - 1$ which leads, by the assumption that $\mu(u) \leqslant 3$, to $\text{wind}_\pi(u) = 0$ and $\text{wind}_\infty(u) = 1$. Therefore,

$$int(u) = 2(1 - e^+) .$$

Since $int(u) \geqslant 0$ and $e^+ \geqslant 1$ we find $int(u) = 0$ and $e^+ = 1$ (hence $h^- = 0$). In view of Theorem 4.6 we therefore conclude that $u(\mathbb{C}) \cap P_u = \emptyset$. In order to complete the argument let us note that $\tilde{u}$ has an injective point. Indeed, otherwise in view of the results in the appendix, $\tilde{u}$ could be written as $\tilde{u} = \tilde{v} \circ q$, where $\tilde{v}$ is somewhere injective and q is a complex polynomial of degree at least two. This implies immediately that $\pi \circ Tu$ vanishes somewhere contradicting the fact that it does not have any zeros (Corollary 4.2). We, therefore, conclude Theorem 4.4, that $u : \mathbb{C} \to M \setminus P$ is an embedding. The proof of Theorem 4.10 is complete. □

Finally, we investigate the intersections of two finite energy planes $\tilde{u}$ and $\tilde{v}$ having the same non-degenerate asymptotic limit x. Define a map $u\sharp v : \mathbb{S}^2 = \mathbb{C} \cup_{S^1_\infty} \overline{\mathbb{C}} \to M$ by gluing the domain $\mathbb{C}$ of v with reversed orientation to the domain $\mathbb{C}$ of u along the circle at infinity. Denote by $c(u, v) \in \mathbb{Z}$ the Chern number of the complex bundle $u(\sharp v)^*\xi \to \mathbb{S}^2$ evaluated on the fundamental class of the two-sphere. We observe that $c(u, v) = -c(v, u)$.

THEOREM 4.11. *Let $\tilde{u} = (a, u)$ and $\tilde{v} = (b, v) : \mathbb{C} \to \mathbb{R} \times M$ be two finite energy planes having the same non-degenerate asymptotic limit $x(t)$. Abbreviating $P = \{x(t) \mid t \in \mathbb{R}\} \subset M$ we assume:*

$$c(u,v) = 0$$
$$cov(u) = cov(v)$$
$$\mu(u) = \mu(v) \leqslant 3$$
$$u(\mathbb{C}) \cap P = \emptyset = v(\mathbb{C}) \cap P\ .$$

Then $u, v : \mathbb{C} \to M \setminus P$ are embeddings and

$$u(\mathbb{C}) = v(\mathbb{C}) \quad \textit{or} \quad u(\mathbb{C}) \cap v(\mathbb{C}) = \emptyset\ .$$

Proof: Observe that $\mathrm{wind}_\infty(u) = \mathrm{wind}_\infty(v) = 1$ if $\mu(x,u) = \mu(x,v) = \mu$ and $\mu \leqslant 3$. Since $\mathrm{wind}_\pi(u) = \mathrm{wind}_\pi(v) = 0$ we have, by Corollary 4.2,

$$\pi Tu \neq 0 \quad \text{and} \quad \pi Tv \neq 0$$

on $\mathbb{C}$. Moreover, $\tilde{u}$ and $\tilde{v}$ are somewhere injective, by the argument at the end of the proof of Theorem 4.10. Therefore, the maps $u, v : \mathbb{C} \to M \setminus P$ are embeddings by Theorem 4.3. We now proceed as in the proof of Theorem 2.3. Let $\tilde{u} = (a, u)$ and $\tilde{v} = (b, v)$ and define $\tilde{u}_\tau = (a + \tau, u)$, for $\tau \in \mathbb{R}$. We claim that there exists $\tau_0 > 0$, such that

$$\tilde{u}_\tau(\mathbb{C}) \cap \tilde{v}(\mathbb{C}) = \emptyset \quad \text{if} \quad |\tau| \geqslant \tau_0\ . \tag{56}$$

Arguing indirectly we find sequences z_k' and z_k in $\mathbb{C}$ and a sequence $\tau_k \in \mathbb{R}$ with $|\tau_k| \to \infty$ such that $\tilde{u}_{\tau_k}(z_k') = \tilde{v}(z_k)$. We may assume that $\tau_k \to +\infty$. Hence $a(z_k') + \tau_k = b(z_k)$ and $u(z_k') = v(z_k)$, with $\tau_k \to \infty$. Since the function a is bounded from below we have $b(z_k) \to \infty$ which implies $|z_k| \to \infty$ and hence $v(z_k) \to P$. Consequently also $u(z_k') \to P$ and since, by assumption $u(\mathbb{C}) \cap P = \emptyset$, we have $|z_k'| \to \infty$. To sum up, we have the following situation:

$$\begin{gathered} u(z_k') = v(z_k) \\ a(z_k') + \tau_k = b(z_k)\ . \\ \tau_k \to \infty, \quad |z_k| \to \infty\ , \quad |z_k'| \to \infty\ . \end{gathered} \tag{57}$$

From the asymptotic description near the limit orbit $x(t)$ we, therefore, find using cylinder coordinates, $z_k = e^{2\pi(s_k + it_k)}$,

$$T\, s_k' + \tau_k = T s_k + c + \varepsilon_k \tag{58}$$

for some $c \in \mathbb{R}$ and $\varepsilon_k \to 0$. Hence

$$\tau_k = T(s_k - s_k') + c + \varepsilon_k\ . \tag{59}$$

Moreover, from the asymptotics of the ϑ-function,

$$t_k' - t_k = \frac{j}{k} + \varepsilon_k\ , \tag{60}$$

where $j \in \{0,1,2,\dots,k-1\}$. "Transversal to the orbit $x(t)$",

$$e^{\int_{s_0}^{s'_k}\gamma(\sigma)d\sigma}\left[e(t'_k)+\varepsilon_k\right] = e^{\int_{s_0}^{s_k}\tilde\gamma(\sigma)d\sigma}\left[f(t_k)+\varepsilon_k\right] . \tag{61}$$

Here $\gamma(s)\to\lambda$ and $\tilde\gamma(s)\to\mu$, as $s\to\infty$, for two negative eigenvalues λ,μ of the selfadjoint operator $A=-J_0\frac{d}{dt}-S(t)$ in $L^2(\mathbf{S}^1,\mathbf{R}^2)$, with eigenvectors $Ae=\lambda e$ and $Af=\mu f$. Clearly $e(t)=e(t+1)\neq 0$ and $f(t)=f(t+1)\neq 0$. From the assumption $\mu\leqslant 3$ on the asymptotic periodic orbit, we conclude that

$$\mathrm{wind}_\infty(u)=\mathrm{wind}_\infty(v)=1 . \tag{62}$$

Since $c(u,v)=0$, the winding numbers of the periodic vectors $e(t)$ and $f(t)\in\mathbf{R}^2$, therefore, agree:

$$\mathrm{wind}\big(e(t)\big)=\mathrm{wind}\big(f(t)\big) . \tag{63}$$

We deduce, perhaps taking a subsequence, that

$$t'_k\to t'_* \quad\text{and}\quad t_k\to t_* \tag{64}$$

as $k\to\infty$. From (60) we conclude that

$$t'_*-t_*=\frac{j}{k} . \tag{65}$$

Consequently we must have, in view of (61),

$$e(t'_*)=bf(t_*)=bf\left(t'_*-\frac{j}{k}\right) , \tag{66}$$

for some positive number b. Assume first that the eigenvalues λ and μ are the same. Then we deduce from (61) that

$$(s_k-s'_k)$$

is a bounded sequence, which, however, contradicts (59). Therefore, we may assume without loss of generality that $\mu<\lambda<0$. Since $S(t)$ in the definition of A is $\frac{1}{k}$-periodic we see that with f also $f(\cdot+\frac{1}{k})$ is an eigenvector. Since by our assumption μ and λ have multiplicity 1 we deduce that $f(t)=f(t+\frac{1}{k})$ for all t and similarly for e. From (66) we therefore deduce

$$e(t''_*)=bf\left(t'_*-\frac{j}{k}\right)=bf(t'_*) , \tag{67}$$

where b is a positive constant. This means that e and f are linearly dependent at the time t'_*. Since, by (63), the winding numbers of $e(t)$ and $f(t)$ agree, we arrive at a contradiction to Lemma 3.5. Hence the claim (56) is proved.

Next we claim that there exists a $\tau^*\in\mathbf{R}$ with the following property: for every $\varepsilon>0$ there exists a compact set $K_\varepsilon\subset\mathbf{C}$ such that if

$$\tilde u_\tau(z')=\tilde v(z) \quad\text{and}\quad |\tau-\tau^*|\geqslant\varepsilon , \tag{68}$$

then $z, z' \in K_\varepsilon$. In order to prove this claim we distinguish two cases. Assume, at first, that the eigenvalues of A satisfy $\lambda \neq \mu$, or $\lambda = \mu$ and then either e and f are linearly independent or $ae = f$ with $a < 0$. Then the claim holds true for $\tau^* = 0$. Indeed, if not, we find sequences $\tau_k \geqslant \varepsilon$ and $|z_k'|, |z_k| \to \infty$ satisfying $\tilde{u}_{\tau_k}(z_k') = \tilde{v}(z_k)$. By (56) the sequence τ_k is bounded and hence we may assume that $\tau_k \to \tau_0^* \geqslant \varepsilon$ as $k \to \infty$, so that, in view of (58), we have $(s_k - s_k') \to \frac{1}{T}\tau_0^*$. This leads to a contradiction to the asymptotic behavior of the ξ-component. In the second case we have $\lambda = \mu$ and $a\, e(t) = f(t)$ for some $a > 0$. In this case the claim (68) follows, if we choose for τ^* the unique $\tau^* > 0$ satisfying

$$\lim_{k\to\infty}\left(\int_{s_0}^{s_k'} \gamma - \int_{s_0}^{s_k} \tilde{\gamma}\right) = \log a$$

if $s_k - s_k' \to \frac{\tau^*}{T}$ and $s_k \to \infty$, $s_k' \to \infty$.

From (68) and (56) we deduce, as in the proof of Theorem 2.3, by using D. McDuff's intersection theory of holomorphic curves ([M]; see also [MiW]), that

$$\tilde{u}_\tau(\mathbb{C}) \cap \tilde{v}(\mathbb{C}) = \emptyset \quad \text{if} \quad \tau \neq \tau^* \ . \tag{69}$$

Recall now, that $T\tilde{u}(z) \neq 0$ and also $T\tilde{v}(z) \neq 0$, as a consequence of $\text{wind}_\pi(u) = \text{wind}_\pi(v) = 0$. Consider the set of points in $\mathbb{C}$:

$$S = \{z \in \mathbb{C} \ ; \ z \text{ satisfies } (*)\}$$

where property $(*)$ requires:

There exists a point $z' \neq z$ and sequences (z_k'), (z_k) satisfying $z_k' \to z'$, $z_k \to z$ and $\tilde{u}(z_k') = \tilde{v}(z_k)$.

This set S is closed. Using $\pi T u \neq 0$ and $\pi T v \neq 0$ one shows, by means of the similarity principle, that the set is also open. Consequently, either $S = \mathbb{C}$ and hence $\tilde{u}(\mathbb{C}) = \tilde{v}(\mathbb{C})$ and $u(\mathbb{C}) = v(\mathbb{C})$, or the intersections of the two holomorphic curves are isolated. However, if there exists an isolated intersection it has a positive intersection index, and therefore $\tilde{u}_\tau(\mathbb{C}) \cap \tilde{v}(\mathbb{C}) \neq 0$ also for $\tau \neq \tau^*$ but close by. This contradicts (25). Consequently there are no intersections and the proof of Theorem 4.11 is complete. □

5. Algebraic Invariants for More General Finite Energy Surfaces

In this section we study more general finite energy surfaces. The methods are the same as in the previous sections. The results will be important in [HWyZ1-2].

Assume $\dot{S} = S \setminus \Gamma$ is a compact Riemannian surface with a finite number of punctures. To be more precise, S is a compact Riemannian surface and Γ

the finite set of punctures. In the following we shall denote by Γ^+ (Γ^-) the set of positive (negative) punctures. Let $\tilde{u}: \dot{S} \to \mathbf{R} \times M$ be a finite energy surface, i.e. it satisfies

$$T\tilde{u} \circ i = \tilde{J} \circ T\tilde{u} \, , \tag{70}$$

and has a finite, but nonzero energy

$$0 < E(\tilde{u}) < \infty \, . \tag{71}$$

In the case $\dot{S} = \mathbf{C}$ we know that $E(\tilde{u}) = \int_{\mathbf{C}} u^* d\lambda$ provided the energy is finite. However, in the general case, this is no longer true. Hence it turns out to be useful to classify all solutions of (70) and (71) satisfying

$$\int_{\dot{S}} u^* d\lambda = 0 \, .$$

This classification is given in the appendix. Observe that the set of singularities of a finite energy surface in non-empty. This has already been proved in the introduction, and we present next a second proof.

LEMMA 5.1. *Let $\tilde{u} : \dot{S} \to \mathbf{R} \times M$ be a non-constant finite energy surface (i.e. $\tilde{u}$ satisfies (70) and (71)). Then $\Gamma \neq \emptyset$.*

Proof: Indeed, otherwise $\tilde{u}$ is a map $S \to \mathbf{R} \times M$, where S is a compact Riemannian surface without boundary. By Stokes' theorem,

$$\int_S u^* d\lambda = 0 \, .$$

Now, observe that the integrand is non-negative. Indeed, one computes in local coordinates $z = s + it$ that $u^* d\lambda = |\pi \frac{\partial u}{\partial s}|^2 (ds \wedge dt)$. Consequently, $\pi T u(z) = 0$ for all $z \in S$, and $u^* d\lambda = 0$. On the other hand,

$$-u^* d\lambda = d(da \circ i) \, .$$

Fix any Riemannian metric g on S such that i acts by multiplication by 90 degrees, i.e. $g(v, iv) = 0$ for all $v \in TS$. Denote by $*$ the Hodge-star operator. For every 1-form we have $(*\alpha)(v) = -\alpha(iv)$. Define by $\|\alpha\|^2 = \int_S \alpha \wedge *\alpha$ the L^2-norm of the 1-form α, associated to g. Then we compute, using $u^* d\lambda = 0$, that

$$\begin{aligned} 0 &= -\int_S a u^* d\lambda = \int_S a d(da \circ i) \\ &= \int_S \left[d(a da \circ i) - da \wedge da \circ i \right] = -\int_S da \wedge da \circ i \\ &= \int_S da \wedge *da = \|da\|^2 \, . \end{aligned}$$

Consequently, a is constant implying that $u^* \lambda = 0$. Hence $\tilde{u}$ is constant contradicting our hypothesis. This completes the proof of the lemma. □

Next we sharpen the result somewhat

LEMMA 5.2. *For a non-constant finite energy surface $\tilde{u} : \dot{S} \to \mathbf{R} \times M$ we have $\Gamma^+ \neq \emptyset$.*

Proof: Arguing indirectly we assume that $\Gamma^+ = \emptyset$. Since $\Gamma \neq \emptyset$ by the previous lemma we have $\Gamma^- \neq \emptyset$. If $\tilde{u} = (a, u)$ we see that a is bounded from above. Let $R < 0$ be a sufficiently negative number, which is a regular value for a. Let $\Omega_R = a^{-1}([R, \infty))$. Then, for every such R, the set Ω_R is a compact manifold with smooth boundary $\partial\Omega_R$. We compute on Ω_R

$$\begin{aligned}\Delta a &= (d\delta + \delta d)a = \delta d a \\ &= - * d * da = - * d(-da \circ i) \\ &= *(-u^* d\lambda) \leqslant 0 \, .\end{aligned}$$

(Observe that in suitable coordinates $\Delta = -\left(\frac{\partial^2}{\partial s^2} + \frac{\partial^2}{\partial t^2}\right)$). Moreover $a = R$ on $\partial\Omega_R$. By the maximum principle this implies that

$$a \leqslant R \text{ on } \Omega_R \, .$$

Since this is true for every regular and negative R we deduce that $a \leqslant R$ for every real number R. This is of course absurd and the lemma is proved. □

In the special case of the plane $\mathbf{C}$, the Conley-Zehnder index of the asymptotic limit with respect to u has been defined in section 3 above. If we have more than one puncture it will turn out that we cannot assign to a single puncture a Conley-Zehnder index with respect to u. However, as we shall show, there will be a well-defined notion of the difference of the total Conley-Zehnder index of the positive and the total Conley-Zehnder index of the negative punctures. This difference then will be called the Conley-Zehnder index of u and it coincides in the case of the plane with the original definition. There will be also a notion of winding number at infinity as well as a notion of π-winding number.

In order to introduce this generalizations we need a result about the trivialization of symplectic plane bundles over punctured Riemannian surfaces. We begin with the following

LEMMA 5.3. *Let S be a compact connected Riemannian surface and $\dot{S} = S \setminus \Gamma$, where Γ is a finite subset. Let Γ be the disjoint union of positive and negative punctures $\Gamma^{\pm}$:*

$$\Gamma = \Gamma^- \cup \Gamma^+$$

and assume $\Phi : \dot{S} \to Sp(1)$ is any map. Then, the integer m denoting the Maslov-index satisfies

$$\sum_{z \in \Gamma^+} \mathrm{m}(\Phi | C_z) = - \sum_{z \in \Gamma^-} \mathrm{m}(\Phi | C_z) \, . \tag{73}$$

Here C_z is a circle around the puncture $z \in \Gamma$ bounding a small disc. The orientation chosen is the one induced from $\tilde{S}$, where $\tilde{S}$ is equal to S with the small open discs around the punctures removed.

Proof: Triangulating $\tilde{S}$, which is equal to S with the union of the open discs removed, we may view $\Phi : \tilde{S} \to Sp(1)$ as an element in the second chain group $C_2(Sp(1), \mathbf{Z})$. The boundary is then precisely

$$\partial\Phi = \sum_{z\in\Gamma} \Phi|C_z \ .$$

The Maslov-index can be viewed as a map from $H_1(Sp(1), \mathbf{Z}) \to \mathbf{Z}$. Since $\partial(\Phi|C_z) = 0$, we see that $\Phi|C_z$ represents a homology class. Therefore, we obtain by the previous discussion

$$0 = \mathrm{m}(\partial\Phi) = \sum_{z\in\Gamma} \mathrm{m}(\Phi|C_z) \ . \tag{74}$$

This proves our result. □

Let us now define the Conley-Zehnder index for a finite energy surface with non-degenerate asymptotic limits and $\int_{\dot{S}} u^* d\lambda > 0$. So assume $\tilde{u} : \dot{S} \to \mathbf{R} \times M$ satisfies the nonlinear Cauchy-Riemann equations, having finite energy, $E(\tilde{u}) < \infty$, so that the asymptotic limits are non-degenerate. Observe now that $u^*\xi$ is trivializable. Indeed, every punctured and connected Riemannian surface has the homotopy type of a wedge of finitely many one-spheres $\mathbf{S}^1$. This can be seen, e.g. by constructing a Morse function having no maximum and only one minimum. Thus the spheres are the connecting orbits issuing from the saddle points, and the claim follows from the fact that a complex vector bundle over $\mathbf{S}^1$ is trivializable. Pick any trivialization $\Phi : \bar{u}^*\xi \to \bar{S} \times \mathbf{C}$ of the pullback of ξ by $\bar{u}$. Here $\bar{S}$ is obtained from $\dot{S}$ by attaching to every puncture an asymptotic $\mathbf{S}^1$ as carried out in the previous section for the plane and $\bar{u}$ is the smooth extension. This induces, in view of the asymptotic behavior of u as proved in [HWyZ3] trivializations of ξ over the asymptotic limits. In order to get all the signs right let us observe the following. At a positive puncture z take a biholomorphic embedding $\phi : [0, \infty) \times \mathbf{S}^1 \to S \setminus \{z\}$ onto a neighbourhood of the puncture. We know that $u \circ \phi(r, \theta) \to x(T\theta)$ as $r \to \infty$, where x is the asymptotic limit at z having period T. The circle C_z used above, properly chosen, may be viewed as the image of $\{r\} \times \mathbf{S}^1$ for some r. With the orientations chosen, the above map is orientation preserving. For a negative puncture we take a biholomorphic embedding $\phi : (-\infty, 0] \to S \setminus \{z\}$ onto a neighbourhood of the puncture. We observe that $u \circ \phi(r, \theta) \to x(T\theta)$, in $C^\infty(\mathbf{S}^1)$, as before. However, this time the map between $\{r\} \times S^1$ and C_z is orientation reversing.

Using the symplectic trivialization chosen above, we can define, for every puncture, the number $\mu^{\Phi}(z)$ which is the Conley-Zehnder index of the associated asymptotic limit using the trivialization associated to Φ.

DEFINITION 5.4. *The Conley-Zehnder index of $\tilde{u}$ with respect to the trivialization Φ is defined by*

$$\mu^{\Phi}(\tilde{u}) = \sum_{z\in\Gamma+} \mu^{\Phi}(z) - \sum_{z\in\Gamma-} \mu^{\Phi}(z)\,, \tag{75}$$

where the integer $\mu^{\Phi}(z)$ is defined in section 3.

The crucial observation is now the following

PROPOSITION 5.5. *With the assumptions above, the integer $\mu^{\Phi}(\tilde{u})$ is independent of the choice of Φ.*

Proof: In order to prove this let Φ and Ψ be two symplectic trivializations. By definition, $\theta \to \Phi(\infty,\theta)\Lambda(\theta T)\Phi(\infty,0)^{-1}$, $\theta \in [0,1]$ is the symplectic arc in $Sp(1)$ whose Conley-Zehnder index is $\mu^{\Phi}(z)$. Here $\Phi(\infty,\cdot)$ is the trivialization over the asymptotic limit and $\Lambda(t)$ is the arc of symplectic maps along $x(t)$. Similarly for Ψ. We shall abbreviate in the following $\Phi(\theta) \equiv \Phi(\infty,\theta)$ and compute using Theorem 3.1,

$$\begin{aligned}\mu^{\Phi}(z) &= \mu\big[\Phi(\theta)\Psi(\theta)^{-1}\Psi(\theta)\Lambda(\theta T)\Psi(0)^{-1}(\Phi(0)\Psi(0)^{-1})^{-1}\big] \\ &= 2\,\mathrm{m}\big[(\Phi(\theta)\Psi(\theta)^{-1})(\Phi(0)\Psi(0)^{-1})^{-1}\big] + \mu^{\Psi}(z)\,.\end{aligned} \tag{76}$$

For a positive puncture we have, in view of the homotopy invariance of the Maslov index m,

$$\mathrm{m}\big[(\Phi(\theta)\Psi(\theta)^{-1})(\Phi(0)\Psi(0)^{-1})^{-1}\big] = \mathrm{m}\big[(\Phi\Psi^{-1})|C_z\big]\,. \tag{77}$$

For a negative puncture we find

$$\mathrm{m}\big[(\Phi(\theta)\Psi(\theta)^{-1})(\Phi(0)\Psi(0)^{-1})^{-1}\big] = -\,\mathrm{m}\big[(\Phi\Psi^{-1})|C_z\big]\,. \tag{78}$$

The latter is true due to the difference of orientations, as remarked before Definition 5.4. Using (77),(78) and Lemma 5.3 we find

$$\begin{aligned}\mu^{\Phi}(\tilde{u}) &= \sum_{z\in\Gamma+} \mu^{\Phi}(z) - \sum_{z\in\Gamma-} \mu^{\Phi}(z) \\ &= \sum_{z\in\Gamma+} \big(\mu^{\Psi}(z) + 2\,\mathrm{m}((\Phi\Psi^{-1})|C_z)\big) \\ &\quad - \sum_{z\in\Gamma-} \big(\mu^{\Psi}(z) + 2(-\,\mathrm{m}((\Phi\Psi^{-1})|C_z))\big) \\ &= \mu^{\Psi}(\tilde{u})\,.\end{aligned}$$

□

Similarly we can define the asymptotic winding number $\text{wind}_\infty(\tilde{u})$, as the difference of the asymptotic winding numbers at positive and negative punctures. Again everything is computed with respect to a trivialization. However, as in the previous proposition it does not depend on the choices involved. We define $\text{wind}_\pi(\tilde{u})$, assuming that $\pi \circ Tu$ does not vanish identically, to be the sum of all local indices of the zeros of the section $\pi \circ Tu$, computed with respect to any trivialization.

We shall prove

PROPOSITION 5.6. *Let $\tilde{u}: \dot{S} \to \mathbf{R} \times M$ be a finite energy surface with non-degenerate asymptotic limits as above, where $\dot{S} = S \setminus \Gamma$. Denote by $\chi = 2 - 2g$ the Euler characteristic of S and by $\sharp\Gamma$ the cardinality of Γ. Then*

$$\text{wind}_\pi(\tilde{u}) = \text{wind}_\infty(\tilde{u}) - \chi + \sharp\Gamma \, .$$

Remark 5.7: We observe that in the case of the plane we have $\chi = \chi(S^2) = 2$ and $\sharp\Gamma = 1$. Hence we recover our result from the previous section:

$$\text{wind}_\pi(\tilde{u}) = \text{wind}_\infty(\tilde{u}) - 1 \, .$$

Proof: Let $\dot{S}$ be a compact Riemannian surface with a nonempty finite set Γ of punctures. Pick a vector field Z on S which is transversal to the zero section in TS and has the points in Γ as zeros. In addition, we choose Z such that the negative points are sources and the positive points are sinks. Denote by $\dot{\sharp}Z$ the algebraic number of zeros contained in $\dot{S}$. Observe that the total algebraic number of zeros of Z is equal to the Euler characteristic χ. Each point in Γ contributes 1 to the index. Hence we find that

$$\dot{\sharp}Z = \chi - \sharp\Gamma \, . \tag{79}$$

We see that the number $\dot{\sharp}Z$ is independent of Z provided Z has the chosen behavior near Γ.

Consider the section $\pi \circ Tu$ of $\text{Hom}(T\dot{S}, u^*\xi) \to \dot{S}$. If $\pi \circ Tu$ does not vanish identically we know that this section only has a finite number of zeros. We know, moreover, that $\pi \circ Tu(z) \neq 0$ near the punctures. Let Y be a non-vanishing section of $T\dot{S}$ and $\Phi : u^*\xi \to \dot{S} \times \mathbf{C}$ a symplectic trivialization of $u^*\xi$. Define $\alpha : \dot{S} \to \mathbf{C}$ by

$$\alpha(z) = \Phi(z) \circ \pi \circ Tu(z)\big(Y(z)\big) \, .$$

Clearly α vanishes at some point iff $\pi \circ Tu$ vanishes there. Moreover, we know that the zeros of $\pi \circ Tu$ have positive index. The integer $\text{wind}(\alpha)$ counts the number of zeros of $\pi \circ Tu$ with multiplicities and is therefore always greater or equal to 0. Clearly this number is equal to $\text{wind}_\pi(\tilde{u})$. Let us write $Z = fY$. Then $f : \dot{S} \to \mathbf{C}$ and does not vanish near the punctures. We note that the zeros of f are non-degenerate (by the assumption on Z). The algebraic count of zeros for f, by (79), gives

$$\sharp f = \sharp \dot{Z} = \chi - \sharp \Gamma \; . \tag{80}$$

Take small circles C_i around the punctures enclosing open discs B_i with smooth boundaries so that $\pi \circ Tu(z) \neq 0$ for all $z \in \bar{B}_i \setminus \{z_i\}$. Denote by $\tilde{S}$ the complement of the union of these discs in S. Equip it with the complex orientation and orient its boundary components (the C_i) in the usual way. Given any map $\eta : \tilde{S} \to \mathbb{C}$, which does not vanish on the boundary, we can define the winding number wind(η) as the sum of the winding numbers of the maps $\eta : C_i \to \mathbb{C} \setminus \{0\}$, using the given orientations. We obtain, therefore,

$$\begin{aligned} \text{wind}(\alpha) &= \text{wind}(\Phi \circ \pi \circ Tu \circ Y) \\ &= \text{wind}(\Phi \circ \pi \circ Tu \circ Z) - \text{wind}(f) \\ &= \text{wind}(\Phi \circ \pi \circ Tu \circ Z) - \chi + \sharp\Gamma \; . \end{aligned} \tag{81}$$

From the asymptotic studies it follows as in section 4 that the contribution to wind$(\Phi \circ \pi \circ Tu \circ Z)$ at a positive puncture is the winding number of the asymptotic eigenvector with repect to the trivialization Φ. At a negative puncture the contribution is the negative of the winding number, due to our orientation convention. This implies that

$$\text{wind}(\Phi \circ \pi \circ Tu \circ Z) = \text{wind}_\infty(\tilde{u}) \; . \tag{82}$$

Hence combining equations (81) and (82) we find the required formula. The proof of the proposition is complete. □

Next we relate wind$_\infty(\tilde{u})$ and $\mu(\tilde{u})$. We fix a trivialization Φ of $\bar{u}^*\xi$ and compute everything with respect to this trivialization. The formula relating the winding numbers to the Conley-Zehnder index is $\mu = 2\alpha + p$ (Theorem 3.10). Let z be a puncture with asymptotic limit x. If z is positive, then

$$\text{wind}_\infty^\Phi(z) \leqslant \alpha^\Phi(z) \; , \tag{83}$$

and if z is negative (now the directional convergence of u at this puncture is given by an eigenvector belonging to the positive part of the spectrum) we find

$$\begin{aligned} &\text{wind}_\infty^\Phi(z) \geqslant \alpha^\Phi(z) \quad \text{if} \quad p = 0 \\ &\text{wind}_\infty^\Phi(z) \geqslant \alpha^\Phi(z) + 1 \quad \text{if} \quad p = 1 \; . \end{aligned} \tag{84}$$

Denote by Γ_p, where $p \in \{0,1\}$, the subset of punctures whose asymptotic limit has parity p, similarly we define $\Gamma_p^\pm$ for the positive respectively negative punctures. Our main results are the following estimates.

THEOREM 5.8. *Let $\tilde{u} : \dot{S} \to \mathbf{R} \times M$ be a finite energy surface with non-vanishing $\pi \circ Tu$. Then*

$$\mu(\tilde{u}) \geqslant 2\mathrm{wind}_\infty(\tilde{u}) + \sharp\Gamma_1 , \tag{85}$$

and

$$\mu(\tilde{u}) \geqslant 2\mathrm{wind}_\pi(\tilde{u}) + 2\chi - 2\sharp\Gamma_0 - \sharp\Gamma_1 . \tag{86}$$

Before we give a proof let us discuss these estimates in more detail. We observe that always $\mathrm{wind}_\pi(\tilde{u}) \geqslant 0$. For the case of a plane we therefore have $\mu \geqslant 4 - 2\sharp\Gamma_0 - \sharp\Gamma_1$. If the asymptotic limit has parity 1 we deduce $\mu \geqslant 3$, and if the asymptotic limit has parity 0 we deduce $\mu \geqslant 2$. For the case of a cylinder, i.e. the case of a sphere with two punctures we deduce $\mu \geqslant 0$, if the punctures both have parity 0. If one puncture has parity 1 and the other one has parity 0 we find $\mu \geqslant 1$. If both have parity 1, we have $\mu \geqslant 2$.

Proof: We deduce from $\mu = 2\alpha + \rho$ using (83) and (84),

$$\begin{aligned}\mu(\tilde{u}) &= \sum_{z\in\Gamma^+} \mu(z) - \sum_{z\in\Gamma^-} \mu(z) \\ &\geqslant \sharp\Gamma_1^+ + \sum_{z\in\Gamma^+} 2\mathrm{wind}_\infty(z) + \sharp\Gamma_1^- - \sum_{z\in\Gamma^-} 2\mathrm{wind}_\infty(z) \\ &= \sharp\Gamma_1 + 2\mathrm{wind}_\infty(\tilde{u}) .\end{aligned} \tag{87}$$

This proves the first part of the theorem. For the second part we combine this with Proposition 5.6 to obtain the desired inequality. □

Finally, we study a "mixed" boundary value problem. Let D be the closed unit disc in $\mathbb{C}$ and let $\Gamma \subset D \setminus \partial D$ be a finite set of punctures. Assume $F \subset M$ is an embedded closed disc with boundary ∂F. Assume F has only one singular point q (where $T_qF = \xi_q$), which is elliptic and belongs to the interior of F, i.e. $F \setminus \partial F$. Suppose F is oriented in such a way that q is positively elliptic. In addition, we assume that the interior of F is transversal to the Reeb vector field X. Such surfaces will be important in the applications given in [HWyZ4].

Let us write $\dot{F}$ for the set $F \setminus (\partial F \cup \{q\})$. We consider the following boundary value problem for a function $\tilde{u} : D \setminus \Gamma \to \mathbb{R} \times M$

$$\begin{gathered} T\tilde{u} \circ i = \tilde{J} \circ T\tilde{u} \\ a|\partial D = 0 \quad \text{and} \quad u(\partial D) \subset \dot{F} , \\ \text{where the winding number of } u|\partial D \to F \setminus \{q\} \text{ is } 1 , \\ 0 < E(\tilde{u}) < \infty . \end{gathered} \tag{88}$$

Let us write $\dot{D}$ for $D \setminus \Gamma$. We need some more assumptions. First of all we assume:

$$\text{All punctures are negative .} \tag{89}$$

This seems to contradict Lemma 5.1, however, observe that in this lemma $\tilde{u}$ is defined on a manifold without boundary. Moreover,

$$\text{The asymptotic limits } x_z, \text{ for all } z \in \Gamma \text{ are non-degenerate and contractible loops.} \tag{90}$$

Assume moreover that for every asymptotic orbit x_z a disc v_z is given such that the Conley-Zehnder index $\mu(x_z, v_z)$ is defined. Gluing these discs to u we can construct a disc map $\hat{u} : D \to M$ with $\hat{u}(\partial D) \subset F$ such that $\hat{u}$ and u coincide on ∂D. The curve $u(\partial D)$ encloses a disc $\hat{F} \subset F$ since the characteristic starts at the singularity q and leaves through the boundary transversally, see Proposition 5.9 below. Let $\hat{v} : D \to \hat{F}$ be a parametrisation of this disc. We observe that $\hat{v}$ and $\hat{u}$ coincide on ∂D. Our last assumption is the following

$$\text{The maps } \hat{u} \text{ and } \hat{v} \text{ are homotopic in } M \text{ with boundary values fixed}. \tag{91}$$

The first consequence of the hypotheses is

PROPOSITION 5.9. *Let the data be as described above. Then* $\pi \circ Tu(z) \neq 0$ *for all* $z \in \partial D$ *and for all* z *near the punctures.*

Proof: From the results in [HWyZ3] we know that $\pi \circ Tu(z) \neq 0$ for z near the punctures. Moreover we know by the asymptotic studies the behavior of $\pi \circ Tu$ near the punctures. To see that $\pi \circ Tu(z) \neq 0$ on ∂D we argue as follows. Let $\nu(z)$ be the the unit vector in $T_z\partial D$ compatible with the orientation of the boundary. Let $n(z)$ the outward pointing normal vector. Then $\nu(z) = in(z)$. By assumption we must have $u^*d\lambda \neq 0$. Denoting by $s + it$ the complex coordinates on D we have, by (88),

$$(-\Delta a) ds \wedge dt = -u^*d\lambda = -|\pi u_s|^2 ds \wedge dt \tag{92}$$

$$a|\partial D = 0\,. \tag{93}$$

Using that all punctures are negative we deduce by the strong maximum principle that

$$0 < \frac{\partial a}{\partial n}(z) = da(z)n(z) = \lambda\big(u(z)\big)\big(Tu(z)\nu(z)\big)\,.$$

Hence $Tu(z)\nu(z) \notin \xi_{u(z)}$. Moreover,

$$0 = \frac{\partial a}{\partial \nu}(z) = \lambda\big(u(z)\big)\big(Tu(z)(i\nu(z)\big) = -\lambda\big(u(z)\big)\big(Tu(z)n(z)\big)\,.$$

This implies that $Tu(z)n(z) \in \xi_{u(z)}$. If $\pi \circ Tu(z) = 0$ for some $z \in \partial D$ we must have $\mathrm{im}(Tu(z)) \subset \ell_{u(z)}$ implying that

$$Tu(z)h = b(h)X\big(u(z)\big)\,,$$

for some real linear map $b : \mathbf{C} \to \mathbf{R}$. However, $u(\partial D) \subset F \setminus (\partial F \cup \{q\})$, hence we deduce that $X(u(z)) \in T_{u(z)}F$, contradicting our assumption that the interior of F is transversal to X. This completes the proof of the proposition. □

Next we would like to estimate $\sum_{z\in\Gamma}\mu(x_z,v_z)$. This is the sum of the Conley Zehnder indices at the punctures z, with respect to the unique homotopy-class of trivializations of $v_z^*\xi \to D$.

Consider the map $\beta : \mathbf{S}^1 \ni \tau \to \pi u_s(\tau)$. This is a section of $\xi|u(S^1)$. We trivialize this bundle symplectically in such a way that the trivialization extends over $\hat{u}^*\xi$. Since $\hat{v}$ and $\hat{u}$ are homotopic with boundaries fixed we obtain the same homotopy class of trivializations of $\xi|u(S^1)$ if we require the trivialization to be extendable over $\hat{v}^*\xi$. Denote by $\text{wind}(\beta,\hat{u}) = \text{wind}(\beta,\hat{v})$ the winding number of β with respect to such trivializations. Let us show that this winding number is 0. By construction we have $\hat{u} = \hat{v}$ on ∂D. Moreover $u, \hat{u}$ and $\hat{v}$ coincide on ∂D. Denote by $\nu(\tau)$ the unit tangent vector to ∂D at the point τ, inducing the boundary orientation. We have

$$Tu(\tau)\nu(\tau) = T\hat{v}(\tau)\nu(\tau) . \tag{94}$$

We observe that by our assumption of transversality of the interior of F with X the map $\pi : \xi|F \to TF$ is a vector bundle isomorphism. Since $\hat{v} : D \to \hat{F}$ is an orientation preserving diffeomorphism we conclude from (94) that $\tau \to \pi \circ Tu(\tau)\frac{\partial}{\partial s}$ and $\tau \to \pi \circ T\hat{v}\frac{\partial}{\partial s}$ have the same winding number. Now, the map $\pi \circ T\hat{v} : TD \to \hat{v}^*\xi$ is an orientation preserving diffeomorphism since X is transversal to F. Therefore, the winding number has to be zero. Hence we note the following

$$\text{wind}(\beta,\hat{u}) = 0 . \tag{95}$$

Next we observe since the zeros of $\pi \circ Tu$ are isolated and have positive index that

$$0 \leqslant \text{wind}(\beta,\hat{u}) - \sum_{z\in\Gamma}\text{wind}_\infty(\pi u_s, v_z, z) . \tag{96}$$

Here $\text{wind}_\infty(\pi u_s, v_z, z)$ denotes the winding number on a small circle near the puncture z, where the orientation is chosen in such a way that the image of the circle under u follows the asymptotic limit x in positive direction. A calculation as done for the previous theorem gives, at a puncture z, abbreviating the parity $p = p(z)$,

$$\begin{aligned} \alpha(z,v_z) &\leqslant \text{wind}_\infty(z,v_z) = \text{wind}_\infty(\pi u_s, v_z, z) + 1 \text{ if } p = 0 \\ \alpha(z,v_z) + 1 &\leqslant \text{wind}_\infty(z,v_z) = \text{wind}_\infty(\pi u_s, v_z, z) + 1 \text{ if } p = 1 . \end{aligned} \tag{97}$$

Hence we find

$$\begin{aligned} \mu(z,v_z) &\leqslant 2 + 2\text{wind}_\infty(\pi u_s, v_z, z) \text{ if } p = 0 \\ \mu(z,v_z) &\leqslant 1 + 2\text{wind}_\infty(\pi u_s, v_z, z) \text{ if } p = 1 . \end{aligned} \tag{98}$$

Define the total Conley-Zehnder index of the punctures by

$$\mu^{\Gamma}(\tilde{u}) = \sum_{z\in\Gamma} \mu(z, v_z) .$$

Denote by Γ_p the punctures of parity p. Then combining (95),(96),(97) and (98) we obtain

$$\mu^{\Gamma}(\tilde{u}) \leqslant \sharp\Gamma + \sharp\Gamma_0 + \sum_{z\in\Gamma} 2\text{wind}_{\infty}(\pi u_s, v_z, z)$$
$$\leqslant \sharp\Gamma + \sharp\Gamma_0 .$$

Hence we have proved the following theorem:

THEOREM 5.10. *Under the assumptions (88),(89),(90) and (91) we have the following inequality (note that in this case we only have negative punctures)*

$$\sum_{z\in\Gamma} \mu(z, v_z) \leqslant \sharp\Gamma_0 + \sharp\Gamma . \tag{99}$$

We apply this result to prove a theorem which has an application in our papers [HWyZ1,2]. Let us first introduce a formalism to describe certain arrays of finite energy surfaces.

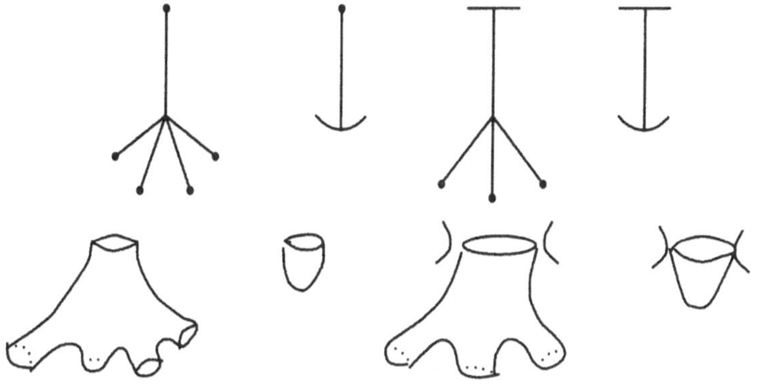

type 1 type 2 type 3 type 4

The first figure describes a finite energy sphere with one positive and four negative punctures. The second figure describes a finite energy plane and the third picture a punctured disc with boundary on a totally real surface F (i.e. $\tilde{J}(T_xF) \oplus T_xF = T_x(\mathbf{R} \times M)$ for $x \in F$) and having 3 punctures all being negative, the fourth picture represents a disc. So top vertices correspond to positive singularities the bottom vertices correspond to negative singularities.

Consider a graph which is composed of these "subgraphs". Assume it contains precisely one subgraph of the third type, which is at the top, whereas the bottom subgraphs are of type two, so that the whole graph is a tree. The following picture illustrates three examples

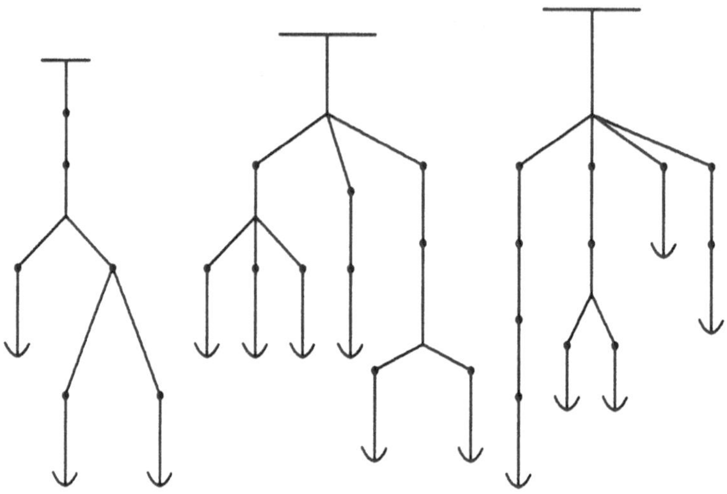

A geometric realization of such a tree is a map associating to each subgraph a corresponding finite energy surface, so that at common vertices the asymptotic limits coincide. Moreover each finite energy surface has positive $d\lambda$-energy. The following figure shows a tree together with a geometric realization.

Next assume we are given a tree together with a geometric realization in a compact oriented connected three-manifold equipped with a contact form λ. Assume we are given an embedded disc F bounded by a non-degenerate closed integral curve P_0 having Conley-Zehnder index 3 with respect to this disc. Assume the interior of F is transversal to the Reeb vector field X and that F has precisely one singular point q which is elliptic. The finite energy disc in our geometric realization has its boundary on F, with winding number 1 with respect to q. Moreover we assume that all asymptotic limits occurring for our finite energy surfaces have parity 1. If we take the geometric realization and glue the different parts along the asymptotic limit together we obtain a set which can be parametrised by a disc so that the parametrisation above ∂D coincides with the boundary values of the finite energy disc. We call this set the disc associated to the tree. Assume that this disc associated to the tree is homotopic with boundary values in $F \setminus \{q\}$ to a small disc around q.

THEOREM 5.11. *Under the above assumptions, the tree consists of one single element of type* 4, *i.e. it represents a disc.*

Remark 5.12: The importance of this theorem, which should be viewed together with a bubbling-off analysis as a compactness theorem, is the following. Using the above F let us apply the disc filling method, see [E2],[H],[Y], in the symplectisation $\mathbb{R} \times M$. If bubbling off occurs, see [HWyZ1], one obtains a geometric realization of a tree precisely as described above. Assume now that all periodic orbits for the Reeb vector field of period $T \leqslant \int_F d\lambda$ are non-degenerate and have parity 1. Then the application in [HWyZ1] of the above theorem shows that bubbling off does not occur, since bubbling off would create a finite energy plane, together with a punctured disc of type 3 so that the graph would contain at least 2 subgraphs, contradicting Theorem 5.11.

Proof: In order to prove the theorem, observe that the finite energy surfaces together with the tree define homotopy classes of trivializations for the asymptotic limits. Let $\tilde{u}$ be a finite energy plane. Then we know that the Conley-Zehnder index is at least 2. If it would be 2 the parity of the asymptotic limit would be 0. Hence we see that $\mu(\tilde{u}) \geqslant 3$. Remove now all the subgraphs of the tree representing a finite energy plane and label the lower vertices by their Conley-Zehnder indices which are at least 3, but always odd since the parity is 1. Consider now all lower vertices belonging to the same subgraph. To this subgraph there belongs a unique positive vertex at the top, unless it is of type 3. In the first case we apply Theorem 5.8 to find that the Conley-Zehnder index of the top vertex of the subgraph is at least $3 + 2N^-$, where N^- is the number of negative vertices of our subgraph. Indeed, using $\text{wind}_\pi \geqslant 0$ and $\sharp\Gamma_0 = 0$, we conclude from (86) that $\mu(\text{top}) - \mu(\text{bottom}) \geqslant 2\chi(S^2) - \sharp\Gamma_1$, and, with $\sharp\Gamma_1 = N^- + 1$, we find $\mu(\text{top}) \geqslant 3N^- + 4 - (N^- + 1) = 3 + 2N^- \geqslant 5$. Now remove all subgraphs of the present type and label the bottom vertices by their Conley-Zehnder indices which we know are at least 5 and then we proceed gaining at each step at least Conley-Zehnder index 2. In any case we finally end up with the subgraph of type 3, where we have at least one bottom vertex. Moreover, all the lower vertices are labeled by odd numbers of value at least 3. Hence, if N^- is the number of negative vertices we must have for the associated total Conley-Zehnder index the estimate $\mu \geqslant 3N^-$. By Theorem 5.10 we have, however, the estimate $\mu \leqslant N^-$. This contradiction shows that there cannot be any negative puncture. Hence we are in case 4. The proof of Theorem 5.11 is complete. □

Let us end this section with a theorem which will be extremely useful in [HWyZ1,2]. The proof is similar to the proof above. Recall that on $\mathbb{S}^3$ the Conley-Zehnder index is always well-defined. Knowing that the manifold is $\mathbb{S}^3$, all quantities introduced, like the winding numbers, are defined absolutely, since there is only one homotopy class of discs bounded by a periodic orbit.

Assume we are given a tree together with a geometric realization in $\mathbb{S}^3$ equipped with a contact form λ. Assume we are given an embedded disc F bounded by a non-degenerate closed integral curve P_0 having Conley-Zehnder index 3. Assume that the interior of F is transversal to the Reeb vector field and that F has precisely one singular point q which is positively elliptic. The finite energy disc in our geometric realization has his boundary on F, having winding number 1 with respect to q. Moreover, we assume that all asymptotic limits occurring for our finite energy surfaces have Conley-Zehnder index at least 3. Then

THEOREM 5.13. *Under the above assumptions, the tree consists of one single element of type 4, i.e. it represents a disc.*

Proof: The proof is similar to the proof of the previous theorem and hence will be omitted. □

6. Appendix

6.1 Factorization through somewhere injective finite energy planes.

We shall study the somewhere injective and multiply-covered curves in more detail. We begin with the following

LEMMA 6.1. *Assume $\tilde{u}$ and $\tilde{v}$ are finite energy planes and suppose that $p : \mathbb{C} \to \mathbb{C}$ is a continuous map satisfying*

$$\tilde{u} = \tilde{v} \circ p \, .$$

Then p is a complex polynomial. If the asymptotic limits are non-degenerate we have, in addition,

$$\operatorname{cov}(\tilde{u}) = \deg(p) \operatorname{cov}(\tilde{v}) \, .$$

Proof: Given a compact subset K of $\mathbb{C}$ we pick $R \in \mathbb{R}$ satisfying $K \subset b^{-1}((-\infty, R])$, where $\tilde{v} = (b, v)$. We know from the results in [H] that b is coercive ($b(z) \to \infty$ as $|z| \to \infty$). Let us write $\tilde{u} = (a, u)$. Then we deduce

$$a^{-1}((-\infty, R]) = p^{-1}(b^{-1}((-\infty, R])) \supset p^{-1}(K) \, .$$

Since a is coercive as well we see that p is proper. The points where $\pi \circ Tu$ or $\pi \circ Tv$ vanishes are isolated as we have proved above by using the similarity principle, see also [HWyZ3]. Hence the hypothesis implies by locally inverting $\tilde{v}$ that p is holomorphic away from a set of isolated points. Since p is bounded near an isolated singularity it is holomorphic on $\mathbb{C}$, by the removable singularity theorem. A proper holomorphic self-map of $\mathbb{C}$ is a polynomial. The second assertion concerning the non-degenerate case follows. This completes the proof of the lemma. □

Next we prove the following

THEOREM 6.2. *Let $\tilde{u} : \mathbb{C} \to \mathbb{R} \times M$ be a finite energy plane. Then there exists a somewhere injective finite energy plane $\tilde{v}$ and a complex polynomial $p : \mathbb{C} \to \mathbb{C}$ satisfying*

$$\tilde{u} = \tilde{v} \circ p .$$

The construction is quite similar to constructions in Riemannian surface theory. We follow more or less the arguments in [MS]. However, we give more details.

On $\mathbb{C}$ we define the relation $\Gamma \subset \mathbb{C} \times \mathbb{C}$ as follows. We say that (z, z') belongs to Γ if there exists a sequence $((z_k, z_k')) \subset (\mathbb{C} \times \mathbb{C}) \setminus \{(z, z')\}$ such that $(z_k, z_k') \to (z, z')$ and $\tilde{u}(z_k) = \tilde{u}(z_k')$. Clearly Γ is symmetric. We shall show later on that that Γ is an equivalence relation and that $\mathbb{C}/\sim$ carries in a natural way the structure of a Riemannian surface so that the projection $\mathbb{C} \to \mathbb{C}/\sim$ is holomorphic. Moreover, the quotient space turns out to be biholomorphic to the complex plane, so that with that identification the projection is the required polynomial map.

Let us denote by $\mathcal{S}$ the set of all points $z \in \mathbb{C}$ such that $T\tilde{u}(z) = 0$. We know by the results in [HWyZ3] that $\mathcal{S}$ is a finite set. This is a consequence of the asymptotic analysis in [HWyZ3] and results in [AH],[CFH],[HZ],[MS]. Further, we introduce the sets

$$\begin{aligned} X &= \mathbb{C} \setminus \mathcal{S} \\ Z &= \tilde{u}(\mathbb{C}) \setminus \tilde{u}(\mathcal{S}) \\ Y &= X \cap \tilde{u}^{-1}(Z) . \end{aligned} \tag{100}$$

If $z \in \mathbb{C}$, we denote by $[z]$ the subset of $\mathbb{C}$ consisting of all points z' such that $(z, z') \in \Gamma$. Clearly $[z] \subset Y$ if $z \in Y$.

LEMMA 6.3. *The set $\mathbb{C} \setminus Y$ consists of isolated points in $\mathbb{C}$.*

Proof: Let $z \in \mathbb{C} \setminus Y$. As a consequence of the similarity principle we have $\tilde{u}(\tilde{z}) \neq \tilde{u}(z)$ for $0 < |z - \tilde{z}| < \varepsilon$ for a suitable small positive number ε and in addition $T\tilde{u} \neq 0$ on the same punctured ball, provided ε is small enough. Since $\tilde{u}(\mathcal{S})$ is a finite set we deduce immediately that

$$\tilde{u}(B_\varepsilon(z) \setminus \{z\}) \cap \tilde{u}(S) = \emptyset .$$

Consequently

$$B_\varepsilon(z) \setminus \{z\} \subset Y ,$$

as claimed in the Lemma. □

DEFINITION 6.4. *For $z \in \mathbb{C}$ the multiplicity of z, denoted by $M(z)$, is defined by*

$$M(z) = \sharp[z] .$$

Assume $(z, z') \in \Gamma$ and $z \in Y$. Then $T\tilde{u}(z) \neq 0$ and $T\tilde{u}(z') \neq 0$. Moreover there exists a sequence (z_k, z'_k), $z_k \neq z$ and $z'_k \neq z'$ with $z_k \to z$, $z'_k \to z'$ such that $\tilde{u}(z_k) = \tilde{u}(z'_k)$. From the similarity principle we can, therefore, conclude that there exist open neighbourhoods U, U' of z and z' respectively such that

$$\tilde{u}(U) = \tilde{u}(U') \tag{101}$$

and

$$(\tilde{u}|U)^{-1} \circ \tilde{u}|U' : U' \to U \tag{102}$$

is a diffeomorphism.

The next lemma shows that $M(z)$ is finite and constant on Y.

LEMMA 6.5. *The number $M(z)$ is finite and independent of z for all $z \in Y$.*

Proof: Take points z_0 and z_1 in Y. In view of Lemma 6.3 the set Y is pathwise connected. Take a path γ in Y connecting z_0 and z_1, say $\gamma(i) = z_i$, $i = 0, 1$. Since the $\mathbb{R}$-component of $\tilde{u}$ is coercive the set

$$\bigcup_{t \in [0,1]} [\gamma(t)] \tag{103}$$

is bounded. Fix $t_0 \in [0, 1]$. Let us first show that $M(\gamma(t_0))$ is finite. Otherwise we find a sequence (z_k) of mutually different points in Y with $\tilde{u}(z_k) = \tilde{u}(z_\ell)$. In view of (103) this sequence is bounded. Hence eventually taking a subsequence we may assume that $z_k \to z_*$. From this we immediately obtain a contradiction to the fact that $T\tilde{u}(z_*) \neq 0$ for points in Y. This contradiction shows that $M(z) < \infty$ for $z \in Y$. Let $[z_0] = \{y_1 = z_0, y_2, \ldots, y_n\}$, where $n = M(z_0)$. Define

$$t_0 = \sup \{ t \in [0, 1] \mid M(\gamma(\tau)) = n \text{ for all } \tau \in [0, t] \} .$$

In view of (101) and (102) we have $t_0 > 0$. We claim that $M(\gamma(t_0)) = n$. If $M(\gamma(t_0)) > n$ we immediately obtain a contradiction using (101) and (102). If $M(\gamma(t_0)) < n$, then we find n sequences $(y^n_k)_k$ such that

$$[y^1_k] = \{y^1_k \mid y^n_k\} ,$$

where $\gamma(t_n) = y_k^1$ for some sequence $(t_k) \subset [0, t_0)$. Moreover taking a subsequence we may assume that $y_k^i \to y^i$ for $k \to \infty$. Since $y^i \in Y$ it follows immediately that the y^i are pairwise different and obviously $\{y^1, \dots, y^n\} \subset [y^1]$ implying that $M(\gamma(t_0)) \geqslant n$ contrary to our assumption $M(\gamma(t_0)) < n$. We have proved that $M(\gamma(t_0)) = n$. If $t_0 < 1$ the above argument shows that $M(\gamma(t)) = n$ for t near to t_0. Hence we must have $t_0 = 1$ by the maximality of t_0 in $[0,1]$. By the first step, this implies $M(\gamma(1)) = n$. □

LEMMA 6.6. *Given $z_0 \in \mathbb{C}$ there exists a disc B around z_0 such that given a sequence $z_k \to z_0$ with $z_k \in Y$ and a sequence $(z_k') \subset B$ with $(z_k, z_k') \in \Gamma$ we have $z_k' \to z_0$.*

Proof: Arguing indirectly we find a sequence (y_k) satisfying $y_k \to z_0$, $y_k \neq z_0$ and $\tilde{u}(y_k) = \tilde{u}(z_0)$. This, however, contradicts the similarity principle which implies that the image under the map $\tilde{u}$ of a small punctured ball around z_0 does not contain $\tilde{u}(z_0)$.

PROPOSITION 6.7. *Γ defines an equivalence relation $\sim$ on $\mathbb{C}$.*

Proof: Clearly $\Gamma|(Y \times Y)$ is an equivalence relation. This follows immediately using (101) and (102). Assume now $(z, z'), (z', z'') \in \Gamma$. We pick a sequence $((z_k, z_k')) \subset Y \times Y$ converging to (z, z') and satisfying $\tilde{u}(z_k) = \tilde{u}(z_k')$, and similarly, a sequence $((y_k, z_k'')) \subset Y \times Y$ converging to (z', z''). For k large enough let us homotope y_k in a suitable annular domain around z' to z_k'. This gives a homotopy of z_k'' to some new point y_k''. As an application of the similarity principle this new point y_k'' is again close to z'' provided y_k is close enough to z'. Indeed the similarity principle shows that during the homotopy the point moves in an annular domain around z''. equivalent to the former and still close to z''. Hence, we have a sequence $((z_k, y_k'')) \subset \Gamma$ converging to (z, z''). This shows that Γ is an equivalence relation on $\mathbb{C}$. □

Equip $\mathbb{C}/\sim$ with the quotient topology and let $p : \mathbb{C} \to \mathbb{C}/\sim$ be the canonical projection. We have

LEMMA 6.8. *$\mathbb{C}/\sim$ is Hausdorff and p is open.*

Proof: That the quotient space is Hausdorff is easily verified. Indeed let $[z]$ and $[z']$ be two different points. By definition $\tilde{u}(z) \neq \tilde{u}(z')$. Take disjoint open neighbourhoods W and W' of $\tilde{u}(z)$ and $\tilde{u}(z')$ respectively and define

$$U = \{[y] \mid \tilde{u}(y) \in W\} \tag{104}$$

$$U' = \{[y'] \mid \tilde{u}(y') \in W'\} . \tag{105}$$

Clearly $U \cap U' = \emptyset$. Moreover $p^{-1}(U) = \tilde{u}^{-1}(W)$. Since $\tilde{u}$ is continuous this set is open. Hence U is open. Similarly for U'. Next let us show that p is open. Given an open set $V \subset \mathbb{C}$ we have to show that $p^{-1}(p(V))$ is open. If $V \subset Y$ this is trivial. So assume that V is an open disc around a point $z_0 \in \mathbb{C} \setminus Y$. Then, assuming V to be small enough, $p^{-1}(p(V \setminus \{z\}))$ is a finite union of open punctured discs (contained in Y) as is easily verified. Hence $p^{-1}(p(V))$ is a union of open discs. This completes the proof of the lemma. □

Next we define a complex structure on $\mathbb{C}/\sim$ for which p is a holomorphic map. Take first a point $[z_0]$ with $z_0 \in Y$. We find an open neighbourhood U of z_0 such that $[z] \cap U = \{z\}$ for all $z \in U$. Define $\phi : p(U) = \tilde{U} \to \mathbb{C}$ by $\phi([z]) = z'$, where z' is the unique point in $U \cap [z]$. This defines an atlas on $Y/\sim$. Moreover if $\tilde{U} \cap \tilde{V} \neq \emptyset$ we see that $\phi_{\tilde{U}} \circ (\phi_{\tilde{V}})^{-1}$ satisfies on its domain of definition

$$\phi_{\tilde{U}} \circ (\phi_{\tilde{V}})^{-1}(w) = (\tilde{u}|U)^{-1} \circ \tilde{u}(z) .$$

These maps are holomorphic. Moreover, $\phi_{\tilde{U}} \circ p = Id$. Hence p is holomorphic. The interesting part of the construction happens at points not belonging to Y. Let $[z_0] \in \mathbb{C}/\sim$ and assume that $z_0 \notin Y$. This implies the existence of a point $z_0' \in \mathbb{C}$ equivalent to z_0 such that $T\tilde{u}(z_0') = 0$. Hence without loss of generality we may assume that $T\tilde{u}(z_0) = 0$. We have to construct a compatible chart. Without loss of generality we may assume that $z_0 = 0$. Take a small closed ball D around 0 and consider the components of $p^{-1}(p(D))$. In view of Lemma 6.6 we see that if the disc B in Lemma 6.6 is small enough the component containing 0 is as small as we wish. Denote the component containing 0 by A. By construction, Γ induces an equivalence relation Γ_A on A. Away from 0, an equivalence class has the same number of points, say $1 < N \leqslant n$ where n is the number of points in an equivalence class of Γ on $\mathbb{C}$ (as introduced above). If A is small enough it is diffeomorphic to a closed disc. Denote by $\dot{A}$ the set $A \setminus \{0\}$. Then $\dot{A}/\sim$ is a Riemannian surface with the complex structure defined above and

$$\dot{A} \to \dot{A}/\sim \tag{106}$$

is a N-sheeted unbranched covering. Consider the set G of Deck transformations, i.e. holomorphic maps $\phi : A \to A$ preserving the fibration (106). The boundary of A is invariant under Γ_A. Let ∂A be equipped with the induced orientation of A. Fix a point z_0 on the boundary of A and denote by γ_0 a closed positively oriented arc on ∂A connecting z_0 with a point z_1 such that all points on $\gamma_0 \setminus \{z_1\}$ are not equivalent. With other words z_1 is the first point on ∂A (counter clockwise) equivalent to z_0. Similarly construct points z_2 equivalent to z_1 and corresponding arcs γ_1 and so on. Let us construct a Deck transformation $\phi A \to A$ mapping z_0 to z_1. Take any loop $\beta(t)$, $t \in [0,1]$ in $\dot{A}$ with winding number k with respect to 0 starting at z_0. We find numbers $t_0 = 0 < t_1 < t_2 < \dots t_m = 1$, open discs $B(\beta(t_i))$ and holomorphic maps $\phi_i : B(\beta(t_i)) \to \dot{A}$ such that $\phi_0(z_0) = z_1$ and

$$\phi_i \mid \left(B(\beta(t_i)) \cap B(\beta(t_{i+1})\right) = \phi_{i+1} \mid \left(B(\beta(t_i)) \cap B(\beta(t_{i+1}))\right) .$$

We claim that

$$\phi_m(z_0) = z_1 . \tag{107}$$

We note that the above construction is an analytic continuation along β. Obviously, by homotopy invariance, the above assertion is true provided it is true for any loop through z_0 homotopic to β. Therefore we can take any loop in ∂A with winding number k. We may assume without loss of generality that $k \neq 0$ and that $\beta : \mathbf{S}^1 \to \partial A$ is an immersion. Here $\mathbf{S}^1 = \mathbf{R}/\mathbf{Z}$. Clearly, it suffices to prove the assertion for $k = 1$. Hence, let $\beta(\frac{1}{m}) = z_1$, $\beta(\frac{2}{m}) = z_2$, etc. and finally $\beta(1) = z_0$. The analytic continuation produces a map $\psi_\beta : [0,1] \to \partial A$, where $\psi_\beta(t)$ is the point associated to $\beta(t)$. Clearly $\psi_\beta(0) = z_1$, $\psi_\beta(\frac{1}{m}) = z_2$, etc. Hence $\psi_\beta(1) = z_1 = \varphi_m(z_0)$, which is equivalent to the desired assertion (107). In view of the above discussion we have the following

PROPOSITION 6.9. *Given any two equivalent points z_0 and z_1 in ∂A there exists a Deck transformation $\phi \in G$ satisfying $\phi(z_0) = z_1$. Moreover, $G \simeq \mathbf{Z}_N$, where $N = \sharp[z]_{A'}$.*

Proof: The above discussion shows that the Deck transformations can be defined by analytic continuation along arcs, since the continuation along a loop produces the identity. Since $[z_0]_A$ consists of precisely N elements we must have $\sharp G = N$. Moreover the above discussion showed that if ϕ is the element mapping z_0 to z_1 then $\phi^N = Id$. This completes the proof.

By analytic continuation (Riemann removable singularity theorem) we can extend the Deck transformations on $\dot{A}$ to A. Hence we have a free holomorphic action of $\mathbf{Z}_N$ on A commuting with the projection $A \to A/\sim$. By the Riemannian mapping theorem we find a biholomorphic map Ψ : $D \to A$ with $\Psi(0) = 0$ and $\Psi(1) = z_0$. Hence we obtain an induced free holomorphic $\mathbf{Z}_N$-action on D fixing 0. The biholomorphic self-maps of D fixing 0 are the rotations. Hence the action has to be multiplication by a N-th root of unity.

In order to complete our construction it suffices to define a chart on $D/\sim$, where the equivalence relation in D is defined by $a \sim b$ iff $a^N = b^N$. Define $D \to D : z \to z^N$. Then we obtain an induced homeomorphism $D/\sim \to D$. Equip $D/\sim$ with the unique complex structure which turns the map $[z] \to z^N$ into a biholomorphic diffeomorphism. The proof is finished if we can show that the map $(A \setminus \{0\})/\sim \to D/\sim$ induced by Ψ is holomorphic.

In that case the map $\alpha : A/\sim \to D : [z] \to \eta(z)^N$ is a chart compatible with those chosen already on $Y/\sim$. This can be seen as follows. We have the equality $\Psi(z)^N = \alpha \circ p$ on $A \setminus \{0\}$ and want to show that α is biholomorphic. Since $p : A \setminus \{0\} \to (A \setminus \{0\})/\sim$ is a local biholomorphic diffeomorphism and the same is true for $z \to \Psi(z)^N$ away from 0 our assertion follows.

Next define $\tilde{v} : \mathbb{C}/\sim \to \mathbb{R} \times M$ by

$$\tilde{v}([z]) = \tilde{u}(z) .$$

Since $\tilde{u} = \tilde{v} \circ p$ and p is a local biholomorphic diffeomorphism away from the isolated points in $\mathcal{S}$ we see that $\tilde{v}|(Y/\sim)$ is pseudoholomorphic. However the complement of $Y/\sim$ in $\mathbb{C}/\sim$ consists of isolated points and we can apply Gromov's removable singularity theorem to find that $\tilde{v}$ is pseudo holomorphic everywhere.

Since $\tilde{u} = (a, u)$ and a is coercive we see that $p : \mathbb{C} \to \mathbb{C}/\sim$ is proper. Consider the map $z \to \tilde{u}(\frac{1}{z})$ near 0. The equivalence relation is nicely behaved in a punctured neighbourhood of 0 since $\tilde{u}$ is asymptotically an immersion. In the non-degenerate case this follows from the asymptotics. However, it is also true in general according to the following lemma.

LEMMA 6.10. *Let $\tilde{u} = (a, u)$ be a finite energy plane. Then there exists an $R > 0$ such that $T\tilde{u}(z) \neq 0$ for $|z| \geqslant R$.*

Proof: We argue by contradiction. Working in cylindrical coordinates we assume that there are sequences $s_k \to \infty$ and $t_k \to t$ such that for the tangent map

$$T\tilde{u}(s_k, t_k) = 0 .$$

Defining $\tilde{u}_k(s, t) = \tilde{u}(s + s_k, t)$ we observe that

$$\int_{\{0\} \times S^1} u_k^* \lambda \to \int_{\mathbb{C}} u^* d\lambda > 0$$

as $k \to \infty$, and, given, $r > 0$,

$$\int_{[-r,r] \times S^1} u_k^* d\lambda \to 0$$

as $k \to \infty$. Moreover, $E(\tilde{u}_k) = E(\tilde{u}) < \infty$. Therefore, after taking a subsequence, we may assume, in view of the C^∞ estimates for $\tilde{u}$, that $\tilde{u}_k \to \tilde{v}$ in $C^\infty_{loc}(\mathbb{R} \times S^1, \mathbb{R} \times M)$. Hence $\tilde{v}$ satisfies:

$$\bar{\partial} \tilde{v} = 0$$

$$E(\tilde{v}) < \infty$$
$$\int_{\mathbf{R}\times\mathbb{S}^1} v^* d\lambda = 0$$
$$\int_{\{0\}\times\mathbb{S}^1} v^*\lambda = \int_{\mathbf{C}} u^* d\lambda > 0$$

and, for the tangent map,

$$T\tilde{v}(0,t) = 0 \; . \tag{108}$$

Using Theorem 6.11 we have $\tilde{v}(s,t) = (Ts+d, x(Tt))$ for a suitable constant d, and a T-periodic solution x of $\dot{x} = X(x)$, where

$$T = \int_{\mathbf{C}} u^* d\lambda > 0 \; .$$

Consequently, $T\tilde{v}(s,t) \neq 0$, for every $(s,t) \in \mathbf{R} \times \mathbb{S}^1$, contradicting (108). This finishes the proof of Lemma 6.10. □

Back to the proof of Proposition 6.9, we can define a small punctured disc-like neighbourhood of 0 and construct a map ϕ there which is conjugated to a multiplication on $D \setminus \{0\}$ by a root of unity. Hence carrying out the same procedure as before $(\mathbb{C}/\sim) \cup \{\infty\} = (\mathbb{C} \cup \{\infty\})/\sim$ can be given a natural complex structure such that the projection map with $\mathbb{S}^2 = \mathbb{C} \cup \{\infty\}$ induces a holomorphic map $p : \mathbb{S}^2 \to \mathbb{S}^2/\sim$. Here $\mathbb{S}^2/\sim$ is a compact Riemannian surface and p, being non-constant, has positive degree. By Poincaré duality $p^* : H^1(\mathbb{S}^2/\sim) \to H^1(\mathbb{S}^2)$ has to be injective. Therefore $H^1(\mathbb{S}^2/\sim) = 0$ which implies that $\mathbb{S}^2/\sim$ is the topological 2-sphere. Taking a biholomorphic diffeomorphism to the Riemannian sphere, which exists by the uniformization theorem, we may view p as a map $\mathbb{S}^2 \to \mathbb{S}^2$ with $p^{-1}(\infty) = \{\infty\}$. That however means p is the extension of a polynomial to the Riemannian sphere. Summing up we have $\tilde{u}$ factorized as $\tilde{v} \circ p$, where p is a complex polynomial and $\tilde{v}$ is a somewhere injective map. This finishes the proof of Proposition 6.9. □

6.2 Characterization of finite energy surfaces with vanishing $d\lambda$-energy. Assume $\tilde{u} : \dot{S} \to \mathbf{R} \times M$ satisfies

$$\pi \circ Tu \circ i = J \circ \pi \circ Tu* \tag{109}$$
$$da = (u^*\lambda)i \; .$$

Here $\dot{S} = S \setminus \Gamma$ is a punctured compact Riemannian surface without boundary, i.e. a closed Riemannian surface with a finite set Γ of points removed. Assume the solution $\tilde{u}$ of equation (109) satisfies, in addition,

$$\int_{\dot{S}} u^* d\lambda = 0 \;\text{ and }\; 0 < E(\tilde{u}) < \infty \; . \tag{110}$$

We note that the assertion $0 < E(\tilde{u})$ is equivalent to the assumption that $\tilde{u}$ is not constant. By the results in [HWyZ3] the following three cases can occur at a singularity $z \in \Gamma$ as already explained in section 1.

removable singularity: The $\mathbf{R}$–component is bounded near z.
positive singularity: The $\mathbf{R}$–component is unbounded but bounded from below near z.
negative singularity: The $\mathbf{R}$–component is unbounded but bounded from above near z.

Without loss of generality we may assume that we do not have removable singularities. Denote the positive and negative singularities by $\Gamma^{\pm}$, respectively. We recall that $\Gamma = \emptyset$ is impossible.

THEOREM 6.11. *Let $\tilde{u}$ satisfy* (109) *and* (110). *Then there exists a periodic orbit x of period $T = E(\tilde{u})$ and a non-constant holomorphic map $\phi : S \to \mathbb{S}^2 = \mathbf{C} \cup \{+\infty\}$ satisfying*

$$\phi^{-1}(+\infty) = \Gamma^{+} \quad \text{and} \quad \phi^{-1}(0) = \Gamma^{-} \ , \tag{111}$$

such that, in addition,

$$\tilde{u}(z) = \tilde{\eta} \circ \phi(z) \quad \text{for} \quad z \in \dot{S} \ , \tag{112}$$

where the trivial pseudo holomorphic map $\tilde{\eta} : \mathbf{C} \setminus \{0\} \to \mathbf{R} \times M$ is defined by $\tilde{\eta}(e^{2\pi(s+it)}) = (Ts + c, x(Tt))$ for a suitable constant $c \in \mathbf{R}$.

Proof: Denote by $\tilde{\dot{S}}$ the universal covering of $\dot{S}$ and by $p : \tilde{\dot{S}} \to \dot{S}$ the (holomorphic) covering projection. Since (by (110)) $\pi \circ Tu = 0$, we find that the image of u must be contained in an orbit $x(\mathbf{R})$ of the differential equation $\dot{x} = X(x)$. Since $\tilde{\dot{S}}$ is simply connected we find a smooth map $f : \tilde{\dot{S}} \to \mathbf{R}$ such that $x(f(z)) = u(p(z))$ for all $z \in \tilde{\dot{S}}$. Consider the map $h : \tilde{\dot{S}} \to \mathbf{C}$ defined by $h = a \circ p + if$. It must be non-constant because otherwise $\tilde{u}$ would be constant. We claim that h is holomorphic. Indeed, we compute

$$\begin{aligned} dh \circ i &= da \circ Tp \circ i + i \circ df \circ i \\ &= (u^{*}\lambda) \circ i \circ Tp \circ i + i \circ df \circ i \\ &= -p^{*}u^{*}\lambda + i \circ df \circ i \\ &= i \circ i \circ df + i \circ df \circ i \\ &= i \circ (i \circ df + df \circ i) \\ &= i \circ \big(d(a \circ p) + i \circ df\big) \\ &= i \circ dh \ . \end{aligned} \tag{113}$$

The fundamental group G of $\dot{S}$ acts on $\tilde{\dot{S}}$. Let us denote this action by $G \times \tilde{\dot{S}} \to \tilde{\dot{S}} : (g, z) \to gz$. Consider for a point $z \in \tilde{\dot{S}}$ the map $g \to h(gz)$. Since we have $u(p(z)) = u(p(gz)) = x(f(gz))$ we must have $f(gz) = f(z)$ if the orbit x is not periodic, which of course would imply that f can be lifted to $\dot{S}$. In that case denote the lift by f. We know the asymptotic behavior of $\tilde{u}$ near its singular points. From that it follows immediately that $\int_C u^*\lambda$ is nonzero for a small loop C of winding 1 around a singular point. This contradicts the fact that $u = x \circ f$ and the image of f is in $\mathbb{R}$. Hence x is periodic and $f(gz) - f(z) = \eta(g)T$, where $T > 0$ is the period, and $\eta : G \to \mathbb{Z}$ is a nontrivial representation of G in the Abelian group $\mathbb{Z}$. Also near a puncture f remains bounded, whereas a converges to $\pm\infty$ depending whether the point is positive or negative. Denote by $N \in \mathbb{N}$ the positive integer defined by image$(\eta) = N\mathbb{Z} \subset \mathbb{Z}$. Clearly f induces a map, again denoted by f, from $\dot{S}$ into $\mathbb{R}/(NT\mathbb{Z})$. Hence h induces a holomorphic map, again denoted by h, from $\dot{S}$ into $\mathbb{C}/(NT\mathbb{Z})$. We can compactify this cylinder to $\mathbb{S}^2$ by adding the points 0 and $+\infty$ and obtain a holomorphic map again denoted by h from $S \to \mathbb{S}^2$ such that $h^{-1}(0) = \Gamma^-$ and $h^{-1}(+\infty) = \Gamma^+$. The proof of Theorem 6.11 is complete. □

References

[AH] C. ABBAS, H. HOFER, Holomorphic curves and global questions in contact geometry, to appear in Lectures in Mathematics, ETH Zürich, Birkhäuser.

[Au] AUDIN ET AL., Holomorphic Curves in Symplectic Geometry, Progress in Mathematics 117, Birkhäuser, 1994.

[B] D. BENNEQUIN, Entrelacements et équations de Pfaff, Astérisque 107-108 (1983), 83–161.

[CFH] K. CIELIEBAK, A. FLOER, H. HOFER, Sympilectic homology II: general symplectic manifolds, to appear in Math. Zeit.

[CoZ] C. CONLEY, E. ZEHNDER, An index theory for periodic solutions of a Hamiltonian system, Geometric Dynamics, Springer Lecture Notes in Mathematics 1007 (1983), 132-145.

[E1] Y. ELIASHBERG, Contact 3–manifolds, twenty year since J. Martinet's work, Ann. Inst. Fourier 42 (1992), 165–192.

[E2] Y. ELIASHBERG, Filling by holomorphic discs and its applications, London Math. Society Lecture Notes, series 151 (1991), 45–67.

[E3] Y. ELIASHBERG, Legendrian and transversal knots in tight contact manifolds, In "Topological Methods in Modern Mathematics", Publish or Perish, 1993.

[FHS] A. FLOER, H. HOFER, D. SALAMON, Transversality results in the elliptic Morse theory for the action functional, to appear in Duke.

[G] E. GIROUX, Convexité en topologie de contact, Comm. Math. Helvetici 66 (1991), 637–677.

[Gr] M. GROMOV, Pseudo Holomorphic curves in symplectic manifolds, Invent. Math. 82 (1985), 307–347.

[H] H. Hofer, Pseudo Holomorphic curves in symplectisations with application to the Weinstein conjecture in dimension three, Invent. Math. 114 (1993), 515–563.
[HWyZ1] H. Hofer, K. Wysocki, E. Zehnder, A characterization of the tight three-sphere, preprint.
[HWyZ2] H. Hofer, K. Wysocki, E. Zehnder, The dynamics on a strictly convex energy surface in $\mathbb{R}^4$. preprint.
[HWyZ3] H. Hofer, K. Wysocki, E. Zehnder, Properties of pseudo holomorphic curves in symplectisations I: Asymptotics, preprint.
[HWyZ4] H. Hofer, K. Wysocki, E. Zehnder, Properties of pseudo holomorphic curves in symplectisations III: Fredholm theory, preprint.
[HZ] H. Hofer, E. Zehnder, Hamiltonian Dynamics and Symplectic Invariants, Birkhäuser, 1994.
[K] T. Kato, Perturbation Theory for Linear Operators, Springer, grundlehren edition, 1976.
[M] D. McDuff, The local behavior of J–holomorphic curves in almost complex 4–manifolds, J. Diff. Geom. 34 (1991), 143–164.
[MS] D. McDuff, D. Salamon, J-holomorphic Curves and Quantum Cohomology, AMS, 1994.
[MiW] M. Micallef, B. White, The structure of branch points in minimal surfaces and in pseudo holomorphic curves, preprint 1994.
[PdM] J. Palis, W. de Melo, Geometric Theory of Dynamical Systems, Springer, Berlin, 1982.
[RS] J.W. Robbin, D. Salamon, The spectral flow and the Maslov index, Journal of the LMS, to appear.
[Ro] D. Rolfson, Knots, Publish or Perish, 1976.
[SZ] D. Salamon, E. Zehnder, Morse theory for periodic solutions of Hamiltonian systems and the Maslov index, Comm. Pure Appl. Math. 45 (1992), 1303–1360.
[Y] R. Ye, Filling by holomorphic discs in symplectic 4-manifolds, preprint.

H. Hofer, K. Wysocki and E. Zehnder
Matematik
ETH Zentrum
CH-8092 Zurich
Switzerland

Submitted: December 1994

Geometric And Functional Analysis 1016-443X/95/0200329-35$1.50+0.20/0
Vol. 5, No. 2 (1995)

BILINEAR FORMS ON EXACT OPERATOR SPACES AND $B(H) \otimes B(H)$

M. JUNGE AND G. PISIER

Abstract

Let E, F be exact operator spaces (for example subspaces of the C^*-algebra $K(H)$ of all the compact operators on an infinite dimensional Hilbert space H). We study a class of bounded linear maps $u : E \to F^*$ which we call tracially bounded. In particular, we prove that every completely bounded (in short *c.b.*) map $u : E \to F^*$ factors boundedly through a Hilbert space. This is used to show that the set OS_n of all n-dimensional operator spaces equipped with the *c.b.* version of the Banach Mazur distance is not separable if $n > 2$.

As an application we show that there is more than one C^*-norm on $B(H) \otimes B(H)$, or equivalently that

$$B(H) \otimes_{\min} B(H) \neq B(H) \otimes_{\max} B(H) ,$$

which answers a long standing open question.

Finally we show that every "maximal" operator space (in the sense of Blecher-Paulsen) is not exact in the infinite dimensional case, and in the finite dimensional case, we give a lower bound for the "exactness constant". In the final section, we introduce and study a new tensor product for C^*-albegras and for operator spaces, closely related to the preceding results.

0. Introduction and Background

Following the remarkable results of Kirchberg on exact C^*-algebras (*cf.* [K1,2]), the notion of "exact operator space" was studied in the paper [Pi1]. In this paper we continue the investigation started in [Pi1].

Let E, F be operator spaces. We denote

$$d_{cb}(E, F) = \inf \{ \|u\|_{cb} \|u^{-1}\|_{cb} \}$$

where the infimum runs over all possible isomorphisms $u : E \to F$. If E, F are not isomorphic, we set, by convention, $d_{cb}(E, F) = \infty$. We denote by $B(H)$ (resp. $K(H)$) the algebra of all bounded (resp. compact) operators on a Hilbert space H. (See below for unexplained notation and terminology.)

In [Pi1], the following characteristic of an operator space, E with $\dim E = n$, was studied:

$$d_{SK}(E) = \inf \{ d_{cb}(E, F) \mid F \subset K(\ell_2) \} , \tag{0.1}$$

The second author is partially supported by the NSF.

or equivalently

$$d_{SK}(E) = \inf\left\{d_{cb}(E,F) \mid F \subset M_N,\ N \geq n\right\}. \tag{0.1)'}$$

For an infinite dimensional operator space X, we define

$$d_{SK}(X) = \sup\left\{d_{SK}(E)\right\}$$

where the supremum runs over all possible finite dimensional subspaces $E \subset X$.

We say that X is exact if $d_{SK}(X) < \infty$. This is equivalent (see [Pi1]) to the exactness of a certain sequence of morphisms in the category of operator spaces, whence the terminology.

Let E, F be exact operator spaces. We will prove below an inequality for completely bounded linear maps $u : E \to F^*$. Here F^* means the "operator space dual" of F in the sense of [ER1], [BlP].

This key inequality can be viewed as a form of Grothendieck's inequality for exact operator spaces. It implies that every *c.b.* map $u : E \to F^*$ can be factorized (in the Banach space sense) through a Hilbert space with a norm of factorization $\leq 4\|u\|_{cb}$. Actually, in this inequality "complete boundedness" can be replaced by a more general notion which we call "tracial boundedness" which has already been considered by previous authors (*cf.* [I], [Bl3]). Using this key inequality and a somewhat surprising application of Baire's theorem, we prove that the metric space OS_n of all n-dimensional operator spaces equipped with the distance

$$\delta_{cb}(E,F) = \log d_{cb}(E,F)$$

is not separable as soon as $n > 2$. We can even show that the subset of all isometrically Hilbertian and homogeneous (in the sense of [Pi3]) n-dimensional operator spaces, is non-separable.

This has a surprising application to C^*-algebra theory. Recall that a C^*-algebra A is called nuclear if $A \otimes_{\min} B = A \otimes_{\max} B$ for any C^*-algebra B (see [L]). For a long time it remained an open problem whether it suffices for the nuclearity of A to assume that

$$A \otimes_{\min} A^{op} = A \otimes_{\max} A^{op} \tag{*}$$

where A^{op} is the opposite C^*-algebra (with the product in reverse order). In [K2], Kirchberg gave the first counterexamples. However, he pointed out that it remained unknown whether $(*)$ holds in the (non-nuclear) case of $A = B(H)$. (Note that $B(H)$ is isomorphic to its opposite.) Kirchberg also proposed an approach to this question together with a series of equivalent conjectures. One of his conjectures was our main motivation to investigate the non-separability of the metric space OS_n, and as a result, we obtain a negative answer to the above mentioned question: the identity (*) fails for $A = B(H)$. In other words, we have

$$B(H) \otimes_{\min} B(H) \neq B(H) \otimes_{\max} B(H) .$$

Equivalently, there is more than one C^*-norm on $B(H) \otimes B(H)$ whenever H is an infinite dimensional Hilbert space. This is proved in section 3.

In the same section, we also show that if E is an n-dimensional Banach space then the operator space max(E) in the sense of [BlP] satisfies

$$d_{SK}(\max(E)) \geq \frac{\sqrt{n}}{4} .$$

In particular, for any infinite dimensional Banach space X the operator space max(X) is not exact.

After circulating a first version of this paper, we observed that the non-separability of OS_n for $n > 2$ can be alternately deduced from properties of Kazhdan-groups, following the ideas of Voiculescu [Vo] (see Remark 2.10 below). However, this approach does not seem to give the same auxiliary information as our original one.

In the final section 4, we introduce a "new" tensor product obtained as follows. Let $E \subset B(H)$ and $F \subset B(K)$ be two operator spaces. We denote by $E \otimes_M F$ the completion of the linear tensor product with respect to the norm induced on it by $B(H) \otimes_{\max} B(K)$. This tensor product makes sense both in the category of operator spaces and in that of C^*-algebras. Our previous results show that it differs in general from the minimal tensor product. We include several properties of this tensor product, based mainly on [K3].

In the rest of this introduction we give some background and explain our notation. We refer to [S], [T] for operator algebra theory. Let H, K be Hilbert spaces. We will denote by $H \otimes_2 K$ their Hilbertian tensor product. By an operator space we mean a closed subspace of $B(H)$ for some Hilbert space H. If $E_1 \subset B(H_1)$, $E_2 \subset B(H_2)$ are operator spaces, we will denote by $E_1 \otimes_{\min} E_2$ their minimal (or spatial) tensor product equipped with the minimal (or spatial) tensor norm induced by the space $B(H_1 \otimes_2 H_2)$. When E_1 and E_2 are C^*-subalgebras then this norm is a C^*-norm, and actually it is the smallest C^*-norm on the algebraic tensor product $E_1 \otimes E_2$. In the case of C^*-algebras, we will denote by $E_1 \otimes_{\max} E_2$ the completion of $E_1 \otimes E_2$ with respect to the largest C^*-norm on $E_1 \otimes E_2$.

We recall that if $E \subset B(H)$ and $F \subset B(K)$ are operator spaces, then a map $u : E \to F$ is completely bounded (in short *c.b.*) if the maps $u_m = I_{M_m} \otimes u : M_m(E) \to M_m(F)$ are uniformly bounded when $m \to \infty$, i.e. if we have $\sup_{m \geq 1} \|u_m\| < \infty$. The *c.b.* norm of u is defined as

$$\|u\|_{cb} = \sup_{m\geq 1} \|u_m\| \ .$$

We will say that u is completely isometric (resp. completely contractive) or is a complete isometry (resp. a complete contraction) if the maps u_m are isometries (resp. of norm ≤ 1) for all m. This is the same as saying that for any operator space X the map

$$I_X \otimes u : X \otimes_{\min} E \to X \otimes_{\min} F$$

is an isometry (resp. a contraction). We also recall that u is called completely positive (in short *c.p.*) if all the maps u_m are positive.

If E_1, F_1, E_2, F_2 are operator spaces and if

$$u_1 : E_1 \to F_1 \qquad u_2 : E_2 \to F_2$$

are completely bounded, then

$$u_1 \otimes u_2 : E_1 \otimes_{\min} F_1 \longrightarrow E_2 \otimes_{\min} F_2$$

is *c.b.* and we have

$$\|u_1 \otimes u_2\|_{cb} = \|u_1\|_{cb} \|u_2\|_{cb} \ . \tag{0.2}$$

If $u : E_1 \to F_1$ and $u_2 : E_2 \to F_2$ are completely isometric then $u_1 \otimes u_2 : E_1 \otimes_{\min} E_2 \longrightarrow F_1 \otimes_{\min} F_2$ also is completely isometric. In particular, we note the completely isometric identity

$$M_n(E) = M_n \otimes_{\min} E \ .$$

We will use (0.2) repeatedly in the sequel with no further reference. We refer the reader to [P1] for more information.

It is known that the analogue of (0.2) fails for the max-tensor product. Instead we have:

(0.3) If E_1, F_1 and E_2, F_2 are C^*-algebras (or merely operator systems) and if $u_1 : E_1 \to F_1$ and $u_2 : E_2 \to F_2$ are completely positive maps, then the map $u_1 \otimes u_2$ defined on the algebraic tensor product extends to a completely positive and *c.b.* map

$$u_1 \otimes u_2 : E_1 \otimes_{\max} F_1 \longrightarrow E_2 \otimes_{\max} F_2$$

satisfying

$$\|u_1 \otimes u_2\|_{cb(E_1 \otimes_{\max} F_1, E_2 \otimes_{\max} F_2)} = \|u_1\| \|u_2\| \ .$$

See e.g. [T, Proposition 4.23], [P1, Proposition 3.5 and Proposition 10.11] or [W2, Proposition 1.11] for more details.

We will use the duality theory for operator spaces, which was introduced in [BlP], [ER1] using Ruan's "abstract" characterization of operator spaces ([R]). This can be summarized as follows: Let E be an operator space and let E^* be the dual Banach space. Then, for some Hilbert space K, there is an isometric embedding $E^* \subset B(K)$ such that for any operator space F, the minimal (=spatial) norm on $E^* \otimes F$ coincides with the norm induced on it by the space $cb(E, F)$. Moreover, this property characterizes the operator space

$$E^* \subset B(K)$$

up to complete isometry. We will refer to this operator space as the "operator space dual" of E. We refer to [BlP], [ER1,2], [Bl1,2], [R] for detailed information.

Consider arbitrary Banach spaces E, Y and an operator $u : E \to Y$.

Recall that u is called 2-absolutely summing if there is a constant C such that for all n for all $(x_1, \ldots, x_n)$ in E we have

$$\sum \|ux_i\|^2 \le C^2 \sup \left\{ \sum |\xi(x_i)|^2 \;\middle|\; \xi \in E^* \ , \ \|\xi\| \le 1 \right\} .$$

We denote by $\pi_2(u)$ the smallest constant C for which this holds.

It is easy to check that for any bounded operators $v : Y \to Y_1$ and $w : E_1 \to E$ we have

$$\pi_2(vuw) \le \|v\| \pi_2(u) \|w\| . \tag{0.4}$$

Notation. Let (x_i) be a finite sequence in a C^*-algebra A. We will denote for brevity

$$RC((x_i)) = \max \left\{ \left\| \sum x_i x_i^* \right\| \ , \ \left\| \sum x_i^* x_i \right\| \right\} .$$

Note that if A is commutative and if $E \subset A$ is any subspace equipped with the induced norm, we have for all $(x_1, \ldots, x_n)$ in E

$$RC((x_i)) = \sup \left\{ \sum |\xi(x_i)|^2 \mid \xi \in E^* \ , \ \|\xi\| \le 1 \right\} . \tag{0.5}$$

DEFINITION 0.3. *Consider a C^*-algebra A and a Banach space Y. Let $E \subset A$ be subspace. A linear map $u : E \to Y$ will be called $(2, RC)$-summing if there is a constant C such that for all n and for all $(x_1, \ldots, x_n)$ in E we have*

$$\sum \|ux_i\|^2 \le C^2 RC((x_i)) .$$

We denote by $\pi_{2,RC}(u)$ the smallest constant C for which this holds. We refer to [Pi3] for a more systematic treatment of $(2, w)$-summing operators when w is a "weight" in the sense of [Pi3]. (See also [Pi4, §2] or [Pi2].) By a well known variant of a Pietsch's factorization theorem for 2-absolutely summing operators, we have

PROPOSITION 0.4. *Consider $u : E \to Y$ as in the preceding definition. The following are equivalent.*

(i) *u is $(2, RC)$-summing and $\pi_{2,RC}(u) \le C$.*

(ii) *There are states f, g on A and $0 \le \theta \le 1$ such that*

$$\forall x \in E \qquad \|u(x)\| \le C\{\theta f(x^*x) + (1-\theta)g(xx^*)\}^{1/2} .$$

(iii) *The map $u : E \to Y$ admits an extension $\tilde{u} : A \to Y$ such that $\pi_{2,RC}(\tilde{u}) \le C$.*

Proof: This is, by now, a well known application of the Hahn-Banach theorem. For more details we refer the reader to, e.g. [Pi4, Lemma 1.3] or [Pi3, Prop. 5.1].

For bilinear forms, we have the following known analogous statement. (In the commutative case, this can be found in [Kw].)

PROPOSITION 0.5. *Let A, B be C^*-algebras and let $E \subset A$, $F \subset B$ be closed subspaces. Let $C > 0$ be a fixed constant. The following properties of a linear map $u : E \to F^*$ are equivalent.*

(i) *For any n, any $(x_1, \ldots, x_n)$ in E and any $(y_1, \ldots, y_n)$ in F, we have*

$$\left|\sum \langle u(x_i), y_i\rangle\right| \le C\left[RC((x_i))RC((y_i))\right]^{1/2} .$$

(ii) *There are states f_1, g_1 on A, f_2, g_2 on B and $0 \le \theta_1, \theta_2 \le 1$ such that*

$$\forall (x,y) \in E \times F \quad |\langle u(x), y\rangle| \le C\left[\theta_1 f_1(x^*x) + (1-\theta_1)g_1(xx^*)\right]^{1/2}$$
$$\left[\theta_2 f_2(y^*y) + (1-\theta_2)g_2(yy^*)\right]^{1/2} .$$

(iii) *For some Hilbert space H, u admits a factorization of the form*

$$E \xrightarrow{a} H \xrightarrow{b^*} F^*$$

with operators $a : E \to H$ and $b : F \to H^*$ such that

$$\pi_{2,RC}(b)\pi_{2,RC}(a) \le C .$$

Proof: See e.g. [Pi3, Theorem 6.1] or [Pi4, Lemma 1.3].

Notation. Let $u : X \to Y$ be an operator between Banach spaces. Assume that u factors through a Hilbert space H, i.e. we have $u = \alpha\beta$ with $\alpha : H \to Y$ and $\beta : X \to H$. Then we will denote

$$\gamma_2(u) = \inf\{\|\alpha\| \, \|\beta\|\}$$

where the infimum runs over all possible factorization. This is the "norm of factorization through Hilbert space" of u. See [Pi2] for more on this theme.

COROLLARY 0.6. *Let u be as in Proposition 0.5. Then if* (iii) *holds there are operators $\tilde{a} : A \to H$ and $\tilde{b} : B \to H^*$ such that $\tilde{b}^*\tilde{a}$ (viewed as a bilinear form on $A \times B$) extends u and*

$$\pi_{2,RC}(\tilde{a})\pi_{2,RC}(\tilde{b}) \leq C \ .$$

In particular, if we let $\tilde{u} = \tilde{b}^*\tilde{a}$, then the operator $\tilde{u} : A \to B^*$ satisfies $\|\tilde{u}\| \leq C$ and $\langle \tilde{u}(x), y\rangle = \langle u(x), y\rangle$ for all (x, y) in $E \times F$. Moreover we have $\gamma_2(u) \leq C$.

Proof: This follows from Propositions 0.4 and 0.5.

The following fact is well known to specialists. (See, e.g. [K1, Lemma 3.9])

LEMMA 0.7. *Let C be a C^*-algebra and let $I \subset C$ be a closed ideal. Let E, X be operator spaces with $E \subset X$. Consider the canonical (complete) contraction $C \otimes_{\min} X \to (C/I) \otimes_{\min} X$. Since this map vanishes on $I \otimes_{\min} X$, we clearly have a canonical (complete) contraction*

$$T_X : C \otimes_{\min} X / I \otimes_{\min} X \to (C/I) \otimes_{\min} X \ .$$

If T_X is an isomorphism, then T_E also is an isomorphism. Moreover,

$$\|T_E^{-1}\| \leq \|T_X^{-1}\| \ .$$

Proof: It is well known (*cf.* [T, p. 27]) that I possesses an approximate unit formed of elements $p_i \in I$ such that $0 \leq p_i \leq 1$ (hence $\|1 - p_i\| \leq 1$) and $p_i x \to x \ \forall x \in I$. Then the proof can be completed exactly as in [Pi1, Lemma 3].

Remark: Equivalently, if we consider the complete isometry

$$C \otimes_{\min} E \to C \otimes_{\min} X$$

then this map defines after passing to the quotient spaces a complete isometry

$$C \otimes_{\min} E / I \otimes_{\min} E \to C \otimes_{\min} X / I \otimes_{\min} X \ .$$

We will also invoke the following elementary fact which follows easily (like the preceding result) from the existence of an approximate unit in any ideal of a C^*-algebra. (Recall that $A \otimes B$ denotes the algebraic tensor product.)

LEMMA 0.8. *Let A, B, C be C^*-algebras. Let $\pi : C \to A$ be a surjective $*$-homomorphism. Let $\mathcal{I} = Ker(\pi)$. Then (viewing the three sets appearing below as subsets of $C \otimes_{\min} B$) we have*

$$[C \otimes B] \cap [\mathcal{I} \otimes_{\min} B] = \mathcal{I} \otimes B.$$

Equivalently, let

$$q : C \otimes_{\min} B \to [C \otimes_{\min} B]/[\mathcal{I} \otimes_{\min} B]$$

be the quotient map and let

$$T : [C \otimes_{\min} B]/[\mathcal{I} \otimes_{\min} B] \to A \otimes_{\min} B$$

be the morphism associated to $\pi \otimes I_B$. *Then* T *induces a linear and* $*$-*algebraic isomorphism between* $q(C \otimes B)$ *and* $A \otimes B$.

Acknowledgement. The second author would like to thank E. Kirchberg for introducing him to the questions considered in this paper and B. Maurey for a stimulating conversation. We are grateful to Alain Valette for communicating to us the connection with the work of Lubotzky-Phillips-Sarnak (see Remark 2.12).

1. Factorization of Bilinear Forms on Exact Operator Spaces

Let $n \geq 1$ be an integer. We will denote by

$$J_n : M_n \to M_n^*$$

the map defined by

$$\forall x, y \in M_n \quad \langle J_n(x), y \rangle = \frac{1}{n} tr({}^t yx) \ .$$

The following notion is natural for our subsequent results. It has already been considered in [I].

DEFINITION 1.1. *Let* E, F *be operator spaces. Let* $u : E \to F^*$ *be a linear map. We will say that* u *is tracially bounded (in short t.b.) if*

$$\sup_{n \geq 1} \|J_n \otimes u\|_{M_n(E) \to M_n(F)^*} < \infty$$

and we denote

$$\|u\|_{tb} = \sup_{n \geq 1} \|J_n \otimes u\|_{M_n(E) \to M_n(F)^*} \ .$$

Equivalently, $u : E \to F^*$ *is t.b. iff the bilinear forms*

$$u_n : M_n(E) \times M_n(F) \to \mathbb{C}$$

defined by

$$u_n\big((x_{ij}), (y_{ij})\big) = \frac{1}{n} \sum_{ij} \langle u(x_{ij}), y_{ij} \rangle$$

are bounded uniformly in n *and we have*

$$\|u\|_{tb} = \sup_{n \geq 1} \|u_n\| \ .$$

We immediately observe

LEMMA 1.2. *For a linear map $u : E \to F^*$*

complete boundedness $\Rightarrow$ tracial boundedness $\Rightarrow$ boundedness ,

and we have

$$\|u\| \leq \|u\|_{tb} \leq \|u\|_{cb} \ .$$

Proof: If u is *c.b.* then for any u and any $(x_{ij}) \in M_n(E)$

$$\|(u(x_{ij}))\|_{M_n(F^*)} \leq \|u\|_{cb}\|x\|_{M_n(E)} \ .$$

We have $M_n(F^*) = cb(F, M_n)$, hence

$$\|(u(x_{ij}))\|_{M_n(F^*)} = \sup\left\{\|\langle u(x_{ij}), y_{k\ell}\rangle\|_{M_n(M_n)} \;\middle|\; \|(y_{k\ell})\|_{M_n(F)} \leq 1\right\}$$

$$= \sup_{y \in B_{M_n(F)}} \left\{\left|\sum_{ijk\ell} \langle u(x_{ij}), y_{k\ell}\rangle \alpha_{j\ell}\beta_{ik}\right| \;\middle|\; \sum_{j\ell} |\alpha_{j\ell}|^2 \leq 1 \ \sum_{ik} |\beta_{ik}|^2 \leq 1\right\} .$$

Taking $\alpha, \beta = \frac{I}{\sqrt{n}}$ we get

$$\|(u(x_{ij}))\|_{M_n(F^*)} \geq \sup\left\{\frac{1}{n}\left|\sum \langle u(x_{ij}), y_{ij}\rangle\right| \;\middle|\; \|(y_{ij})\|_{M_n(F)} \leq 1\right\}$$

hence $\|u\|_{tb} \leq \|u\|_{cb}$. The inequality $\|u\| \leq \|u\|_{tb}$ is clear by taking $n = 1$.□

The following consequence of the non-commutative Grothendieck inequality is known ([Bl3]).

LEMMA 1.3. *Let A, B be C^*-algebras. Then any bounded linear operator $u : A \to B^*$ is tracially bounded and we have*

$$\|u\| \leq \|u\|_{tb} \leq K\|u\|$$

for some numerical constant K. Let $E \subset A$ and $F \subset B$ be closed subspaces and let $i_E : E \to A$, $i_F : F \to B$ be the inclusion maps. Then the restriction $i_F^* u i_E : E \to F^*$ satisfies

$$\|i_F^* u i_E\|_{tb} \leq K\|u\| \ .$$

Proof: Consider $x = (x_{ij}) \in M_n(A)$ and $y = (y_{ij}) \in M_n(B)$.

Let $\alpha(x) = \max\left\{n^{-1/2}\left\|\sum_{ij} x_{ij}^* x_{ij}\right\|^{1/2}, n^{-1/2}\left\|\sum_{ij} x_{ij} x_{ij}^*\right\|^{1/2}\right\}$. It is easy to check that

$$\alpha(x) \leq \|x\|_{M_n(A)} \ .$$

By the non-commutative Grothendieck inequality (cf. [H1], see also [Pi2]) we have for some numerical absolute constant K

$$\left|\frac{1}{n}\sum_{ij} \langle u(x_{ij}), y_{ij}\rangle\right| \leq K\|u\|\alpha(x)\alpha(y)$$

hence a fortiori

$$\leq K\|u\|\|x\|_{M_n(A)}\|y\|_{M_n(B)} \ .$$

Therefore $\|u\|_{tb} \le K\|u\|$. The second part is obvious since

$$\|i_F^* u i_E\|_{tb} \le \|u\|_{tb} . \qquad \square$$

Remark: The preceding argument actually shows the following: Consider $u : A \to B^*$ of the form $u = a^*b$, where $b : E \to H$ and $a : F \to H^*$ are $(2, RC)$–summing operators. Then we have

$$\|u\|_{tb} \le \pi_{2,RC}(a)\pi_{2,RC}(b) .$$

We will now show that if E and F are exact operator spaces, the converse to the second part of Lemma 1.3 also holds, that is to say the bilinear form on $E \times F$ associated to a tracially bounded map $u : E \to F^*$ is the restriction of a bounded bilinear form on $A \times B$. This is the key result for this paper. We denote by F_∞ the free group with countably many generators denoted by $g_1, g_2, \dots$. Let $\lambda : F_\infty \to B(\ell_2(F_\infty))$ be the left regular representation and let us denote simply by C_λ the reduced C^*-algebra of F_∞. Let E be an operator space. Let $(x_t)_{t\in F_\infty}$ be a finitely supported family of elements of E. For simplicity we will denote simply by

$$\left\|\sum \lambda(t) \otimes x_t\right\|_{\min}$$

the norm of $\sum_{t\in F_\infty} \lambda(t) \otimes x_t$ in $C_\lambda \otimes_{\min} E$.

The following inequality (cf. [HPi, Prop. 1.1]) plays an important rôle in the sequel.

For any finite sequence (x_i) in a C^*-algebra we have (1.1)

$$\left\|\sum \lambda(g_i) \otimes x_i\right\|_{\min} \le 2 \max\left\{\left\|\sum x_i^* x_i\right\|^{1/2}, \left\|\sum x_i x_i^*\right\|^{1/2}\right\} .$$

THEOREM 1.4. *Let E,F be exact operator spaces. Let $C = d_{SK}(E)d_{SK}(F)$. Let $u : E \to F^*$ be a tracially bounded linear map. Let $(x_t)_{t\in F_\infty}$ (resp. $(y_t)_{t\in F_\infty}$) be a finitely supported family of elements of E (resp. F). Then we have*

$$\left|\sum \langle u(x_t), y_t\rangle\right| \le C\|u\|_{tb} \left\|\sum \lambda(t) \otimes x_t\right\|_{\min} \left\|\sum \lambda(t) \otimes y_t\right\|_{\min} . \quad (1.2)$$

In particular, for all n and all $x_i \in E$, $y_i \in F$ $(i = 1, 2, \dots, n)$ we have

$$\left|\sum \langle u(x_i), y_i\rangle\right| \le C\|u\|_{tb} \left\|\sum \lambda(g_i) \otimes x_i\right\|_{\min} \left\|\sum \lambda(g_i) \otimes y_i\right\|_{\min} , \quad (1.3)$$

hence

$$\le 4C\|u\|_{tb} \max\left\{\left\|\sum x_i^* x_i\right\|^{1/2}, \left\|\sum x_i x_i^*\right\|^{1/2}\right\} \cdot \max\left\{\left\|\sum y_i^* y_i\right\|^{1/2}, \left\|\sum y_i y_i^*\right\|^{1/2}\right\} . \quad (1.3)'$$

Furthermore, if A, B are C^*-algebras such that $E \subset A$, $F \subset B$ (completely isometrically) then u admits an extension $\tilde{u} : A \to B^*$ such that

$$\|\tilde{u}\| \leq \|\tilde{u}\|_{tb} \leq 4C\|u\|_{tb}$$

and $\langle \tilde{u}(x), y\rangle = \langle u(x), y\rangle \ \forall (x, y) \in E \times F$.

Proof: The proof is essentially the same as that of [Pi1, Theorem 8]. An essential ingredient in the proof is Wassermann's construction of a specific embedding of C_λ into an ultrapower (in the von Neumann sense) of matrix algebras, cf. [W1]. This is based on the residual finiteness of the free group. More precisely, consider the family

$$\{M_\alpha \mid \alpha \geq 1\}$$

of all matrix algebras. Let $L = \{(x_\alpha)_{\alpha\geq 1} \mid x_\alpha \in M_\alpha, \sup_{\alpha\geq 1} \|x_\alpha\|_{M_\alpha} < \infty\}$ equipped with the norm $\|(x_\alpha)\|_L = \sup_{\alpha\geq 1} \|x_\alpha\|_{M_\alpha}$. Let $\mathcal{U}$ be an ultrafilter on $\mathbf{N}$. Let us denote by τ_α the normalized trace on M_α, and let

$$I_{\mathcal{U}} = \left\{(x_\alpha)_{\alpha\geq 1} \in L \mid \lim_{\mathcal{U}} \tau_\alpha(x_\alpha^* x_\alpha) = 0\right\} .$$

We then set $\mathcal{N} = L/I_{\mathcal{U}}$. It is a well known fact (cf. [S]) that $\mathcal{N}$ is a finite von Neumann algebra. Let VN_λ be the von Neumann algebra generated by λ. (Note that $C_\lambda \subset VN_\lambda$.) Wassermann proved that for a suitable $\mathcal{U}$ one can find for each g in F_∞ a sequence $(u_\alpha^g)_{\alpha\geq 1}$ such that:

(i) u_α^g is unitary in M_α and has real entries ($\alpha \geq 1, i = 1, 2, \ldots$),

(ii) $\lim_{\mathcal{U}} \tau_\alpha(u_\alpha^{s\,*} u_\alpha^g) = 0$ if $g \neq s$, or equivalently since the entries are real $\lim_{\mathcal{U}} \tau_\alpha({}^t u_\alpha^s u_\alpha^g) = 0$ if $g \neq s$.

(iii) The mapping $\Phi : VN_\lambda \to L/I_{\mathcal{U}}$ which takes $\lambda(g)$ to the equivalence class of $(u_\alpha^g)_{\alpha\geq 1}$ is an isometric representation mapping VN_λ onto a von Neumann subalgebra of $\mathcal{N} = L/I_{\mathcal{U}}$. A fortiori it is completely isometric.

The last point implies that we can write

$$\left\| \sum_{t\in F_\infty} \lambda(t) \otimes x_t \right\|_{\min} = \left\| \sum_{t\in F_\infty} \Phi[\lambda(t)] \otimes x_t \right\|_{(L/I_{\mathcal{U}})\otimes_{\min} E} . \tag{1.4}$$

Without loss of generality we may assume $\dim E = \dim F = n$. Hence, for any $\varepsilon > 0$, for some integer N there is $E_1 \subset M_N$ such that

$$d_{cb}(E, E_1) < d_{SK}(E)(1 + \varepsilon) . \tag{1.5}$$

Clearly we have completely isometrically

$$L/I_{\mathcal{U}} \otimes_{\min} M_N = (L \otimes_{\min} M_N)/I_{\mathcal{U}} \otimes_{\min} M_N .$$

By Lemma 0.7, since $I_{\mathcal{U}}$ is an ideal this remains true with E_1 in the place of M_N. By (1.5), it follows that the natural (norm one) map

$$T_E : (L \otimes_{\min} E)/(I_{\mathcal{U}} \otimes_{\min} E) \to (L/I_{\mathcal{U}}) \otimes_{\min} E$$

has an inverse with norm $\|T_E^{-1}\| < d_{SK}(E)(1+\varepsilon)$, and since ε is arbitrary and the same is true for F, we actually have

$$\|T_E^{-1}\| \le d_{SK}(E) \qquad \|T_F^{-1}\| \le d_{SK}(F)\ . \tag{1.6}$$

Since u is tracially bounded, for any α, the linear map $V_\alpha : M_\alpha(E) \to M_\alpha(F)^*$ defined by

$$\langle V_\alpha(a_\alpha \otimes x), b_\alpha \otimes y\rangle = \tau_\alpha({}^t b_\alpha a_\alpha)\langle u(x), y\rangle\ , \quad \forall\, x \in E,\ \forall\, y \in F,\ \forall\, a_\alpha,\ b_\alpha \in M_\alpha$$

is bounded with

$$\|V_\alpha\| \le \|u\|_{tb}\ . \tag{1.7}$$

For $a \in L$, let us denote by (a_α) its coordinates. Similarly consider an element z in $L \otimes E$. Clearly z can be identified to a family (z_α) with $z_\alpha \in M_\alpha \otimes E$. Note that

$$\|z\|_{L \otimes_{\min} E} = \sup_\alpha \|z_\alpha\|_{M_\alpha(E)}\ . \tag{1.8}$$

With this notation, we can define a linear map $V : L \otimes_{\min} E \to (L \otimes_{\min} F)^*$ by setting

$$\langle V(a \otimes x), b \otimes y\rangle = \lim_{\mathcal{U}} \tau_\alpha({}^t b_\alpha a_\alpha)\langle u(x), y\rangle \qquad \forall\, a, b \in L\ ,\ \forall\, x \in E\ ,\ \forall\, y \in F\ .$$

Clearly, by (1.7) and (1.8) we have $\|V\| \le \sup_\alpha \|V_\alpha\| \le \|u\|_{tb}$. Moreover, it is clear that for all ξ in $I_{\mathcal{U}} \otimes_{\min} E$ and all η in $I_{\mathcal{U}} \otimes_{\min} F$ we have $\langle V(\xi), \eta\rangle = 0$. Therefore V defines canonically a map

$$\tilde{V} : (L \otimes_{\min} E)/(I_{\mathcal{U}} \otimes_{\min} E) \to \big[(L \otimes_{\min} F)/(I_{\mathcal{U}} \otimes_{\min} F)\big]^*$$

such that $\|\tilde{V}\| = \|V\| \le \|u\|_{tb}$. By (1.6), $\tilde{V}$ also defines a map

$$\hat{V} : (L/I_{\mathcal{U}}) \otimes_{\min} E \to \big[(L/I_{\mathcal{U}}) \otimes_{\min} F\big]^*$$

such that

$$\|\hat{V}\| \le C\|u\|_{tb}\ . \tag{1.9}$$

Let (x_t) and (y_t) be as in Theorem 1.4, and let $T_1 = \sum \lambda(t) \otimes x_t$, $T_2 = \sum \lambda(t) \otimes y_t$, and let

$$\hat{T}_1 = (\Phi \otimes I_E)(T_1) \in (L/I_{\mathcal{U}}) \otimes_{\min} E\ , \quad \hat{T}_2 = (\Phi \otimes I_F)(T_2) \in (L/I_{\mathcal{U}}) \otimes_{\min} F\ .$$

By (ii) above, we clearly have $\sum \langle u(x_t), y_t\rangle = \langle \hat{V}(\hat{T}_1), \hat{T}_2\rangle$; hence

$$\Big|\sum \langle u(x_t), y_t\rangle\Big| \le \|\hat{V}\| \|\hat{T}_1\|_{\min} \|\hat{T}_2\|_{\min}\ ,$$

and this together with (1.4) and (1.9) implies (1.2). This proves (1.2). Clearly (1.3) is but a particular case of (1.2) and (1.3)′ follows, using (1.1). Finally, the last assertion follows from Corollary 0.6 and the remark after Lemma 1.3. □

Remark: The preceding proof of (1.2) remains valid if we replace F_∞ by any residually finite discrete group G. Actually, we only use the fact that there is a completely isometric embedding into an ultraproduct, say $\Phi : C^*_\lambda(G) \to L/I_{\mathcal{U}}$ such that

$$\forall\, s,g \in G \quad \lim_{\mathcal{U}} \tau_\alpha\big({}^t\Phi_\alpha(\lambda(s))\Phi_\alpha(\lambda(g))\big) = \delta_{sg}\ .$$

A corollary of the non-commutative Grothendieck theorem says that every bounded linear operator $u : A \to B^*$ factors through a Hilbert space when A, B are C^*-algebras and we have $\gamma_2(u) \le K\|u\|$ for some absolute constant K. In the same vein, we have

COROLLARY 1.5. *Let E, F be as in Theorem 1.4 with $C = d_{SK}(E)d_{SK}(F)$. Then, if $u : E \to F^*$ is tracially bounded there is a Hilbert space H and a factorization $u = a^*b$*

$$E \xrightarrow{b} H \xrightarrow{a^*} F^*$$

with a, b such that

$$\pi_{2,RC}(a)\pi_{2,RC}(b) \le 4C\|u\|_{tb}\ .$$

Conversely, if there is such a factorization we have

$$\|u\|_{tb} \le \pi_{2,RC}(a)\pi_{2,RC}(b)\ . \tag{1.10}$$

Proof: The direct implication follows from Theorem 1.4 and Proposition 0.5. The converse follows from the remark after Lemma 1.3. □

COROLLARY 1.6. *Let E,F be exact operator spaces. Let C=$d_{SK}(E)d_{SK}(F)$. Then every completely bounded map $u : E \to F^*$ factors through a Hilbert space and we have*

$$\gamma_2(u) \le 4C\|u\|_{cb}\ . \tag{1.11}$$

Proof: First recall that $\|u\|_{tb} \le \|u\|_{cb}$ by Lemma 1.2. Then this is deduced from Theorem 1.4 using Proposition 0.5 and Corollary 0.6. □

COROLLARY 1.7. *Let E be a closed subspace of a commutative C^*-algebra and let F be an exact operator space. Then every completely bounded $u : E \to F^*$ is 2-absolutely summing and we have $\pi_2(u) \le 4d_{SK}(F)\|u\|$.*

Proof: Since a commutative C^*-algebra is nuclear we have $d_{SK}(E) = 1$. The result then follows from Theorem 1.4 taking (0.5) into account. □

In particular we have the following corollary which is already known. It was proved independently by V. Paulsen and the second author on one hand (using Clifford matrices, this gives the better constant 2, see [P3]) and by the first author on the other (using random matrices, this yields a worse constant).

COROLLARY 1.8. *Consider C^*-algebras A, B and subspaces $E \subset A$ and $F \subset B$. If A, B are assumed commutative, then any c.b. map $u : E \to F^*$ can be written as $u = a^*b$ with $\pi_2(a)\pi_2(b) \leq 4\|u\|_{cb}$.*

Proof: Again, by the nuclearity of A, B we have $d_{SK}(E) = d_{SK}(F) = 1$. Note that by (0.5) we have $\pi_2(a) = \pi_{2,RC}(a)$ for all $a : E \to H$ and similarly for F. Hence this follows from Theorem 1.4. □

COROLLARY 1.9. *Let E be an exact operator space. Consider a linear map $v : E \to H$ into a Hilbert space. Let $C > 0$ be a constant. Assume that for all n and all (x_{ij}) in $M_n(E)$ we have*

$$\left(\frac{1}{n}\sum_{ij}\left\|v(x_{ij})\right\|^2\right)^{1/2} \leq C\left\|(x_{ij})\right\|_{M_n(E)} . \tag{1.12}$$

Then v is $(2, RC)$-summing and

$$\pi_{2,RC}(v) \leq 2C d_{SK}(E) . \tag{1.13}$$

Proof: Consider the mapping $u = \overline{v}^* v : E \to \overline{E}^*$, obtained by identifying H with its antidual $\overline{H}^*$. From (1.12) it is easy to deduce by Cauchy-Schwarz that $\|\overline{v}^*v\|_{tb} \leq C^2$. Hence by (1.2) we have for all (x_i) in E

$$\sum \|vx_i\|^2 = \sum \langle \overline{v}^* v x_i, x_i\rangle \leq 4C^2 d_{SK}(E)^2 RC\big((x_i)\big)^2$$

and (1.13) follows. □

We need to recall some elementary facts on ultraproducts of operator spaces. Let $(E_i)_{i\in I}$ be a family of operator spaces with $E_i \subset B(H_i)$. Then their ultraproduct $\hat{E} = \Pi E_i/\mathcal{U}$ embeds isometrically into $\Pi B(H_i)/\mathcal{U}$. The latter being a C^*-algebra, this embedding defines an operator space structure on $\hat{E}$. It is easy to check that we have isometrically

$$M_n(\hat{E}) = \Pi M_n(E_i)/\mathcal{U} .$$

Equivalently

$$M_n \otimes_{\min} \hat{E} = \Pi(M_n \otimes_{\min} E_i)/\mathcal{U} .$$

This identity clearly remains valid with M_n replaced by any subspace $F \subset M_n$, therefore we also have the following isometric identity, valid if $d_{SK}(F) = 1$ $(\dim F < \infty)$.

$$F \otimes_{\min} \hat{E} = \Pi(F \otimes_{\min} E_i)/\mathcal{U} . \tag{1.14}$$

This yields

COROLLARY 1.10. *Let I by any set. Let $(E_i)_{i\in I}$, $(F_i)_{i\in I}$ be exact operator spaces with*

$$C = \sup_{i\in I} d_{SK}(E_i) d_{SK}(F_i) < \infty .$$

Let $u_i : E_i \to F_i^*$ be tracially bounded maps with $\sup_{i \in I} \|u_i\|_{tb} \leq 1$. Let $\mathcal{U}$ be an ultrafilter on I and let $\hat{E}$ (resp. $\hat{F}$) be the ultraproduct of $(E_i)_{i \in I}$ (resp. $(F_i)_{i \in I}$) modulo $\mathcal{U}$. Let $\hat{u} : \hat{E} \to (\hat{F})^*$ be the map associated to the family $(u_i)_{i \in I}$. Then for all finite sets (x_k) in $\hat{E}$ and (y_k) in $\hat{F}$ we have

$$\left|\sum \langle \hat{u}(x_k), y_k \rangle\right| \leq C \left\|\sum \lambda(g_k) \otimes x_k\right\|_{\min} \left\|\sum \lambda(g_k) \otimes y_k\right\|_{\min} . \tag{1.15}$$

Proof: We use the fact that $C_\lambda^*(F_\infty)$ is an exact C^*-algebra since it has the slice map property (*cf.* [dCH, Corollary 3.12] and see [Kr] for more details). Therefore if $F = \text{span}(\lambda(g_1), \ldots, \lambda(g_n))$ we have $d_{SK}(F) = 1$. Then it is easy to derive (1.15) from (1.2) (applied to each u_i) taking (1.14) into account.□

2. The Non-separability of OS_n

We will denote by OS_n the set of all n dimensional operator spaces. We identify two elements $E, F \in OS_n$ if they are completely isometric. For $E, F \in OS_n$, let

$$d_{cb}(E,F) = \inf \left\{ \|u\|_{cb} \|u^{-1}\|_{cb} \mid u : E \to F, u \text{ complete isomorphism} \right\} .$$

Then it can be shown (see [Pi1] for the easy details) that this infimum is actually attained and that

$$\delta_{cb}(E,F) = \text{Log } d_{cb}(E,F)$$

is a distance on OS_n for which it is a complete metric space.

We will need a weaker metric structure on the space OS_n. To introduce it we need the following notation: For any linear map $u : E \to F$ between operator spaces we denote

$$\|u\|_k = \|I_{M_k} \otimes u\|_{M_k(E) \to M_k(F)} .$$

Note that

$$\|u\|_{cb} = \sup_{k \geq 1} \|u\|_k .$$

Now consider $E, F \in OS_n$. We define

$$d_k(E,F) = \inf \left\{ \|u\|_k \|u^{-1}\|_k \mid u : E \to F, u \text{ linear isomorphism} \right\} .$$

Then by a simple compactness argument (the unit ball of the space $\mathcal{L}(E,F)$ of all linear maps is compact for any norm on $\mathcal{L}(E,F)$) one can check that

$$d_{cb}(E,F) = \sup_{k \geq 1} d_k(E,F) . \tag{2.1}$$

We set

$$\delta_w(E,F) = \sum_{k \geq 1} 2^{-k} \text{ Log } d_k(E,F) .$$

Then, δ_w is a distance on OS_n. Let $\{E_i\}$ be a sequence in OS_n. Then

$\delta_w(E_i, E) \to 0$ iff

$$d_k(E_i, F) \to 1 \quad \text{for all} \quad k \geq 1 .$$

In that case we will write simply $E_i \xrightarrow{w} E$. It was observed in [Pi1] that $E_i \xrightarrow{w} E$ iff for any non-trivial ultrafilter $\mathcal{U}$ on $\mathbb{N}$ the ultraproduct $\Pi E_i/\mathcal{U}$ is completely isometric to E.

We will need the following known fact.

PROPOSITION 2.1. *For any E, F in OS_n and any $k \geq 1$ we have*

$$d_{cb}(E, F) = d_{cb}(E^*, F^*)$$

and

$$d_k(E, F) = d_k(E^*, F^*)$$

for all $k \geq 1$. Hence in particular

$$\delta_{cb}(E, F) = \delta_{cb}(E^*, F^*) \quad \text{and} \quad \delta_w(E, F) = \delta_w(E^*, F^*) .$$

Proof: It clearly suffices to know that for any $u : E \to F$ we have

$$\|u\|_{cb} = \|u^*\|_{cb} \tag{2.2}$$

and

$$\forall k \geq 1 \qquad \|u\|_k = \|u^*\|_k . \tag{2.3}$$

The identity (2.2) was proved in [BlP], ER1], while (2.3) is easy to check using the definition of $\|u\|_k$. We have (by [Sm])

$$\|u\|_k = \sup \left\{ \|bua\|_{cb} \;\middle|\; a : M_k^* \to E \ , \ b : F \to M_k \ , \ \|a\|_{cb} \leq 1 \ , \ \|b\|_{cb} \leq 1 \right\} . \tag{2.4}$$

Clearly (2.4) implies (2.3). Then the above proposition is obvious. □

The following was proved in [Pi1].

PROPOSITION 2.2. *Let $E \in OS_n$. The following are equivalent*

(i) *For any sequence $\{E_i\}$ in OS_n tending weakly to E we have $d_{cb}(E, E_i) \to 1$ when $i \to \infty$.*
(ii) *Same as (i) with each E_i isometric to E.*
(iii) $d_{SK}(E) = d_{SK}(E^*) = 1$.

Remark: For any fixed integer $k \geq 1$, these are also equivalent to the same property as (i) restricted to E_i k-isometric to E.

Proof: We only prove (i) $\Rightarrow$ (iii) which is what we use in the sequel.
Assume (i). Then $E \subset B(\ell_2)$. Let $P_m : B(\ell_2) \to M_m$ be the projection which maps e_{ij} to itself if $1 \leq i, j \leq m$ and to zero otherwise. Let $E_m = P_m(E) \subset M_m$. It is very easy to check that $E_m \xrightarrow{w} E$. Hence if (i) holds we

have

$$\begin{aligned} d_{SK}(E) &\leq d_{cb}(E, E_m) d_{SK}(E_m) \\ &\leq d_{cb}(E, E_m) \end{aligned}$$

hence
$$d_{SK}(E) \leq \lim d_{cb}(E, E_m) = 1 \,.$$

Now if we apply the same to E^* (equipped with the dual operator space structure) we obtain by Proposition 2.1 that $d_{SK}(E^*) = 1$.
Reformulated in more concise terms, the proof reduces to this: let $OS_n(m)$ be the subset of OS_n formed of all n-dimensional subspaces of M_m. Then the union $\bigcup_{m \geq n} OS_n(m)$ is weakly dense in OS_n. Hence if (i) holds, E (and also E^* by Proposition 2.1) must be in the strong closure of $\bigcup_{m \geq n} OS_n(m)$, which means that $d_{SK}(E) = 1$ (and $d_{SK}(E^*) = 1$).

We can now prove

THEOREM 2.3. *The metric space (OS_n, δ_{cb}) is non-separable if $n > 2$.*

Proof: Let $f : (OS_n, \delta_w) \to (OS_n, \delta_{cb})$ be the identity mapping. Note that f^{-1} is continuous, but in general f is not.
However, if we assume OS_n strongly separable then we claim that f is in the first Baire class. Indeed, by (2.1) for any closed ball β in (OS_n, δ_{cb}), $f^{-1}(\beta)$ is weakly closed. Hence if OS_n is strongly separable, for any U strongly open in OS_n $f^{-1}(U)$ must be an F_σ-set in weak topology, hence f is in the first Baire class. Note that the domain of f is compact, hence is a Baire space. By Baire's classical theorem (cf. [B], see also [Ku, §31, X, Th. 1, p. 394]), if the range of f is separable, the set of points of continuity of f must be dense in the domain of f, i.e. dense for the weak topology.
This implies by Proposition 2.2 that for any E in OS_n there is a sequence $\{E_i\}$ in OS_n such that $d_{SK}(E_i) = d_{SK}(E_i^*) = 1$ which tends weakly to E. Equivalently, E can be viewed as the ultraproduct of (E_i) with respect to a non-trivial ultrafilter $\mathcal{U}$.
By Corollary 1.10 applied to the identity of E (with $u_i = I_{E_i}$), this implies that for any biorthogonal system $(x_1, \ldots, x_n)$ $(x_1^*, \ldots, x_n^*)$ in E we have

$$n \leq \left\| \sum_1^n \lambda(g_i) \otimes x_i \right\|_{C_\lambda \otimes_{\min} E} \left\| \sum \lambda(g_i) \otimes x_i^* \right\|_{C_\lambda \otimes_{\min} E^*} . \tag{2.5}$$

Now let $E_n^\lambda = \operatorname{span}(\lambda(g_1), \ldots, \lambda(g_n)) \subset C_\lambda$.
Let $\lambda_*(g_i)$ be the biorthogonal functionals in $(E_n^\lambda)^*$. Then, if $n > 1$, we have by [AO]

$$\left\| \sum_1^n \lambda(g_i) \otimes \lambda(g_i) \right\| = 2\sqrt{n-1} \,,$$

and since $t = \sum_{i=1}^n \lambda(g_i) \otimes \lambda_*(g_i)$ represents the inclusion map $j : E_n^\lambda \to C_\lambda$, we have

$$\|t\|_{C_\lambda \otimes_{\min} (E_n^\lambda)^*} = \|j\|_{cb(E_n^\lambda, C_\lambda)} = 1 .$$

Hence taking $x_i = \lambda(g_i)$, $x_i^* = \lambda_*(g_i)$ in (2.5) we obtain

$$n \le 2\sqrt{n-1}$$

or equivalently $n \le 2$. □

Remark 2.4: By a simple modification of the preceding proof, one can prove that the subset $HOS_n \subset OS_n$ formed of all the n-dimensional operator spaces which are isometric to ℓ_2^n is non-separable if $n > 2$. Our original argument here gave only $n > 4$, the improvement is due to Timur Oikhberg. Here is briefly the argument: By the proof of Theorem 2.3, if HOS_n is separable, then any E in HOS_n must satisfy (2.5). Consider then the operator space $\min(\ell_2^n)$ obtained by embedding ℓ_2^n isometrically into a commutative C^*-algebra. Let (e_i) be the basis of ℓ_2^n. Assume $n > 1$. Let $\gamma_n = 2(1-n^{-1})^{1/2}$. Consider the subspace

$$E \subset E_n^\lambda \oplus \min(\ell_2^n)$$

spanned by the vectors $x_i = \lambda(g_i) \oplus \gamma_n e_i$. Then by [AO, p. 1038] we have $\|\sum \alpha_i x_i\| = \gamma_n (\sum |\alpha_i|^2)^{1/2}$, $\forall\, (\alpha_i) \in \mathbb{C}^n$, hence $E \in HOS_n$. Furthermore, we have

$$\left\| \sum_1^n \lambda(g_i) \otimes x_i \right\|_{C_\lambda \otimes_{\min} E} = \max\{2\sqrt{n-1}, \gamma_n^2\}$$

and (2.6)

$$\left\| \sum \lambda(g_i) \otimes x_i^* \right\|_{C_\lambda \otimes_{\min} E^*} \le 1 ,$$

so that we conclude again from (2.5) that $n \le 2$.

Remark 2.5: (i) An operator space E is called homogeneous if for any $u : E \to E$ we have $\|u\| = \|u\|_{cb}$. This notion seems particularly interesting in the Hilbertian case (see [Pi3]). Consider an arbitrary n-dimensional operator space E given with a basis $(e_1, \ldots, e_n)$. We can define its "homogeneous hull" $\hat{E}$ as follows. Let $U(n)$ be the unitary group. We view the coordinates (u_{ij}) of a unitary matrix u as a continuous function on $U(n)$, so that $u_{ij} \in C(U(n))$. In the space $C(U(n)) \otimes_{\min} E$ we consider the elements

$$\hat{e}_i = \sum_j u_{ij} \otimes e_j \in C(U(n)) \otimes_{\min} E .$$

Let $\hat{E}$ be the operator space spanned by $(\hat{e}_1, \ldots, \hat{e}_n)$. Observe that $U(n)$ acts isometrically (and actually completely isometrically) on $\hat{E}$, therefore it is easy to check that $\hat{E}$ is Hilbertian and homogeneous. Moreover, we have $d_{SK}(\hat{E}) \le d_{SK}(E)$. Now let F be another operator space and let

$u : E \to F$ be an isomorphism. Let $f_i = u(e_i)$ and let $\hat{F}$ be the operator space associated to F and this basis. Then it is easy to check that

$$d_{cb}(\hat{E}, \hat{F}) \leq \|u\|_{cb}\|u^{-1}\|_{cb} \quad \text{and} \quad d_k(\hat{E}, \hat{F}) \leq \|u\|_k\|u^{-1}\|_k \ \ \forall k .$$

(ii) Let us denote by HH_n the subset of OS_n formed of all the Hilbertian homogeneous spaces. Using the first part of this remark, it is easy to check that any space E in HH_n is the weak limit of a net (E_i) in HH_n such that $d_{SK}(E_i) = 1$ for all i. Then a simple modification of the proof of Theorem 2.3 shows that HH_n is a non-separable subset of OS_n for the (strong) distance δ_{cb}, if $n > 2$. Indeed, we can replace the space E and its basis (x_i) in Remark 2.4 by $\hat{E}$ and $\hat{x}_i$. Using an inequality due to Haagerup [H3, Lemma 2.4], one can check that (2.6) remains valid. This gives us a space in HH_n which satisfies (2.5) only if $n \leq 2$. (Alternately, one could replace E by the linear span of a circular system in the sense of Voiculescu [VoDN], but this seems to yield non-separability only for $n > 4$.)

PROPOSITION 2.6. *Let A be any separable C^*-algebra (or any separable operator space) and let $n \geq 1$.*

(i) *The subset $S_n(A) \subset OS_n$ formed of all the n-dimensional subspaces of A is separable in (OS_n, δ_{cb}).*
(ii) *For any $n > 2$, there is an operator space $E_0 \in OS_n$ and $\varepsilon_0 > 0$ such that for any n-dimensional subspace $E \subset A$ we have $d_{cb}(E, E_0) \geq 1 + \varepsilon_0$.*

Proof: The first part is proved by a standard perturbation argument. We merely sketch it: Let $D \subset A$ be a dense countable subset. For any n-tuple $x = (x_1, ..., x_n)$ of linearly independent elements of A, let $E_x \subset A$ be their linear span. Let $x_1^*, ..., x_n^*$ be functionals in A^* which are biorthogonal to x_i. Fix $\varepsilon > 0$. Pick $y_1, ..., y_n$ in D such that $\sum \|x_i - y_i\| < \varepsilon$. Consider then the operator $u : A \to A$ associated to $\sum x_i^* \otimes (y_i - x_i)$. Clearly $\|u\|_{cb} \leq f(\varepsilon)$ with $f(\varepsilon) \to 0$ when $\varepsilon \to 0$. Moreover, $(I + u)(x_i) = y_i$, and (say, when $f(\varepsilon) < 1$) $x_i = (I+u)^{-1}(y_i)$. This immediately yields $d_{cb}(E_x, E_y) \leq 1+g(\varepsilon)$ with $g(\varepsilon) \to 0$ when $\varepsilon \to 0$, whence (i).
The second part follows from the first one, since it merely expresses the fact that (by Theorem 2.3) $S_n(A)$ is not dense in OS_n for $n > 2$. □

We now give a more precise version of Theorem 2.3 based on the following well known variant of Baire's Theorem.

LEMMA 2.7. *Let S, T be metric spaces and let $f : S \to T$ be a mapping such that for every closed ball $B \subset T$, $f^{-1}(B)$ is closed in S. Fix a number $\varepsilon > 0$. Let $C_\varepsilon(f)$ be the set of all points of ε-continuity of f, i.e. all points s in S such that whenever $s_i \to s$ we have $\limsup d(f(s_i), f(s)) \leq \varepsilon$. Now* *assume that T is ε-separable, i.e. there exists a sequence $\{B_n\}$ of closed balls of radius ε in T such that $T = \bigcup B_n$. Then, if S is a Baire space, the set $C_{2\varepsilon}(f)$ of points of (2ε)-continuity of f is dense in S.*

Proof: Let $S' = \bigcup_n [f^{-1}(B_n) \setminus \overset{\circ}{\widehat{f^{-1}(B_n)}}] \subset S$. Since S' is a countable union of closed sets with empty interior in a Baire space, its complement $S \setminus S'$ is dense in S. But it is easy to check (recall $S = \bigcup f^{-1}(B_n)$), that $S \setminus S' \subset C_{2\varepsilon}(f)$. □

THEOREM 2.8. *For each $n \geq 1$, let ε_n be the infimum of all numbers $\varepsilon > 0$ such that (OS_n, δ_{cb}) is ε-separable. Let $\delta_n = \exp(\varepsilon_n)$. Then we have for all $n \geq 3$*

$$\left(\frac{n}{2\sqrt{n-1}}\right)^{1/4} \leq \delta_n \ . \tag{2.7}$$

and there is a constant $c > 0$ such that for all $n \geq 3$

$$cn^{1/8} < \delta_n \leq n^{1/2} \ . \tag{2.8}$$

Proof: We apply Lemma 2.7 to the same map f as in the proof of Theorem 2.3. Consider $\delta > \delta_n$ and let $\varepsilon = \log(\delta)$ so that (OS_n, δ_{cb}) is ε-separable. By Lemma 2.7, $C_{2\varepsilon}(f)$ is δ_w-dense. Then, by a simple modification of the proof of Proposition 2.2, any E in OS_n is the weak limit of a net E_i such that $\limsup d_{SK}(E_i) \leq e^{2\varepsilon} = \delta^2$ and $\limsup d_{SK}(E_i^*) \leq e^{2\varepsilon} = \delta^2$. Then (recalling Corollary 1.10) we conclude, as in the proof of Theorem 2.3, that

$$n \leq \delta^4 2\sqrt{n-1}$$

whence (2.7).
It was proved in [Pi3, Theorem 9.6] that $d_{SK}(E) \leq n^{1/2}$ for all E in OS_n, therefore $\delta_n \leq n^{1/2}$ and (2.8) follows. □

COROLLARY 2.9. *There is a constant $c > 0$ such that for each $n \geq 3$ there is an uncountable collection $(E_i)_{i \in I}$ in OS_n satisfying*

$$\forall i \neq j \qquad d_{cb}(E_i, E_j) > cn^{1/8} \ .$$

Remark 2.10: Two or three months after this paper had been circulated as a preprint, Simon Wassermann mentioned to us that he conjectured that the linear spans of the n-tuples of operators considered by Voiculescu in [Vo] should yield a more explicit non-separable family of finite dimensional operator spaces. Some form of this conjecture is indeed correct. This shows that groups with Kazhdan's property T (see [dHaV]) can be used to prove the non-separability of OS_n. Here are the details. In [Vo], to each subset Ω of the integers, Voiculescu associates an n-tuple $\tau_\Omega = (T_1^\Omega, \ldots, T_n^\Omega)$ of operators in $B(H)$ (with say $H = \ell_2$), in the following way. Let G be any discrete group with Kazhdan's property T admitting a countable collection of pairwise disjoint finite dimensional representations (π_k). (For instance we can take $G = SL_3(\mathbb{Z})$). Let $(t_1, \ldots, t_{n-1})$ be a finite set of generators and let t_n be equal to the unit element. Then for any $\Omega \subset \mathbb{N}$, we define

$$T_j^\Omega = \oplus_{k\in\Omega}\pi_k(t_j) \ , \quad j = 1, 2, \ldots, n \ .$$

Let M_Ω be the finite dimensional operator space spanned by τ_Ω. We claim these form a non-separable collection of operator spaces. This does not seem to follow from Voiculescu's stated results but it does follow easily from his ideas, as follows:

First, in case M_Ω is not n-dimensional we consider an n-dimensional operator space E_Ω containing τ_Ω.

Following [Vo], we use property T through the following: There is a fixed number $\varepsilon > 0$ such that if Ω and Ω' are subsets of $\mathbb{N}$ with $\Omega \not\subset \Omega'$, then there are unitary operators $u_1, \ldots, u_n$ satisfying

$$n = \left\|\sum_1^n T_j^\Omega \otimes u_j\right\|_{\min} \quad \text{and} \quad \left\|\sum_1^n T_j^{\Omega'} \otimes u_j\right\|_{\min} < n - \varepsilon \ . \tag{2.9}$$

(More precisely, if we pick $k \in \Omega - \Omega'$ then we can take $u_j = \overline{\pi_k(t_j)}$.)

Now fix $\delta > 1$ and assume that the metric space (OS_n, δ_{cb}) contains a sequence (E_m) such that for any E in OS_n there is an m such that $d_{cb}(E, E_m) < \delta$. Fix a number $\eta > 0$. Then there is an integer m and a continuous collection $\mathcal{C}$ of subsets of $\mathbb{N}$ such that for each Ω in $\mathcal{C}$ there is a map $v_\Omega \colon E_\Omega \to E_m$ such that $\|v_\Omega\|_{cb} < \delta$ and $\|v_\Omega^{-1}\|_{cb} = 1$. Now consider the continuous family $(v_\Omega(T_j^\Omega))_{j\le n}$ of n-tuples of elements of E_m. Since $(E_m)^n$ (the space of n-tuples of elements of E_m) is norm-separable, there must exist a continuous subcollection $\mathcal{C}_1 \subset \mathcal{C}$ such that for all Ω, Ω' in $\mathcal{C}_1$ we have

$$\sum_1^n \left\|v_\Omega(T_j^\Omega) - v_{\Omega'}(T_j^{\Omega'})\right\| < \eta \ . \tag{2.10}$$

A fortiori, $\mathcal{C}_1$ has cardinality > 1, hence we can find Ω, Ω' in $\mathcal{C}_1$ satisfying $\Omega \not\subset \Omega'$. By (2.9) this implies

$$\left\|\sum v_{\Omega'}(T_j^{\Omega'}) \otimes u_j\right\|_{\min} \le \|v_{\Omega'}\|_{cb}\left\|\sum T_j^{\Omega'} \otimes u_j\right\|_{\min} \le \delta\ (n - \varepsilon) \ ,$$

and

$$n = \left(\|v_\Omega^{-1}\|_{cb}\right)^{-1}\left\|\sum_1^n T_j^\Omega \otimes u_j\right\|_{\min} \le \left\|\sum v_\Omega(T_j^\Omega) \otimes u_j\right\|_{\min} \ .$$

This gives by (2.10)

$$n \leq \left\| \sum v_{\Omega'}(T_j^{\Omega'}) \otimes u_j \right\|_{\min} + \sum_1^n \| v_{\Omega'}(T_j^{\Omega'}) - v_\Omega(T_j^\Omega) \| \leq \delta(n-\varepsilon) + \eta \,.$$

When $\eta > 0$ and $\delta - 1$ are small enough, this is impossible (since $\varepsilon > 0$ remains fixed). This contradiction shows that the spaces (E_Ω) form a non separable collection. Since we may obviously ensure that $d_{cb}(E_\Omega, E_{\Omega'}) \leq d_{cb}(M_\Omega, M_{\Omega'})$ holds (by choosing each (E_Ω) simply as a suitable direct sum) this completes the proof of our claim that the spaces (M_Ω) themselves form a non separable collection. (Alternatively, if g_j are the generators of the free group, we can replace T_j^Ω by $T_j^\Omega \oplus \lambda(g_j)$ for $j = 1, \ldots, n-1$ and T_n^Ω by $T_n^\Omega \oplus I$, this guarantees dimension n for the span and, by [AO], it does not spoil the estimates.) □

Actually, Simon Wassermann conjectured that the whole family of spaces (M_Ω) is uniformly d_{cb}-separated, i.e. that for some $\eta > 0$ we have $d_{cb}(M_\Omega, M_{\Omega'}) > 1 + \eta$ whenever $\Omega \neq \Omega'$. As far as we know this is still open.

It is known ([Tr]) that $G = SL_3(\mathbb{Z})$ admits two generators so that (recall that g_n is the unit) we obtain by this reasoning a continuous collection of 3-dimensional operator spaces (E_t) such that for some $\varepsilon > 0$ we have $d_{cb}(E_t, E_s) > 1+\varepsilon$ for all $t \neq s$. By a simple modification, we can make sure that the spaces we obtain are spanned by three unitaries. The same cannot be achieved with spans of two unitaries. Indeed, the span of two unitaries (U_1, U_2) is completely isometric to the span of (I, U) with $U = U_1^* U_2$, which itself embeds completely isometrically into a commutative (hence nuclear) C^*-algebra.

Remark 2.11: (Added March 3 1995) Let us denote by c_n the infimum of all numbers c with the following property:

there is an infinite collection $\{u_i^m | 1 \leq i \leq n\}$, $m \in \mathbb{N}$, of n-tuples of unitary matrices (more precisely u_i^m is a unitary matrix of size say $N_m \times N_m$) such that

$$\sup_{m \neq m'} \left\| \sum_{i=1}^{i=n} u_i^m \otimes \overline{u_i^{m'}} \right\|_{\min} \leq c \,.$$

Note that if $m = m'$ the preceding norm is equal to n. Then a close look at the preceding argument shows that the number δ_n defined in Theorem 2.8 satisfies, for any $n \geq 1$,

$$n / c_n \leq \delta_n \,. \tag{2.11}$$

From the upper bound in (2.8) we deduce $c_n \geq n^{1/2}$. By Remark 2.10, we know that $c_n < n$ for all $n > 2$ and by the preceding observation about pairs of unitaries, we know that $c_2 = 2$ (but whether $\delta_2 > 1$ remains open).

Remark 2.12: (Added March 3 1995) Very recently, Alain Valette has kindly showed us that the striking work of Lubotzky-Phillips-Sarnak (see A. Lubotzky's recent book "Discrete Groups, Expanding Graphs and Invariant Measures", Birkhauser 1994, or P. Sarnak's book "Some Applications of Modular Forms" Cambridge Univ. Press 1990) can be used to show that $c_n \leq 2(n-1)^{1/2}$ at least for infinitely many integers n; by (2.11) this implies the following (asymptotically sharp) estimate

$$n/[2(n-1)^{1/2}] \leq \delta_n.$$

Here is a sketch of Valette's argument:
Let G be the free group with k generators denoted by $g_1, \ldots, g_k$, where $k = (p+1)/2$ with $p > 2$ a prime number. Then (from Lubotzky-Phillips-Sarnak's work) one can find a decreasing sequence $\ldots H_q \subset H_{q-1} \subset \ldots \subset H_1 \subset G$ of normal subgroups with finite index and intersection reduced to the unit element, with the following key property:
Let $G_q = G/H_q$ and let $\sigma_q : G \to G_q$ be the quotient mapping. Let λ_q^0 be the compression of the left regular unitary representation of G_q to the orthogonal of the constant functions on G_q. Let $\Lambda_q = \lambda_q^0 \circ \sigma_q$. Then Λ_q is a finite dimensional unitary representation of G, and its key property is (for each q)

$$\left\| \sum_1^k \Lambda_q(g_i) + \Lambda_q(g_i^{-1}) \right\| \leq 2p^{1/2} = 2(2k-1)^{1/2} . \tag{2.12}$$

We refer the reader in particular to Theorem 7.4.3 in Lubotzky's book (in the non-bipartite case).

Now for each q we decompose Λ_q into a direct sum of irreducible subrepresentations and we form the union over q of all the (finite dimensional unitary) irreducible representations of G obtained in this way. Since the sequence (Λ_q) separates the points of G and G is infinite, this union must contain an infinite sequence of mutually inequivalent representations, which we denote by (π^m). Then, setting by convention $g_{k+j} = g_j^{-1}$ for $j = 1, \ldots, k$, we define $\forall m \in \mathbb{N}$, $\forall i = 1, \ldots, 2k \quad u_i^m = \pi^m(g_i)$. From (2.12) we deduce that for all $m \neq m'$

$$\left\| \sum_{i=1}^{i=2k} u_i^m \otimes \overline{u_i^{m'}} \right\|_{\min} = \left\| \sum_{i=1}^{i=2k} [\pi^m \otimes \overline{\pi^{m'}}](g_i) \right\| \leq 2(2k-1)^{1/2} , \tag{2.13}$$

indeed, note that since π^m and $\pi^{m'}$ are inequivalent and irreducible, $\pi^m \otimes \overline{\pi^{m'}}$ does not contain the trivial representation, hence is a direct sum of subrepresentations of Λ_q for some q. Finally, (2.13) clearly implies, as announced, that $c_{2k} \leq 2(2k-1)^{1/2}$.

3. Applications to $B(H) \otimes B(H)$ and Maximal Operator Spaces

By Kirchberg's results in [K2], Theorem 2.3 implies

COROLLARY 3.1. *If* $\dim H = \infty$, *there is more than one* C^**-norm on* $B(H) \otimes B(H)$. *In other words we have*

$$B(H) \otimes_{\min} B(H) \neq B(H) \otimes_{\max} B(H) .$$

Proof: This follows from the equivalence of the conjectures (A7) and (A2) in [K2, p. 483]. For the convenience of the reader, we include a direct argument, as follows. For any discrete group G we denote by $C^*(G)$ the full C^*-algebra of G. By Proposition 2.6 (ii) (applied with $A = C^*(F_\infty)$), there is an operator space E_0 such that for some $\varepsilon_0 > 0$ we have

$$d_{cb}(E, E_0) \geq 1 + \varepsilon_0$$

for all n-dimensional subspaces $E \subset C^*(F_\infty)$. Now let F_I be a free group associated to a set of generators $\{g_i \mid i \in I\}$ where I is any set with infinite cardinality. Observe that for any finite dimensional (or merely separable) subspace $E \subset C^*(F)$ there is a countable infinite subset $J \subset I$ such that $E \subset C^*(F_J)$. (Indeed only countably many "letters" are being used.) Hence we also have

$$d(E, E_0) \geq 1 + \varepsilon_0$$

for all n-dimensional subspaces $E \subset C^*(F_I)$. Let $\pi : C^*(F) \to B(H)$ be a C^*-algebra representation of the full C^*-algebra of a big enough free group F onto $B(H)$. We have

$$B(H) \approx C^*(F)/\text{Ker } \pi .$$

Let $\mathcal{I} = \text{Ker } \pi$. Then (by Lemma 0.8) the quotient norm of the space

$$Q = \left[C^*(F) \otimes_{\min} B(H)\right] / \left[\mathcal{I} \otimes_{\min} B(H)\right]$$

induces on $B(H) \otimes B(H)$ a C^*-norm. Assume that there is only one such norm. Then we have isometrically

$$Q = B(H) \otimes_{\min} B(H) . \tag{3.1}$$

Now we may clearly assume that E_0^* is a subspace of $B(H)$, i.e. we have $E_0^* \subset B(H)$ completely isometrically, so that the completely isometric inclusion $j : E_0 \to B(H)$ can be viewed as an element j_0 in $B(H) \otimes_{\min} E_0^* \subset B(H) \otimes_{\min} B(H)$. Note that $\|j_0\|_{\min} = \|j\|_{cb} = 1$. By Lemma 0.7 and by (3.1), for any $\varepsilon > 0$ there is a lifting $\tilde{j}_0$ in $C^*(F) \otimes_{\min} E_0^*$ with $\|\tilde{j}_0\|_{\min} < 1 + \varepsilon$ and $(\pi \otimes I_{E_0^*})(\tilde{j}_0) = j_0$. Now, let $\tilde{j} : E_0 \to C^*(F)$ be the associated linear operator. We have $\pi \tilde{j} = j$ and $\|\tilde{j}_0\|_{\min} = \|\tilde{j}\|_{cb}$, hence $d_{cb}(E_0, \tilde{j}(E_0)) \leq \|\pi\|_{cb} \|\tilde{j}\|_{cb} < 1 + \varepsilon$. When $\varepsilon < \varepsilon_0$ this is impossible. □

Remark: By [K2, §8], Corollary 3.1 has the following consequences

(i) There is a separable unital C^*-algebra with the WEP in the sense of Lance [L] for which $\mathrm{Ext}(A)$ is not a group.

(ii) There are separable unital C^*-algebras A, B with WEP such that

$$A \otimes_{\min} B \neq A \otimes_{\max} B .$$

(iii) The WEP does not imply the local lifting property (in short LLP) in the sense of [K2].

(iv) There is a separable unital C^*-algebra A with WEP which is not approximately injective in the sense of [EH].

(v) The identity $A^{op} \otimes_{\min} A = A^{op} \otimes_{\max} A$ does not imply the approximate injectivity of A.

(vi) There is a unital separable C^*-algebra B with WEP which is not a quotient C^*-algebra of an approximately injective C^*-algebra.

By [K2, p. 484], Corollary 3.1 also implies a negative answer to Kirchberg's conjecture (C1) in [K2]. Thus we have:

$C^*(F_\infty)$ and $C^*(SL_2(\mathbb{Z}))$ are not approximately injective in $B(H)$.

Finally, by [K2, p. 487], Corollary 3.1 implies a negative answer to the conjecture (P2) in [K2], hence we have: Let $K = K(H)$, $B = B(H)$. Consider the canonical morphism

$$\Phi : B \otimes_{\min} B \mapsto (B/K) \otimes_{\min} (B/K) .$$

Then the kernel of Φ is strictly larger than the set $F(K, B, B \otimes_{\min} B) + F(B, K, B \otimes_{\min} B)$ where $F(.,.,.)$ denotes the Fubini product.

In a different direction, we can give more examples of non-exact operator spaces, completing those of [Pi1]. Following Blecher and Paulsen [BlP], given a Banach space E, we denote by $\min(E)$ the operator space obtained by embedding E into a commutative C^*-algebra, or equivalently by embedding E into the space $C(K)$ of all continuous functions on K with $K = (B_{E^*}, \sigma(E^*, E))$. Similarly, let I be a suitable set and let $g : \ell_1(I) \to E$ be a metric surjection (i.e. q^* is an isometry). We view the space $\ell_1(I) = c_0(I)^*$ as an operator space with the dual operator space structure. Then (cf. [BlP],[P2]) we denote by $\max(E)$ the operator space obtained by equipping E with the operator space structure (in short o.s.s.) of the quotient space $\ell_1(I)/\mathrm{Ker}(q)$. Equivalently, we have a complete isometry

$$\max(E) \to \left(\min(E^*)\right)^* ,$$

and $\max(E)$ is characterized by the isometric identity

$$cb\left(\max(E), M_n\right) = B(E, M_n) . \tag{3.2}$$

More generally, for any operator space F we have isometrically

$$cb\left(\max(E), F\right) = B(E, F) . \tag{3.2$'$}$$

We refer to [BlP],[P2] for more information. It will be convenient to introduce the following characteristic for an operator space E,

$$d_{QSK}(E) = \inf\{d_{cb}(E,F)\}$$

where the infimum runs over all operator spaces F of the form $F = E_1/E_2$ where $E_2 \subset E_1 \subset K$ and $\dim E = \dim F$.

THEOREM 3.2. *Let E be any n-dimensional Banach space. Then*

$$d_{QSK}(\max(E)) \geq \frac{\sqrt{n}}{4}\ .$$

Proof: By a well known result in Banach space theory (cf. e.g. [Pi2, Theorem 1.11, p.15]) we have $\pi_2(I_{E^*}) = \sqrt{n}$. Let $u : X \to Y$ be a 2-absolutely summing operator between Banach spaces and let $J : Y \to Y_1$ be an isometric embedding. Then it is easy to see that

$$\pi_2(u) = \pi_2(Ju)\ . \tag{3.3}$$

Consider an isomorphism $v : E_1/E_2 \to \max(E)$ with $E_2 \subset E_1 \subset K$. Observe $(E_1/E_2)^* = E_2^\perp \subset E_1^*$. Let $J : E_2^\perp \to E_1^*$ be the (isometric) canonical inclusion. We will now apply Corollary 1.7 to the map

$$Jv^* : \min(E^*) \xrightarrow{v^*} E_2^\perp \xrightarrow{J} E_1^*\ .$$

Note that $d_{SK}(\min(E^*)) = d_{SK}(E_1) = 1$. Hence Corollary 1.7 yields by (3.3)

$$\pi_2(v^*) = \pi_2(Jv^*) \leq 4\|Jv^*\|_{cb} \leq 4\|v^*\|_{cb} = 4\|v\|_{cb}\ ,$$

but this implies by (0.4)

$$n^{1/2} = \pi_2(I_{E^*}) \leq \|v^{-1*}\|\pi_2(v^*) \leq 4\|v^{-1}\|\ \|v\|_{cb}\ ,$$

hence $n^{1/2} \leq 4d_{QSK}(\max(E))$. □

Remark: In particular the preceding result answers a question raised by Vern Paulsen (private communication): the space $\max(\ell_2^n)$ is quite different (when n is large enough) from the linear span of n Clifford matrices, i.e. matrices (u_i) in M_{2^n} satisfying the relations

$$u_i = u_i^* \qquad u_iu_j^* + u_j^*u_i = 2\delta_{ij}I\ .$$

Indeed, if we denote $Cl_n = \mathrm{span}(u_1,\ldots,u_n)$ then by Theorem 3.2 we have

$$d_{cb}(\max(\ell_2^n), Cl_n) \geq d_{QSK}(\max(\ell_2^n)) \geq \sqrt{n}/4\ .$$

4. A New Tensor Product for C^*-algebras or Operator Spaces

Let $E_1 \subset B(H_1), E_2 \subset B(H_2)$ be arbitrary operator spaces.
Let us denote by $\| \ \|_M$ the norm induced on the algebraic tensor product $E_1 \otimes E_2$ by $B(H_1) \otimes_{\max} B(H_2)$. Recalling (0.3), and using the injectivity of $B(H_1)$ and $B(H_2)$ as well as the decomposition property of *c.b.* maps into $B(H)$, it is easy to check (see Lemma 4.1 below) that this norm is independent of the choice of the completely isometric embeddings $E_1 \subset B(H_1)$, $E_2 \subset B(H_2)$. In other words, the norm $\| \ \|_M$ on $E_1 \otimes E_2$ depends only on the operator space structures of E_1 and E_2.
We will denote by $E_1 \otimes_M E_2$ the completion of $E_1 \otimes E_2$ under this norm. We equip the space $E_1 \otimes_M E_2$ with the natural operator space structure induced by the C^*-algebra $B(H_1) \otimes_{\max} B(H_2)$ via the isometric embedding $E_1 \otimes_M E_2 \subset B(H_1) \otimes_{\max} B(H_2)$.
Clearly, if A, B are C^*-algebras, then $\| \ \|_M$ is a C^*-norm on $A \otimes B$ and $A \otimes_M B$ also is a C^*-algebra.

LEMMA 4.1. *Let F_1, F_2 be two operator spaces. Consider c.b. maps $u_1 : E_1 \to F_1$ and $u_2 : E_2 \to F_2$. Then $u_1 \otimes u_2$ defines a c.b. map from $E_1 \otimes_M E_2$ to $F_1 \otimes_M F_2$ with*

$$\|u_1 \otimes u_2\|_{cb(E_1 \otimes_M E_2, F_1 \otimes_M F_2)} \leq \|u_1\|_{cb} \|u_2\|_{cb} \ . \tag{4.1}$$

Proof: Indeed, note that if $F_1 \subset B(\mathcal{H}_1)$, $F_2 \subset B(\mathcal{H}_2)$ then by the extension property of *c.b.* maps (*cf.* [P1, p. 100]) u_1, u_2 admit extensions $\tilde{u}_1 : B(H_1) \to B(\mathcal{H}_1)$ and $\tilde{u}_2 : B(H_2) \to B(\mathcal{H}_2)$ with $\|\tilde{u}_1\|_{cb} = \|u_1\|_{cb}$ and $\|\tilde{u}_2\|_{cb} = \|u_2\|_{cb}$. Hence it suffices to check this in the case when each of E_1, E_2, F_1, F_2 is $B(H)$ for some H. Then the idea is to use the decomposition property of *c.b.* maps on $B(H)$ as linear combinations of completely positive maps to reduce checking (4.1) to the case of completely positive maps. In the completely positive case, the relevant point here is of course (0.3). This idea leads to a simple proof of (4.1) with some additional numerical factor. However, this factor can be removed at the cost of a slightly more technical argument based on [H3]. For lack of a suitable reference, we now briefly outline this (straightforward) argument to check (4.1).
Let A, B be C^*-algebras. We will denote by $CP(A, B)$ the set of all completely positive maps $u : A \to B$, and by $D(A, B)$ the set of all decomposable maps $u : A \to B$, *i.e.* maps which can be written as $u = u_1 - u_2 + i(u_3 - u_4)$ with $u_1, \ldots, u_4 \in CP(A, B)$.
In [H3], Haagerup defines the norm $\|u\|_{dec}$ on $D(A, B)$ as follows. Consider all possible mappings S_1, S_2 in $CP(A, B)$ such that the map $v : A \to M_2(B)$ defined by

$$v(x) = \begin{pmatrix} S_1(x) & u(x^*)^* \\ u(x) & S_2(x) \end{pmatrix}$$

is completely positive. Then we set $\|u\|_{dec} = \inf\{\max\{\|S_1\|, \|S_2\|\}\}$ where the infimum runs over all possible such mappings.
In [H3, Proposition 1.3 and Theorem 1.6] the following results appear:

$$\forall\, u \in D(A,B) \qquad \|u\|_{cb} \le \|u\|_{dec}\ , \tag{4.2}$$

$$\forall\, u \in CP(A,B) \qquad \|u\| = \|u\|_{cb} = \|u\|_{dec}\ , \tag{4.3}$$

(see also [P1, p. 28] for the first equality)

if C is any C^*-algebra, if $u \in D(A,B)$ and $v \in D(B,C)$, then $vu \in D(A,C)$ and (4.4)

$$\|vu\|_{dec} \le \|v\|_{dec}\|u\|_{dec}\ ,$$

$$\forall\, u \in cb(B(H), B(\mathcal{H})) \qquad \|u\|_{cb} = \|u\|_{dec}\ . \tag{4.5}$$

Now let A_1, A_2, B_1, B_2 be arbitrary C^*-algebras, and let $u_1 \in D(A_1, B_1)$, $u_2 \in D(A_2, B_2)$. We claim that Haagerup's results imply that $u_1 \otimes u_2$ extends to a decomposable map from $A_1 \otimes_{\max} A_2$ into $B_1 \otimes_{\max} B_2$ (still denoted by $u_1 \otimes u_2$) satisfying

$$\|u_1 \otimes u_2\|_{dec} \le \|u_1\|_{dec}\|u_2\|_{dec}\ . \tag{4.6}$$

By (4.2) we have a fortiori

$$\|u_1 \otimes u_2\|_{cb(A_1 \otimes_{\max} A_2, B_1 \otimes_{\max} B_2)} \le \|u_1\|_{dec}\|u_2\|_{dec}\ . \tag{4.7}$$

To verify (4.6) (hence also (4.7)) we may assume (using (4.4) and $u_1 \otimes u_2 = (u_1 \otimes I)(I \otimes u_2)$) that $A_2 = B_2$ and u_2 is the identity on A_2. Consider then $v_1 : A_1 \to M_2(B_1)$ of the form

$$v_1(x) = \begin{pmatrix} S_1(x) & u_1(x^*)^* \\ u_1(x) & S_2(x) \end{pmatrix}$$

with v_1 and S_1, S_2 all completely positive.
By (0.3), the associated map $v_1 \otimes I_{A_2}$ as well as $S_1 \otimes I_{A_2}$ and $S_2 \otimes I_{A_2}$ are (bounded and) completely positive from $A_1 \otimes_{\max} A_2$ to $B_1 \otimes_{\max} A_2$. Therefore (by the definition of $\|\ \|_{dec}$)

$$\|u_1 \otimes I_{A_2}\|_{dec} \le \max\{\|S_1 \otimes I_{A_2}\|, \|S_2 \otimes I_{A_2}\|\}$$

hence by (0.3)

$$\le \max\{\|S_1\|, \|S_2\|\}\ .$$

It follows that

$$\|u_1 \otimes I_{A_2}\|_{dec} \le \|u_1\|_{dec}\ ,$$

and this is enough to verify (4.6) (and a fortiori (4.7)).

The proof of (4.1) is now easy: let $u_1, \widetilde{u}_1$ and $u_2, \widetilde{u}_2$ be as explained above, by (4.7)

$$\|u_1 \otimes u_2\|_{cb(E_1\otimes_M E_2, F_1\otimes_M F_2)} \leq \|\widetilde{u}_1 \otimes \widetilde{u}_2\|_{cb(B(H_1)\otimes_{\max} B(H_2), B(\mathcal{H}_1)\otimes_{\max} B(\mathcal{H}_2))}$$
$$\leq \|\widetilde{u}_1\|_{dec}\|\widetilde{u}_2\|_{dec}$$

hence by (4.5) $\leq \|\widetilde{u}_1\|_{cb}\|\widetilde{u}_2\|_{cb} = \|u_1\|_{cb}\|u_2\|_{cb}$.

This completes the proof of (4.1). □

We will use several times the following obvious consequence of (4.1) and the definition of $E_1 \otimes_M E_2$:

If u_1 and u_2 (as above) are complete isometries, then $u_1 \otimes u_2 : E_1 \otimes_M E_2 \to F_1 \otimes_M F_2$ also is a complete isometry . (4.8)

By Corollary 3.1, we know that there are operator spaces E, F such that $E \otimes_{\min} F \neq E \otimes_M F$. It is natural to try to understand the meaning of this new tensor norm $\|\ \|_M$ and to characterize the operator spaces E, F for which the equality holds. For that purpose the following result due to Kirchberg [K3] will be crucial:
For any free group F_I and any H, we have an isometric identity

$$C^*(F_I) \otimes_{\min} B(H) = C^*(F_I) \otimes_{\max} B(H) . \tag{4.9}$$

Using this, we have

PROPOSITION 4.2. *Let E, F be operator spaces, let $u \in E \otimes F$ and let $U : F^* \to E$ be the associated finite rank linear operator. Consider a finite dimensional subspace $S \subset C^*(F_\infty)$ and a factorization of U of the form $U = ba$ with bounded linear maps $a : F^* \to S$ and $b : S \to E$, where $a : F^* \to S$ is weak-$*$ continuous. Then*

$$\|u\|_M = \inf\left\{\|a\|_{cb}\|b\|_{cb}\right\}$$

where the infimum runs over all such factorizations of U.

Proof: Assume $E \subset B(H)$ and $F \subset B(K)$ with H, K Hilbert. It clearly suffices to prove this in the case when E and F are both finite dimensional. Assume U factorized as above with $\|a\|_{cb}\|b\|_{cb} < 1$. Then by Kirchberg's theorem (4.9) the min and max norms are equal on $C^*(F_\infty)\otimes B(K)$, hence, by (4.8), we have isometrically $S\otimes_{\min} F = S\otimes_M F$, so that if $\hat{a}$ is the element of $S \otimes_{\min} F$ associated to a, we have $\|\hat{a}\|_M = \|a\|_{cb}$ and $u = (b \otimes I_F)(\hat{a})$. Therefore, by (4.1) we have $\|u\|_M \leq \|b\|_{cb}\|\hat{a}\|_M \leq \|a\|_{cb}\|b\|_{cb} < 1$.
The proof of the converse is essentially the same as for Corollary 3.1 above. We skip the details. □

Let A be any C^*-algebra. For any finite dimensional operator space E, let

$$d_{SA}(E) = \inf \left\{ d_{cb}(E,F) |\ F \subset A \right\} .$$

Then the preceding result immediately implies

COROLLARY 4.3. *Let E be a finite dimensional operator space. Let $i_E \in E \otimes E^*$ be the tensor associated to the identity on E. Then*

$$\|i_E\|_M = d_{SC^*(F_\infty)}(E) .$$

In particular we have

$$d_{SC^*(F_\infty)}(E) = d_{SC^*(F_\infty)}(E^*) . \tag{4.10}$$

Proof: The first part is clear by Proposition 4.2. To check (4.10), observe more generally that for any operator spaces E, F, the "flip isomorphism" $(x \otimes y \to y \otimes x)$ is a complete isometry between the spaces $E \otimes_M F$ and $F \otimes_M E$. Hence (4.10) follows by symmetry. □

Remark: The preceding argument shows the following: If $A \otimes_{\min} B(H) = A \otimes_{\max} B(H)$ then for any finite dimensional operator space E we have

$$d_{SC^*(F_\infty)}(E) \le d_{SA}(E^*) . \tag{4.11}$$

In particular, this holds (by definition of nuclearity) if A is nuclear. (Hence (4.11) still holds if A is exact, since $d_{SK}(E^*) \le d_{SA}(E^*)$ in that case). The proof of Corollary 3.1 shows that (4.11) is false in general for $A = B(H)$. (Observe that $d_{SA}(E^*) = 1$ for any E if $A = B(H)$ and $\dim(H) = \infty$.)

Remark: We can now give a quantitative version of Corollary 3.1. For any n let

$$\lambda(n) = \sup \left\{ \frac{\|u\|_{\max}}{\|u\|_{\min}} \right\} \tag{4.12}$$

where the supremum runs over all u in $B(H) \otimes B(H)$ with $\operatorname{rank}(u) \le n$. We claim that (with the notation of Theorem 2.8 and Remark 2.12)

$$\delta_n \le \lambda(n) \le \sqrt{n} . \tag{4.13}$$

To verify this, first observe that

$$\lambda(n) = \sup \left\{ \frac{\|u\|_M}{\|u\|_{\min}} \right\} \tag{4.12$'$}$$

where the supremum runs over all operator spaces E, F in OS_n and all u in $E \otimes F$. Equivalently, we have

$$\lambda(n) = \sup \left\{ \|i_E\|_M \right\} \tag{4.12$''$}$$

where the supremum runs over all E in OS_n and where $i_E \in E \otimes E^*$ represents (as above) the identity on E.

Indeed, any u as in (4.12)$'$ can be rewritten as $u = (I_E \otimes \widetilde{u})(i_E)$ where $\widetilde{u} : E^* \to F$ is the linear map corresponding to u, hence by (4.1) we have

$\|u\|_M \le \|\tilde{u}\|_{cb} \|i_E\|_M = \|u\|_{\min} \|i_E\|_M$, and (4.12)'' follows.
By Corollary 4.3, this implies

$$\lambda(n) = \sup \left\{ d_{SC^*(F_\infty)}(E) \mid E \in OS_n \right\} . \tag{4.12)'''}$$

Now since $C^*(F_\infty)$ is separable, with the notation of Theorem 2.8, we clearly have (recall Proposition 2.6 (i)) $\delta_n \le \sup\{d_{SC^*(F_\infty)}(E) \mid E \in OS_n\}$, hence the left side of (4.13) follows from (4.12)'''. For the other side, note that by (4.4) and (4.5) we have for any E in OS_n, $d_{SC^*(F_\infty)}(E) \le d_{SK}(E)$ and by [Pi3, Theorem 9.6] this is $\le \sqrt{n}$. □

THEOREM 4.4. *Let A be a C^*-algebra and let $H = \ell_2$. The following are equivalent.*

(i) $A \otimes_{\min} B(H) = A \otimes_M B(H)$.

(ii) *For any $\varepsilon > 0$ and any finite dimensional subspace $E \subset A$, there is a subspace $\hat{E} \subset C^*(F_\infty)$ such that $d_{cb}(E, \hat{E}) < 1 + \varepsilon$.*

(iii) *Same as (ii) with $\varepsilon = 0$. of A is completely isometric to a subspace of $C^*(F_\infty)$.*

Proof: Assume (i). Then for any finite dimensional subspace $E \subset A$, consider a completely isometric embedding $E^* \subset B(H)$ and view $E \otimes E^*$ as a subspace of $A \otimes_{\min} B(H)$. If we apply Proposition 4.2 when $u \in E \otimes E^*$ represents the identity on E, we have $\|u\|_{cb} = 1$, hence $\|u\|_M = 1$ and we immediately obtain (ii). Conversely, (ii) clearly implies (i) by Proposition 4.2 and (4.8). The fact that (i) implies (iii) follows from [EH, Theorem 3.2] (this was kindly pointed out to us by Kirchberg). Indeed, it suffices to prove (iii) for finite dimensional operator systems $E \subset A$. Then, assuming (i) the second condition in [EH, Theorem 3.2] must hold by Lemma 0.7 and the short exactness property of the max-tensor product. Therefore, if we represent A as a quotient of $C^*(F_I)$ for some free group F_I, by [EH, Theorem 3.2] any unital completely positive map $v : E \to A$ has a unital completely positive lifting $\hat{v} : E \to C^*(F_I)$. In particular E embeds completely isometrically into $C^*(F_I)$, hence into $C^*(F_\infty)$ (see the proof of Corollary 3.1). This shows (i)⇒(iii). Finally, (iii)⇒(ii) is trivial. □

Remark: The notion appearing in (ii) above is analogous to that of "finite representability" in Banach space theory.

THEOREM 4.5. *Let X be an operator space and let $c \ge 1$ be a constant. The following are equivalent.*

(i) *For any operator space F, we have $X \otimes_{\min} F = X \otimes_M F$ and $\|u\|_M \le c \|u\|_{\min}$ for any u in $X \otimes F$.*

(ii) *The same as* (i) *with $F = B(\ell_2)$.*

(iii) *For any finite dimensional subspace $E \subset X$ we have*

$$d_{SC^*(F_\infty)}(E) \le c\ .$$

Proof: (i)⇒(ii) is trivial and (ii)⇒(iii) is clear by Proposition 4.2, taking $F = E^*$ and $u \in X \otimes E^*$ associated to the inclusion $E \subset X$. Finally assume (iii). Consider u in $X \otimes F$. We have $u \in E \otimes F$ for some finite dimensional subspace $E \subset X$. By (iii), for each $\varepsilon > 0$, there is a subspace $\tilde{E} \subset C^*(F_\infty)$ such that $d_{cb}(E, \tilde{E}) \le c + \varepsilon$. By Proposition 4.2, this implies $\|u\|_M \le (c+\varepsilon)\|u\|_{\min}$, whence (i). □

Remark: It can be shown that the operator Hilbert space OH introduced in [Pi3] satisfies the equivalent conditions in Theorem 4.5 for $c = 1$. In particular we have

$$d_{SC^*(F_\infty)}(OH) = 1\ .$$

This is a consequence of some unpublished work by U. Haagerup, namely the inequality (4.14) below (itself a consequence of [Pi3, Corollary 2.7]). To explain this inequality, let $\overline{B(H)}$ be the complex conjugate of $B(H)$, *i.e.* the same space but with the conjugate complex multiplication. We denote by $x \to \overline{x}$ the canonical anti-isomorphism between $B(H)$ and $\overline{B(H)}$. Note that $\overline{B(H)} \approx B(\overline{H}) \approx B(H)$. In the sequel, we simply denote by $\|\ \|_{\max}$ (resp. $\|\ \|_{\min}$) the max-norm (resp. the min-norm) on the space $B(H) \otimes \overline{B(H)}$.

Then, for any $x_1, \ldots, x_n$ and $y_1, \ldots, y_n$ in $B(H)$ we have

$$\left\|\sum x_i \otimes \overline{y}_i\right\|_{\max} \le \left\|\sum x_i \otimes \overline{x}_i\right\|_{\min}^{1/2} \left\|\sum y_i \otimes \overline{y}_i\right\|_{\min}^{1/2}\ . \tag{4.14}$$

To check (4.14), we first recall an entirely elementary fact: for any $a_1, \ldots, a_n$, $b_1, \ldots, b_n$ in a C^*-algebra A, we have

$$\left\|\sum a_i b_i\right\| \le \left\|\sum a_i a_i^*\right\|^{1/2} \left\|\sum b_i^* b_i\right\|^{1/2}\ .$$

Hence if $a_i b_i = b_i a_i$ we also have

$$\left\|\sum a_i b_i\right\| \le \left\|\sum a_i^* a_i\right\|^{1/2} \left\|\sum b_i b_i^*\right\|^{1/2}\ .$$

Applying these inequalities in the case $a_i = x_i \otimes 1$, $b_i = 1 \otimes \overline{y}_i$ we get

$$\left\|\sum x_i \otimes \overline{y}_i\right\|_{\max} \le \left\|\sum x_i x_i^*\right\|^{1/2} \left\|\sum y_i^* y_i\right\|^{1/2} \tag{4.14.0}$$

$$\left\|\sum x_i \otimes \overline{y}_i\right\|_{\max} \le \left\|\sum x_i^* x_i\right\|^{1/2} \left\|\sum y_i y_i^*\right\|^{1/2}\ . \tag{4.14.1}$$

Let us denote by A_0 (resp. A_1) the space $B(H)^n$ equipped with the norm $\|(x_i)\| = \left\|\sum x_i^* x_i\right\|^{1/2}$ (resp. $\|\sum x_i x_i^*\|^{1/2}$). Moreover for any (x_i) in $B(H)^n$, we denote by $\|(x_i)\|_{\frac{1}{2}}$ the norm in the complex interpolation space $(A_0, A_1)_{\frac{1}{2}}$. Then, by the complex interpolation theorem applied to the sesquilinear map

$$(x_i),(y_i)\to\sum x_i\otimes\overline{y}_i\,,$$

(4.14.0) and (4.14.1) imply that we have

$$\left\|\sum x_i\otimes\overline{y}_i\right\|_{\max}\le\|(x_i)\|_{\frac{1}{2}}\|(y_i)\|_{\frac{1}{2}}\,. \tag{4.15}$$

On the other hand, by Corollary 2.7 in [Pi3] we have for all (x_i) in $B(H)^n$

$$\|(x_i)\|_{\frac{1}{2}}=\left\|\sum x_i\otimes\overline{x}_i\right\|_{\min}^{1/2}. \tag{4.16}$$

Hence (4.15) implies (4.14).

Finally, let $(T_i)_{i\ge1}$ be an orthonormal basis in the operator Hilbert space OH introduced in [Pi3]. We may assume $OH\subset B(H)$ with (say) $H=\ell_2$. Recall (see [Pi3]) that $\|\sum_1^n T_i\otimes\overline{T}_i\|_{\min}=1$ and for any $x_1,\ldots,x_n$ in $B(H)$ we have

$$\left\|\sum T_i\otimes x_i\right\|_{\min}=\left\|\sum x_i\otimes\overline{x}_i\right\|_{\min}^{1/2}.$$

Therefore, by (4.14) for any $x_1,\ldots,x_n$ in $B(H)$ we have

$$\left\|\sum T_i\otimes x_i\right\|_{\max}\le\left\|\sum x_i\otimes\overline{x}_i\right\|_{\min}^{1/2}=\left\|\sum T_i\otimes x_i\right\|_{\min}.$$

Equivalently, we conclude that $\|\ \|_{\max}$ and $\|\ \|_{\min}$ coincide on $OH\otimes B(H)$ so that $X=OH$ satisfies the equivalent properties in Theorem 4.5. Note that if $x_i=y_i$ in (4.14) we have

$$\left\|\sum x_i\otimes\overline{x}_i\right\|_{\max}\le\left\|\sum x_i\otimes\overline{x}_i\right\|_{\min}.$$

In other words, we obtain that $\|\ \|_{\max}$ and $\|\ \|_{\min}$ coincide on the "positive" cone in $B(H)\otimes\overline{B(H)}$ formed of all the tensors of the form $\sum_1^n x_i\otimes\overline{x}_i$.

In [Pi5], a modified version of the identity (4.16) is proved with an arbitrary semi-finite von Neumann algebra in the place of $B(H)$. We refer the reader to a possibly forthcoming paper by U. Haagerup and the second author for extended versions of (4.14) and (4.15).

References

[AO] C. Akemann, P. Ostrand, Computing norms in group C^*-algebras, Amer. J. Math. 98 (1976), 1015-1047.

[B] R. Baire, Sur les fonctions des variables réelles, Ann. di Mat. 3:3 (1899), 1-123.

[Bl1] D. BLECHER, Tensor products of operator spaces II, Canadian J. Math. 44 (1992), 75-90.

[Bl2] D. BLECHER, The standard dual of an operator space,. Pacific J. Math. 153 (1992), 15-30.

[Bl3] D. BLECHER, Tracially completely bounded multilinear maps on C^*-algebras, Journal of the London Mathematical Society 39 (1989), 514-524.

[BlP] D. BLECHER, V. PAULSEN, Tensor products of operator spaces, J. Funct. Anal. 99 (1991) 262-292.

[dCH] J. DE CANNIÈRE, U. HAAGERUP, Multipliers of the Fourier algebras of some simple Lie groups and their discrete subgroups, Amer. J. Math. 107 (1985), 455-500.

[dHaV] P. DE LA HARPE, A. VALETTE, La Propriété T de Kazhdan pour les Groupes Localement Compacts, Astérisque, Soc. Math. France 175 (1989).

[EH] E. EFFROS, U. HAAGERUP, Lifting problems and local reflexivity for C^*-algebras, Duke Math. J. 52 (1985), 103-128.

[ER1] E. EFFROS, Z.J. RUAN, A new approach to operator spaces, Canadian Math. Bull. 34 (1991), 329-337.

[ER2] E. EFFROS, Z.J. RUAN, On the abstract characterization of operator spaces, Proc. Amer. Math. Soc. 119 (1993), 579-584.

[H1] U. HAAGERUP, The Grothendieck inequality for bilinear forms on C^*-algebras, Advances in Math. 56 (1985), 93-116.

[H2] U. HAAGERUP, An example of a non-nuclear C^*-algebra which has the metric approximation property, Invent. Math. 50 (1979), 279-293.

[H3] U. HAAGERUP, Injectivity and decomposition of completely bounded maps, in "Operator algebras and their connection with Topology and Ergodic Theory", Springer Lecture Notes in Math. 1132 (1985), 170-222.

[HPi] U. HAAGERUP, G. PISIER, Bounded linear operators between C^*-algebras, Duke Math. J. 71 (1993), 889-925.

[I] T. ITOH, On the completely bounded maps of a C^*-algebra to its dual space, Bull. London Math. Soc. 19 (1987), 546-550.

[K1] E. KIRCHBERG, On subalgebras of the CAR-algebra, to appear in J. Funct. Anal.

[K2] E. KIRCHBERG, On non-semisplit extensions, tensor products and exactness of group C^*-algebras, Invent. Math. 112 (1993), 449-489.

[K3] E. KIRCHBERG, Commutants of unitaries in UHF algebras and functorial properties of exactness, to appear in J. reine angew. Math.

[Kr] J. KRAUS, The slice map problem and approximation properties, J. Funct. Anal. 102 (1991), 116-155.

[Ku] W. KURATOWSKI, Topology, Vol. 1. (New edition translated from the French), Academic Press, New-York 1966.

[Kw] S. KWAPIEŃ, On operators factorizable through L_p-spaces, Bull. Soc. Math. France, Mémoire 31-32 (1972), 215-225.

[L] C. LANCE, On nuclear C^*-algebras, J. Funct. Anal. 12 (1973), 157-176.

[P1] V. PAULSEN, Completely bounded maps and dilations, Pitman Research Notes 146. Pitman Longman (Wiley) 1986.

[P2] V. PAULSEN, Representation of function algebras, Abstract operator spaces and Banach space geometry, J. Funct. Anal. 109 (1992), 113-129.

[P3] V. PAULSEN, The maximal operator space of a normed space, to appear.

[Pi1] G. PISIER, Exact operator spaces, Colloque sur les algèbres d'opérateurs, Astérisque, Soc. Math. France, to appear.

[Pi2] G. PISIER, Factorization of Linear Operators and the Geometry of Banach Spaces, CBMS (Regional conferences of the A.M.S.) 60, (1986); Reprinted with corrections 1987.

[Pi3] G. PISIER, The operator Hilbert space OH, complex interpolation and tensor norms, submitted to Memoirs Amer. Math. Soc.
[Pi4] G. PISIER, Factorization of operator valued analytic functions, Advances in Math. 93 (1992), 61-125.
[Pi5] G. PISIER, Projections from a von Neumann algebra onto a subalgebra, Bull. Soc. Math. France, to appear.
[R] Z.J. RUAN, Subspaces of C^*-algebras, J. Funct. Anal. 76 (1988), 217-230.
[S] S. SAKAI, C^*-algebras and W^*-algebras, Springer Verlag New-York, 1971.
[Sm] R.R. SMITH, Completely bounded maps between C^*-algebras, J. London Math. Soc. 27 (1983), 157-166.
[T] M. TAKESAKI, Theory of Operator Algebras I, Springer-Verlag New-York 1979.
[Tr] S. TROTT, A pair of generators for the unimodular group, Canad. Math. Bull. 3 (1962), 245-252.
[Vo] D. VOICULESCU, Property T and approximation of operators, Bull. London Math. Soc. 22 (1990), 25-30.
[VoDN] D. VOICULESCU, K. DYKEMA, A. NICA, Free random variables, CRM Monograph Series, 1, Amer. Math. Soc., Providence RI.
[W1] S. WASSERMANN, On tensor products of certain group C^*-algebras, J. Funct. Anal. 23 (1976), 239-254.
[W2] S. WASSERMANN, Exact C^*-algebras and related topics, Lecture Notes Series 19, Seoul National University, 1994.

M. Junge
Mathematisches Seminar
CAU Kiel
24098 Kiel, Germany

G. Pisier
Texas A&M University and Université Paris 6
College Station, TX 77843
USA

Equipe d'Analyse
75252 Paris Cedex 05
France

Submitted: March 1994

Geometric And Functional Analysis 1016-443X/95/0200364-23$1.50+0.20/0
Vol. 5, No. 2 (1995)

LOCAL NON-SQUEEZING THEOREMS AND STABILITY

F. LALONDE AND D. MCDUFF

Dedicated to Misha Gromov on the occasion of his 50th birthday.

1. Introduction

One of the most fundamental results in symplectic topology is the non-squeezing theorem which asserts that there is no symplectic embedding which takes a standard $2n+2$-ball of radius 1 into a cylinder $(M\times D^2(a), \omega\oplus\sigma)$ whose base $D^2(a)$ is a closed 2-disc of σ-area $a<\pi$. This was first proved by Gromov ([G]) for a range of manifolds including standard Euclidean space, and was generalized to all manifolds by Lalonde–McDuff ([LM1]). In this paper we consider "local" versions of this theorem. The word local can here be interpreted in two ways. Sometimes we localize in space and think of embedding not a whole set such a ball or ellipsoid but just its germ along a central 2-disc $0\times D$. Sometimes we localize in time and look for embeddings which are close to a given inclusion.

Our problem can be formulated as follows. Let $(W,\Omega)=(M\times D,\omega\oplus\sigma)$ be a symplectic cylinder, where D is a closed 2-disc of σ-area π and (M,ω) is some symplectic manifold. Suppose that S is a compact subset of W, whose boundary is a smooth hypersurface. When can S be moved symplectically to lie strictly inside W? The main Theorem below gives an essentially complete answer to this question. As one might expect, the answer lies in the geometry of S near the points which meet the boundary ∂W of the cylinder. The interesting case is when S meets ∂W along some closed characteristic $x\times\partial D$, and we will see that our problem is closely connected to the properties of the linearization of the characteristic flow around this closed orbit. As a corollary, we prove the sufficiency of the condition for the stability of geodesics in Hofer's metric.

We wish to acknowledge the hospitality of the Newton Institute, Cambridge, where this paper was completed. We also wish to thank Lisa Traynor for explaining symplectic homology to us, and Helmut Hofer for suggesting

---

The work of the first author is partially supported by NSERC grant OGP 0092913 and FCAR grant ER-1199. The work of the second author is partially supported by NSF grant DMS 9401443.

its application in Proposition 2.1. We are also grateful to Leonid Polterovich for carefully reading the paper and making useful suggestions.

1.1 Local squeezing and characteristic flows. To state our results precisely, we will need the following definitions.

DEFINITION 1.1. *We say that S is* **squeezable by isotopy** *in W if there is a smooth 1-parameter family $\psi_{t\in[0,1]}$ of symplectic embeddings of S in W starting at the inclusion such that*

$$\psi_1(S) \subset \operatorname{Int} W ,$$

where $\operatorname{Int} W = M \times \operatorname{Int} D$. S is **locally squeezable by isotopy** *in W if in addition $\psi_t(S) \subset \operatorname{Int} W$ for all $t > 0$. Finally, S is* **locally squeezable** *in W if there is a sequence ψ_i, $i \geq 1$, of symplectic embeddings $S \hookrightarrow \operatorname{Int} W$ such that ψ_i converges C^1 to the inclusion as $i \to \infty$.*

Clearly, a set which is locally squeezable by isotopy is both locally squeezable and squeezable by isotopy. However, the exact relationship between the latter two concepts is somewhat complicated because an isotopy of S in W usually will not extend to an isotopy from W to W. This point is discussed further below.

From now on, we assume that S is a compact subset of W whose boundary is a smooth hypersurface. Observe that all characteristics on the boundary ∂W of W are flat circles $pt \times \partial D$. The following observation is a well-known fact:

LEMMA 1.2. *If $S \cap \partial W$ contains no closed characteristics $pt \times \partial D$ then S is locally squeezable by isotopy in W.*

Proof: To prove this, one constructs a Hamiltonian H on ∂W whose flow points into W at all points of $\partial S \cap \partial W = S \cap \partial W$. For this, we need $\partial H/\partial t < 0$ at all points of $S \cap \partial W$, which is possible if this set contains no closed characteristic (here $t \in [0,1]$ is the angle coordinate of the boundary of the 2-disc D, the base of the cylinder). To be complete, here is a more detailed proof.

Because S is compact, the set $P = S \cap \partial W$ is compact too. Thus $P^c = (M \times S^1) - P$ is an open subset of the manifold $N = M \times S^1$, which contains $(M - K) \times S^1$ for some compact subset $K \subset M$. Because there is no closed loop $\{pt\} \times S^1$ in P, P^c contains some non-empty open interval $\{p\} \times I$ for each $p \in M$. Now choose any smooth function $f : N \to (-\infty, 0]$ which is strictly negative on $K \times S^1$, and has compact support. Let $g : M \to \mathbf{R}$ be its integral over the S^1-factor, and $h : N \to [0, \infty)$ be a smooth positive function with compact support inside the open set P^c, whose integral on each S^1-factor $\{p\} \times S^1$ is equal to $-g(p)$. (The existence of h is obvious: simply take a finite open covering $U_i \subset M$ of the projection on M of supp$f$, and a partition of unity ϕ_i associated to it. Choose open subsets $O_i = U_i \times I_i \subset P^c$, refining the open covering $\{U_i\}$ if needed. Then define for each O_i a positive function h_i with compact support in O_i whose S^1-integral is $-\phi_i g$, and set $h = \sum_i h_i$.)

Hence $f + h$ is a smooth function with compact support in N whose integral over each $\{pt\} \times S^1$ vanishes, and which is negative on P. Fixing a point $t_0 \in S_1$, the integral $\int_{t_0}^{t_0+2\pi}(f+g)$ is a smooth function on N whose derivative with respect to $t \in S^1 = \mathbf{R}/\mathbf{Z}$ is negative everywhere in P, and which vanishes outside some compact set in N. Finally, let $H : M \times \mathbf{R}^2 \to \mathbf{R}$ be any extension of $f+g$, with compact support. Then the flow ϕ_t induced by H sends S strictly inside W for sufficiently small times $t > 0$. □

Therefore the interesting case is when ∂S contains a closed characteristic $x_0 \times \partial D$. Then, because S is a subset of W, ∂S must be tangent to ∂W at all points y of this circle. Since the characteristics of ∂S point along the null direction of $\Omega|_{T_y \partial S} = \Omega|_{T_y \partial W}$, this circle is a characteristic on ∂S as well. What turns out to be crucial is the linearization of the characteristic flow of ∂S around this circle. If we identify all the tangent spaces $T_y M \subset T_y \partial W$ with $\mathbf{R}^{2n} = T_{x_0 M}$ in the obvious way, this linearization is a family of symplectic linear maps

$$A_t : \mathbf{R}^{2n} \to \mathbf{R}^{2n}, \quad t \in [0,1].$$

We will say that A_t has a non-constant closed orbit at time T if there is some fixed point $v \in T_{x_0} M$ of A_T which is not fixed by all the A_t, $t \in [0,T]$. Thus v is in the kernel of $A_t - \mathbb{1}$ for $t = T$ but not for all $t < T$. Here is our main result.

THEOREM 1.3. *Let S be a compact subset of a symplectic cylinder (W, Ω) which intersects ∂W along at least one closed characteristic $x \times \partial D$.*

(i) If $S \cap \partial W$ contains only finitely many closed characteristics $x \times \partial D$, and if the linearized flow around each has a non-constant closed orbit in time < 1, then S is locally squeezable by isotopy in W.

(ii) If there is a closed characteristic in $S \cap \partial W$ along which the linearized flow has no non-constant closed orbit in time ≤ 1, then S is neither locally squeezable nor squeezable by isotopy in W.

Remark 1.4: (*i*) Recall that when $\det(A_1 - \mathbb{1}) \neq 0$, the index of the closed characteristic $x_0 \times \partial D$ is given by counting (with multiplicities) the number of times $T < 1$ at which $\det(A_T - \mathbb{1}) = 0$: see Ekeland [E2] for example. Therefore, the hypothesis in part (*ii*) is equivalent to assuming that the closed characteristic $x_0 \times \partial D$ has the same index on S as it does on W. This suggests that symplectic homology might be directly applicable to prove part (ii). Unfortunately, this does not seem to work, although it does work in the special case of ellipsoids: see § 2.

(ii) We will see that any small neighborhood S of the disc $0 \times D$ in the $(2n+2)$-ball of radius 1 in $\mathbf{R}^{2n} \times D$ is squeezable by isotopy but not locally squeezable. This is a rather exceptional example which occurs because the closed characteristic $0 \times \partial D$ is not isolated in the set of all closed characteristics on ∂S. In general one would expect that if S intersects ∂W along an isolated closed characteristic then S is squeezable by isotopy only if it is also locally squeezable. It is not obvious how to prove this in all cases, though Theorem 1.3 shows that this does hold generically. See also Lemma 2.6.

The above theorem is very closely related with the question of the stability of geodesics in Hofer's metric on the group of Hamiltonian symplectomorphisms which we studied in [LM2,3]. In fact, we will show in § 3 that part (i) is essentially just a restatement of Theorem 1.6 in [LM2] where we used the non-constant closed orbit to construct a local squeezing of S near each closed characteristic in $S \cap \partial W$.

In view of this, we now concentrate on explaining the proof of part (ii). Observe that this is again local in S, that is to say it depends only on the germ of S near the flat disc bounded by the closed characteristics in $S \cap \partial W$. Given such a closed characteristic $x_0 \times \partial D$, choose a Darboux chart near $x_0 \in M$ which takes x_0 to $0 \in \mathbf{R}^{2n}$ and write the ε-neighborhood S_ε of the corresponding disc $0 \times D$ in S as

$$S_\varepsilon = \{(x,c,t) : c \leq \pi - H_t(x) \ , \ \|x\| < \varepsilon\} \subset \mathbf{R}^{2n} \times D$$

where $H_t(x) \geq 0$, $H_t(0) = 0$ for all t. Here we have used action-angle coordinates (c,t) on D where $t \in \mathbf{R}/\mathbf{Z}$ as before and $c = \pi r^2$. Thus the area form is $dc \wedge dt$. Therefore, the characteristics of S are given by the Hamiltonian flow of the function $\pi - H_t(x) - c$ and the linearized flow A_t is generated by the 2-jet of $-H_t$. Observe that the linearized flow has a non-constant closed trajectory at time t exactly when A_t has an eigenvalue 1, i.e. $\det(A_t - \mathbb{1}) = 0$. We will call a path $A_{t\in[0,t_0]} \in \mathrm{Sp}(2n,\mathbf{R})$ *short* if $\det(A_t - \mathbb{1}) \neq 0$ for $t > 0$, and *positive* if it is generated by $-Q_t$ where Q_t is a positive definite non-degenerate quadratic Hamiltonian.

Because the conditions considered in part (ii) are open and we may replace S by a slightly smaller set, it suffices to prove the result in the case when H_t is quadratic and non-degenerate for all t. Thus, we will suppose that S_ε is a quadratic slice of the form

$$S_{Q_t,\varepsilon} = \{(x,c,t) : c < \pi - Q_t(x) \ , \ \|x\| < \varepsilon\} \ ,$$

where each $\mathcal{Q}_t$ is a positive definite quadratic form on $\mathbf{R}^{2n}$. If $\mathcal{K} = \mathcal{Q}_t$ is independent of t, the slice $S_{\mathcal{K},\varepsilon}$ is contained in the ellipsoid

$$E_{\mathcal{K}} = \{(x,c,t) : c + \mathcal{Q}(x) \leq \pi\} .$$

In this case part (i) of the above theorem may be proved directly, and part (ii) follows by using the theory of symplectic homology: see §2.

Now suppose that $\mathcal{Q}_t$ depends on time, and let

$$x \mapsto A_t x \ , \quad t \in [0,1]$$

be the corresponding linear flow, where $A_t \in \mathrm{G} = \mathrm{Sp}(2n, \mathbf{R})$. By slightly perturbing the family $\mathcal{Q}_{t\in[0,1]}$ we may suppose that A_1 is diagonalizable. Let $\mathcal{U}$ denote the set of diagonalizable matrices with all eigenvalues on the unit circle, and observe that the flow of a time-independent positive quadratic form lies in $\mathcal{U}$. The following proposition allows us to reduce to the time-independent case provided that the time 1-map A_1 of $\mathcal{Q}_t$ lies in $\mathcal{U}$.

PROPOSITION 1.5. *Suppose that $\mathcal{Q}_t$, $t \in \mathbf{R}/\mathbf{Z}$ is a 1-periodic family of positive definite quadratic forms on $\mathbf{R}^{2n}$ which generates a path $A_t \in \mathrm{Sp}(2n, \mathbf{R})$ such that $A_1 \in \mathcal{U}$. Then there is a time-independent positive definite quadratic form $\mathcal{K}$ and $\varepsilon > 0$ such that there is a symplectic isotopy*

$$\Phi_s : S_{\mathcal{K},\varepsilon} \to W \ , \quad 0 \leq s \leq 1 \ ,$$

which fixes all points of the disc $D_0 = 0 \times D$, starts at the inclusion and ends at a map Φ_1 which takes the ellipsoidal slice $S_{\mathcal{K},\varepsilon}$ into $S_{\mathcal{Q}_t,\varepsilon}$. Moreover, if $A_{t\in[0,1]}$ is short, so is the path generated by $\mathcal{K}$.

Roughly speaking, this proposition says that any slice S which satisfies the conditions of Theorem 1.3(ii) and whose monodromy is conjugate to a unitary matrix, contains an ellipsoidal slice which satisfies the same conditions. A few extra arguments are needed to deduce non-squeezing results for S from those for the ellipsoidal slice: see § 2.2. In the general case we prolong the path A_t to a path with endpoint in $\mathcal{U}$ without introducing any new closed trajectories: see Theorem 1.6(i) below.

The proof of Proposition 1.5 is based on the study of positive paths in G which is carried out in [LM4]. We will only consider paths in G which start at $\mathbb{1}$. We will write

$$\mathcal{S}_1 = \{A \in \mathrm{G} : \det(A - \mathbb{1}) = 0\} ,$$

and denote by $\mathcal{P}_{\mathrm{Aut}}$ the space of short positive paths in G which are generated by time-independent Hamiltonians $\mathcal{K}$. If $\mathcal{K}(x) = -\frac{1}{2}x^T P x$, for some positive definite symmetric matrix P then it is easy to check that the corresponding flow is

$$A_t = e^{JPt} ,$$

where J is multiplication by i in $\mathbf{R}^{2n} = \mathbf{C}^n$. In particular, all elements A_t lie in $\mathcal{U}$, and the corresponding slice $S_{\mathcal{K},\varepsilon}$ is part of an ellipsoid. Let

$$e : \mathcal{P}_{\mathrm{Aut}} \to \mathcal{U}$$

be the endpoint map. It is not hard to see that every element of $\mathcal{U}$ with no multiple eigenvalues is the endpoint of a unique element of $\mathcal{P}_{\mathrm{Aut}}$. However, elements with multiple eigenvalues have larger inverse image in $\mathcal{P}_{\mathrm{Aut}}$ and hence it is not possible to lift every path in $\mathcal{U}$ to a path in $\mathcal{P}_{\mathrm{Aut}}$.

The main result of [LM4] is:

THEOREM 1.6. *(i) Every element of* G *is the endpoint of a positive path. An element of* $\mathrm{G} - \mathcal{S}_1$ *is the endpoint of a short positive path if and only if it has an even number of real eigenvalues* λ *with* $\lambda > 1$.

(ii) Any short positive path may be extended to a short positive path with endpoint in $\mathcal{U}$.

(iii) If $A_{t\in[0,1]}$ *is a short positive path with endpoint in* $\mathcal{U}$*, it is homotopic through short positive paths with endpoint in* $\mathcal{U}$ *to a short autonomous path. Moreover, we may choose this homotopy so that the path formed by its endpoints lifts to* $\mathcal{P}_{\mathrm{Aut}}$.

In the language of Ekeland [E2], this implies that every stable positive linear periodic Hamiltonian system is homotopic through such systems to one generated by an autonomous Hamiltonian. Moreover the homotopy may be chosen so that the index of the system does not change.

1.2 Stability of geodesics in $\mathrm{Ham}^c(M)$. Let $\mathrm{Ham}^c(M)$ be the group of Hamiltonian symplectomorphisms of the symplectic manifold (M, ω), generated by compactly supported Hamiltonian $M \times [0,1] \to \mathbf{R}$, with the Hofer norm $\|\phi\|$ defined by:

$$\|\phi\| = \inf \mathcal{L}(\phi_{t\in[0,1]}) \ ,$$

where the infimum is taken over all paths in $\mathrm{Ham}^c(M)$ from the identity $\mathbb{1}$ to ϕ, and where the length $\mathcal{L}$ of the path generated by the Hamiltonian $H_{t\in[0,1]}$ is defined to be

$$\mathcal{L}(\phi_t) = \mathcal{L}(H_t) = \int_0^1 \big(\max_{x\in M} H_t(x) - \min_{x\in M} H_t(x)\big) dt \ .$$

A path $\gamma = \phi_{t\in[0,1]}$ is said to be a stable geodesic if it is a local minimum for $\mathcal{L}$ on the space of all paths from $\mathbb{1}$ to ϕ_1. (This path space is given the C^1-topology.) It was shown in [BP],[LM2],[U] that a stable geodesic γ is quasi-autonomous, that is it has at least one fixed maximum P (a point at which H_t assumes its maximum for all t) and one fixed minimum p. Moreover, if there are only finitely many such fixed extrema, there must be one such pair P, p at which the linearised flow has no non-trivial closed trajectory in time < 1. We will show below that this statement is essentially equivalent to part (*i*) of Theorem 1.3 above. Similarly, part (*ii*) is equivalent to the converse:

THEOREM 1.7. *Suppose that γ has a fixed maximum and minimum at which the linearised flow A_t has no non-trivial closed trajectory in time ≤ 1. Then γ is a stable geodesic.*

As shown by Ustilovsky in [U], this result is a fairly easy consequence of the second variation formula for $\mathcal{L}$ provided that the Hessian of H_t at the fixed extrema are non-degenerate at all times. We prove the general result in § 3.

2. Local Squeezing

2.1 Ellipsoidal slices. Let Q be a positive definite quadratic form on $\mathbf{R}^{2n}$, and consider the ellipsoidal slice

$$S_{Q,\varepsilon} = \{(x, c, t) \in \mathbf{R}^{2n} \times \mathbf{R}^2 \mid c \leq \pi - Q(x) \text{ and } \|x\| < \varepsilon\} ,$$

where ε is small enough so that $\pi - Q(x)$ is positive over the ball $\|x\| \leq \varepsilon$. In this section we will consider the squeezing properties of $S_{Q,\varepsilon}$ in the cylinder $W = \mathbf{R}^{2n} \times D$, where D is the unit disc in $\mathbf{R}^2$. Since Q may be diagonalized with respect to a symplectic basis of $\mathbf{R}^{2n}$, there is a symplectomorphism of the form $\Psi \times \mathbb{1}$ of W which takes $S_{Q,\varepsilon}$ to the corresponding set defined by the diagonalized form

$$\sum_{i=1}^{n} \pi a_i^2 (x_{2i-1}^2 + x_{2i}^2)$$

where $a_1 \geq \ldots \geq a_n > 0$. Therefore, we will assume that Q has this form. The analog of Theorem 1.3 for these slices is:

PROPOSITION 2.1. *(i) If $a_1 > 1$, then $S_{Q,\varepsilon}$ is locally squeezable by isotopy in W.*

(ii) If $a_1 = 1$, then $S_{Q,\varepsilon}$ is squeezable by isotopy but not locally squeezable in W.

(iii) If $a_1 < 1$, then $S_{Q,\varepsilon}$ is neither locally squeezable nor squeezable by isotopy in W.

Proof: Let (u, v) be rectangular coordinates of the $\mathbf{R}^2$-plane containing the base D of the cylinder W. If $a_1 > 1$ then one can rotate $S = S_{Q,\varepsilon}$ in the x_1, x_2, u, v-plane so that it does not project surjectively onto D for any $t > 0$. For example, one can use the matrix

$$\begin{pmatrix} \cos t & 0 & -\sin t & 0 \\ 0 & \cos t & 0 & -\sin t \\ \sin t & 0 & \cos t & 0 \\ 0 & \sin t & 0 & \cos t \end{pmatrix} .$$

This proves (i). If $a_1 = 1$ then the same rotation keeps S in W and will take S to a position in which the projection to D is not surjective when $\sin(t) > \varepsilon$. However, S is not locally squeezable in this case. To see this, suppose by contradiction that it were and let ψ_i be a sequence of embeddings of S into $\operatorname{Int} W$ which converge C^1 to the inclusion. We may suppose that the ψ_i are so close to the inclusion that their graphs in $-W \times W$ may be considered as sections of the cotangent bundle T^*W. The ψ_i therefore extend to embeddings of the whole ellipsoid

$$E_Q = \left\{(x,r,t) \in \mathbf{R}^{2n} \times \mathbf{R}^2 \mid r^2 \leq 1 - Q(x)\right\} ,$$

which also converge to the inclusion. Hence, for large enough i, these embeddings take E_Q into $\operatorname{Int} W$. But E_Q contains the unit ball, and this cannot be mapped strictly inside W by the Non-Squeezing theorem ([G],[LM1]).

Now consider case (iii). Since the sets S increase when the a_i decrease, S cannot be locally squeezable by (ii). Suppose, by contradiction, that it were squeezable by isotopy, and choose a symplectic isotopy $\phi_t : S \to W$ such that

$$\phi_1(S) \subset W_{2\delta} = M \times D(\pi - 2\delta)$$

for some $\delta > 0$. (Here $D(a)$ denotes the 2-disc in $\mathbf{R}^2$ of area a centered at the origin.) It is easy to see that there is some $\nu > \delta$ such that $S_{Q,\varepsilon}$ contains the product

$$P = B^{2n}(\nu) \times D(\pi - \delta) .$$

We claim that P cannot be isotoped in W to a subset of $W_{2\delta}$. The basic reason for this is that there is an element of the symplectic homology of W coming from a closed characteristic of ∂W which is non-zero in P but which vanishes on any subset of $W_{2\delta}$. To be precise, we will use the formulation of symplectic homology with $\mathbf{Z}_2$ coefficients given by Floer, Hofer, Wysocki ([FH],[FHW]). The result we want can almost be quoted directly from there, but to be complete we will give some details.

Symplectic homology is calculated using a complex of the form

$$\ldots \xrightarrow{d} \oplus_i(C_i, k; b_i) \xrightarrow{d} \oplus_j(C_j, k-1; b_j) \xrightarrow{d} \ldots$$

where C_i, C_j are vector spaces over $\mathbf{Z}_2$ generated by certain periodic orbits, the grading k is given by the index of the orbit and the level b_i is determined by the action of the orbits. The differential is defined, as in Floer homology, by counting connecting orbits satisfying some elliptic PDE. The group $S_k^{[a,b)}$ is then defined as the homology of the truncated quotient complex, obtained by ignoring all boundary operators with domain $(C_i, k; b_i)$ where $b_i \geq b$ and by quotienting out by all groups $(C_i, k; b_i)$ with $b_i < a$. For example, the terms with grading ≤ 4 in the complex for the 2-disc $D(\nu)$ of capacity ν are

$$\xrightarrow{0} (\mathbf{Z}_2, 4; 2\nu) \xrightarrow{\mathrm{Id}} (\mathbf{Z}_2, 3; \nu) \xrightarrow{0} (\mathbf{Z}_2, 2; \nu) \xrightarrow{\mathrm{Id}} (\mathbf{Z}_2, 1; 0) \to 0 .$$

Therefore, if $0 \leq \delta < \nu$ and $0 < \delta' \leq \nu$, the groups

$$S_k^{[\nu-\delta,\nu+\delta')}(D(\nu)) , \qquad k = 2, 3 ,$$

are non-zero. The complex for the $2n$-ball $B^{2n}(\nu)$ of capacity ν is similar except that there are $2n$ contributions at level ν with gradings going from $n+1$ to $3n$ and then further contributions at levels $j\nu$, $j > 1$ with gradings $\geq 3n+1$. Therefore, with δ, δ' as above, the groups

$$S_{n+1}^{[\nu-\delta,\nu+\delta')}(B^{2n}(\nu)) \text{ and } S_{3n}^{[\nu-\delta,\nu+\delta')}(B^{2n}(\nu))$$

are non-zero.

The groups for the product P are calculated by the tensor product of the complexes for the factors where

$$(C, k; b) \otimes (C', k'; b') = (C \otimes C', k + k'; b + b')$$

with the differential $\mathbb{1} \otimes d' + d \otimes \mathbb{1}$. Therefore, if $P = B^{2n}(\nu) \times D(\alpha)$, the complex in degrees $n+1, n+2, n+3$ is

$$\begin{aligned} \dots \to &(\mathbf{Z}_2, n+3, \nu) \oplus (\mathbf{Z}_2, n+3, \nu+\alpha) \\ &\xrightarrow{d_{n+3}} (\mathbf{Z}_2, n+2, \nu) \oplus (\mathbf{Z}_2, n+2, \alpha) \to (\mathbf{Z}_2, n+1; \nu) \oplus (\mathbf{Z}_2, n+1; 0) \to \dots . \end{aligned}$$

Here the terms at level ν are products of entries in the complex for the ball with the lowest term $(\mathbf{Z}_2, 1; 0)$ for the 2-disc, and so the differential has the form $d \otimes \mathbb{1}$ on them. However the term at level $\nu + \alpha$ is the product

$$(\mathbf{Z}_2, n+1, \nu) \otimes (\mathbf{Z}_2, 2; \alpha)$$

and so d_{n+3} takes its generator 1 to the element $1 \oplus 1$. Thus d_{n+3} is surjective if the window $[a, b)$ contains both α and $\nu + \alpha$. However, if

$$0 < a < \alpha \leq \pi , \quad \pi < b < \min(\nu + \alpha, 2\alpha) ,$$

then $S_{n+2}^{[a,b)}(P(\nu, \alpha)) = \mathbf{Z}_2$.

Now, let $\alpha_{t\in[0,1]}$ increase from $\alpha_0 = \alpha$ to $\alpha_1 = \pi$, let a, b be as above, and consider the restriction map

$$S_{n+2}^{[a,b)}(P(\nu, \alpha_t)) \to S_{n+2}^{[a,b)}(P(\nu, \alpha)) .$$

This is the identity map when $t = 0$, and as α_t increases none of the levels in the complexes we are considering crosses the given window. Hence by [FHW, Lemma 18], it is an isomorphism when $t = 1$.

Next let $\nu_{t\in[0,1]}$ increase from $\nu_0 = \nu$ to some large value $\nu_1 = \kappa$, and consider the restriction map

$$S_{n+2}^{[a,b)}(P(\nu_t,\pi)) \to S_{n+2}^{[a,b)}(P(\nu,\alpha)) .$$

As ν_t increases, some levels of the part of the complex coming from $B^{2n}(\nu_t)$ cross the window. However, the part of the complex used to calculate $S_{n+2}^{[a,b)}$ for the given a, b does not change. Therefore, as before, because this is the identity map when $t = 1$ it is the identity map for all t.

Now suppose that $P(\nu, \pi - \delta)$ can be isotoped in W into $W_{2\delta}$. Choose κ so that this isotopy takes place in $P(\kappa, \pi)$. and suppose that

$$0 < \pi - 2\delta < a < \pi - \delta , \quad \pi < b < \pi + \nu .$$

Then, by the above, the restriction map

$$S_{n+2}^{[a,b)}(P(\kappa,\pi)) \to S_{n+2}^{[a,b)}(P(\nu,\pi-\delta))$$

is non-zero. But, because it is invariant under isotopy in $P(\kappa, \pi)$ it equals the restriction map

$$S_{n+2}^{[a,b)}(P(\kappa,\pi)) \to S_{n+2}^{[a,b)}(P(\nu,\pi-2\delta)) .$$

But this is the zero map because $a > \pi - 2\delta$. A contradiction. □

Remark 2.2: The case $Q(x) = \|x\|^2$ is borderline. Although the slice $S_{Q,\varepsilon}$ of the ball considered above is squeezable by isotopy inside the whole cylinder $\mathbf{R}^{2n} \times D$ it is very likely that it does not squeeze by isotopy inside a very short cylinder $B^{2n}(\delta) \times D$ when δ is only just larger than ε. (One could also compactify $B^{2n}(\delta)$ to a complex projective space $\mathbf{C}P^n$ of small volume if one wants to assume that M is in some sense complete.)

2.2 General slices. We now consider quadratic non-negative Hamiltonians Q_t which are 1-periodic in t, and let A_t be the corresponding flow. We write S_A for the germ along the 2-disc $D_0 = 0 \times D$ of the corresponding slice $S_{Q_t,\varepsilon}$.

LEMMA 2.3. *(i) If two positive paths A_t and B_t have the same endpoint $A_1 = B_1$ and are homotopic in* $\mathrm{Sp}(2n, \mathbf{R})$ *rel endpoints, there exists a symplectic diffeomorphism*

$$\Phi : S_A \to S_B .$$

(ii) Suppose that $A^s_{t\in[0,1]}$ where $s \in [0,1]$ is a smooth family of positive paths with fixed endpoints. Then there is a symplectic isotopy

$$\Phi^s : S_{A^0} \to S_{A^s} , \quad 0 \le s \le 1 .$$

Proof: (i) If A_t, B_t are generated by $-\mathcal{Q}_t, -\mathcal{K}_t$ respectively, then

$$S_A = \{(x,c,t) : c \leq \pi - \mathcal{Q}_t(x)\ ,\ \|x\| \text{ small}\}$$
$$S_B = \{(x,c,t) : c \leq \pi - \mathcal{K}_t(x)\ ,\ \|x\| \text{ small}\}\ ,$$

and it is easy to check that the map

$$\Psi : (x,c,t) \mapsto \left(B_t A_t^{-1}x, c - \mathcal{K}_t(B_t A_t^{-1}x) + \mathcal{Q}_t(x), t\right)$$

is a symplectomorphism defined on some neighborhood V_A of ∂S_A in S_A which takes ∂S_A to ∂S_B. (Note that S_A, S_B are germs along D_0, so that we can make them smaller as necessary.) Moreover, since $\mathcal{Q}_t(0) = \mathcal{K}_t(0) = 0$, we may extend Ψ to D_0 by the identity.

Since the paths A_t, B_t are homotopic, Ψ can be extended further to a smooth diffeomorphism of S_A onto S_B. The pull-back $\Omega = \Psi^*(\omega \oplus \sigma)$ coincides with $\omega \oplus \sigma$ on V_A and is such that $\Omega\ |_{D_0} = \sigma$. After a slight perturbation of Ψ in the transversal direction along D_0, we may assume that

$$\Omega = \omega \oplus \sigma$$

on the full tangent space $T_{(x,c,t)}(\mathbf{R}^{2n} \times D)$ at all points of $V_A \cup D_0$. But then the 1-parameter family of closed 2-forms

$$\Omega_\lambda = (1-\lambda)(\omega \oplus \sigma) + \lambda\Omega\ , \quad \lambda \in [0,1]$$

is a symplectic isotopy in some small neighbourhood of $V_A \cup D_0$, which always equals the split form in V_A and on D_0. By Moser's argument, there is a diffeotopy Φ_λ on a neighbourhood N of $V_A \cup D_0$ which is the identity on $V_A \cup D_0$ and is such that Φ_1 pulls Ω back to the split form $\omega \oplus \sigma$.

This proves (i). Statement (ii) is obvious since all choices can be made to depend smoothly on s. □

COROLLARY 2.4. *Proposition 1.5 holds.*

Proof: Let $A_{t\in[0,1]}$ be a short positive path with endpoint $A_1 \in \mathcal{U}$. Then $A_1 \in \mathcal{U}$ is the endpoint of a short positive autonomous path B_t. By Theorem 1.6 (iii), there exists a 1-parameter family of short positive autonomous paths B_t^s and a 1-parameter family of short positive paths A_t^s such that $A_t^1 = A_t$, $B_t^1 = B_t$, $A_1^s = B_1^s (\in \mathcal{U})$ for all $s \in [0,1]$, and $A_t^0 = B_t^0$.

We first wish to define a smooth 1-parameter family of symplectic automorphisms (with respect to the standard symplectic structure) $f^s : \mathbf{R}^{2n} \to \mathbf{R}^{2n}$ with the following property: if Q^s denotes the autonomous Hamiltonian that generates B_1^s, the pull-back $Q^s \circ f^s$ is diagonal in the standard basis for each $0 \leq s \leq 1$. Here of course "diagonal" means that $Q^s \circ f^s$ is of the form

$$Q^s \circ f^s = \sum_{i=1}^{n} \pi a_i^2 (x_{2i-1}^2 + x_{2i}^2) .$$

To do so, note first that since our argument is purely homotopical, we have complete freedom to jiggle or reparametrize the matrices B_1^s along time s. So first jiggle the path $B_1^{s \in [0,1]}$ so that

(i) the endpoints B_1^0 and B_1^1 are fixed, and

(ii) the autonomous positive quadratic forms Q^s corresponding to B_1^s are such that there exists only a finite number of points $s \in [0,1]$ where Q^s is not generic (that is to say: having a $2k$-dimensional space over which Q^s is equal to const$\|v\|^2$, for $k \geq 2$).

This is possible because the set of non-generic quadratic forms is of codimension at least one. Of course, one jiggles accordingly the paths A_t^s so that $A_1^s = B_1^s$ for all s. Let $\mathcal{I} \subset [0,1]$ denote the finite subset referred to in (ii) above.

Now we can reparametrize the path B_1^s so that the following holds: for each $s_k \in \mathcal{I}$, there exists an interval $I_k \subset [0,1]$ of non-zero measure such that all matrices B_1^s with s in this interval are equal. In this case, although the choice of a symplectic ordered basis $\mathcal{B}_s$ of oriented eigen-2-planes of each Q^s is not unique, there exists some choice of $\mathcal{B}_s = (P_{1,s}, \ldots, P_{n,s})$ which varies smoothly with s. The reason is that the unordered basis of nonoriented eigen-2-planes is unique when s is not in one of the intervals I defined above. And when s belongs to one of them, the above condition makes it possible to move the unordered basis smoothly from the one needed at the left end of the interval I_k to the one needed at the other end. This defines a smooth path of unordered and nonoriented bases: we then choose an order and an orientation at time $s = 0$ and extend this uniquely over all $s \in [0,1]$. Finally, since each $P_{i,s}$ is oriented, there exists a continuous lift of $\mathcal{B}_s$ to f_s, obtained by choosing a pair of symplectic vectors in each $P_{i,s}$. Such a lift obviously exists because we need to define it only over a path.

Now we conjugate both paths $A^s_{t \in [0,1]}, B^s_{t \in [0,1]}$ by f^s. This gives new paths, that we still denote in the same way, but which are such that the endpoints $A_1^s = B_1^s$ are all diagonalizable with respect to the standard symplectic basis. By the last lemma, there exist symplectic diffeomorphisms

$$\Phi^s : S_{B^s} \to S_{A^s} , \quad s \in [0,1]$$

with $\Phi^0 = \mathrm{id}$. But since all B^s are generated by diagonalized time-independent quadratic Hamiltonians, the sets S_{B^s} are ellipsoids with principal axes independent of t and s. Hence $\cap_{s \in [0,1]} S_{B^s}$ contains $S_{\mathcal{K},\varepsilon}$ for

$$\mathcal{K} = \sum_{i=1}^{n} \pi a_i^2 (x_{2i-1}^2 + x_{2i}^2)$$

where $a_i = \max_s a_i^s$. Note that the path generated by $\mathcal{K}$ is still short. The restriction of Φ^s to $S_{\mathcal{K},\varepsilon}$ gives the desired isotopy. □

In order to prove part (*ii*) of Theorem 1.3 we need the following lemmas. Recall that if $W = M \times D(\pi)$ we write W_δ for $M \times D(\pi - \delta)$.

LEMMA 2.5. *If S is locally squeezable in W, there is some $\delta > 0$ such that every compact subset X of $S \cap \operatorname{Int} W$ is squeezable by isotopy into W_δ.*

Proof: Given $\psi : S \to W$ consider the associated map $\tilde{\psi} : \partial W \cup X \to W$ which equals ψ on X and the identity on ∂W. If ψ is so close to the inclusion that its graph may be identified with a partial Lagrangian section of T^*W, then $\tilde{\psi}$ extends over W and is isotopic to the identity by maps which fix all points of ∂W. Note that this statement holds whatever X is provided that $X \cap \partial W = \emptyset$. The result now follows by applying this to a local squeezing ϕ_i of S. □

LEMMA 2.6. *Suppose that S is squeezable by isotopy in W and that it intersects ∂W in a closed characteristic $x_0 \times \partial D$ which is isolated among the closed characteristics on ∂S. Then the germ of S along $x_0 \times D$ is isotopic in W by an isotopy which fixes $x_0 \times \partial D$ to a slice germ which is locally squeezable in W.*

Proof: Suppose that $\psi_t : S \to W$ is an isotopy such that $\psi_1(S) \subset M \times \operatorname{Int} D$. Let γ be the closed characteristic $x_0 \times \partial D$ and consider the set

$$\mathcal{I} = \left\{ t \in [0,1] : \psi_t(\gamma) \subset M \times \partial D \right\} .$$

If $T = \inf\{t \in [0,1] : t \notin \operatorname{Int} \mathcal{I}\}$, then $T < 1$ since S is squeezable by isotopy. Further, since $\psi_t(\gamma)$ must always be a flat circle $x \times \partial D$ for $t \leq T$ we may compose ψ_t with maps of the form $g_t \times h_t$ so that ψ_t fixes all points of γ for $t \leq T$. We claim that $\psi_T(S)$ is locally squeezable. For, by hypothesis there is a sequence $\varepsilon_i \to 0^+$ such that $\psi_{T+\varepsilon_i}(S)$ does not intersect $M \times D$ in a closed characteristic, and so these sets can be pushed inside $M \times D$ by Lemma 1.2. □

Let $A_{t\in[0,1]}$ be a positive path, and $A^T = A_{t\in[0,T]}$ any subpath. We define S_{A^T} as the slice associated to the positive path A_{tT}, $t \in [0,1]$.

LEMMA 2.7. *If a positive path $A_{t\in[0,1]}$ is such that S_A is neither locally squeezable nor squeezable by isotopy, the same is true of any subpath $A^T = A_{t\in[0,T]}$.*

Proof: This is obvious. We may extend the subpath by setting $A_t = A_T$ for $t \geq T$, which corresponds to setting $Q_t = 0$ for $t \geq T$. Denote by $A'_{t\in[0,1]}$ that path. Clearly, $S_{A'}$ is locally squeezable (or squeezable by isotopy) if and only if the same is true of S_{A^T}. (The non-smoothness of $S_{A'}$ is irrelevant here.) But the corresponding set $S_{A'}$ clearly contains S_A, and so if it were locally squeezable or squeezable by isotopy, the same would be true of S_A.□

Proof of Theorem 1.3(ii): Let S be any slice germ along $D_0 = x_0 \times D$ whose characteristic flow is given by a short positive path $A_{t\in[0,1]}$. Since we are only looking at the germ we may suppose that M is just a small neighborhood of x_0 and so identify it with an open neighborhood U of $\{0\}$ in $\mathbf{R}^{2n}$. Further, by Theorem 1.6(ii) and Lemma 2.7 we may suppose that $A_1 \in \mathcal{U}_0$ so that we can apply Proposition 1.5. We may therefore isotop S in $U \times D$ by Φ_s, fixing D_0, so that $\Phi_1(S)$ is an ellipsoidal slice S_E, say. By Proposition 2.1 the latter slice is neither locally squeezable nor squeezable by isotopy in $U \times D$. It follows that S is not locally squeezable by isotopy in $M \times D$. However, we want to prove more: namely that S is neither locally squeezable nor squeezable by isotopy.

To see that S is not locally squeezable, observe that for all $\delta > 0$ there are subsets X' of $S_E \cap U \times D$ which cannot be squeezed by isotopy into $U \times D(\pi - \delta)$: namely the polydiscs $P(\varepsilon, \alpha)$ where $\varepsilon + \alpha > \pi$. Therefore, a similar statement holds for $X = \Phi_1^{-1}(X')$, and the desired conclusion follows from Lemma 2.5.

To see that S is not squeezable by isotopy in the original cylinder $M \times D$ we apply Lemma 2.6. Note first that because S is the set associated to a (time-dependent) quadratic positive non-degenerate form, the hypothesis of Theorem 1.3(ii) on the nonexistence of non-trivial closed trajectories of A_t implies in particular that there is no closed characteristic of ∂S near $x_0 \times \partial D$. This is because in the quadratic case, the flow A_t is given by the characteristic foliation of ∂S. Now we can apply Lemma 2.6: it implies that if S were squeezable by isotopy, there would be an isotopic set $S' \subset M \times D$ which was locally squeezable. But the characteristic flow of S' near γ is the same as that of S, and so such a set S' cannot exist by which we have just proved. □

3. Relations with Hofer's Geometry and Proof of Part (i) of the Main Theorem

As we mentioned earlier, the problem of (in)stability in Hofer's metric and the problem of local (un)squeezability of subsets of cylinders are essentially equivalent: the necessary condition for stability is equivalent to part (i) of

Theorem 1.3 and the sufficient condition for stability is equivalent to part (*ii*) of that theorem. We established a necessary condition for stability in [LM2] and we will use it below to prove part (*i*) of the main Theorem on the squeezability of sets $S \subset W$. Concerning the sufficient condition for stability or unsqueezability, we proceed in the reverse direction: we have just proved part (*ii*) of the main theorem and we will show below that it implies the sufficiency of the condition for the stability of geodesics.

3.1 Proof of part (*i*) of the Main Theorem. Here is the rough outline of the proof. Consider the part of ∂S inside a thin cylindrical annulus $W - W_\delta$. There, ∂S is the graph of a partially defined Hamiltonian and we can therefore apply the curve-shortening construction in the proof of Theorem 1.6 of [LM2] to move the graph away from ∂W. Since this construction is localized near the extrema, we can attach the shortened curves to the set $S \cap W_\delta$ and obtain a squeezing isotopy of S inside W.

So choose $\delta > 0$ small enough so that $\partial S \cap (W - W_\delta)$ is the graph of a function $G = \pi - H : E \to \mathbf{R}$, where E is some compact (not necessarily connected) subset of $\mathbf{R}^{2n} \times S^1$. Of course E contains a neighbourhood of the set of all maxima of G. We will suppose that G is normalized so that its maximum value is π. Let $\tilde{G} : \tilde{E} \to \mathbf{R}$ denote the pull-back of G by the map $\mathrm{id} \times h : \mathbf{R}^{2n} \times [0,1] \to \mathbf{R}^{2n} \times S^1$, where h identifies 0 and 1. Thus

$$\mathrm{graph}(\tilde{G}) \subset \tilde{E} \times [0,\pi] \subset \mathbf{R}^{2n} \times [0,1] \times [0,\pi] .$$

Now, by the proof of Theorem 1.6 of [LM2], there exists a smooth isotopy

$$\tilde{\phi}_\lambda : \mathrm{graph}(\tilde{G}) \to \tilde{E} \times \mathbf{R} , \quad \lambda \in [0,\varepsilon]$$

which begins with the identity and satisfies for all $\lambda > 0$:

(*i*) $\tilde{\phi}_\lambda$ is a symplectic diffeomorphism onto its image and the restriction of $\tilde{\phi}_\lambda$ to $\tilde{G}^{-1}([\pi - \delta, \pi - \delta/2])$ is the identity, as well as its restriction to some time intervals containing $\{t = 0\}$ and $\{t = 1\}$;

(*ii*) the set $\tilde{\phi}_\lambda(\mathrm{graph}(\tilde{G}))$ sits inside $\tilde{E} \times [0,\pi]$ and its restriction to some time interval $I \subset (0,1)$ sits inside $\tilde{E} \times [0,\pi)$; here I is independent of λ.

To see this, one first extends $\tilde{G}$ to a compactly supported Hamiltonian $\mathbf{R}^{2n} \times [0,1] \to \mathbf{R}$ whose values outside $\tilde{E}$ are in $[0, \pi - \delta/2)$. One then constructs the desired isotopy as in section 4.2 of [LM2].

Because $\tilde{\phi}_\lambda$ is the identity on some time intervals containing $\{t = 0\}$ and $\{t = 1\}$, it descends to an isotopy

$$\phi_\lambda : \mathrm{graph}(G) = \partial S \cap (W - W_\delta) \to W$$

which trivially extends to an isotopy $\partial S \to W$. By condition (*ii*) above, this isotopy, that we still denote ϕ_λ, has for all $\lambda > 0$ an image whose projection on the base D of the cylinder is *not* onto. Extend this isotopy to a symplectic isotopy defined on some interior collar neighbourhood of ∂S in S. Let $\bar{\phi}_\lambda$ be its lift to S. Then the pull-back of the standard symplectic form $\bar{\phi}^*_\lambda(\Omega)$ is a symplectic isotopy rel ∂S for all sufficiently small λ. The relative Moser argument then yields a symplectic isotopy $\psi_\lambda : S \to W$ beginning with the identity, which is such that $\mathrm{Im}(\psi_\lambda)$ does not project onto the base D^2 of the cylinder W when $\lambda > 0$. Then composing with an appropriate area preserving map of the base gives a local squeezing of S by isotopy. □

3.2 Sufficient condition for the stability of geodesics. We show that Theorem 1.3 (*ii*) proved in § 2 implies the sufficient condition for the stability of geodesics in Hofer's metric.

As in the characterisation of geodesics established in [LM3], we need the gluing-along-monodromy construction. However, we must use a slightly different normalization here, and so we begin by repeating the main constructions of [LM3, §2] in modified form. For now, we will assume that M is closed and that ω has been rescaled so that $\mathrm{vol}\, M = 1$.

Let $H_{t\in[0,1]}$ be a regular path in $\mathrm{Ham}^c(M)$ which has a fixed minimum p and a fixed maximum P at which the linearized flows have no non-trivial closed orbits in time less than or equal to 1. Proceeding by contradiction, we assume that this is not a stable geodesic, and will show that this contradicts Theorem 1.3 (*ii*). By replacing H_t by $H_t - H_t(p)$, we may assume that $\min H_t = 0$ for all t. Since $H_{t\in[0,1]}$ is regular, its maximum value $m_H(t) = H_t(P)$ is strictly greater than its minimum value $H_t(p)$ for all t. We reparametrize all flows which we consider, i.e. both $H_{t\in[0,1]}$ and the nearby flows $K_{t\in[0,1]}$, so that they are generated by Hamiltonians which vanish along with all their derivatives when $t = 0, 1$. Since the reparametrized flow $\phi_{\beta(t)}$ is generated by the Hamiltonian $\beta'(t)H_{\beta(t)}$ this does not change the length of the isotopy. Moreover, if we use the same reparametrization function β for all flows, it is easy to check this will not affect the stability properties of the path $H_{t\in[0,1]}$. The maximum of H_t will be denoted $m_H(t) = H_t(P)$. So our conventions imply that $m_H(t)$ is a smooth function of t which is infinitely tangent to 0 at $t = 0, 1$ and is > 0 on $(0, 1)$.

Denote by $\tilde{R}^-_H$ and $\tilde{R}^+_H$ the parts under and over the graph of H:

$$\tilde{R}^-_H = \left\{(x, s, t) \in M \times \mathbf{R}^2 : 0 \leq s \leq H_t(x)\right\}$$
$$\tilde{R}^+_H = \left\{(x, s, t) \in M \times \mathbf{R}^2 : H_t(x) \leq s \leq m_H(t)\right\},$$

and define

$$U_H = \left\{(s, t) \in \mathbf{R}^2 : 0 \leq s \leq m_H(t)\right\}.$$

By assumption on $m_H(t)$, the set U_H has area $\mathcal{L}(H)$ and lies between two curves $s = 0$ and $s = m_H(t)$ which are infinitely tangent at their endpoints but are otherwise disjoint. Similarly, both $\tilde{R}_H^-$ and $\tilde{R}_H^+$ are regions lying between two hypersurfaces which are infinitely tangent along $M \times 0 \times \{t = 0, 1\}$. One of these hypersurfaces is the graph of H with monodromy ϕ_1, and the other (either $s = 0$ or $s = m_H(t)$) has trivial monodromy.[1] Note that these hypersurfaces will touch at points where $H_t(x)$ is either 0 or $m_H(t)$. However, this does not matter because we will be interested only in the piece of $\tilde{R}_H^-$ near P (or the piece of $\tilde{R}_H^+$ near p). The gluing of both halves $\tilde{R}_H^-$ and $\tilde{R}_H^+$ by the identification of their common side, which matches the characteristic foliations, gives back simply the product $M \times U_H$ of area $\mathcal{L}(H)$.

Now let $K_{t\in[0,1]}$ be another path that we may assume to be homotopic and C^∞-close to $H_{t\in[0,1]}$. As above, $K_t = 0$ when $t = 0, 1$ and is infinitely tangent to 0 there, but, because $K_{t\in[0,1]}$ need not have a fixed minimum, we cannot necessarily normalize K_t, keeping it a smooth function of t, so that $\min K_t = 0$ for each t. Therefore we define ε-thickenings of $\tilde{R}_K^\pm$ as follows:

$$\tilde{R}_{K,\varepsilon}^- = \{(x,s,t) \in M \times \mathbf{R}^2 : \lambda_1(t) \leq s \leq K_t(x)\}$$
$$\tilde{R}_{K,\varepsilon}^+ = \{(x,s,t) \in M \times \mathbf{R}^2 : K_t(x) \leq s \leq \lambda_2(t)\} ,$$

where $\lambda_1(t) < \min K_t$ and $\lambda_2(t) > \max K_t$ are smooth functions which are infinitely tangent to 0 at $t = 0, 1$, and so close to the minimum and maximum of K_t that

$$\mathrm{vol}(\tilde{R}_{K,\varepsilon}^+ \cup \tilde{R}_{K,\varepsilon}^-) = \mathcal{L}(K) + \varepsilon .$$

Since the flow $\psi_{t\in[0,1]}$ of $K_{t\in[0,1]}$ has endpoint $\psi_1 = \phi_1$, we can use the map

$$\Phi_{K,H} : \tilde{R}_{K,\varepsilon}^+ \to \tilde{R}_H^-$$
$$\Phi_{K,H}(x,s,t) = \left(\phi_t \circ \psi_t^{-1}(x), s - K_t(x) + H_t(\phi_t \circ \psi_t^{-1}(x)), t\right) .$$

to glue $\tilde{R}_{K,\varepsilon}^+$ to $\tilde{R}_H^-$ to get a subset of $M \times \mathbf{R}^2$ which we call $\tilde{R}_{H,K,\varepsilon}$. The set

$$\tilde{R}_{K,\varepsilon,H} = \tilde{R}_{K,\varepsilon}^- \cup \tilde{R}_H^+$$

is defined similarly. These sets $\tilde{R}_{H,K,\varepsilon}, \tilde{R}_{K,\varepsilon,H}$ have the same basic shape as $\tilde{R}_H^\pm$, i.e. they are manifolds except for the fact that their front and back faces are infinitely tangent along $M \times \{0\} \times \{t = 0, 1\}$. But now the ε-thickening prevents the front and back faces from touching each other. Recall from [LM3] that the *area* of a set such as $\tilde{R}_H^+$ is the number A defined by:

$$\mathrm{vol}(\tilde{R}_H^+) = A\,\mathrm{vol}(M) .$$

---

[1] The monodromy of a hypersurface diffeomorphic to $M \times [0,1]$ is the (partially defined) map $M \to M$ which takes the point $x \in M$ to y, where $y \times 1$ is the endpoint of the leaf of the characteristic foliation which goes through $x \times 0$.

Since we are assuming that $\mathrm{vol}(M) = 1$, A is just $\mathrm{vol}(\tilde{R}_H^+)$.

Before beginning the proof, we must deal with the smoothing/normalization problem. We need a canonical way to thicken sets like U_H, and have to be careful because there is no extra room to play with. Fix a real number $\lambda > 0$ and denote by

$$\tilde{R}_H^-(\lambda) = \{(x,s,t) : 0 \le s \le \lambda + H_t(x)\}$$
$$\tilde{R}_H^+(\lambda) = \{(x,s,t) : H_t(x) \le s \le m_H(t) + \lambda\}$$
$$U_H(\lambda) = \{(s,t) : 0 \le s \le \lambda + m_t\} .$$

Hence, for instance, $U_0(\lambda)$ is simply the square $[0,\lambda] \times [0,1]$. Correspondingly, we may thicken $\tilde{R}_{H,K,\varepsilon}$ to

$$\tilde{R}_{H,K,\varepsilon}(\lambda) = \tilde{R}_H^-(\lambda) \cup \tilde{R}_{K,\varepsilon}^+ ,$$

where $\tilde{R}_{K,\varepsilon}^+$ is translated by λ in the s direction so that it fits together with $\tilde{R}_H^-(\lambda)$.

All symplectic embeddings

$$\tilde{R}_{H,K,\varepsilon}(\lambda) \to M \times \mathbf{R}^2$$

which we will later consider will be infinitely tangent to the inclusion along the three sides $s = 0$, $t = 0, 1$. We will call such maps *normalized*.

Now let us begin the proof. If H_t is not a stable geodesic, there is a sequence K_t^i of Hamiltonians which converge to H_t and are such that

$$\mathcal{L}(K_t^i) < \mathcal{L}(H_t) = m$$

for all i. Because

$$\begin{aligned}\operatorname{area}\left(\tilde{R}_{H,K,\varepsilon}(\lambda)\right) + \operatorname{area}\left(\tilde{R}_{K,\varepsilon,H}(\lambda)\right) &= \operatorname{area}\left(\tilde{R}_H(\lambda)\right) + \operatorname{area}\left(\tilde{R}_{K,\varepsilon}(\lambda)\right) \\ &= \mathcal{L}(H) + \mathcal{L}(K) + 2(\lambda + \varepsilon) ,\end{aligned}$$

this can happen only if there is a sequence of positive real numbers ε_i converging to 0 such that either

$$\operatorname{area}\left(\tilde{R}_{H,K^i,\varepsilon_i}(\lambda)\right) < \mathcal{L}(H) + \lambda = \operatorname{area}\left(U_H(\lambda)\right)$$

or

$$\operatorname{area}\left(\tilde{R}_{K^i,\varepsilon_i,H}(\lambda)\right) < \mathcal{L}(H) + \lambda = \operatorname{area}\left(U_H(\lambda)\right) .$$

We will suppose the former. (The latter case is of course symmetric and would be handled in the same way.) Denoting by a_i the area of $\tilde{R}_{H,K^i,\varepsilon_i}$, we then have: $a_i + \lambda < m + \lambda$ for all i.

In order to deduce Theorem 1.7 from Theorem 1.3, the key technical lemma needed is the following:

LEMMA 3.1. *Let the regular path $H_{t\in[0,1]}$ satisfy the hypothesis of the Stability Theorem 1.7 and suppose that it is not a stable geodesic from $\mathbb{1}$ to*

ϕ. Assume that K_i is a sequence of Hamiltonians with time-1 map ϕ which converge C^∞ to H_t and are such that

$$\text{area}\left(\tilde{R}_{H,K^i,\varepsilon_i}(\lambda)\right) < \mathcal{L}(H) + \lambda\ .$$

Let B be a closed neighbourhood of P such that $H_t(x) \geq \frac{1}{2}H_t(P)$ for $x \in B$, and define

$$\tilde{S} = \tilde{R}_H^-(\lambda) \cap (B \times \mathbf{R}^2)\ ,$$
$$\tilde{W} = M \times U_H(\lambda)\ .$$

Then there exists a sequence of symplectic embeddings

$$f_i : \tilde{S} \to \tilde{W}\ , \quad i = 1, 2, \ldots$$

which satisfies:

- *(i) each map f_i coincides with the inclusion on the three sides $s = 0$, $t = 0, 1$ and has contact of infinite order with the inclusion there,*
- *(ii) the sequence converges in the C^∞-topology to the inclusion $\tilde{S} \hookrightarrow \tilde{W}$, and*
- *(iii) for each i, the composite*

$$\tilde{S} \xrightarrow{f_i} \tilde{W} \xrightarrow{\pi} U_H(\lambda)$$

is not onto (where the second map is the projection).

Indeed, with this, it is then an easy matter to complete the proof:

COROLLARY 3.2. *Theorem 1.3 (ii) implies the Stability Theorem 1.7.*

Proof: The set $\tilde{S}$ is the region under the graph of a Hamiltonian $G_{t\in[0,1]} : U \to [\lambda, \infty)$ which equals the constant map λ at $t = 0, 1$ and is infinitely tangent to it there. Since this Hamiltonian is obtained from the initial Hamiltonian $H_{t\in[0,1]}$ by reparametrisation and addition of a constant map, its linearised flow at P is the same as the one of H, and therefore it has no non-trivial closed trajectory in time less than or equal to 1. Now let $\Phi : U_H(\lambda) \to [0, m + \lambda] \times [0, 1]$ be an area preserving map of the form $(s,t) \mapsto (a(s,t), b(t))$ (thus preserving the fibers $t = \text{const}$). Then the maps

$$g_i = (\text{id} \times \Phi) \circ f_i \circ (\text{id} \times \Phi)^{-1} : \bar{S} = (\text{id} \times \Phi)(\tilde{S}) \to M \times [0, m + \lambda] \times [0, 1]$$

are symplectic embeddings, converge to the inclusion, are tangent to infinite order to the inclusion on $s = 0$, and $t = 0, 1$, and do not have surjective projection onto $[0, m + \lambda] \times [0, 1]$. Further, $\bar{S}$ is the graph of a Hamiltonian obtained from G by reparametrisation, and therefore its linearised flow at P has no non-trivial closed trajectory in time ≤ 1. But the condition on the tangency of all maps g_i on the three sides of $\bar{S}$ shows that these maps descend to symplectic embeddings

$$h_i : S = (\text{id} \times \phi)(\bar{S}) \to (\text{id} \times \phi)\big(M \times [0, m + \lambda] \times [0, 1]\big) = M \times D^2(m + \lambda)$$

where $D^2(m+\lambda)$ is the standard closed disk of area $m+\lambda$ and where ϕ is the map taking the (s,t) coordinate to the action-angle coordinates $(c = s, t)$ $(c = \pi r^2,\ 2\pi t = \theta)$. Thus the family h_i gives a local squeezing of the set S inside the split round cylinder, although this set does satisfy the hypothesis on non-existence of closed trajectories of Theorem 1.3. This contradicts the latter theorem. □

Proof of Lemma 3.1: We need first the following definition.

DEFINITION 3.3: For $a \geq 0$, choose a smooth family of functions $\mu_a : [0,1] \to [0,\infty)$ which
 (i) increase with a,
 (ii) map $(0,1)$ into $(0,\infty)$ and are infinitely tangent to 0 at $t = 0, 1$, and
 (iii) are such that the set

$$U_a = \{(s,t) : 0 \leq s \leq \lambda + \mu_a(t)\ ,\ 0 \leq t \leq 1\}$$

 has area $\lambda + a$.

Then we will say that $\tilde{R}_{H,K,\varepsilon}(\lambda)$ is a *square cylinder* of area a if there is a smooth normalized symplectomorphism

$$\Phi_{H,K} : \big(\tilde{R}_{H,K,\varepsilon}(\lambda),\ \omega \oplus \sigma\big) \to (M \times U_a,\ \omega \oplus \sigma)\ .$$

This is possible only if $a = \operatorname{vol}\tilde{R}_H^- + \operatorname{vol}\tilde{R}_{K,\varepsilon}^+$ (recall that $\operatorname{vol} M$ has been set equal to 1), and if $\tilde{R}_{H,K}$ has trivial monodromy (or, equivalently, that the time 1 maps of the flows of H_t and K_t are the same).

The *front face* of a square cylinder consists of the points which map onto

$$\{(x, \lambda + \mu_a(t), t) \mid t \in [0,1]\ ,\ x \in M\}\ .$$

The following lemma is an adaptation of [LM3, Lemma 2.6] to the present context.

LEMMA 3.4. *Fix $\phi \in \operatorname{Ham}^c(M)$ and let H_t, K_t be Hamiltonians with flows ϕ_t, ψ_t from id to ϕ normalized as above. Then, there is a C^1-neighbourhood $\mathcal{U}$ of id in $\operatorname{Ham}^c(M)$ such that $\tilde{R}_{H,K,\varepsilon}(\lambda)$ is a square cylinder whenever $\phi_t \circ \psi_t^{-1} \in \mathcal{U}$ for all t.*

Proof: First observe that because $\phi_1 = \psi_1$, the gluing map $\Psi_{K,H}$ defined above by

$$\Psi_{K,H}(x,s,t) = \big(\phi_t \circ \psi_t^{-1}(x),\ s + \lambda - K_t(x) + H_t(\phi_t \circ \psi_t^{-1}(x)),\ t\big)$$

equals the identity when $t = 1$. Therefore, if we extend it by the identity, it defines a normalized map

$$\Psi_K : \partial\tilde{R}_{H,K,\varepsilon}(\lambda) \to M \times \partial U_a$$

where $a = \mathrm{vol}\,\tilde{R}_H^- + \mathrm{vol}\,\tilde{R}_{K,\varepsilon}^+$. We take $\mathcal{U}$ to be a star-shaped neighbourhood of id in $\mathrm{Ham}^c(M)$ consisting of Hamiltonian diffeomorphisms ψ whose graphs lie close enough to the diagonal $diag$ in $(M \times M, -\omega \oplus \omega)$ to correspond to graphs of 1-forms $\rho(\psi)$ in $(T^*M, -d\lambda_{\mathrm{can}})$. Then, if K_t is so close to H_t that the corresponding paths $\{\psi_t\}, \{\phi_t\}$ satisfy $\phi_t \circ \psi_t^{-1} \in \mathcal{U}$ for all t, there is a unique choice of retracting homotopy $f_{c,t}$ from $f_{0,t} = \mathrm{id}$ to $f_{1,t} = \phi_t \circ \psi_t^{-1}$ defined by

$$\rho(f_{c,t}) = c\,\rho(\phi_t \circ \psi_t^{-1}) .$$

It will be convenient to parametrize this homotopy a little differently, by the points of $U_a \cap \{s \geq \lambda\}$ instead of the (c,t) square. In fact, because H_t and K_t are both infinitely tangent to 0 at $t = 0, 1$, there is for each K_t a smooth map

$$g_K : U_a \cap \{s \geq \lambda\} \to \mathrm{Ham}^c(M)$$

such that

$$g_K(0,t) = \mathrm{id}\ , \quad g_K(s,t) = f\big(c(s,t),t\big)\ , \quad g_K\big(\mu_a(t),t\big) = \phi_t \circ \psi_t^{-1} .$$

(Note that the reparametrization map $c(s,t)$ itself need not be smooth at $t = 0, 1$.) Moreover, we may assume that $g_K(s,t)$ is infinitely tangent to the identity along the line $s = 0$. Then g_K may be used to extend Ψ_K to a smooth normalized map

$$\widetilde{\Psi}_K : \tilde{R}_{H,K,\varepsilon}(\lambda) \to M \times U_a\ ,$$

which has the form

$$(x,s,t) \mapsto \big(g_K(s,t)x, s'(x,s,t), t\big)$$

on $\tilde{R}_{H,K,\varepsilon} \cap \{s \geq 0\}$ for some suitable function s'. When $\mathcal{U}$ is sufficiently C^1-small (that is when K_t is C^2-close to H_t), the push-forward

$$\widetilde{\Omega} = (\widetilde{\Psi}_K)_*(\omega \oplus \sigma)$$

on $M \times U_a$ restricts to an area form on each flat disc $pt \times U_a$. The standard Moser method now shows that there is an isotopy f_t of $M \times U_a$ which is the identity on $M \times \partial U_a$ (and is infinitely tangent to the identity on the sides $s = 0$, $t = 0, 1$) such that $f_1^*(\widetilde{\Omega}) = \omega \oplus \sigma$. See [LM1, Lemma 2.3]. To make f_t have the required properties near $M \times \partial U_a$, one should first adjust $\widetilde{\Omega}$ near the front face and then adjust it inside. Further details are left to the reader. □

LEMMA 3.5. *Let H_t be as in Lemma 3.4, and recall that $\tilde{S} = \tilde{R}_H^-(\lambda) \cap B \times \mathbf{R}^2$, where $B \subset M$ is a neighbourhood of the fixed maximum P on which H_t is $\geq m_H(t)/2$. Then, given the sequences K_t^i converging C^∞ to H_t and ε_i converging to 0, we may construct the normalized maps $\widetilde{\Psi}_{K^i}$ so that their* *restrictions to $\tilde{S}$ converge to the inclusion*

$$\iota : \tilde{S} \hookrightarrow \tilde{R}_H^-(\lambda) \subset M \times U_m$$

where $m = \mathcal{L}(H)$.

Proof: The reader may check that at each point of the above construction the distance of the map from ι is dominated by the distance of K_t from H_t. We need to assume that $H_t(x)$ is bounded away from 0 on B in order that the function s' which gives the s-coordinate of $\widetilde{\Psi}_K$ is well-behaved (and independent of ε). Note also that the size of the isotopy provided by Moser's method depends only on the distance between the endpoints $\widetilde{\Omega}$ and $\omega \oplus \sigma$ of the isotopy of forms, which in the given situation can be assumed to tend to 0. □

Lemma 3.1 is a direct consequence of the preceeding two lemmas.

This completes the proof of Theorem 1.7 when M is closed. To get rid of this last hypothesis, assume that M is non-compact and without boundary. If $H_{t\in[0,1]}$ is not stable with respect to the (strong) C^∞-topology, this means that there is a compact set $X \subset M$ containing the support of H and a sequence of Hamiltonians $K^i_{t\in[0,1]}$ with support in X such that $K^i \to H$ on X with, say,

$$\text{area}\left(\tilde{R}_{H,K^i,\varepsilon}(\lambda)\right) < \text{area}\left(U_H(\lambda)\right) ,$$

for all i. Then the proof goes as before.

Finally, if (M, ω) is a manifold with boundary, one may slightly extend both M and ω to a small open collar neighbourhood V of ∂M. Then, if all Hamiltonian isotopies were the identity on some neighbourhood of ∂M, one would first extend them by the identity to V and apply the previous argument.

References

[BP] M. Bialy, L. Polterovich, Geodesics of Hofer's metric on the group of Hamiltonian diffeomorphisms, preprint, Tel Aviv, 1994.

[E1] I. Ekeland, An index theory for periodic solutions of convex Hamiltonian systems, Proc. Symp. Pure Math. 45 (1986), 395–423.

[E2] I. Ekeland, Convexity Methods in Hamiltonian Mechanics, Ergebnisse Math 19, Springer-Verlag, Berlin (1989).

[FH] A. Floer, H. Hofer, Symplectic homology I, preprint, 1992.

[FHW] A. Floer, H. Hofer, K. Wysocki, Applications of symplectic homology, preprint, 1993.

[G] M. Gromov, Pseudo-holomorphic curves in symplectic manifolds, Invent. Math. 82 (1985), 307–347.

[LM1] F. Lalonde, D. McDuff, The geometry of symplectic energy, Annals of Math. to appear.

[LM2] F. LALONDE, D. MCDUFF, Hofer's L^∞-geometry: energy and stability of Hamiltonian flows I, preprint, 1994.

[LM3] F. LALONDE, D. MCDUFF, Hofer's L^∞-geometry: energy and stability of Hamiltonian flows II, preprint, 1994.

[LM4] F. LALONDE, D. MCDUFF, Homotopy properties of stable positive paths and bifurcations of eigenvalues, preprint, 1994.

[U] I. USTILOVSKY, Conjugate points on geodesics of Hofer's metric, preprint, Tel Aviv, 1994.

François Lalonde
Université du Québec
Montréal
Canada
e-mail: flalonde@math.uqam.ca

Dusa McDuff
State University of New York
Stony Brook
USA
e-mail: dusa@math.sunysb.edu

Submitted: November 1994
Revised version: January 1995

Geometric And Functional Analysis
Vol. 5, No. 2 (1995)
1016-443X/95/0200387-15$1.50+0.20/0

ON SELBERG'S EIGENVALUE CONJECTURE

W. Luo, Z. Rudnick and P. Sarnak

1. Introduction

Let $\Gamma \subset SL_2(\mathbf{Z})$ be a congruence subgroup, and $\lambda_0 = 0 < \lambda_1 < \dots$ be the eigenvalues of the non-euclidean Laplacian on $L^2(\Gamma\backslash\mathbf{H}^2)$. A fundamental conjecture of Selberg ([Se]) asserts that the smallest nonzero eigenvalue $\lambda_1(\Gamma) \geq 1/4 = 0.25$. In the same paper Selberg proved that $\lambda_1(\Gamma) \geq 3/16 = 0.1875$. Gelbart and Jacquet ([GJ]), using very different methods, improved this to $\lambda_1(\Gamma) > 3/16$. Iwaniec ([I]) showed that for almost all Hecke congruence groups $\Gamma_0(p)$ with a certain multiplier χ_p, one has $\lambda_1(\Gamma_0(p), \chi_p) \geq 44/225 = 0.19555\dots$. In [I], he also established a density theorem for possible exceptional eigenvalues as above, which while not giving any improvement on 3/16 for an individual Γ, is sufficiently strong to substitute for Selberg's conjecture in many applications to number theory. Selberg's conjecture is the archimedean analogue of the "Ramanujan conjectures" on the Fourier coefficients of Maass forms. For these, much progress has been made in improving the relevant estimates, beginning with Serre ([Ser]) and later on Shahidi ([Sh2]) and Bump-Duke-Hoffstein-Iwaniec ([BDHI]). In this paper we restore the balance and establish in part for the archimedean place what is known at the finite places. The method on the face of it is quite different, but the quality of the results coincide (the reason will be made clear later).

THEOREM 1.1. *For any congruence subgroup $\Gamma \subset SL_2(\mathbf{Z})$, we have*

$$\lambda_1(\Gamma) \geq \frac{21}{100} .$$

There are numerous applications of this result. We present some immediate ones. The first is towards the Linnik-Selberg conjecture ([L],[Se]) on cancellation in sums of Kloosterman sums. For $m, n, c \geq 1$ one defines the Kloosterman sum as

$$S(m,n,c) = \sum_{\substack{x \bmod c \\ x\bar{x}\equiv 1 \bmod c}} e\left(\frac{mx + n\bar{x}}{c}\right) .$$

Supported by NSF grants DMS-9304580, DMS-9400163 and DMS-9102082

COROLLARY 1.1. *Fix N, n, m. Then as $x \to \infty$,*

$$\sum_{\substack{c \leq x \\ c \equiv 0 \bmod N}} \frac{S(m,n,c)}{c} \ll x^{2/5} .$$

Note that Weil's bound ([We]) $S(m,n,c) \ll_\epsilon c^{1/2+\epsilon}$ would give a bound of $x^{1/2+\epsilon}$ for the sum so that Corollary 1.1 indicates that there is considerable cancellation due to the signs of the Kloosterman sums along any progression $c \equiv 0 \bmod N$. Indeed this Corollary is the first such result on cancellation of Kloosterman sums on a general progression.

A second application is to the remainder term in the "Prime Geodesic Theorem" for congruence subgroups Γ of $SL_2(\mathbf{Z})$. Let $\pi_\Gamma(x)$ be the number of prime closed geodesics of length $\ell \leq \log x$ on $\Gamma \backslash \mathbf{H}^2$.

COROLLARY 1.2. *For any congruence group $\Gamma \subset SL_2(\mathbf{Z})$,*

$$\pi_\Gamma(x) = Li(x) + O(x^{7/10})$$

where $Li(x) = \int_2^x \frac{dt}{\log t}$.

A remainder term of the form $O(x^{3/4})$ has been known for a some time (see [S]). The natural conjecture here is that as in the theory of primes, the remainder term is $O_\epsilon(x^{1/2+\epsilon})$ for any $\epsilon > 0$.

Our proof of Theorem 1.1 is based on the Gelbart-Jacquet lift ([GJ]) and so is naturally concerned with the cuspidal spectrum of GL_m (in what follows $m \geq 2$). To describe our results we need to introduce various L-functions. For this we assume some familiarity with the adelic language. Let $\mathbf{A}$ be the adeles of $\mathbf{Q}$, and $\pi = \otimes_{p \leq \infty} \pi_p$ be an irreducible cuspidal automorphic representation of $GL_m(\mathbf{A})$, which we normalize to have unitary central character. Assume that the archimedean component π_∞ is spherical, so that one associates to it a semi-simple conjugacy class $diag(\mu_\infty(1), \ldots, \mu_\infty(m))$ in $GL_m(\mathbf{C})$. The gamma factor for the principal L-function $L(s,\pi)$ associated to π ([GoJ],[J]) is

$$L(s,\pi_\infty) = \prod_{j=1}^{m} \Gamma_{\mathbf{R}}(s - \mu_\infty(j)) \tag{1.1}$$

where $\Gamma_{\mathbf{R}}(s) = \pi^{-s/2}\Gamma\left(\frac{s}{2}\right)$. The analogue of Selberg's conjecture for GL_m is that π_∞ is *tempered*, i.e. for $j = 1, \ldots, m$

$$\operatorname{Re}\left(\mu_\infty(j)\right) = 0 . \tag{1.2}$$

For $m = 2$, Selberg's bound $\lambda_1 \geq 3/16$ is equivalent to $|\operatorname{Re}(\mu_\infty(j))| \leq 1/4$. For $m \geq 3$ the only known bound toward (1.2) is the (local) result of

Jacquet-Shalika ([JSha1,2]), which asserts that for generic unitary representations

$$\left|\mathrm{Re}(\mu_\infty(j))\right| < \tfrac{1}{2} \tag{1.3}$$

(see [BR] for a proof using Vogan's classification of the unitary dual of $GL_m(\mathbf{R})$).

THEOREM 1.2. *Let π be a cuspidal automorphic representation of $GL_m/\mathbf{Q}$ with π_∞ spherical. Then*

$$\left|\mathrm{Re}(\mu_\infty(j))\right| \le \frac{1}{2} - \frac{1}{m^2+1}\,.$$

Theorem 1.1 follows from Theorem 1.2 via the Gelbart-Jacquet lift. Indeed if $\lambda = 1/4 - r^2$, $r > 0$, is an exceptional eigenvalue for $\Gamma\backslash\mathbf{H}^2$ then there is a cuspidal automorphic representation π on $GL_2/\mathbf{Q}$ such that π_∞ is parametrized by $\mu_\infty(1) = r$, $\mu_\infty(2) = -r$. Since π cannot be monomial (as these have $\lambda \ge 1/4$), it lifts to a cuspidal automorphic representation Π on GL_3 whose archimedean component Π_∞ is also spherical and is parametrized by $diag(2r, 0, -2r)$. Now apply Theorem 1.2.

Our proof of Theorem 1.2 makes heavy use of the by now well developed Rankin-Selberg theory on GL_m. The key to the proof is the following observation: If π on GL_m is as above, the Rankin-Selberg L-function $L(s, \pi\times\tilde{\pi})$ has as its gamma factor

$$L(s, \pi_\infty \times \tilde{\pi}_\infty) = \prod_{j,k=1}^{m} \Gamma_{\mathbf{R}}\big(s - \mu_\infty(j) - \overline{\mu_\infty(k)}\big)\,.$$

Let $\beta_0 = 2\max\mathrm{Re}(\mu_\infty(j))$, then $L(s, \pi_\infty\times\tilde{\pi}_\infty)$ is holomorphic for $\mathrm{Re}\, s > \beta_0$ and has a pole at $s = \beta_0$. If χ is a primitive even Dirichlet character then the same is true for the gamma factor of $L(s, (\pi\otimes\chi)_\infty \times\tilde{\pi}_\infty)$ - in fact the gamma factor is still equal to $L(s, \pi_\infty\times\tilde{\pi}_\infty)$. For χ even primitive of sufficiently large (prime) conductor q we have $\pi\otimes\chi \not\simeq \pi$ and so $L(s, \pi_\infty\times\tilde{\pi}_\infty)L(s, (\pi\otimes\chi)\times\tilde{\pi})$ is entire. Hence β_0 is a *trivial* zero of $L(s, (\pi\otimes\chi)\times\tilde{\pi})$, that is

$$L\big(\beta_0, (\pi\otimes\chi)\times\tilde{\pi}\big) = 0 \tag{1.4}$$

for all such χ. In this way the problem becomes the familiar one of proving that certain twists of L-functions do not vanish at a given point. Theorem 1.2 follows from

$$\sum_{q\sim Q}\ \sum_{\substack{\chi\neq\chi_0\\ \text{even}}} L\big(\beta, (\pi\otimes\chi)\times\tilde{\pi}\big) \gg \frac{Q^2}{\log Q} \tag{1.5}$$

for $\mathrm{Re}\,\beta > 1 - \frac{2}{m^2+1}$, with Q large, the implied constants depending only on π and β.

To prove (1.5) we use the functional equation for $L(s,(\pi\otimes\chi)\times\tilde{\pi})$ to approximate $L(\beta,(\pi\otimes\chi)\times\tilde{\pi})$. This brings in Gauss sums and in order to optimize the analysis we use Deligne's bounds on hyper-Kloosterman sums ([D]); these arise for similar reasons in Duke-Iwaniec ([DuI]), Rohrlich ([R]) and Barthel-Ramakrishnan ([BR]). We note however that an improvement of (1.3) and hence of Selberg's 3/16 bound would result even without use of Deligne's bound.

To end the Introduction we make some further remarks. Firstly one can treat the finite places in a similar way and reduce the problem of bounding the size of Fourier coefficients to one of non-vanishing of twists. That is, fix a prime p at which π is unramified. The local L-factor $L(s,\pi_p\times\tilde{\pi}_p)=\prod_{j,k=1}^{m}(1-\alpha_j(p)\overline{\alpha_k(p)}p^{-s})^{-1}$ has a pole at the point β_0 defined via $p^{\beta_0}=\max_j|\alpha_j(p)|^2$. Hence the partial L-function $L^{(p)}(s,\pi\times\tilde{\pi}):=L(s,\pi_p\times\tilde{\pi}_p)^{-1}L(s,\pi\times\tilde{\pi})$ has a "trivial" zero at $s=\beta_0$. The same is true for the twists $L^{(p)}(s,(\pi\otimes\chi)\times\tilde{\pi})$ for any χ of conductor q for which $\chi(p)=1$ (this being the analogue of $\chi(-1)=1$ which we needed in (1.4)). By choosing special q's (as in Rohrlich ([R])) one can modify the arguments in this paper and obtain similar results towards Ramanujan at p. This puts the finite and infinite places on the same footing and explains the comment before Theorem 1.1. In this connection we note that the full Selberg conjecture (or as above, the Ramanujan conjecture) would follow from the following statement: Given π an irreducible cuspidal automorphic representation and β with $\mathrm{Re}(\beta)>0$, there is an even Dirichlet character such that $L(\beta,\pi\otimes\chi)\neq 0$. Such problems have been studied by many authors ([Shi], [R], [BR]).

In the special case of $m=3$ an improvement of Theorem 1.2 would result from the theory of the symmetric square L-function $L(s,\pi,\mathrm{Sym}^2)$ on GL_3. The point is that by using $L(s,\pi\otimes\chi,\mathrm{Sym}^2)$ instead of $L(s,(\pi\otimes\chi)\times\tilde{\pi})$, the conductor dependence in χ is reduced from q^9 to q^6. On the other hand the location of the trivial zeros remains unchanged. The result would be an improvement in the RHS of Theorem 1.2, with 2/5 being replaced by 5/14 and correspondingly $\lambda_1(\Gamma)\geq\frac{171}{784}=0.21811\ldots$ in Theorem 1.1. Unfortunately, the archimedean theory (even in the unramified case) for $L(s,\pi,\mathrm{Sym}^2)$ is not well understood at present and so we cannot carry out the above analysis.[1] However, in view of [PP-S],[BuGi], this is not a problem at the finite unramified places, and so for these one can carry out the above. This is the analogue of [BDHI].

The results above, both at the infinite and finite places, can be estab-

[1]D. Ramakrishnan has pointed out to us a device using [BuGi] and the functional equation in [Sh1] to overcome this difficulty.

lished with $\mathbf{Q}$ replaced by a number field F with no loss in the quality of the estimates. The point being that the size of the conductor (in the character aspect) of $L(s,(\pi\otimes\chi)\times\tilde{\pi})$ is independent of the number field. The analysis is made more difficult by the presence of units which restrict the choice of χ. A similar difficulty appears, and is overcome, in the work of Rohrlich ([R]; see also [BR]). A complete proof of the results with $\mathbf{Q}$ replaced by a number field will appear in a forthcoming article.

Acknowledgement. We thank H. Iwaniec for several illuminating discussions on this work.

2. Background on Rankin-Selberg L-functions

2.1 Rankin-Selberg theory. We recall the Rankin-Selberg theory as developed by Jacquet, Piatetski-Shapiro and Shalika ([JP-SS]), Shahidi ([Sh1]) and Mœglin-Waldspurger ([MW]). The Rankin-Selberg L-function associated to a pair of cuspidal automorphic representations π' on GL_m, π'' on GL_n is given by an Euler product

$$L(s,\pi'\times\pi'')=\prod_{p<\infty}L(s,\pi'_p\times\pi''_p)\ .$$

For primes p where both π' and π'' are unramified the local factors are given by

$$L(s,\pi'_p\times\pi''_p)=\det(I-A'_p\otimes A''_p p^{-s})^{-1} \tag{2.1}$$

where A'_p (respectively A''_p) are the Satake parameters associated to π'_p (respectively to π''_p). At finite primes where one of π', π'' are ramified, the local factor is still of the form $L(s,\pi'_p\times\pi''_p)=P_p(p^{-s})^{-1}$, where $P_p(x)$ is a polynomial of degree at most mn with $P(0)=1$. The Euler product is absolutely convergent for $\operatorname{Re}s>1$ [JSha1].

In case π'_∞, π''_∞ are spherical, the local factor at infinity is given by

$$L(s,\pi'_\infty\times\pi''_\infty)=\prod_{j=1}^{m}\prod_{k=1}^{n}\Gamma_{\mathbf{R}}(s-\mu'_\infty(j)-\mu''_\infty(k))\ . \tag{2.2}$$

The completed L-function $\Lambda(s,\pi'\times\pi'')=L(s,\pi'_\infty\times\pi''_\infty)L(s,\pi'\times\pi'')$ has a meromorphic continuation and satisfies a functional equation

$$\Lambda(s,\pi'\times\pi'')=\epsilon(s,\pi'\times\pi'')\Lambda(1-s,\tilde{\pi}'\times\tilde{\pi}'') \tag{2.3}$$

where the ϵ-factor is of the form

$$\epsilon(s,\pi'\times\pi'')=\tau(\pi'\times\pi'')\mathfrak{f}(\pi'\times\pi'')^{-s} \tag{2.4}$$

with $\mathfrak{f}(\pi'\times\pi'')>0$ and $\tau(\pi'\times\pi'')\in\mathbf{C}^*$. It can be written as a product of local factors by fixing an additive character $\psi=\prod\psi_p$ of $\mathbf{A}/\mathbf{Q}$ (which we will assume to be everywhere normalized):

$$\epsilon(s,\pi'\times\pi'')=\prod_p \epsilon_p(s,\pi'_p\times\pi''_p,\psi_p) \tag{2.5}$$

and each local factor is 1 if both π'_p and π''_p are unramified and ψ_p is normalized, and otherwise it is of the form

$$\epsilon_p(s,\pi'_p\times\pi''_p,\psi_p)=\tau(\pi'_p\times\pi''_p)p^{-c(\pi'_p\times\pi''_p)s} \tag{2.6}$$

with $c(\pi'_p\times\pi''_p)\in \mathbf{Z}$ and $\tau(\pi'_p\times\pi''_p)$ a "Gauss sum". The "conductor" $\mathfrak{f}(\pi'\times\pi'')$ is the product $\prod_p p^{c(\pi'_p\times\pi''_p)}$. At infinity, since we are assuming that both π_∞ and π''_∞ are unramified, one has $\epsilon(s,\pi'_\infty\times\pi''_\infty,\psi_\infty)=1$.

The completed L-function $\Lambda(s,\pi'\times\pi'')$ is entire unless $\pi''\simeq\tilde{\pi}'\otimes|\cdot|^t$ for some $t\in\mathbf{C}$, and $\Lambda(s,\pi\times\tilde{\pi})$ is holomorphic except for simple poles at $s=0,1$ ([MW]). Moreover it is easily deduced from [JSha1,2] and [MW] that $L(s,\pi'\times\pi'')$ is of order one, see [RuS]. If we express $L(s,\pi\times\tilde{\pi})$ as a Dirichlet series

$$L(s,\pi\times\tilde{\pi})=\sum_{n=1}^{\infty}\frac{b(n)}{n^s} \tag{2.7}$$

then the coefficients $b(n)\geq 0$ (see [RuS] for the verification at the ramified primes) and the following is a simple consequence:

$$\sum_{n\leq x} b(n)\sim c_\pi x, \qquad x\to\infty \tag{2.8}$$

for some $c_\pi>0$.

2.2 Twists. Let χ be a primitive Dirichlet character mod q. As is well known, χ corresponds to a Hecke character of the idele class group $\mathbf{A}^\times/\mathbf{Q}^\times$, trivial on $\mathbf{R}_+^\times$, so χ is of the form $\chi=\otimes\chi_p$. The Dirichlet character being even (i.e. $\chi(-1)=1$) is equivalent to $\chi_\infty\equiv 1$. For $q>2$ prime, there are $(q-1)/2$ such characters mod q.

We apply the Rankin-Selberg theory described above to the following situation: Fix π on GL_m, and let χ be an even *primitive* Dirichlet character mod q, where q is a prime not dividing the conductor $\mathfrak{f}(\pi)$ of π. Take $\pi'=\pi(\chi):=\pi\otimes\chi$ and $\pi''=\tilde{\pi}$. To describe the exact functional equation in this case, we recall the Gauss sum

$$\tau(\chi)=\sum_{x \bmod q}\chi(x)e(\frac{x}{q})\ . \tag{2.9}$$

LEMMA 2.1. *Let q be a prime, $q\nmid\mathfrak{f}(\pi)$, and let χ be a primitive even Dirichlet character mod$q$.*

i) *If we write $L(s,\pi\times\tilde{\pi})=\sum_{n=1}^{\infty}b(n)n^{-s}$, then (recall $\chi(n)=0$ if $(q,n)\neq 1$) :*

$$L(s,\pi(\chi)\times\tilde{\pi})=\sum_{n=1}^{\infty}\frac{\chi(n)b(n)}{n^s}$$

ii) $\Lambda(s,\pi(\chi)\times\tilde{\pi}) = L(s,\pi_\infty\times\tilde{\pi}_\infty)L(s,\pi(\chi)\times\tilde{\pi})$ *satisfies the functional equation*

$$\Lambda(s,\pi(\chi)\times\tilde{\pi})=\epsilon(s,\pi(\chi)\times\tilde{\pi})\Lambda(1-s,\pi(\bar{\chi})\times\tilde{\pi}) \tag{2.10}$$

where the global ϵ-factor is given by

$$\begin{aligned}\epsilon(s,\pi(\chi)\times\tilde{\pi}) &= \chi(\mathfrak{f}(\pi\times\tilde{\pi}))\epsilon(s,\pi\times\tilde{\pi})\epsilon(s,\chi)^{m^2}\\ &= \chi(\mathfrak{f}(\pi\times\tilde{\pi}))\tau(\chi)^{m^2}q^{-m^2 s}\epsilon(s,\pi\times\tilde{\pi})\ .\end{aligned} \tag{2.11}$$

Proof: If $p\not| q\mathfrak{f}(\pi)$ then

$$\begin{aligned}&L(s,\pi(\chi)_p\times\tilde{\pi}_p)=\det\left(I-\chi(p)A(p)\otimes\bar{A}(p)p^{-s}\right)^{-1}\\ &\epsilon(s,\pi(\chi)_p\times\tilde{\pi}_p,\psi_p)=1\ .\end{aligned} \tag{2.12}$$

To describe the local factors in the case $p\mid q\mathfrak{f}(\pi)$, we begin with $p=q$: If $\chi\neq 1$ then the local L-factors are given by

$$L(s,\pi(\chi)_q\times\tilde{\pi}_q)=1\ .$$

Indeed, at the prime q, $\pi_q = Ind(GL_m,B;\mu_1,\dots,\mu_m)$ is an unramified principal series representation, where $\mu_j(x) = |x|^{u_j}$ are unramified characters. Likewise $\tilde{\pi}_q = Ind(GL_m,B;\mu_1^{-1},\dots,\mu_m^{-1})$ is unramified. Then $\pi_q\otimes\chi = Ind(GL_m,B;\chi\mu_1,\dots,\chi\mu_m)$. Hence (see [JP-SS])

$$\begin{aligned}L(s,\pi(\chi)_q\times\tilde{\pi}_q) &= \prod_{j=1}^{m}L(s,\pi_q\otimes\chi_q\otimes\mu_j^{-1})\\ &= \prod_{j,k=1}^{m}L(s,\chi\mu_k\mu_j^{-1})\end{aligned}$$

and since χ_q is ramified, each factor above is 1. As for the epsilon factor, we have by [JP-SS]

$$\begin{aligned}\epsilon(s,\pi(\chi)_q\times\tilde{\pi}_q,\psi_q) &= \prod_{j=1}^{m}\epsilon(s,\pi_q\otimes\chi\mu_j^{-1},\psi_q)\\ &= \prod_{j,k=1}^{m}\epsilon(s,\chi\mu_k\mu_j^{-1},\psi_q)\\ &= \prod_{j,k=1}^{m}\epsilon(s+u_k-u_j,\chi,\psi_q)\end{aligned}$$

where the abelian ϵ-factor (for χ primitive) is given by

$$\epsilon(s,\chi,\psi_q)=\tau(\chi)q^{-s}\,.$$

Therefore we have

$$\begin{aligned}\epsilon\big(s,\pi(\chi)_q\times\tilde{\pi}_q,\psi_q\big)&=\prod_{j,k=1}^{m}\tau(\chi)q^{-(s+u_k-u_j)}\\&=\tau(\chi,\psi_q)^{m^2}q^{-m^2s}\,.\end{aligned}$$

Since the local ϵ-factor $\epsilon(s,\pi_q\times\tilde{\pi}_q,\psi_q)=1$, we see that

$$\epsilon\big(s,\pi(\chi)_q\times\tilde{\pi}_q,\psi_q\big)=\epsilon(s,\chi,\psi_q)^{m^2}\epsilon(s,\pi_q\times\tilde{\pi}_q,\psi_q)\,. \tag{2.13}$$

Now suppose that $p\mid\mathfrak{f}(\pi)$. Then (with $P_p(p^{-s}):=L(s,\pi_p\times\tilde{\pi}_p)$)

$$\begin{aligned}L\big(s,\pi(\chi)_p\times\tilde{\pi}_p\big)&=P_p\big(\chi(p)p^{-s}\big)^{-1}\\\epsilon\big(s,\pi(\chi)_p\times\tilde{\pi}_p,\psi_p\big)&=\chi(p^{c(\pi_p\times\tilde{\pi}_p)})\epsilon(s,\pi_p\times\tilde{\pi}_p,\psi_p)\end{aligned} \tag{2.14}$$

Indeed, $\chi_p(x)=|x|^{v_p}$ is unramified. We claim that $L(s,\pi(\chi)_p\times\tilde{\pi}_p)=L(s+v_p,\pi_p\times\tilde{\pi}_p)$ and similarly for the ϵ-factor. This can be seen from the local Rankin-Selberg integrals of [JP-SS]. With this given, we have

$$L\big(s,\pi(\chi)_p\times\tilde{\pi}_p\big)=P_p(p^{-s-v_v})^{-1}=P_p\big(\chi(p)p^{-s}\big)^{-1}$$

while

$$\begin{aligned}\epsilon\big(s,\pi(\chi)_p\times\tilde{\pi}_p,\psi_p\big)&=\tau(\pi_p\times\tilde{\pi}_p)p^{-c(\pi_p\times\tilde{\pi}_p)(s+v_p)}\\&=\chi(p^{c(\pi_p\times\tilde{\pi}_p)})\tau(\pi_p\times\tilde{\pi}_p)p^{-c(\pi_p\times\tilde{\pi}_p)s}\\&=\chi(p^{c(\pi_p\times\tilde{\pi}_p)})\epsilon(s,\pi_p\times\tilde{\pi}_p,\psi_p)\end{aligned}$$

Since $\chi_\infty=1$, $\epsilon(s,\chi_\infty,\psi_\infty)=1$ and so we find for the global ϵ-factor

$$\begin{aligned}\epsilon\big(s,\pi(\chi)\times\tilde{\pi}\big)&=\prod_p\epsilon\big(s,\pi(\chi)_p\times\tilde{\pi}_p,\psi_p\big)\\&=\epsilon(s,\pi_\infty\times\tilde{\pi}_\infty)\epsilon(s,\chi,\psi_q)^{m^2}\epsilon(s,\pi_q\times\tilde{\pi}_q,\psi_q)\\&\quad\cdot\prod_{p\mid\mathfrak{f}(\pi)}\chi(p^{c(\pi_p\times\tilde{\pi}_p)})\epsilon(s,\pi_p\times\tilde{\pi}_p,\psi_p)\\&=\chi\big(\mathfrak{f}(\pi\times\tilde{\pi})\big)\tau(\chi)^{m^2}q^{-m^2s}\epsilon(s,\pi\times\tilde{\pi})\end{aligned}$$

as required. □

3. The Proofs

For $\chi \neq \chi_0$ a primitive even Dirichlet character mod$q$, $q \nmid \mathfrak{f}(\pi)$, let

$$L(s,\chi) := L\big(s,\pi(\chi)\times\tilde{\pi}\big) = \sum_{n=1}^{\infty} \frac{b(n)\chi(n)}{n^s} . \tag{3.1}$$

We write the functional equation for $L(s,\chi)$ as

$$L(s,\chi) = \epsilon\big(s,\pi(\chi)\times\tilde{\pi}\big)G(s)L(1-s,\bar{\chi}) \tag{3.2}$$

with $\epsilon(s,\pi(\chi)\times\tilde{\pi})$ is given by Lemma 2.1 and where we set

$$G(s) = \frac{L(1-s,\pi_\infty\times\tilde{\pi}_\infty)}{L(s,\pi_\infty\times\tilde{\pi}_\infty)} . \tag{3.3}$$

We investigate the averages

$$\sum_{q\sim Q}\sum_{\substack{\chi\neq\chi_0\\ \text{even}}} L(\beta,\chi) \tag{3.4}$$

where $\sum_{q\sim Q}$ means we sum over primes $Q \leq q \leq 2Q$.

PROPOSITION 3.1. *For* $0 < \operatorname{Re}\beta < 1$, *and* $\epsilon > 0$

$$\sum_{q\sim Q}\sum_{\substack{\chi\neq\chi_0\\ \text{even}}} L(\beta,\chi) = \tfrac{1}{2}\sum_{q\sim Q} q + O_{\beta,\epsilon}\big(Q^{1+\frac{m^2+1}{2}(1-\operatorname{Re}\beta)+\epsilon}\big) . \tag{3.5}$$

Proof of Theorem 1.2: As noted in the introduction, Theorem 1.2 follows from noting that if π_∞ is spherical and parametrized by $diag(\mu_\infty(1),\ldots,\mu_\infty(m))$ then for all even Dirichlet characters χ the Rankin-Selberg L-function $L(s,\pi(\chi)\times\tilde{\pi})$ has a trivial zero at $\beta_0 = 2\max\operatorname{Re}\mu_\infty(j)$. Note that since π_∞ is unitary, $\{\mu_\infty(j)\} = \{-\mu_\infty(k)\}$ and so to prove Theorem 1.2 it suffices to show that $\beta_0 \leq 1-2/(m^2+1)$. However if $\operatorname{Re}\beta > 1-2/(m^2+1)$ then in (3.5) the O-term is of smaller order than $\frac{1}{2}\sum_{q\sim Q} q \sim \frac{3}{4}Q^2/\log Q$ while the left-hand side is zero. This gives a contradiction and so proves Theorem 1.2.

AN APPROXIMATE FUNCTIONAL EQUATION. To prove Proposition 3.1, we need an appropriate series representation of $L(\beta,\chi)$. The following is such a representation which is gotten by a well known use of the functional equation (3.2). For $f \in C_c^\infty(0,\infty)$ with $\int_0^\infty f(x)dx = 1$, set

$$k(s) = \int_0^\infty f(y)y^s\frac{dy}{y} . \tag{3.6}$$

Thus $k(s)$ is entire, rapidly decreasing in vertical strips and $k(0) = 1$. For $x > 0$ set

$$\begin{aligned} F_1(x) &= \frac{1}{2\pi i}\int_{\mathrm{Re}\, s=2} k(s)x^{-s}\frac{ds}{s} \\ F_2(x) &= \frac{1}{2\pi i}\int_{\mathrm{Re}\, s=2} k(-s)G(-s+\beta)x^{-s}\frac{ds}{s}\ . \end{aligned} \tag{3.7}$$

Recall that $\beta_0 = 2\max \mathrm{Re}\, \mu_\infty(j)$ and we assume $0 < \mathrm{Re}\, \beta < 1$.

LEMMA 3.1. i) *$F_1(x)$ and $F_2(x)$ are rapidly decreasing as $x \to \infty$.*
ii) *$F_1(x) = 1 + O(x^N)$ for all $N \geq 1$ as $x \to 0$.*
iii) *$F_2(x) \ll 1 + x^{1-\beta_0-\mathrm{Re}\,\beta-\epsilon}$ as $x \to 0$.*

Proof: The asymptotics of $F_1(x)$ follow upon shifting the contour of integration to the right (for $x \to \infty$) and left (for $x \to 0$). As for $F_2(x)$, by Stirling's formula, $G(s)$ is of moderate growth in vertical strips and so we may shift contours. To get the behaviour as $x \to \infty$, shift the contour to the right. For the behaviour as $x \to 0$, shift to the left. If $\mathrm{Re}\,\beta + \beta_0 - 1 < 0$ then we pick up a simple pole at $s = 0$ which gives $F_2(x) = O(1)$; otherwise we pick up the first pole at $s = \beta + \beta_0 - 1$ and none to its right. In this case we get the bound

$$F_2(x) \ll x^{1-\beta-\mathrm{Re}\,\beta}(-\log x)^{d-1}\ , \qquad \text{as } x \to 0$$

where $d \leq m^2$ is the maximal order of a pole of $L(s, \pi_\infty \times \tilde{\pi}_\infty)$ on the line $\mathrm{Re}\, s = \beta_0$. □

In the rest of this section we set

$$\mathfrak{f} = \mathfrak{f}(\pi \times \tilde{\pi})\ . \tag{3.8}$$

LEMMA 3.2 [Approximate Functional Equation]. *If $\chi \neq \chi_0$ is an even primitive Dirichlet character mod$q$, with $q \nmid \mathfrak{f}(\pi)$, and $0 < \mathrm{Re}\,\beta < 1$ then for any $Y > 1$,*

$$\begin{aligned} L(\beta,\chi) &= \sum_{n=1}^{\infty} \frac{b(n)\chi(n)}{n^\beta} F_1\left(\frac{n}{Y}\right) \\ &\quad + \tau(\pi\times\tilde{\pi})(q^{m^2}\mathfrak{f})^{-\beta} \sum_{n=1}^{\infty} \frac{b(n)\bar{\chi}(n)}{n^{1-\beta}} \chi(\mathfrak{f})\tau(\chi)^{m^2} F_2\left(\frac{nY}{\mathfrak{f}q^{m^2}}\right)\ . \end{aligned} \tag{3.9}$$

Proof: Consider the integral

$$\begin{aligned} \frac{1}{2\pi i}\int\limits_{\mathrm{Re}\, s=2} k(s)L(s+\beta,\chi)Y^s\frac{ds}{s} &= \sum_{n=1}^{\infty} \frac{b(n)\chi(n)}{n^\beta}\frac{1}{2\pi i}\int\limits_{\mathrm{Re}\, s=2} k(s)(\frac{Y}{n})^s\frac{ds}{s} \\ &= \sum_{n=1}^{\infty} \frac{b(n)\chi(n)}{n^\beta} F_1\left(\frac{n}{Y}\right)\ . \end{aligned} \tag{3.10}$$

Both the fact that this converges absolutely and the justification of the contour shifts follow from the comments at the end of section 2.1. On the other hand, shifting the contour to $\mathrm{Re}\, s = -1$, since $L(s,\chi)$ is entire for $\chi \neq \chi_0$,

$$\frac{1}{2\pi i}\int_{\mathrm{Re}\, s=2} k(s)L(s+\beta,\chi)Y^s\frac{ds}{s} = L(\beta,\chi)+\frac{1}{2\pi i}\int_{\mathrm{Re}\, s=-1} k(s)L(s+\beta,\chi)Y^s\frac{ds}{s}\,.$$

On applying the functional equation (3.2), this gives

$$L(\beta,\chi)$$
$$+\frac{1}{2\pi i}\int_{\mathrm{Re}\, s=-1} k(s)\tau(\pi\times\tilde{\pi})\chi(\mathfrak{f})\tau(\chi)^{m^2}(\mathfrak{f}q^{m^2})^{-s-\beta}G(s+\beta)L(1-s-\beta,\bar{\chi})Y^s\frac{ds}{s}\,.$$

On changing variable $s \to -s$ this gives

$$= L(\beta,\chi)$$
$$-\frac{1}{2\pi i}\int_{\mathrm{Re}\, s=1} k(-s)\tau(\pi\times\tilde{\pi})\chi(\mathfrak{f})\tau(\chi)^{m^2}(\mathfrak{f}q^{m^2})^{s-\beta}$$
$$\cdot\, G(-s+\beta)L(s+1-\beta,\bar{\chi})Y^{-s}\frac{ds}{s}$$
$$= L(\beta,\chi)-\tau(\pi\times\tilde{\pi})\chi(\mathfrak{f})\tau(\chi)^{m^2}(\mathfrak{f}q^{m^2})^{-\beta}\sum_{n=1}^{\infty}\frac{b(n)\bar{\chi}(n)}{n^{1-\beta}}F_2\left(\frac{nY}{\mathfrak{f}q^{m^2}}\right)\,.$$

Comparing with (3.10) we recover (3.9). □

Proof of Proposition 3.1: We study the average (3.4) by using the approximate functional equation (3.9) with $Q \ll Y \ll Q^{m^2}$. On using

$$\sum_{\substack{\chi\neq\chi_0\\ \text{even}}}\chi(n) = \begin{cases} 0\,, & n\equiv 0 \bmod q\\ \frac{q-1}{2}-1\,, & n\equiv \pm 1 \bmod q\\ -1\,, & \text{otherwise}\end{cases} \tag{3.11}$$

we find that the contribution of the first sum on the RHS of (3.9) to the average is

$$\sum_{q\sim Q}\sum_{\substack{\chi\neq\chi_0\\ \text{even}}}\sum_{n}\frac{b(n)\chi(n)}{n^\beta}F_1\left(\frac{n}{Y}\right) = \sum_{q\sim Q}\frac{q-1}{2}\sum_{n\equiv\pm1 \bmod q}\frac{b(n)}{n^\beta}F_1\left(\frac{n}{Y}\right)$$
$$-\sum_{q\sim Q}\sum_{\substack{(n,q)=1\\ n\not\equiv\pm1 \bmod q}}\frac{b(n)}{n^\beta}F_1\left(\frac{n}{Y}\right)\,. \tag{3.12}$$

We single out the contribution from $n=1$ in the first term above:

$$\sum_{q\sim Q}\frac{q-1}{2}F_1\left(\frac{1}{Y}\right) = \frac{1}{2}\sum_{q\sim Q}q + O(Q) + O(Q^2Y^{-N})\,. \tag{3.13}$$

We will choose $Y \sim Q^{\frac{m^2+1}{2}}$ and so we use $F_1(x) \to 1$ as $x \to 0$. Note that $\sum_{q\sim Q} q \sim \frac{3}{2}Q^2/\log Q$.

The sum over $n \equiv 1 \bmod q$, $n \neq 1$ contributes

$$\sum_{q\sim Q} \frac{q-1}{2} \sum_{d\geq 1} \frac{b(1+dq)}{(1+dq)^\beta} F_1\left(\frac{1+dq}{Y}\right) \ll Q \sum_m \frac{b(m)m^\epsilon}{m^{\mathrm{Re}\,\beta}} \left|F_1\left(\frac{m}{Y}\right)\right| \quad (3.14)$$

where we use the fact that for $n \neq 1$, the number of different representations $n = 1 + dq = 1 + d'q'$ is $O(n^\epsilon)$. Now apply (2.8) and $F_1(x) \sim 1$ as $x \to 0$ to find that

$$\sum_{q\sim Q} \frac{q-1}{2} \sum_{\substack{n\equiv 1 \bmod q \\ n\neq 1}} \frac{b(n)}{n^\beta} F_1\left(\frac{n}{Y}\right) \ll QY^{1-\mathrm{Re}\,\beta+\epsilon} \quad (3.15)$$

(recall that $\mathrm{Re}\,\beta < 1$). Similarly we find that

$$\sum_{q\sim Q} \frac{q-1}{2} \sum_{n\equiv -1 \bmod q} \frac{b(n)}{n^\beta} F_1\left(\frac{n}{Y}\right) \ll QY^{1-\mathrm{Re}\,\beta+\epsilon}\,. \quad (3.16)$$

The last sum in (3.12) is bounded by

$$\sum_{q\sim Q} \sum_{(n,q)=1} \frac{b(n)}{n^{\mathrm{Re}\,\beta}} \left|F_1\left(\frac{n}{Y}\right)\right| \ll QY^{1-\mathrm{Re}\,\beta+\epsilon}\,. \quad (3.17)$$

To treat the contribution of the second term in (3.9) we first note that if $q \nmid n$ then

$$\sum_{\substack{\chi\neq\chi_0 \\ \text{even}}} \bar\chi(n)\chi(\mathfrak{f})\tau(\chi)^{m^2} \ll q^{\frac{m^2+1}{2}}\,. \quad (3.18)$$

Indeed, setting $r \equiv n\bar{\mathfrak{f}} \bmod q$ (with $\mathfrak{f}\bar{\mathfrak{f}} \equiv 1 \bmod q$) we have

$$\sum_{\substack{\chi\neq\chi_0 \\ \text{even}}} \bar\chi(r)\tau(\chi)^{m^2} = \frac{q-1}{2}\{\mathrm{Kl}_{m^2}(r,q) + \mathrm{Kl}_{m^2}(-r,q)\} - (-1)^{m^2} \quad (3.19)$$

where for $r \neq 0 \bmod q$ the hyper-Kloosterman sum $\mathrm{Kl}_n(r,q)$ is defined by

$$\mathrm{Kl}_n(r,q) = \sum_{x_1\cdot\ldots\cdot x_n \equiv r \bmod q} e\left(\frac{x_1+\ldots+x_n}{q}\right)\,. \quad (3.20)$$

Using Deligne's bound $\mathrm{Kl}_n(r,q) \ll q^{(n-1)/2}$ ([D]), we get (3.18).

Now sum over q to find

$$\sum_{q\sim Q}(\mathfrak{f}q^{m^2})^{-\beta}\sum_{\substack{\chi\neq\chi_0\\ \text{even}}}\sum_{n}\frac{b(n)\bar{\chi}(n)}{n^{1-\beta}}\chi(\mathfrak{f})\tau(\chi)^{m^2}F_2\left(\frac{nY}{\mathfrak{f}q^{m^2}}\right)$$

$$=\sum_{q\sim Q}(\mathfrak{f}q^{m^2})^{-\beta}\sum_{(n,q)=1}\frac{b(n)}{n^{1-\beta}}\Big[\frac{q-1}{2}\operatorname{Kl}_{m^2}(n\bar{\mathfrak{f}},q)$$

$$+\frac{q-1}{2}\operatorname{Kl}_{m^2}(-n\bar{\mathfrak{f}},q)-(-1)^{m^2}\Big]F_2\left(\frac{nY}{\mathfrak{f}q^{m^2}}\right)$$

$$\ll\sum_{q\sim Q}(\mathfrak{f}q^{m^2})^{-\operatorname{Re}\beta}\sum_{(n,q)=1}\frac{b(n)}{n^{1-\operatorname{Re}\beta}}q^{\frac{m^2+1}{2}}\left|F_2\left(\frac{nY}{\mathfrak{f}q^{m^2}}\right)\right|$$

$$\ll\sum_{q\sim Q}(\mathfrak{f}q^{m^2})^{-\operatorname{Re}\beta}q^{\frac{m^2+1}{2}}\int_1^\infty\left|F_2\left(\frac{xY}{\mathfrak{f}q^{m^2}}\right)\right|\frac{dx}{x^{1-\operatorname{Re}\beta}}\ll Q^{1+\frac{m^2+1}{2}}Y^{-\operatorname{Re}\beta}$$

on using the bound for $F_2(x)$ in Lemma 3.1 and $\operatorname{Re}\beta>0$, $\beta_0<1$. Thus

$$\sum_{q\sim Q}(\mathfrak{f}q^{m^2})^{-\beta}\sum_{\substack{\chi\neq\chi_0\\ \text{even}}}\sum_{n}\frac{b(n)\bar{\chi}(n)}{n^{1-\beta}}\chi(\mathfrak{f})\tau(\chi)^{m^2}F_2\left(\frac{nY}{\mathfrak{f}q^{m^2}}\right)\ll Q^{1+\frac{m^2+1}{2}}Y^{-\operatorname{Re}\beta}. \tag{3.21}$$

Collecting together (3.13), (3.15), (3.16), (3.17) and (3.21) we find

$$\sum_{q\sim Q}\sum_{\substack{\chi\neq\chi_0\\ \text{even}}}L(\beta,\chi)=\sum_{q\sim Q}\frac{q-1}{2}+O(QY^{1-\operatorname{Re}\beta+\epsilon}+Q^{1+\frac{m^2+1}{2}}Y^{-\operatorname{Re}\beta})\,. \tag{3.22}$$

On taking $Y\sim Q^{(m^2+1)/2}$ we prove Proposition 3.1.

Applications. We sketch how Corollaries 1.1 and 1.2 follow from Theorem 1.1. For Corollary 1.1, we use the result of Goldfeld and Sarnak ([GolS]) which asserts that if $\lambda_j=s_j(1-s_j)<\frac{1}{4}$, are the exceptional eigenvalues for $\Gamma_0(N)\backslash\mathbf{H}^2$, then

$$\sum_{\substack{1\leq c\leq x\\ c\equiv 0 \bmod N}}\frac{S(m,n,c)}{c}=\sum_{s_j}\tau_j(m,n)x^{2s_j-1}+O_\epsilon(x^{\frac{1}{6}+\epsilon})\,. \tag{3.23}$$

Since Theorem 1.1 gives $\frac{1}{2}\leq s_j<\frac{7}{10}$, we recover Corollary 1.1.

Corollary 1.2 was established in the recent work of Luo and Sarnak ([LuS]) for the full modular group $\Gamma=SL_2(\mathbf{Z})$. It was pointed out there that the only obstruction to establishing Corollary 1.2 for any congruence subgroup is the presence of small eigenvalues $\lambda_j=s_j(1-s_j)$ with $s_j>\frac{7}{10}$. Theorem 1.1 asserts precisely that these do not exist.

References

[BR] L. Barthel, D. Ramakrishnan, A nonvanishing result for twists of L-functions of $GL(n)$, Duke Math. Jour. 74 (1994), 681–700.

[BDHI] D. Bump, W. Duke, J. Hoffstein, H. Iwaniec, An estimate for the Hecke eigenvalues of Maass forms, Inter. Math. Res. Notices 4 (1992), 75–81.

[BuGi] D. Bump, D. Ginzburg, Symmetric square L-functions on $GL(r)$, Ann. of Math. 136 (1992), 137–205.

[D] P. Deligne, SGA $4\frac{1}{2}$ Cohomologie Etale, Lecture Notes in Math. 569, Springer-Verlag 1977.

[DuI] W. Duke, H. Iwaniec, Estimates for coefficients of L-functions, Proc. C.R.M. Conf. on Automorphic Forms and Analytic Number Theory, Montréal (1989), 43–48.

[GJ] S. Gelbart, H. Jacquet, A relation between automorphic representations of $GL(2)$ and $GL(3)$, Ann. Sci. Ecole Norm. Sup. 4e série 11 (1978), 471–552.

[GoJ] R. Godement, H. Jacquet, Zeta Functions of Simple Algebras, Lecture Notes in Math. 260, Springer-Verlag, 1972.

[GolS] D. Goldfeld, P. Sarnak, Sums of Kloosterman sums, Invent. Math. 71 (1983), 243–250.

[I] H. Iwaniec, Small eigenvalues of the Laplacian on $\Gamma_0(N)$, Acta Arith. 56 (1990), 65–82.

[J] H. Jacquet, Principal L-functions of the linear group, Proc. Symp. Pure Math. 33:2, 63–86, American Math. Soc., Providence 1979.

[JP-SS] H. Jacquet, I.I. Piatetski-Shapiro, J.A. Shalika, Rankin-Selberg convolutions, Amer. Jour. of Math 105 (1983), 367–464.

[JSha1] H. Jacquet, J.A. Shalika, On Euler products and the classification of automorphic representations I, Amer. Jour. of Math. 103 (1981), 499–558.

[JSha2] H. Jacquet, J.A. Shalika, Rankin-Selberg convolutions: Archimedean theory, Israel Math. Conf. Proceedings 2 (1990), 125–208.

[L] Yu.V. Linnik, Additive problems and eigenvalues of modular operators, Proc. I.C.M. Stockholm (1962), 270–284.

[LuS] W. Luo, P. Sarnak, Quantum Ergodicity of Eigenfunctions on $PSL_2(\mathbf{Z})\backslash\mathbf{H}^2$, preprint (1994).

[MW] C. Mœglin, J.-L. Waldspurger, Le spectre résiduel de $GL(n)$, Ann. Sci. Ecole Norm. Sup. 4e série 22 (1989), 605–674.

[PP-S] S.J. Patterson, I.I. Piatetski-Shapiro, The symmetric square L-function attached to a cuspidal automorphic representation of $GL(3)$, Math. Ann. 283 (1989), 551–572.

[R] D. Rohrlich, Nonvanishing of L-functions for $GL(2)$, Invent. Math. 97 (1989), 383-401.

[RuS] Z. Rudnick, P. Sarnak, Zeros of Principal L-functions and Random Matrix Theory, preprint.

[S] P. Sarnak, Prime Geodesic Theorems, Ph.D. Thesis, Stanford University (1980).

[Se] A. Selberg, On the estimation of Fourier coefficients of modular forms, Proc. Symp. in Pure Math. 8, Amer. Math. Soc., Providence (1965), 1–15.

[Ser] J.-P. Serre, letter to J.-M. Deshouillers (1981).

[Sh1] F. Shahidi, On certain L-functions, Amer. Jour. of Math. 103 (1981), 297–355.

[Sh2] F. Shahidi, On the Ramanujan conjectures and finiteness of poles for certain L-functions, Ann. of Math. 127 (1988), 547–584.

[Shi] G. SHIMURA, On the periods of modular forms, Math. Ann. 229 (1977), 211–221.

[We] A. WEIL, On some exponential sums, Proc. Nat. Acad. Sci. 34 (1948), 204–207.

Wenzhi Luo
School of Mathematics
Institute for Advanced Studies
Princeton, NJ 08540
USA
Current address:
MSRI
Berkeley, CA 94720
USA

Zeév Rudnick
Dept. of Math.
Princeton University
Princeton, NJ 08544
USA
Current address:
School of Math. Sci.
Tel Aviv University
Tel Aviv 69978
Israel

Peter Sarnak
Dept. of Math.
Princeton University
Princeton, NJ 08544
USA

Submitted: September 1994

Geometric And Functional Analysis
Vol. 5, No. 2 (1995)
1016-443X/95/0200402-32$1.50+0.20/0

THE DIFFERENTIAL OF A QUASI-CONFORMAL MAPPING OF A CARNOT-CARATHEODORY SPACE

G.A. Margulis and G.D. Mostow

1. Introduction

The theory of quasi-conformal mappings has been used to prove rigidity theorems on hyperbolic n space over the division algebras $\mathbf{R}, \mathbf{C}, \mathbf{H}$, and $\mathbf{O}$, by studying quasi-conformal mappings on their boundary spheres S^{kn-1} at infinity, where k is the dimension of the division algebra. The notion of quasi-conformal mappings for such spaces, first introduced in [Mo2], was subsequently reformulated by Pansu in terms of Carnot-Caratheodory spaces M, and Pansu studied quasi-conformal mappings for the special case of graded nilpotent groups M. Subsequently, Pansu's definition was simplified in [Mo3], and this simpler definition was employed by Koranyi-Reimann in their study of quasi-conformal mappings of the nilpotent Heisenberg group operating on the boundary of complex hyperbolic n-space and transitive on the complement of one point.

In this paper, we carry over to quasi-conformal mappings over general Carnot-Caratheodory spaces some of the absolute continuity theorems that are basic for the case of Riemannian spaces.

As a consequence, we can prove in Theorem 10.5 that the differential of a quasi-conformal map between Carnot-Caratheodory spaces exists and is a group isomorphism at almost all points. Our proof, applied to the special case of Riemannian spaces, is new, since we make no use of the Rademacher-Stepanoff theorem. In the special case that M and M' are graded nilpotent groups, our result has been proved by Pansu in [P] by his beautiful method of "Lie algebra development", which is not available in the more general context.

The work of the first author is supported in part by NSF Grant DMS-9204270 and that of the second author by NSF Grant DMS-91036-8.

2. Preliminaries on *cc* Spaces

2.1. Let M be a smooth connected manifold, T its tangent bundle, and H a subbundle. For any $x \in M$, let $T^i(x)$ denote the subspace of $T(x)$ spanned by a local basis of smooth vector fields $X_1, \dots, X_h$ around x for the subbundle H, together with all commutators of these vector fields of order $\leq i$. The subbundle H is called *generic* if for all $x \in X$, $\dim T^i(x)$ is independent of x, and *horizontal* if $T^n(x) = T(x)$ for some n. We call the pair $(M; H)$ a *Carnot-Caratheodory* (abbreviated "cc") *space of depth* n if H is generic and horizontal, and $n = \inf\{k; T^k(x) = T(x)\}$.

Let $(M; H)$ be a *cc* space. We define a *cc norm* on $(M; H)$ to be a continuous real valued function defined on the total space of the bundle H (i.e. for any point $x \in M$, and for any $v \in H_x$, a map $v \to \|v\|$ such that $\| \ \|$ is a norm on each H_x).

Such a pair with a fixed *cc* norm is called a *cc normed space.* Given a *cc* normed space, a piecewise-smooth curve p mapping the interval $a \leq t \leq b$ of $\mathbf{R}$ into M is called *horizontal* if $\frac{dp}{dt} \in H_{p(t)}$ for almost all $t \in [a, b]$. The length of such a horizontal curve is defined as $\int_a^b \|\frac{dp}{dt}\| dt$ and is denoted $\|p\|$.

One defines the *cc distance* between two points x_0, x_1 in X as (cf. [Mi]) $d(x_0, x_1) = \inf\{\|p\|; p$ a horizontal curve with initial point x_0 and final point $x_1\}$. This *cc* distance is a metric on M compatible with the given topology on M. If the depth of the *cc* space is 1 the restriction of this metric to any open subset of M with compact closure is equivalent to the metric given by any Riemannian metric on M.

For any $x \in M$, $r > 0$, set $B(x, r) = \{y \in X; d(x, y) < r\}$. If the depth of the *cc* space is n, then the Hausdorff dimension of M with respect to a *cc* metric is

$$\delta = \sum_{k=1}^{n} k\big(\dim T^k(x) - \dim T^{k-1}(x)\big) \ . \tag{2.1.1}$$

for any $x \in M$ (cf. [Mi]).

On any compact set F of M and for all $x \in F$

(2.1.2) the Riemannian volume of the *cc* ball $B(x, r)$ for small r, considered as an open set in the Riemannian space M satisfies

$$c' r^\delta \leq \ \text{vol}\, B(x, r) \leq c'' r^\delta$$

for some constants c', c'' called the "volume constants of the compact set F". Moreover

(2.1.3) Let μ denote Hausdorff δ-dimensional measure relative to the *cc* metric; then on any compact subset of M, μ is commensurable with any Lebesgue measure.

The *cc* metric on $(M; H)$ depends on the *cc* norm; however, the restriction to compact sets of two *cc* metrics are equivalent.

(2.1.4) We say, as usual, that $x \in M$ is a *point of density* of a measurable subset $A \subset M$ if
$$\lim_{r \to 0} \frac{\mu(A \cap B(x,r))}{\mu(B(x,r))} = 1 .$$

PROPOSITION 2.2. (a) (Density point theorem) *For any measurable subset $A \subset M$, almost all points of A are points of density.*

(b) *If x is a point of density of A then*
$$\lim_{y \to x} \frac{dist(y, A)}{d(y,x)} = 0 .$$

Statement (b) follows directly from (2.1.2). Statement (a) is proved in the same way as the classical density point theorem (here again, we have to use (2.1.2)).

We will use the following lemma.

LEMMA 2.3. *Let $\varphi : M \to Z$ be an injective measurable map of a cc space M into a space Z with a finite measure σ. Set*
$$q(x) = \liminf_{r \to 0} \frac{\sigma(\varphi(B(x,r)))}{vol(B(x,r))} , \qquad x \in M . \tag{2.3.1}$$

Then
$$\int_M q(x) d\mu(x) \leq \sigma(Z)$$
where μ denotes Lebesgue measure on M.

Proof: In view of the density point theorem, there exists a subset $X \subset M$ such that $\mu(M - X) = 0$ and for every $x \in X$
$$\liminf_{r \to 0} \frac{\int_{B(x,r)} q(x) d\mu(x)}{\mu(B(x,r))} = q(x) . \tag{2.3.2}$$

Fix $\varepsilon > 0$. From (2.3.1) and (2.3.2) we get that for any $x \in X$, there exists a sequence $\{r_i(x)\}$ of positive numbers such that $\lim_{i \to \infty} r_i(x) = 0$ and
$$\sigma(\varphi(B(x, r_i(x)))) > (1 - \varepsilon) \int_{B(x,r)} q(x) d\mu(x) . \tag{2.3.3}$$
for every i. Let Ω denote the family $\{B(x, r_i(x)); x \in X, i \in \mathbf{N}^+\}$. In view of (2.1.2) we can apply the Vitali covering theorem to the family Ω and find a sequence $\{B_i\} \subset \Omega$ such that $B_i \cap B_j = \emptyset$ if $i \neq j$ and $\mu(M - \bigcup_{i \in \mathbf{N}^+} B_i) = 0$. Then in view of (2.3.4)

$$\sigma(Z) > (1-\varepsilon)\int_M q(x)d\mu(x) .$$

Since $\varepsilon > 0$ is arbitrary, the lemma is proved.

LEMMA 2.4. *Let M be a cc space. Given a fibration of a compact subset C of M by smooth horizontal curves in M, then there is a positive constant $\alpha \geq 1$ such that for any two points p, q on any fiber F, the ratio of the arc length $\widehat{pq}$ to the distance $d(p,q)$ lies between 1 and α.*

Proof: For any *cc* space M, we denote by Riem(M) the *cc* space obtained from M by enlarging the horizontal subbundle to the entire tangent bundle and extending the norm of M to a norm on the tangent bundle. Then Riem(M) is a Riemannian space in which the arc length of curves that are horizontal in the original M are preserved. Clearly distances in Riem(M) are majorized by those in M. The result now follows from its validity in Riemannian spaces.

In the sequel we shall denote by M a smooth manifold with a fixed Riemannian metric. The horizontal sub-bundle of the tangent bundle that defines a *cc* space will be allowed to vary, and the *cc* norm on each of these *cc* spaces will be given by the fixed Riemannian metric restricted to the horizontal sub-bundle.

3. Some Measure Comparisons

3.1. Let X be a smooth horizontal nowhere vanishing vector field defined on an open subset $\mathcal{O}$ of the *cc* space $(M; H)$, i.e. $0 \neq X(x) \in H_x$ for all $x \in \mathcal{O}$. We call a trajectory $\exp tX(p)$ an X-line. Assume that $\mathcal{O}$ has compact closure $\bar{\mathcal{O}}$. Furthermore, we assume that $\mathcal{O}$ contains a codimension 1 submanifold S transversal to X and that

$$\varphi : (s,t) \to \exp tX(s) , \qquad -a < t < a ,$$

is a diffeomorphism of $S \times (-a,a)$ to $\mathcal{O}$. Set $F_s = \varphi(s \times (-a,a))$ for any $s \in S$ and let $\pi : \mathcal{O} \to S$ denote the fibration $pr_1 \circ \varphi^{-1}$, i.e. $\pi^{-1}(s) = F_s$ for all $s \in S$.

For any $\varepsilon > 0$ and any $x \in \mathcal{O}$, let $f_{\varepsilon,x}$ denote the characteristic function of $B(F_{\pi(x)}, \varepsilon)$, the tube in $\mathcal{O}$ of radius ε around the fiber $F_{\pi(x)}$. Set

$$h_\varepsilon(y) = \int_{\mathcal{O}} f_{\varepsilon,x}(y)d\mu(x) .$$

Then we can write $h_\varepsilon = \int_{\mathcal{O}} f_{\varepsilon,x}d\mu(x)$. Let us note that

$$h_\varepsilon(y) = \mu(\pi^{-1}(\pi(B(y,\varepsilon)))) . \qquad (3.1.1)$$

We write $\mu(f)$ instead of $\int_M f(x)d\mu(x)$.

LEMMA 3.2. *For every compact set $\mathcal{C} \subset \mathcal{O}$, there exists constants $c'_{\mathcal{C}}$ and $c''_{\mathcal{C}}$ such that for all $\varepsilon > 0$ and $y \in \mathcal{C}$*

$$c'_{\mathcal{C}}\varepsilon^{d-1} < \mu(B(F_{\pi(y)}, \varepsilon)) = \mu(f_{\varepsilon,y}) < c''_{\mathcal{C}}\varepsilon^{d-1} \tag{3.2.1}$$

and

$$c'_{\mathcal{C}}\varepsilon^{d-1} < h_\varepsilon(y) < c''_{\mathcal{C}}\varepsilon^{d-1} ; \tag{3.2.2}$$

that is, the tube of radius ε around the connected component in $\mathcal{O}$ of the orbit $\exp tX(y)$ on the one hand, and the orbit $\exp tX(B(y,\varepsilon))$ on the other, have μ measure comparable to ε^{d-1}.

Proof: No generality is lost in assuming that $\mathcal{C} \approx Q \times [-b, b] = \{\exp tX(s);\ s \in Q, -b \le t \le b\}$ where $Q \subset S$ and $0 < b < a$. Given $y \in \mathcal{C}$, we can choose points $p_1, \dots, p_N$ on the line $F_{\pi(y)}$ such that arc length $p_i p_j = |i-j|\varepsilon$, for all $i, j = 1, \dots, N$ and arc length $(p_1, p_N) \ge 2a - \varepsilon$.

$B(p_i, \varepsilon/\alpha) \cap B(p_j, \varepsilon/\alpha)$ is empty for all $i, j = 1, \dots, N$, where α is as in Lemma 2.4 and taken > 1, as we may. Clearly

$$\cup_i B(p_i, \varepsilon) \subset B(F_{\pi(y)}, \varepsilon) \subset \cup_i B(p_i, 2\varepsilon) .$$

By (2.1.2), we have for the volume constants $\gamma'_{\mathcal{C}}$ and $\gamma''_{\mathcal{C}}$ of the compact set $\mathcal{C}$,

$$\gamma'_{\mathcal{C}} N(\varepsilon/\alpha)^d \le \mu\big(\cup_i B(p_i, \varepsilon/\alpha)\big) \le \mu(B(F_{\pi(y)}, \varepsilon)) \le \gamma''_{\mathcal{C}} N(2\varepsilon)^d .$$

From $2a - \varepsilon \le N\varepsilon \le 2a$ if we infer (3.2.1).

The measure μ on M being commensurable with Lebesgue measure, it decomposes with respect to the fibering of $\mathcal{O}$ into the product of arc length measure on the fibers $\{F_s; s \in S\}$ and a quotient measure ν on S. Assertion (3.2.2) says that for suitable constants $c'_{\mathcal{C}}$ and $c''_{\mathcal{C}}$

$$c'_{\mathcal{C}}\varepsilon^{d-1} < \nu\big(\pi(B(y,\varepsilon))\big) < c''_{\mathcal{C}}\varepsilon^{d-1} .$$

From Lemma 2.4, we infer that the arc length of $F_{\pi(x)} \cap B(y,\varepsilon)$ is at most $2\alpha\varepsilon$ for any $x, y \in \mathcal{C}$. Hence

$$\begin{aligned}\int_{B(y,\varepsilon)} 1 d\mu &= \int_{\pi(B(y,\varepsilon))} \left(\int_{F_s} dt\right) d\nu \\ &\le \int_{\pi(B(y,\varepsilon))} 2\varepsilon\alpha d\nu \le 2\varepsilon\alpha\nu\big(\pi(B(y,\varepsilon))\big) .\end{aligned}$$

Thus $c'\varepsilon^d \le \mu(B(y,\varepsilon)) \le 2\varepsilon\alpha\nu(\pi(B(y,\varepsilon)))$ yielding

$$(c'/2\alpha)\varepsilon^{d-1} \le \nu\big(\pi B(y,\varepsilon)\big) .$$

On the other hand, for any $x \in B(y,\varepsilon)$, the arc length of $F_{\pi(x)} \cap B(y, 2\varepsilon)$ is at least 2ε. Hence

$$\int_{\pi(B(y,\varepsilon))} \varepsilon d\nu \leq \int_{B(y,2\varepsilon)} 1 d\mu \leq c''(2\varepsilon)^d$$

yielding $\nu(\pi(B(y,\varepsilon))) \leq 2^d c'' \varepsilon^{d-1}$. This proves (3.2.2).

LEMMA 3.3. *Let $(M;H), \mu, X, S, F_s, \pi, C$ be as above. let M' be a metric space and $\varphi : M \to M'$ a homeomorphism. Let μ' be a finite measure on M'. Then*

$$\liminf_{\varepsilon \to 0} \frac{\mu'(\varphi B(F_{\pi(x)}, \varepsilon))}{\varepsilon^{d-1}} < \infty$$

for almost all x in C.

Proof: No generality is lost in assuming that $\bar{O} = \mathcal{C} = \bigcup_{s \in S} F_s$. Set

$$^{\varphi}f_{\varepsilon,x} = f_{\varepsilon,x} \circ \varphi^{-1}, {}^{\varphi}h_\varepsilon = h_\varepsilon \circ \varphi^{-1} .$$

From (3.2.2) we infer from definitions,

$$c'_C \varepsilon^{d-1} \leq {}^{\varphi} h_\varepsilon(y') \leq c''_C \varepsilon^{d-1}, \text{ for all } y' \in \varphi(C) . \tag{3.3.1}$$

Hence

$$\int_{\varphi(C)} {}^{\varphi}h_\varepsilon(y') d\mu'(y') = \bar{c}(\varepsilon)\varepsilon^{d-1}$$

where $c'_C \mu'(\varphi(\mathcal{L})) \leq \bar{c}(\varepsilon) \leq c''_C \mu'(\varphi(\mathcal{L}))$.

Let μ denote a Lebesgue measure on M. From $L_2 = h_\varepsilon := \int_C f_{\varepsilon,x} d\mu(x)$ we get ${}^{\varphi}h_\varepsilon = \int_C {}^{\varphi}f_{\varepsilon,x} d\mu(x)$. Therefore

$$\begin{aligned} \bar{c}(\varepsilon)\varepsilon^{d-1} &= \int_{\varphi(C)} {}^{\varphi}h_\varepsilon(y') d\mu'(y') = \int_{\varphi(C)} \int_C {}^{\varphi}f_{\varepsilon,x}(y') d\mu(x) d\mu'(y') \\ &= \int_C \int_{\varphi(C)} {}^{\varphi}f_{\varepsilon,x} d\mu'(y') d\mu(x) = \int_C \mu'(\varphi(B_{\pi(x)}, \varepsilon)) d\mu(x) . \end{aligned}$$

Hence

$$\frac{\mu'(\varphi(B(F_{\pi(x)}, \varepsilon)))}{\varepsilon^{d-1}} < {}_1c < \infty$$

for all $x \in C - C({}_1c, \varepsilon)$ with $\mu(C({}_1c, \varepsilon)) \to 0$ as ${}_1c \to \infty$. It follows that

$$\frac{\mu'(\varphi(B(F_{\pi(x)}, \varepsilon)))}{\varepsilon^{d-1}} < \infty$$

for almost all $x \in C$. From this the conclusion of the lemma follows.

Remark: In the proof of Lemma 3.3, we used only that M' is a Borel space and φ an injective Borel map.

§4. Hausdorff Measures

We collect here some classical facts about Hausdorff measure (cf. Saks "Theory of the Integral", referred to in this section by [STI]).

Let M be a metric space and let E be a subset of M. Hausdorff k-dimensional measure is defined by

$$m_k(E) = \lim_{a\to 0}\left(\inf \sum_U \beta_k(\text{radius } U)^k\right) := \lim_{a\to 0}\Lambda_k(E,a)$$

where U ranges over a countable covering of E by balls of radius less than a and β_k is the volume of the unit ball in Euclidean k-space:

$$\beta_k = 2^k\Gamma\left(\frac{1}{2}\right)^{k-1}\Gamma\left(\frac{k+1}{2}\right)\Gamma(k+1)^{-1}\,.$$

In an n-dimensional Riemannian manifold, m_n coincides with Lebesgue measure. For arbitrary k, m_k is an "outer Caratheodory measure", i.e. $C := m_k$ is a set function defined for all subset X of M taking values in $\mathbf{R}\cup\{\infty,-\infty\}$ satisfying

(C_1) $C(X)\le C(Y)$ whenever $X\subset Y$

(C_2) $C(\sum_i X)\le \sum_i C(X_i)$ for each sequence $\{X_i\}$ of sets

(C_3) $C(X\cup Y) = C(X)+C(Y)$ whenever $\mathrm{dist}(X,Y)>0$

One calls a set $E\subset M$ *measurable* with respect to the outer measure C if

$$C(X) = C(X\cap E) + C(X\cap(M-E))$$

for every set X. The class of all sets measurable with respect to C is denoted $\mathcal{L}_C$. $\mathcal{L}_C$ contains all sets X with $C(X)=0$, all Borel sets, and indeed $\mathcal{L}_C$ is a countably additive family of sets and C is countably additive on the class $\mathcal{L}_C$ (cf. [STI, Theorems 4.1 and 4.5 pp. 44-45]) that is $C(\sum_n X_n) = \sum_n C(X_n)$ for every sequence $\{X_n\}$ of disjoint sets in $\mathcal{L}_C$.

A set function defined on a family of sets is called *finite* if it takes on only finite values. The upper and lower bounds of a countably additive set function $\Phi(X)$ on subsets of a set E are called the *upper* and *lower* variation and denoted $\bar{V}$ and $\underline{V}$ respectively, $\bar{V}-\bar{V}$ is called the *total variation*. The total variation of a finite countably additive set function is finite; the total variation is a bound for the set function.

A countably additive set function Φ on a space M with a measure μ is called *absolutely continuous with respect to* μ if for any $E\subset M$, $\mu(E)=0$ implies that $\Phi(E)=0$. For a measure space (M,μ) with μ finite, Φ is absolutely continuous with respect to μ if and only if given any $\varepsilon>0$, there is a $\delta>0$ such that for any subset $E\subset M$, $\mu(E)<\delta$ implies that $\Phi(E)<\varepsilon$; moreover such a Φ is finite. ([STI, Theorem 13.2, p. 31])

Let I denote the interval $[0,1]$ in $\mathbf{R}$ and let $\varphi : I \to M$ be an injective continuous map. For any subset $E \subset I$, set

$$\Phi(E) = m_1\big(\varphi(E)\big)$$

and

$$\|\varphi\| = \sup_P \sum m_1\big(\varphi(a_n, b_n)\big)$$

where P varies over all partitions of I into disjoint intervals (open, closed, or half-open). Then $\|\varphi\|$ is the total variation of Φ. The countably additive set function on the class of subsets $\mathcal{L}_\Phi$ that is obtained from the Caratheodory outer measure Φ is countably additive. We define φ to be *absolutely continuous* if and only if Φ is absolutely continuous.

It may be verified directly that if M is a manifold the map φ is absolutely continuous if and only if for any coordinate system $\{f_1, \dots, f_n\}$ on an open set in M, each set function Φ_i formed from the function $f_i \circ \varphi$ $(i = 1, \dots, m)$ is absolutely continuous.

5. Absolute Continuity on Horizontal Lines

Given the *cc* normed spaces $(M; H)$ and $(M'; H')$, and given a homeomorphism $\varphi : M \to M'$, set

$$L_\varphi(x,r) = \big\{\inf s; \varphi(B(x,r)) \subset B'(\varphi(x), s)\big\}$$
$$\ell_\varphi(x,r) = \big\{\sup s; B'(\varphi(x), s) \subset \varphi(B(x,r))\big\}\ .$$

It follows at once from these definitions that

$$\ell_{\varphi^{-1}}\big(\varphi(x), L_\varphi(x,r)\big) = r = L_{\varphi^{-1}}\big(\varphi(x), \ell_\varphi(x,r)\big)\ .$$

We call φ *K-quasiconformal* if

$$\limsup_{r\to 0} \frac{L_\varphi(x,r)}{\ell_\varphi(x,r)} \le K \text{ for all } x \in M$$

We call φ *quasi-conformal* if it is K-quasiconformal for some K.

Set

$$I_\varphi = \limsup_{r\to 0} \frac{L_\varphi(x,r)}{r}$$
$$J_\varphi = \limsup_{r\to 0} \frac{\mu'(\varphi(B(x,r)))}{\mu(B(x,r))}\ .$$

LEMMA 5.1. *Let $\varphi : M \to M'$ be a homeomorphism of (normed) cc spaces, let d, d' denote their Hausdorff dimensions, and let μ, μ' denote their Hausdorff measures m_d and $m'_{d'}$. Let X be a horizontal smooth vector field on*

M whose flow induces a fibering of M, and let π denote the map of M to the quotient space of the fibering. Let $x \in M$. Assume that

(1) $A := \liminf_{r \to 0} \frac{\mu'(\varphi(B(F_{\pi(x)}, r)))}{\mu(B(F_{\pi(x)}, r))} < \infty$

(2) $d' \leq d$

(3) φ *or* φ^{-1} *is quasiconformal*

Then φ is absolutely continuous on the line $F_{\pi(x)}$ and $d' = d$

Proof: We must prove: given $\varepsilon > 0$, there is a δ such that for any measurable subset $E \subset F_{\pi(x)}$, $m_1(E) < \delta$ implies that $m_1(\varphi(E)) < \varepsilon$ (here, m_1 denotes Hausdorff length). No generality is lost in assuming that E is compact.

CASE (i). φ is K-quasi-conformal.

No generality is lost in assuming that there is a number $a > 0$ such that $\frac{L_\varphi(p,r)}{\ell_\varphi(p,r)} < K$ for all $0 < r < a$. For if we let E_a denote the subset of $p \in E$ for which the above holds, then E_a increases to E as $a \to 0$ and the result for E follows from the result for E_a. By uniform continuity of φ on E, we can choose $a_1 > 0$ so that $L(p, a_1) < a$ for all $p \in E_1$.

Choose $t < \inf\{a, a_1\}$ with $\frac{\mu'(\varphi(B(F_{\pi(x)}, t)))}{\mu(B(F_{\pi(x)}, t))} \leq 2A$. We can choose points $\{p_i \in E; i = 1, \ldots, N\}$ such that the arc length of $p_i p_j$ along $F_{\pi(x)}$ is at least $(|i-j|-1)t$ for all i, j, $R := \bigcup_{i=1}^{i=N} B(p_i, t) \supset E$, and $Nt \leq m_1(E) + a$. Set $s_i = L_\varphi(p_i, t)$, $i = 1, \ldots, N$. For a constant c depending on the volume constants of M', and by Hölder's inequality, we have

$$\Big(\sum_i s_i\Big)^{d'} \leq N^{d'-1}\Big(\sum s_i^{d'}\Big) \leq cN^{d'-1} \sum_i \mu'(B(\varphi(p_i), t)) \qquad (*)$$

and by Lemma 2.4,

$$\sum_i \mu'(B(\varphi(p_i), t)) \leq 3\alpha\mu'(\varphi(R))$$

because each point of $R \cap F_{\pi(x)}$ lies in at most three $B(p_i, t) \cap F_{\pi(x)}$. Thus, for a constant c_1 depending on the volume constants of M and M' and on the fiber $F_{\pi(x)}$. By definition, $\Lambda_1(\varphi(E), a) \leq \sum_i 2s_i$,

$$\Lambda_1(\varphi(E), a)^{d'} \leq 2^{d'} c_1 (Nt)^{d'-1} \frac{\mu'(\varphi(B(F_{\pi(x)}, t)))}{t^{d-1}} t^{d-d'}$$

$$m_1(\varphi(E))^{d'} \leq 2^{d'-1} \cdot c_1 \big(m_1(E) + a\big)^{d'-1} A\, t^{d-d'}\ .$$

If $d' < d$, then upon letting $t \to 0$, we conclude that $m_1(\varphi(E)) = 0$. In particular, we can apply this reasoning to $E = F_{\pi(x)}$ to conclude that $m_1(\varphi(F_\pi(x))) = 0$ - contradicting that φ is a homeomorphism. Hence $d' = d$ and we conclude that

$$m_1(\varphi(E))^d \leq 2^{d'-1} A c_1 m_1(E)^{d-1} .$$

This implies that φ is absolutely continuous on $F_{\pi(x)}$.

CASE (ii). φ^{-1} is K-quasi-conformal. Here we can assume that

$$L_{\varphi^{-1}}(\varphi(p), r) \leq K\ell_{\varphi^{-1}}(\varphi(p), r) \qquad 0 < r < a\ , \quad \text{for all } p \in E\ .$$

Again we fix a, and select a_1 so that $L(p, a_1) < a$. Again we choose $t < \inf\{a, a_1\}$ and the points $p_1, \ldots, p_N$ so that the balls $B(p_k, t)$ cover E, the arc length $p_i p_j$ along $F_{\pi(x)}$ is at least $(|i-j|-1)t$ for all i, j, and $Nt \leq m_1(E) + a$.

Again we set $s_i = L_\varphi(p_i, t)$ $(i = 1, \ldots, N)$. Then

$$\ell_{\varphi^{-1}}(\varphi(p_i), s_i) = t\ , \quad L_{\varphi^{-1}}(p_i, s_i) < Kt\ .$$

Hence

$$\varphi(B(p_i, t)) \subset B'(\varphi(p_i), s_i) \subset \varphi(B(p_i, Kt))\ .$$

As before

$$\Big(\sum s_i\Big)^{d'} \leq N^{d'-1}\Big(\sum s_i^{d'}\Big) \leq cN^{d'-1} \sum_i \mu'(B'(\varphi(p_i), s_i))\ .$$

Here we observe that $B'(\varphi(p_i), s_i)$ meets no more $B'(\varphi(p_j), s_j)$ than $B(p_i, Kt)$ meets $B(p_j, Kt)$, i.e. at most $3K\alpha$, where α is as in Lemma 2.4. Consequently,

$$\sum_i \mu'(B'(\varphi(p_i), s_i)) \leq 3K\alpha\mu'\Big(\bigcup_i B'(\varphi(p_i), s_i)\Big)$$

$$\Lambda_1(\varphi(E), a)^{d'} \leq 2^{d'} c_1 3K\alpha N^{d'-1} \mu'(\varphi(B(F_{\pi(x)}, Kt)))\ .$$

By the same argument as in Case (i), we conclude that $d = d'$ and

$$m_1(\varphi(E))^{d'} \leq 2^d \cdot 3 \cdot c_1 K^{d+1} A m_1(E)^{d-1}\ .$$

The proof of Lemma 5.1 is now complete.

Remark: Inasmuch as F has finite measure, $\varphi(F_s)$ has finite length if φ is absolutely continuous on F_s; cf. §4.

6. The Inverse of a Quasi-conformal Map

LEMMA 6.1. *Let $\varphi : M \to M'$ be K quasi-conformal, let L' be a horizontal line in M' on which φ^{-1} is absolutely continuous, and let x' be a point at which the restriction of φ^{-1} to L' is differentiable. Set $L = \varphi^{-1}(L')$. Let t be an arc length parameter along L', i.e. $x'(t)$ is the point on L' where $|t|$ is the arc length of the L' arc from x' to $x(t)$ and set $x(t) = \varphi(x'(t))$, $x = x(0) = \varphi^{-1}(x')$. Assume that*

$$C = \lim_{t\to 0} \frac{d(x, x(t))}{t}$$

exists. Then

(i) $\limsup_{s\to 0} \frac{L_{\varphi^{-1}}(x',s)}{s} \le K^2 \liminf_{s\to 0} \frac{L_{\varphi^{-1}}(x',s)}{s}$. *If moreover, $C \neq 0$, then*

(ii) $K^{-1}C \le I_{\varphi^{-1}}(x') \le KC$ *and*

(iii) $c'K^{-2d} I_{\varphi^{-1}}(x')^d \le J_{\varphi^{-1}}(x') \le c''K^{2d} I_{\varphi^{-1}}(x')^d$ *where c' and c'' are the volume constants of M as in (2.2).*

Proof: For all sufficiently small $t > 0$, a point on L at a distance ct from x is of the form $\varphi^{-1}(x(t + o(t)))$ with $o(t)/t \to 0$ as $t \to 0$, i.e. at a distance $t + o(t)$. We can assume without loss of generality that $K > 1$. Hence for sufficiently small t,

$$B'(\varphi(x), K^{-1}C^{-1}t) \subset \varphi(B(x,t)) \subset B'(\varphi(x), KC^{-1}t)$$

and hence

$$B(x, K^{-2}t) \subset \varphi^{-1}(B'(\varphi(x), K^{-1}C^{-1}t)) \subset B(x,t)$$

or equivalently

$$B(x, K^{-1}C^{-1}t) \subset B(\varphi(x), t) \subset B(x, KCt)\ .$$

It follows at once that

$$K^{-1}C \le \frac{L_{\varphi^{-1}}(x',t)}{t} \le KC\ .$$

From this (i) follows, as well as

$$\text{(ii)} \qquad K^{-1}C \le I_{\varphi^{-1}}(x') \le KC$$

and

$$c'(K^{-1}C)^d \le J_{\varphi^{-1}}(x')$$

where c' and c'' are the volume constants of M. From this (iii) follows.

LEMMA 6.2. *Let $\varphi : M \to M'$ be a homeomorphism of cc spaces and assume that φ^{-1} is K quasi-conformal. Let $r_2 > r_1$, and set $D_{r_1,r_2} = B(x,r_2) - \overline{B(x,r_1)}$, $a = L_\varphi(x,r_1)$,$b = \ell_\varphi(x,r_2)$. Then $\log\frac{b}{a} \le k\log\frac{r_2}{r_1}$ for some constant k depending on K and the volume constants, if r_1 is sufficiently small and $r_2/r_1 \le K^2$.*

Proof: We can assume without loss of generality that $b > a$. Proof is by contradiction. Set $A = \frac{r_2}{r_1}$. Set $D'_{a,b} = B'(\varphi(x), b) - \overline{B'(\varphi(x), a)}$. Suppose $\frac{b}{a} = kA$ with $\log k$ large in relation to $\log A$, i.e. $k > A^e$ with e large.

We modify the cc norm in the subspace $M' - \varphi(B(x, r_1))$ via multiplying the norm at each of its points q by the distance $d(q, \varphi(x))^{-1}$; denote by d^* the distance of the resulting cc space, and by $L^*_\varphi(x, s)$ the radius of the circumscribed ball of $\varphi(B(x, s))$ with respect to the new distance.

Let m^* denote the resulting measure. Let X be a nowhere vanishing horizontal vector field on the ball $B(x, r_2)$, and let R denote the orbit in $B(x, r_2)$ of $B(x, r_1)$ under the flow $\{\exp tX; t \in \mathbf{R}\}$. Taking r_1 sufficiently small, we may assume that R is fibered by the orbits of its points and that the measure in R decomposes into a product of arc length along the fiber and the quotient measure, the measure of R with respect to the latter being of the order r_1^{d-1}. Set $R_0 = \varphi^{-1}(D'_{a,b}) \cap R$. R_0 has shorter fibers than R but the same quotient. Since φ maps any fiber into a curve which crosses from the inner boundary of $D'_{a,b}$ to the outer boundary, we find for the modified $I_\varphi := \limsup_{s \to 0} \frac{L^*_\varphi(x,s)}{s}$ at any $y \in R$ and for any orbit L on which φ is absolutely continuous, setting $L_0 = L \cap R_0$,

$$\int_{L_0} I_\varphi \int_a^b \frac{dt}{t} = \log \frac{b}{a} > \log(kA) = \log k + \log A \, .$$

Let f denote the characteristic function of the set R_0. Then $\int_L f I_\varphi = \int_{L_0} I_\varphi$ for each fiber of R and thus

$$\int_{R_0} I_{\varphi_0} = \int_R f I_\varphi \geq c(\log k) r_1^{d-1}$$

for some constant C depending on the volume constants in M. By Lemma 8.1, $J_\varphi \geq c_1 I_\varphi^d$ for some constant c_1 depending on K and the volume constants of M'. Applying Hölder's inequality, we get

$$m^*(\varphi(R_0)) \geq \int_{R_0} J_\varphi \geq c_1 \int_{R_0} I_\varphi^d \geq c_1 \int_{R_0} \left(\frac{(\log k) r_1^{d-1}}{\mu(R_0)} \right)^d \geq \frac{c_1((\log k) r_1^{d-1})^d}{\mu(R_0)^{d-1}} \, .$$

Thus

$$m^*(D'_{a,b}) = m^*(\varphi(R_0)) \geq \frac{c_1((\log k) r_1^{d-1})}{\mu(R)^{d-1}} = \frac{c_2 (\log k)^d r_1^{d(d-1)}}{r_1^{d(d-1)}}$$

where c_2 depends on c_1 and on A and hence on c_1 and K if $A \leq K^2$. Therefore $m^*(D'_{a,b}) \geq c_2 (\log k)^d$. On the other hand

$$\bigcup_{n=0}^{[\log_2 k]} D'_{2^n a, 2^{n+1} a} \subset D'_{a,b} \subset \bigcup_{n=0}^{[\log_2 k]+1} D'_{2^n a, 2^{n+1} a}$$

yielding that $m^*(D'_{a,b})$ is approximately $\sum_{n=0}^{[\log_2 k]} m^*(D'_{2^n a, 2^{n+1} a})$ which is $c_3 \log_2 k$, where c_3 is a constant depending on the volume constants of M'. This gives

$$c_3 \log_2 k \geq c_2 (\log k)^d$$

which is impossible for $d > 1$ and k large; more precisely, if $\log k / \log A$ is large.

THEOREM 6.3. *Let $\varphi : M \to M'$ be a homeomorphism of normed cc spaces with φ^{-1} quasi-conformal. Then φ is quasi-conformal.*

Proof: Let $x \in M$. For any homeomorphism $\varphi : M \to M'$ and for any $r > 0$, we observe that by definition $B(x,r)$ is the inscribed ball of $\varphi^{-1}(B'(\varphi(x), L_\varphi(x,r))$ and the circumscribed ball of $\varphi^{-1}(B'(\varphi(x), \ell_\varphi(x,r))$; that is

$$\ell_{\varphi^{-1}}(\varphi(x), L_\varphi(x,r)) = r = L_{\varphi^{-1}}(\varphi(x), \ell_\varphi(x,r)) .$$

Set

$$a = \ell_\varphi(x,r) , \quad b = L_\varphi(x,r) .$$

Applying the foregoing observation with φ, φ^{-1}, r replaced by φ^{-1}, φ, a and φ^{-1}, φ, b respectively, we get

$$a = L_\varphi(x, \ell_{\varphi^{-1}}(\varphi(x), a)) , \quad b = \ell_\varphi(x, L_{\varphi^{-1}}(\varphi(x), b)).$$

Set

$$r_1 = \ell_{\varphi^{-1}}(\varphi(x), a) , \quad r_2 = L_{\varphi^{-1}}(\varphi(x), b) .$$

Then

$$a = \ell_\varphi(x, r_1) , \quad b = L_\varphi(x, r_2) .$$

We have that φ^{-1} is K-quasi-conformal for some $K \geq 1$. Inasmuch as $\ell_{\varphi^{-1}}(\varphi(x), b) = r$, we infer that $r_2/r \leq K$ for all sufficiently small r. Inasmuch as $\ell_{\varphi^{-1}}(\varphi(x), a) = r$, we infer that $r/r_1 \leq K$ for all sufficiently small r. Hence $r_2/R_1 \leq K^2$ for all sufficiently small r_1. All the hypotheses of the preceding lemma being satisfied, we conclude that $\log \frac{b}{a} \leq k \log \frac{r_2}{r_1}$ where k is a constant depending on K and the volume constants, i.e. $b/a \leq K^{2k}$. This proves that φ is K^{2k}-quasi-conformal.

COROLLARY 6.4. *Let $\varphi : M \to M'$ be a quasi-conformal map between normed cc spaces of Hausdorff dimension d, d' respectively. Then*

(i) *φ^{-1} is quasi-conformal*

(ii) *$d = d'$*

Proof: (i) follows at once from Theorem 6.3. Interchanging M and M' if necessary, we can assume that $d' \leq d$. The hypotheses of Lemma 5.1 then are valid for almost all $x \in M$ by Lemma 3.3, and by Lemma 5.1, $d = d'$.

COROLLARY 6.5. *Let X be a horizontal smooth vector field m the cc space M and let $\varphi : M \to M'$ be a quasiconformal map of cc spaces. Then for almost all $x \in M$, $t \to \varphi(\exp tX(x))$ is an absolutely continuous path.*

Proof: By Corollary 6.4, M and M' have equal Hausdorff dimension. By Lemma 3.3, the hypotheses of Lemma 5.1 are satisfied. Our corollary is now an immediate consequence of Lemma 5.1.

By Corollary 6.5, one can define a *measurable* vector field X' at almost all points of M' by the formula

$$X'(\varphi(x)) = \frac{d}{dt}_{|t=0} \varphi(\exp tX(x)) \, .$$

The vector field X' on M' is *integrable* in the sense that $t \to \varphi(\exp tX(x))$ is its exponential, i.e. a family of trajectories to X', the trajectory through almost every point $x' \in M'$ being absolutely continuous.

COROLLARY 6.6. *Let $\varphi, M, M', I_\varphi, \mu'$ be as above and let P be a measurable subset of M with $I_\varphi(x) = 0$ for all $x \in P$. Then $\mu'(\varphi(P)) = 0$.*

Proof: No generality is lost in assuming that P is small enough to be contained in a subset $\mathcal{O}$ fibered by horizontal lines as in §3. By the Fubini theorem applied to the product fibering of $\varphi(\mathcal{O})$

$$\mu'(\varphi(P)) = \int_{\pi'(\varphi(P))} m_1(\varphi(P \cap F_s))d\nu' \leq \int_{\pi'(\varphi(P))} \left(\int_{P \cap F_s} I_\varphi \right) d\nu = 0$$

7. Absolute Continuity

THEOREM 7.1. *Let $\varphi : M \to M'$ be quasi-conformal. Then $J_\varphi > 0$ a.e. in M, the set function $E \to \mu'(\varphi(E))$ is absolutely continuous in M, and*

$$\mu'(\varphi(E)) = \int_E J_\varphi$$

for any measurable set E in M.

Proof: Let Z denote the set of zeros of J_φ. We prove that $\mu(Z) = 0$ by contradiction. Suppose $\mu(Z) > 0$. Let x be a point of density of Z, i.e. $\limsup_{r \to 0} \frac{\mu(B(x,r) \cap Z)}{\mu(B(x,r)} = 1$. Given any $\varepsilon > 0$, we can select arbitrarily small r with $\frac{\mu(B(x,r) \cap Z)}{\mu(B(x,r))} > 1 - \varepsilon$. The map φ is K-quasi-conformal for some $K \geq 1$. Fix c_0 so that $\ell_\varphi(x, c_0 r) - L_\varphi(x, r) \geq r$. As in §3 we choose a nowhere vanishing horizontal vector field X on $B(x, c_0 r)$ and let R denote the orbits in $D_{r,cr} := B(x, c_0 r) - \overline{B(x,r)}$ of $B(x,r)$ via the flows $\{\exp tX; t \in \mathbf{R}\}$, i.e. $R = \pi^{-1}(\pi(B(x,r)) \cap D_{r,cr}$. For r sufficiently small, R is fibered by the orbits and its measure decomposes into the product of arc length along fibers and a quotient measure on the base, as above.

For any fiber L, we have

$$\int_L I_\varphi > r \ , \quad \int_R I_\varphi > c_1 r \cdot r^{d-1} = c_1 r^d$$

where c_1 depends on the volume constants. Thus

$$c_1 r^d = \int_R I_\varphi = \int_{R-Z} I_\varphi \ ,$$

since $I_\varphi^d \sim J_\varphi$ by Lemma 6.1(iii). By the Hölder inequality

$$\int_{R-Z} I_\varphi^d \geq \left(\int_{R-Z} I_\varphi\right)^d \Big/ \big(m(R-Z)\big)^{d-1} \geq c_2 (r^d)^d/(\varepsilon r^d)^{d-1} \geq c_2 \frac{r^d}{\varepsilon^{d-1}}$$

with c_2 a constant depending on the volume constants.

By Lemma 6.2, $\log \frac{\ell_\varphi(x,c_0 r)}{L_\varphi(x,r)} \leq k \log c_0$, for some constant k depending on K and the volume constants, and therefore $L_\varphi(x, c_0 r)/\ell_\varphi(x,r) < c_3$, where c_3 is a constant depending on K and the volume constants. Consequently,

$$\mu(\varphi(R)) = \int_R J_\varphi \geq c_4 \int_{R-Z} I_\varphi^d \geq c_5 \frac{r^d}{\varepsilon^{d-1}} \ .$$

Clearly $L_\varphi(x, c_0 r)^d \geq \mu(\varphi(R))$ and $L_\varphi(x, c_0 r)^d \leq (c_0 r)^d$ because $I_\varphi(x) = 0$. Consequently $(c_0 r)^d \geq \frac{c_2 r^d}{\varepsilon^{d-1}}$, i.e. $c_0^d \geq \frac{c_2}{\varepsilon^{d-1}}$ – a contradiction since ε can be arbitrarily small. This proves that $\mu(Z) = 0$.

Let $P = \{x \in M; J_\varphi(x) = \infty\}$. In any $x \in P$, $\infty = I_\varphi(x) = \limsup_{r \to 0} \frac{L_\varphi(x,r)}{r}$. Hence

$$0 = \liminf_{r \to 0} \frac{r}{L_{\varphi(x,r)}} = \liminf_{s \to 0} \frac{\ell_{\varphi^{-1}}(\varphi(x), s)}{s} = \liminf_{s \to 0} \frac{L_{\varphi^{-1}}(\varphi(x), s)}{s} \ .$$

At any point x' in $\varphi(P)$ which lies on a horizontal line L' on which φ is absolutely continuous and is moreover differentiable at x', we have that $K^2 \liminf_{s \to 0} \frac{L_{\varphi^{-1}}(x',s)}{s} \geq \limsup_{s \to 0} \frac{L_{\varphi^{-1}}(x',s)}{s}$ by Lemma 6.1(i). Consequently $I_{\varphi^{-1}} = 0$ on P except for a set of measure zero. Since φ^{-1} is quasi-conformal by Theorem 6.3, we can apply Corollary 6.4 to get $\mu(P) = 0$. By the deVallee Poussin decomposition $E \to \mu'(\varphi(E))$ is absolutely continuous and $\mu'(\varphi(E)) = \int_E J$ for any measurable set E.

8. Preliminaries on the Tangent Cone

8.1. Let M be a *cc* space whose horizontal subbundle H has a norm coming from a fixed smooth Riemannian metric g. We denote by d_M, the resulting *cc* metric on M. When $(M;H)$ is fixed in a discussion, we set $d = d_M$ and denote by (M,d) the corresponding *cc* space. We shall have occasion to consider varying horizontal subbundles H_t $(t \in \mathbf{R}^+)$ on U, a coordinate neighborhood of the underlying Riemannian space M; in the resulting *cc* structure, we use the norm on H_t coming from the fixed Riemannian metric g, and we denote by (U,d_t) the resulting *cc* space with metric d_t. In addition to such a one parameter family of *cc* spaces, we shall consider the family of metric *cc* spaces with the same underlying manifold M, and the same horizontal subbundle H but with metric td.

Gromov has defined the tangent cone $(\bar{M}_x, \bar{d})$ to M at a point $x \in M$ as

$$\lim_{\substack{t\to\infty \\ x=\text{base point}}} (M, td) \ .$$

The limit signifies that for each $r > 0$, the ball of radius r in (M,td) about the base point x *converges* to the ball of radius r about a fixed point in $\bar{M}_x$ in the sense that the infimum of the Hausdorff distance between these compact abstract metric spaces approach o as $t \to \infty$. Here Hausdorff distance $H(B_1,B_2)$ between two metric spaces B_1 and B_2 means $\inf_C H_C(B_1,B_2)$, where $H_C(B_1,B_2)$ is the Hausdorff distance between the images under isometric embeddings of B_1 and B_2 into the metric space C.

8.2. Mitchell has given the following description of the tangent cone in [Mi]. Let $X_1,\dots,X_h$ be a base of horizontal C^∞ vector fields on M. For any multi-index $I = \{i_1,\dots,i_m\}$, let X_I denote the m-fold commutator $[X_{i_1}\cdots,[X_{i_{m-1}},X_{i_m}]\cdots]$. Enlarge the base of horizontal vector fields to a base of vector fields $X_1,\dots,X_h,\dots,X_n$ which at each point of M spans the tangent space of the underlying manifold M and with each $X_j = X_{I_j}$ for some I and is moreover *coherent* with the increasing filtration of the tangent bundle $(0) \subset T^1 \subset T^2 \subset \cdots \subset T^r = T$ of depth r that was introduced in §2; "coherent" is defined as follows: Letting $[i]$ denote $\{\inf j; X_i(x) \in T^j(x)\}$, then for each $i = 1,\dots,r$, $\{X_j; [j] = i\}$ spans $T^i(x)$ modulo $T^{i-1}(x)$. We call the base $X_1,\dots,X_n$ *coherent with the horizontal sub-bundle* H that defines the *cc* space. Introduce "normal coordinates" (cf. Mitchell) on any sufficiently small neighborhood U of x_0 in M via

$$x = \left(\exp\sum_{i=1}^{n} a_i X_i\right)(x_0)$$

and define for any $t < 0$ the homothety map ${}_{x_0}h_t : U \to M$ in terms of normal coordinates $(a_1,\dots,a_n)$ via

$$({}_{x_0}h_t x)_i = t^{[i]} a_i$$

We shall denote ${}_{x_0}h_t$ by h_t when the point x_0 remains fixed in its context; when the a_i are small, h_t makes sense for large t.

Let $h_t(X)$ denote the transform of X by h_t; i.e. for any function $U \in C^\infty(M)$.

$$(h_t X)(u) = (h_t^{*^{-1}} X h_t^*)(u) = \big(X(u \circ h_t)\big) \circ h_t^{-1} \,. \tag{8.2.1}$$

Then as is known (cf. Metivier p. 487 (3.2), Mitchell (iii) of Theorem of §3, p. 38)

$$\lim_{t\to\infty} t^{-[i]} h_t(X_i) = \hat{X}_i \,, \qquad i = 1, \ldots, n \,. \tag{8.2.2}$$

Clearly $h_t \hat{X}_i = t^{[i]} \hat{X}_i$ $(i = 1, \ldots, n)$; i.e. each $\hat{X}_i$ is a homogeneous vector field with respect to the normal coordinates $(a_1, \ldots, a_n)$, of degree $[i]$, and has degree 1 for $1 \le i \le h$. At each point x in the coordinate neighborhood U of the point x_0 in M, the vectors $\hat{X}_1(x), \ldots, \hat{X}_n(x)$ span the tangent space $T_x(M)$. Finally $\hat{X}_1, \ldots, \hat{X}_h$ are the Lie algebra generators of a graded nilpotent Lie algebra; they induce homogeneous vector fields on $\mathbf{R}^n$ with respect to the graded nilpotent Lie group structure on $\mathbf{R}^n$ resulting from its graded nilpotent Lie algebra structure.

Mitchell's Lemma 3.1 and 3.2 can be reformulated as follows.

Set $H_t = h_t(H)$ and let H_∞ be the subtangent bundle spanned by $\hat{X}_1, \ldots, \hat{X}_h$ on the normal coordinate neighborhood U $(0 < t \le \infty)$. Let d_t $(0 < t \le \infty)$ denote the *cc* metric on $(U; H_t)$ induced from the underlying Riemannian metric of M. Then

(8.2.3) (U, d_t) converges in the sense of Hausdorff to (U, d_∞) as $t \to \infty$.

(8.2.4) The quasi-isometric distance between (U, td) and (U, d_t) tends to zero as $t \to \infty$, in fact

$$\lim_{t\to\infty} \big\{\text{metric distortion of } h_t : (U, td) \to (U, d_t)\big\} = 1$$

(cf. ibid p. 43).

As a consequence of (8.2.2), (8.2.3) and (8.2.4), Mitchell gets,

THEOREM 8.3. *The tangent cone to M at x_0 can be identified via an isometry with the cc Lie group $(\hat{G}_{x_0}, \hat{d})$ whose Lie algebra is n dimensional, graded, and generated by the homogeneous vector fields of degree one of (8.2.2). Moreover, the horizontal subbundle of the cc space $(\hat{G}_{x_0}, \hat{d})$ is given by the space of homogeneous vector fields of degree 1 and has the same dimension as the horizontal subbundle of M.*

8.4. It is convenient to reformulate the definition of the tangent cone to M at a point x_0. Namely, $(\bar{M}_{x_0}, \bar{d})$ can be identified with equivalence classes of certain curves. We consider continuous (parametrized) curves $\{c(s) \mid 0 \leq s \leq 1\} \subset M$ such that $c(0) = x_0$. Two curves $c_1(s)$ and $c_2(s)$ are called *equivalent* if $\lim_{s\to 0} \frac{1}{s} d(c_1(s), c_2(s)) = 0$. Then $(\bar{M}_{x_0}, \bar{d})$ can be considered as the space of equivalence classes of curves equivalent to orbits $\{_{x_0}h_s x\}, x \in M$, with the distance

$$\bar{d}(\{c_1(s)\}, \{c_2(s)\}) = \lim_{s\to 0} \frac{1}{s} d(c_1(s), c_2(s)) \ .$$

The above interpretation generalizes a standard interpretation of the tangent space to a differentiable manifold as the space of equivalence classes of differentiable curves.

Let us also note that for different "normal coordinates" we get the same set of equivalence classes.

In order to verify this, it suffices to prove the following.

Let $(a_1, \ldots, a_n)$ be a set of normal coordinates corresponding to a choice of base of horizontal C^∞ vector fields $X_1, \ldots, X_h$ and certain of their successive commutators coherent with the increasing filtration $0 \subset T^1 \subset \cdots \subset T^r = T$ of the tangent bundle to the *cc* spaces M at the point x_0 (cf. (8.2)); let $X_1, \ldots X_h, \ldots X_n$ denote one such choice of base of vector fields for the tangent bundle of M on some fixed coordinate neighborhood U of x_0. Let $X'_1, \ldots, X'_n$ be another such choice of base of vector fields on U with $X'_i(x_0) = X_i(x_0)$ for $i = 1, \ldots, n$. Let Ψ denote the map of a neighborhood U of x_0 given by $\psi\big((\exp \sum a_i X_i)(x_0)\big) = \big(\exp \sum a_i X'_i\big)(x_0)$, i.e.

$$\begin{aligned} \psi &= \Big(\exp \sum a_i X'_i\Big)\Big(\exp \sum a_i X_i\Big)^{-1} : U \to M \\ &= \Big(\exp \sum a_i X'_i\Big)\Big(\exp - \sum a_i X_i\Big) \ . \end{aligned}$$

Let $Z = Z(a_1 \ldots, a_n)$ denote the formal Campbell-Hausdorff expansion in the vector fields $\sum_1^n a_i X'_i$ and $-\sum_1^n a_i X_i$. Then

$$Z = \sum_{k=1}^{\infty} f_k Y_k$$

where $f_k = f_k(a_1, \ldots, a_n)$ is a homogeneous polynomial of degree d_k and Y_k is a d_k-fold commutator of $\{X_1, \ldots, X_n, X'_1, \ldots, X'_n\}$ with $d_k \leq d_j$ if $k \leq j$ and with $d_k = d_j$ for only a finite number of j. Then for any C^∞ real-valued function g on U, the partial derivatives $\frac{\partial^s}{\partial a_1^{i_1} \cdots \partial a_n^{i_n}}$ of degree s of $g\big((\exp \sum_1^n a_i X_i)(x)\big)$ and of

$$\Big(\sum_{k=0}^{m} \Big(\Big(\sum_1^n a_i X_i\Big)^k g\Big) \Big/ k!\Big)(x) \ , \quad x \in U \ ,$$

coincide for $s = 0, 1, \ldots, m$. A similar relation holds for $g\big(\exp \sum_1^n a_i X'\big)(x)$ and its truncated formal power series, and for $g(\psi(x))$ and the truncated formal power series for $g(\exp Z(x))$. Consequently, given any $a_1, \ldots, a_n$, we can use the Campbell-Hausdorff formula to compute the map $\big(\exp \sum_1^n t a_i^{[i]} X_i'\big)$ $\big(\exp \sum_1^n t^{[i]} a_i X_i\big)^{-1}$ up to terms of order t^s for arbitrary s.

We have by definition of the X_i',

$$X_i' = X_i + \sum_{j=1}^{k} \alpha_{ij} X_i \qquad \alpha_{ij}(x_0) = 0\ , \quad \alpha_{ij} \in C^\infty(U)\ .$$

Hence

$$X_i' = X_i + \sum_{j=1}^{n} \beta_{ij} Y_i \qquad \beta_{ij}(x_0) = 0\ , \quad \beta_{ij} \in C^\infty(U)$$

degree $Y_j = [i]$ for $[i] = 2, \ldots, r$, $r = [n] =$ depth of the *cc* structure. Hence, up to terms of degree t^s

$$\Big(\exp \sum_{i=1}^{n} t^{[i]} a_i X_i'\Big)\Big(\exp - \sum_{i=1}^{n} t^{[i]} a_i X_i\Big)(x) = \exp\Big(\sum f_k Y_k\Big)(x)$$

where the coefficient of each monomial Poisson bracket of total *cc* degree k on the right hand side has the form $t^k \gamma_k(x)$ where $\gamma_k(x_0) = 0$ and $x = \big(\exp \sum_{i=1}^n t^{[i]} a_i X_i\big)(x_0)$ depends on t. Hence $\sum f_k Y_k$ has the form $\sum_{k=1}^r t^{j_k} Z_k'$ where Z_k' is a linear combination of Poisson brackets of degree k (i.e. degree k with respect to the *cc* degree) and $j_k > [k]$.

Consider now $d(h_t x, h_t' x)$ in the *cc* metric. We have $\psi(h_t x) = h_t' x$ and hence up to terms of degree $t^{\frac{r+1}{r}}$ we have $d(h_t x, h_t' x) \le \sum_{k=1}^r c_k t^{[k+1]/[k]}$, with c_k constant. It follows at once that

$$\lim_{t\to 0} \frac{d(h_t x, h_t' x)}{t} = 0\ .$$

In the argument above, we have used the estimate: There exist constants $c_1, \ldots, c_r$ such that for all y in some compact neighborhood of x_0,

$$d\bigg(\exp\Big(\sum_{k=1}^{r} s_k Z_k'\Big)(y), y\bigg) \le \sum_{k=1}^{r} c_k s_k^{1/k}\ . \tag{8.4.1}$$

Proof: Let h_s denote the homothety centered at the point x_0, with respect to a normal coordinate system coherent with the *cc* structure on a neighborhood N of the point x_0. Let Z be a non-zero vector field on N homogeneous of degree k with respect to the homothety. Set

$$f(t) = d\big((\exp tZ)(x_0), x_0\big)\ .$$

Then

$$d\big(h_s(\exp tZ(x_0))(x_0), x_0\big) = d\big((\exp ts^k Z)(x_0), x_0\big) = f(ts^k) \ .$$

On the other hand, we have for any $x \in N$ and s^{-1} large,

$$d(h_{s^{-1}} h_s x, x_0) = d(h_{s^{-1}} h_s x, h_{s^{-1}} h_s x_0) \sim s^{-1} d(h_s x, x_0) \ .$$

Hence asymptotically as $s \to 0$ through positive values,

$$d(h_s x, x_0) \sim s d(x, x_0) \ .$$

No generality is lost in assuming that $(\exp Z)(x_0) \in N$. Hence $f(ts^k) \sim sf(t)$ and therefore $f(ts) \sim s^{1/[k]} f(t)$; in particular

$$f(s) \sim s^{1/[k]} f(1) \ . \tag{8.4.2}$$

Repeated use of this estimate, as x_0 varies over a compact neighborhood, yields (8.4.1).

Remark 8.4.3: If Z_k is a smooth vector field on an open subset A of M with compact closure such that

$$Z_k(y) \notin T^{k-1}(y) \quad \text{(cf. (2.1)) for all} \quad y \in \bar{A} \ ,$$

then by (8.4.2) there is a constant c such that for all $y \in A$,

$$d\big(\exp tZ_k(y), y\big) \geq ct^{1/k} \quad \text{for } t \text{ sufficiently small} \ .$$

In particular, if $k \geq 1$, the smooth path $t \to \exp tZ_k$ is not *cc* rectifiable and hence not *cc* absolutely continuous. Conversely, if a smooth path is rectifiable, it is horizontal. In particular, a smooth *cc* absolutely continuous path is horizontal.

Remark: The notion of Hausdorff convergence of metric spaces does not uniquely define convergence of subsets because of the ambiguity introduced by automorphisms of the metric spaces. Our use in (8.5) of non-invariant normal coordinates and a non-invariant homothety in defining the convergence of a family of points x_t in (M, td) to a point in $(\bar{M}_{x_0}, d)$ is unavoidable.

8.5. The family of homotheties $\{h_s\}$ in a neighborhood of a point x_0 induces a one parameter group of homotheties $\{\bar{h}_s\}$ on $(\bar{M}_{x_0}, \bar{d})$ via change of parameters:

$$h_s \cdot (h_t x) = h_{st} x \ .$$

Although h_s depends on the choice of "normal coordinates", $\bar{h}_s$ does not; it can be described in terms of the grading of $(\bar{M}_{x_0}, \bar{d})$, which is canonical.

As a model for the convergence of the *cc* spaces (M, td) to the tangent cone at x_0, we introduce the direct product space

$$\mathcal{E}_{x_0} = (\bar{M}_{x_0}, \bar{d}) \times [0,1]$$

and we fix a neighborhood U of x_0 on which h_s is defined for every s with $0 < s \leq 1$. For any $x \in U$, we denote by $\bar{x}$ the element in the tangent cone represented by the parametrized orbit $\{h_s x; 0 < s \leq 1\}$. We define the embedding $\alpha_t : (U, td) \to \mathcal{E}_{x_0}$, $t > 1$ by $\alpha_t(x) = (\bar{h}_t \bar{x}, t^{-1})$. It follows from (8.2.3) and (8.2.4) that the embedding α_t, though not an isometry, is *asymptotically isometric*, in the sense that:

for every compact subset K in the tangent cone, the restriction of α_t to $\alpha_t^{-1}(K, t^{-1})$ is bi-Lipshitz with Lipshitz coefficient tending to 1 as $t \to \infty$.

We can now define the convergence of a family of points $x_t \in (M, td)$ to a point $y \in \bar{M}_{x_0}$ by:

$$x_t \underset{x_0}{\longrightarrow} y \quad \text{if and only if} \quad \alpha_t(x_t) \to (y, 0) \quad \text{as} \quad t \to \infty\,. \tag{8.5.1}$$

Given mappings $\rho_t : U \times [0,1] \to (M, td)$; $t \geq 1$, and $\rho_\infty : U \times [0,1] \to (\bar{M}_{x_0}, \bar{d})$, we can use (8.5.1) to make sense of

$$\lim_{t \to \infty} \rho_t = \rho_\infty \tag{8.5.2}$$

uniformly on compact sets.

Set $\bar{1}_{x_0} = \lim_{t \to \infty} x_0$; this is the identity element of the tangent cone.

8.6. It will be convenient to have a map of a suitable neighborhood of the identity of the tangent cone onto a neighborhood of x_0 in M, albeit non-canonical, generalizing the exponential map in Riemannian manifolds.

If $y \in \bar{M}_{x_0}$ is represented by $\{h_s x; 0 < s \leq 1\}$, with $x \in M$, we set $x = \in_{x_0}(y)$. Then $\in_{x_0}$ is a homeomorphism of some neighborhood $\bar{U}$ of $\bar{1}_{x_0}$ in $\bar{M}_{x_0}$ onto a neighborhood U_{x_0} of x_0 in M; we define λ_{x_0} to be the inverse of $\in_{x_0}$ on U_{x_0}. The definition of α_t can be rewritten

$$\alpha_t(x) = \left(\bar{h}_t \lambda_{x_0}(x), t^{-1}\right)\,.$$

If $\lambda_{x_0}(x)$ is represented by the parametrized orbit $\{h_s x; s \leq 1\}$, then $\bar{h}_t \lambda_{x_0}(x)$ is represented by the parametrized orbit $\{h_{ts} x; 0 \leq s \leq 1\}$. It follows at once that

(8.6.1) For any point x in U and for any $c > 0$ for which $\alpha_{ct}(x)$ is defined, we have

$$\alpha_{ct}(x) = \left(\bar{h}_{ct} \bar{x}, (ct)^{-1}\right) = \left(\bar{h}_c \bar{h}_t \bar{x}, (ct)^{-1}\right)\,.$$

Suppose now that

$$x_t \underset{x_0}{\longrightarrow} y \quad \text{as} \quad t \to \infty \quad \text{in the sense of (8.5.1)} .$$

Then $\alpha_t(x_t) \to (y, 0)$ as $t \to \infty$. Hence $\alpha_{ct}(x_t) \to (\bar{h}_c y, 0)$. We apply this to the family of parametrized curves $c_s(\mu) := c(t_0 + s\mu)$, $\frac{-\varepsilon}{s} \leq \mu \leq \frac{\varepsilon}{s}$, which is defined for $0 < s\mu < \varepsilon$ with ε fixed, and with $c_s(\mu)$ in $(M, s^{-1}d)$.

If for some sequence $s_n \to 0$ we have $\theta := \lim_{s_n \to 0} c_{s_n}$ exists, then for any $a > 0$, $\lim_{s_n \to 0} c_{as_n}$ exists and equals $\bar{h}_a \circ \theta$, in view of

$$\alpha_{(as)^{-1}}(c_{as}(\mu)) = (\bar{h}_{a^{-1}} \bar{h}_{s^{-1}} \lambda_{x_0}(c_{as}(\mu)), as) .$$

In particular, if $\lim_{s \to 0} c_s$ exists, then the limiting curve is a full orbit of the one parameter group of homotheties $\{h_a; a \in \mathbf{R}^+\}$ acting on curves in $\bar{M}_{x_0}$, i.e.

$$\theta(as) = \bar{h}_a \theta(s) . \tag{8.6.2}$$

8.7. Consider now a smooth horizontal vector field X on the cc space (M, d). Then for any $x_0 \in M$, there is a neighborhood U of x_0 and an interval $-a < s < a$ such that $\rho : (s, x) \to (\exp sX)(x)$ is a smooth map of $(-a, a) \times U \to M$. By (8.1),

$$\hat{\rho}_t : (s, x) \to \left(\exp \frac{s}{t} h_t(X)\right)(x)$$
$$\hat{\rho}_\infty : (s, x) \to (\exp s\hat{X})(x)$$

are well defined maps of $(-a, a) \times V \to (M, d_t)$ for some normal coordinate neighborhood V of x_0 for all t, $N \leq t \leq \infty$ and for N sufficiently large. Moreover, $\hat{\rho}_t$ converges to $\hat{\rho}_\infty$ as $t \to \infty$.

Let ρ_t denote the map of $(-a, a) \times (U \cap B(x_0, \frac{1}{t}))$ into (M, td) given by the restriction of the domain of ρ. By (8.2.4), the map $\hat{\rho}_t$ approximates the map ρ_t.

With the identification $(\bar{M}_{x_0}, \bar{d}) = (\hat{G}_{x_0}, \hat{d}) \equiv (\mathbf{R}^n, d_\infty)$ of Theorem 8.3, we find

$$\lim_{t \to \infty} \rho_t = \lim_{t \to \infty} \hat{\rho}_t = \hat{\rho}_\infty : (-a, a) \times U \to \hat{G}_{x_0} . \tag{8.7.1}$$

Let J be a subset of $\mathbf{R}$ which is unbounded above. Let A be an open subset of M containing x_0. For each $t \in J$, let A_t denote the subset A in (M, td) and let X_t be a smooth vector field on A_t. Then $\lim_{t \to \infty} A_t = \bar{M}_{x_0}$. Let $\bar{X}$ be a smooth vector field on $\bar{M}_{x_0}$. Set

$$\rho_t : (s, x) \to (\exp sX_t)(x) , \quad (s, x) \in I_{x,t} \times A_t$$
$$\rho_\infty : (s, X) \to (\exp s\bar{X})(x) , \quad (s, x) \in \mathbf{R} \times \bar{M}_{x_0}$$

where $I_{x,t} := (a_{x,t}, b_{x,t}) \subset \mathbf{R}$, and this interval is maximal for the trajectories to the vector field X_t; we assume that ρ_∞ is defined for all $s \in \mathbf{R}$.

DEFINITION 8.8: $\lim_{\substack{t\to\infty \\ t\in J}} X_t = \bar{X}$ if and only if

$$\lim_{\substack{t\to\infty \\ t\in J}} \rho_t = \rho_\infty$$

in the sense of convergence (8.5.2).

Explicitly, if X is a smooth horizontal vector field, the image of the family of trajectories $\{\exp sX(x) : (s,x) \in (-a,a) \times U\}$ in $\lim_{t\to\infty}(M, td)$ is $\{(\exp s\bar{X})(y); -a < s < a, y \in \bar{M}_{x_0}\}$ where $\bar{X}$ is a vector field on the tangent cone $\bar{M}_{x_0}$ corresponding to the vector field $\hat{X} = \lim_{t\to\infty} t^{-1}h_t(X)$ on a neighborhood of x_0; $\hat{X}$ is the unique homogeneous vector field of degree 1 with respect to the normal coordinates $a_1, \dots, a_n$ for which $\hat{X}(x_0) = X(x_0)$. By Theorem 8.3, the vector field $\bar{X}$ is horizontal on the tangent cone $\bar{M}_{x_0}$ and is in the Lie algebra of $\bar{M}_{x_0}$.

For any sequence of horizontal vector fields $Y_1, \dots, Y_k$ on M, let ρ_t : $(s_1, \dots, s_k; x) \to (\exp t^{-1}s_1Y_1 \cdot \exp t^{-1}s_2Y_2 \cdots \exp t^{-1}s_kY_k)(x)$ denote the map of $(-a,a)^k \times U \to (M, td)$ with t large. By the identification of $(\bar{M}_{x_0}, \bar{d})$ with $(\mathbf{R}^n, d_\infty)$, we get by (8.7.1)

$$\lim_{t\to\infty} \rho_t = \rho_\infty : (s_1, \dots, s_k) \to (\exp s_1\bar{Y}_1 \cdots \exp s_k\bar{Y}_k)(1) .$$

The path $s \to \exp s\bar{Y} \cdot \exp s_2\bar{Y}_2 \cdots \exp s_k\bar{Y}_k(1)$ is horizontal in $\bar{M}_{x_0}$ whenever Y is a horizontal vector field on U. Thus

(8.9) The curve $s \to \exp s\bar{Y} \cdot g$ is horizontal in $(\bar{M}_{x_0}, \bar{d})$ for any $g \in \bar{G}$; i.e. $\bar{Y}$ is a right-invariant vector field on $\bar{M}_{x_0}$.

To summarize:

PROPOSITION 8.10. *Let X be a smooth horizontal vector field on the cc space (M,d), let $x_0 \in M$, and let X_t denote the same vector field regarded as a vector field in (M, td). Then*

$$\lim_{t\to\infty} X_t = \bar{X}$$

where $\bar{X}$ is the right invariant horizontal vector field on $\bar{M}_{x_0}$ whose value at the identity element is $X(x_0)$, upon identifying horizontal elements in the tangent space at the identity element of $\bar{M}_{x_0}$ with elements of $T^1_{x_0}(M)$.

Remark: By (8.9), the metric $\bar{d}$ on $\bar{M}_{x_0}$ is right invariant; in Mitchell's paper, $\bar{d}$ is asserted to be left-invariant – because of a different convention about left and right: for Mitchell $\exp X$ acts on the right.

LEMMA 8.11. *For any compact subset $K \subset M$, and $a > 0$ and any $\varepsilon > 0$ one can find $\delta = \delta(K, a, \varepsilon) > 0$ and $p = p(K, a, \varepsilon) > 0$ such that if $0 < s < p$, $x \in K$, v_1 and v_2 are horizontal tangent vectors at x, $\|v_1\| < a$, $\|v_2\| < a$*

and $\|v_1 - v_2\| < \delta$ *then*

$$\frac{1}{s} d(\in_x (sv_1), \in_x (sv_2)) < \varepsilon .$$

If K consists of one point, the above lemma follows from the discussion in (8.6). For general K it is enough to note that estimates obtained in Mitchell's paper are uniform on compact subsets.

Using Proposition 8.10 and Lemma 8.11, we get the following.

PROPOSITION 8.12. *Let* $x_n \in M$, *let* v_n *be a horizontal tangent vector at* x_n *and let* $t_n > 0$. *Assume that* $x_n \to x$, $v_n \to v$ *and* $t_n \to \infty$. *Let us also assume that* $(x_n, t_n d) \to (y, \bar{d})$ *where* $(x_n, t_n d)$ *is* x_n *considered as a point in* $(M, t_n d)$. *Then the curves* $\{s \to \in_{x_n}(sv_n)\}$ *converge to a right coset relative to a one-parameter subgroup in* $(\bar{M}_x, \bar{d})$ *corresponding to* v.

9. *cc* Differentiability

In any metric space M one can define the *length* of a continuous curve $c : [a, b] \to M$ to be $\sup \sum_i d(c(t_i), c(t_{i+1}))$ as mesh $P \to 0$ where P is any partition $a = t_0 < t_1 < \cdots < t_n = b$, and mesh $P = \sup_i(t_{i+1} - t_i)$. If the length is finite, the curve is called *rectifiable*. A metric space is called a *length space* if the distance $d(p, q)$ between any two points p and q is the infimum of the length of continuous curves joining p and q.

LEMMA 9.1. *Let* $c(t)$ *be a rectifiable curve in a length space. Assume that* $c(t)$ *is parametrized by arc length. Then for almost all* t

$$\lim_{\Delta t \to 0} \frac{d(c(t + \Delta t), c(t - \Delta t))}{2\Delta t} = 1 .$$

Proof: Assume that this is false. Then there is an $\varepsilon > 0$ such that $\liminf_{\Delta t \to 0}$ of the above quotient is less than $1 - \varepsilon$ on a set S_ε of positive measure. By the Vitali covering theorem, we can assert: For any positive η, there is a finite set of disjoint open intervals, each of length less than η, which cover S_ε up to a subset of measure less than η; and for each of these open intervals the above ratio of $\frac{\text{chord length}}{\text{arc length}}$ is less than $1 - \varepsilon$. Moreover, this finite set of intervals can be regarded as a sub-collection of intervals of a partition $t_0, t_1, \ldots, t_n$ of the domain of the curve $c(t)$.

For such partitions the sum of the length of the chords $d(c(t_i), c(t_{i+1}))$ is at most length of the curve $c(t)$ minus $\varepsilon[(\text{measure } S_\varepsilon) - \eta]$.

For $\eta < \frac{1}{2}$ (measure S_ε) and arbitrarily small, we find a contradiction.

LEMMA 9.2. *Let* A_n *be a sequence of compact metric spaces which converges to a compact space* B. *Let* $c_n(t)$, $0 \leq t \leq 1$, *be a rectifiable curve* *in A_n, parametrized by arc length. Then there is a subsequence $c_{n_i}(t)$ ($i = 1, 2, \dots$) which converges for each t. Moreover, the limit of any convergent subsequence is a rectifiable curve in B.*

Proof: By a diagonal process, we can select a subsequence c_{n_i} of curves and a denumerable dense subset S of $[0,1]$ such that $\lim_{i\to\infty} c_{n_i}(s)$ exists for each $s \in S$.

Let $t \in [0,1]$. Consider now any limit point b of the sequence $\{c_{n_i}(t); i = 1, 2, \dots\}$. For any $s \in S$, we have

$$d_B\Big(b, \lim_{i\to\infty} c_{n_i}(s)\Big) = \lim_{i\to\infty} d_{A_{n_i}}\big(c_{n_i}(t), c_{n_i}(s)\big) \le |t-s| \,. \qquad (*)$$

Hence for any sequence of points $s_k \in S$ with $s_k \to t$, we have $b = \lim_{k\to\infty} \lim_{i\to\infty} c_{n_i}(s_k)$. it follows at once that $c_{n_i}(t)$ has a unique limit point in B as $i \to \infty$. Thus $c_{n_i}(t)$ converges for all t. By a similar argument, one gets that the limit curve of the subsequence is a rectifiable curve of length at most 1.

9.3. In equation $(*)$ above, the inequality $\le$ may be replaced by equality if each curve c_n is a geodesic; that is, if $d_{A_n}(c_n(t), c_n(s)) = |t-s|$ for each $s, t \in [0,1]$. In that case, the limit curve is also a geodesic. That is, if c_n is a geodesic in A_n with $\lim_{n\to\infty} c_n(0) = b_0 \in B$, then a subsequence of c_n converges to a geodesic in B.

9.4. Let $c : [a,b] \to M$ denote a continuous rectifiable curve in the *cc* space (M,d), parametrized by arc length. Let t_0 be a point in the interval (a,b). We let $c_s : \mu \to c(t_0 + s\mu)$, $-\varepsilon \le s\mu \le \varepsilon$, which is well defined for small positive ε, denote the indicated parametrized curve on the interval $\left[\frac{-\varepsilon}{s}, \frac{\varepsilon}{s}\right]$. In the *cc* metric space $\left(M, \frac{1}{s}d\right)$, the curve c_s is parametrized by arc length.
DEFINITION: The parametrized rectifiable curve c is *cc differentiable* in the *cc* space (M,d) if and only if the family of curves c_s in the *cc* space $\left(M, \frac{1}{s}d\right)$ converges as $s \to 0$ to a limit curve in the tangent cone to $(\bar{M}, \bar{d})$ at $c(t_0)$.

If such a limit exists, it lies on a curve parametrized by μ in the tangent cone and by (8.6.2) invariant under the homothety group; the *cc derivative* of $c(t)$ at t_0 is the point in the tangent cone corresponding to $\mu = s$. On the other hand, since c is parametrized by arc length, the limit curve is rectifiable by Lemma 9.2. Therefore the limit in the tangent cone is a one parameter horizontal subgroup of the graded nilpotent tangent cone. Thus the above definition of a derivative agrees with the usual definition of derivative in Riemannian spaces. We can now make the following remark.

9.4.1 Remark: A rectifiable curve c parametrized by arc length is *cc* differentiable at t_0 if its Riemannian derivative $c'(t_0)$ exists and is horizontal, and moreover

$$\lim_{s\to 0} \frac{1}{s} d\big(c(t_0+s), \epsilon_{c(t_0)}\,(sc'(t_0))\big) = 0 \,.$$

The same is also true for rectifiable curves not necessarily parametrized by arc length but whose arc length is differentiable at t_0.

Let M be a Riemannian space and let (M,d) denote the *cc* space formed from a horizontal subbundle of the tangent bundle of M. Then the length of a piecewise smooth horizontal curve in both the Riemannian metric and the *cc* metric coincide. Furthermore, for any two points p and q in M, the *cc* length of a continuous rectifiable curve joining p and q is minorized by the *cc* length of a piecewise smooth horizontal curve joining p and q. Hence the *cc* space (M,d) is a length space.

Let (M,d) be a *cc* space with metric d. Let $a_1,\dots,a_n$ be a normal coordinate system centered at the point x_0 coherent with the *cc* space M. Let $\pi : (a_1,\dots,a_h) \to (a_1,\dots,a_n)$ denote the projection onto the first h coordinates, $h = \dim T^1$ = the dimension of the horizontal subbundle H.

Let $c(t)$ $(-1 \le t \le 1)$ be a curve in M parametrized by arc length. By Lemma 9.1,

$$\lim_{\Delta t\to 0} \frac{d(c(t-\Delta t), c(t+\Delta t))}{2\Delta t} = 1 \qquad ([\mathrm{G}])$$

for almost all t.

LEMMA 9.5. *If at t_0 condition [G] is satisfied and $c(t)$ is differentiable at t_0 in the Riemannian sense, then $c(t)$ is cc differentiable at t_0.*

Proof: Let x_0 be a point in the *cc* space M, and let $X_1,\dots,X_n$ be a base of vector fields on a neighborhood U of x_0 coherent with the horizontal subbundle H of the tangent bundle, and let $a_1,\dots,a_n$ be the normal coordinates on U given by $x = \big(\exp\sum_{i=1}^n a_iX_i\big)(x_0)$. Define the homothety map ${}_{x_0}h_t : U \to M$ as in (8.2) via

$$a_i({}_{x_0}h_t x) = t^{[i]}a_i$$

where

$$[i] = \inf\left\{j; X_i(x) \in T^j(x)\right\} .$$

In order to prove Lemma 9.5, it suffices by (8.2.4) to prove that

$${}_{c(t_0)}h_{s^{-1}}c(t_0+s\mu) \qquad (9.5.0)$$

approaches a limit in the tangent cone as $s \to 0$.

For convenience, set $h_s = {}_{c(t_0)}h_s$.

Set $c_i(t) = a_i(c(t))$, $1 \le i \le n$. We prove first

$$\lim_{s\to 0} a_i\big(h_{s^{-1}}c(t_0+s\mu)\big) = \frac{d}{dt}_{|t=t_0} c_i(t)\cdot\mu \,, \quad 1 \le i \le h \,. \qquad (9.5.1)$$

Proof: By hypothesis, the map c of $(-1,1)$ to the underlying Riemannian space M is differentiable at $t = t_0$ in the usual sense. By definition of h_s, we have

$$a_i\big(h_{s^{-1}}c(t_0 + s\mu)\big) = s^{-[i]}c_i(t_0 + s\mu) \qquad 1 \le i \le n$$
$$c_i(t_0) = a_i\big(c(t_0)\big) = 0 \ .$$

Hence for $1 \le i \le h$, i.e. for $[i] = 1$, we have

$$\lim_{s\to 0} a_i\big(h_{s^{-1}}(c(t_0 + s\mu))\big) = \lim_{s\to 0} c_i \frac{(t_0 + s\mu) - c_i(t_0)}{s} = \mu \frac{dc_i}{dt}\Big|_{t_0} \ .$$

Set $x_0 = c(t_0)$ and let $\bar{M}_{x_0}$ denote the tangent cone at x_0. Let B denote a ball in M of center x_0. For any $s > 0$, define the curve c_s in $(M, s^{-1}d)$ by $c_s(\mu) : \mu \to c(t_0 + s\mu)$, $-\frac{\varepsilon}{s} \le \mu \le \frac{\varepsilon}{s}$, for some fixed positive ε.

The curve c_s is parametrized by arc length in $(M, s^{-1}d)$ and has length $\frac{2\varepsilon}{s}$. Hence the curve c_s lies in B for s small, by (8.2.4).

9.5.2. The family of curves

$$\alpha_{s^{-1}} \circ c_s : \left[\frac{-\varepsilon}{s}, \frac{\varepsilon}{s}\right] \to \bar{M}_{x_0} \times [0,1] \ , \qquad 0 < s \le a$$

when restricted to any bounded interval of $\mathbf{R}$ is precompact.

Proof: We have $\alpha_{s^{-1}}(c(t_0)) \to \alpha_\infty(c(t_0))$ as $s \to 0$. The family of curves $\{c_s; 0 < s \le a\}$ is equicontinuous and lies in a ball in the metric space $(B, s^{-1}d)$ of center x_0, and diameter $\frac{2\varepsilon}{2}$. Inasmuch as α_t is asymptotically isometric as $t \to \infty$, we may conclude that the family of curves $\{\alpha_{s^{-1}} \circ c_s; 0 < s \le a\}$ is precompact on any bounded interval of $\mathbf{R}$.

Next we claim:

9.5.3. Any convergent subsequence of $\{h_{s^{-1}}c_s\}$ converges as $s \to 0$ to a geodesic in the tangent cone.

Proof: By the hypothesis of our Lemma $\frac{d(c(t_0+s\mu),c(t_0-s\mu))}{2s\mu} \to 1$ as $s\mu \to 0$. For any bounded set of μ in $\mathbf{R}$,

$$\frac{(s^{-1}d)(c(t_0 + s\mu), c(t_0 - s\mu))}{2\mu} \to 1 \quad \text{as} \quad s \to 0 \ .$$

Suppose now that $h_{s_n^{-1}}c_{s_n}$ is a convergent subsequence. By Lemma 9.2, the limit $c_\infty := \lim_{n\to\infty} h_{s_n^{-1}}c_{s_n}$ is a rectifiable curve. By (8.2) and (8.3) we have $d_\infty(c_\infty(t_0 + \mu), c_\infty(t_0 - \mu)) = 2\mu$, the cc arc length of the curve $c_\infty(t)$, $-\mu \le t \le \mu$. It follows at once that $\mu \to c_\infty(t_0 + \mu)$ is a geodesic in the tangent cone $\bar{M}_{x_0}$; in particular, it is horizontal.

9.5.4. All convergent subsequences of the family of parametrized curves $\{h_s^{-1}c_s; s \to 0\}$ have the same limit.

Proof: By (9.5.1), the projection of $h_{s^{-1}}c_s$ on the first h coordinates yields a family of curves in $\mathbf{R}^h$ which converge to a segment of the one parameter subgroup $\mu \to \frac{d}{dt}\big|_{t=t_0} c_i(t) \cdot \mu$ $(i = 1, \ldots, h)$, $-1 \leq \mu \leq 1$.

Let $\pi : \bar{M}_{x_0} \to \mathbf{R}^h$ denote the projection of the graded nilpotent group onto its vector subspace of degree 1; i.e. its one dimensional vector subspaces are horizontal. Then any convergent subsequence of $h_{s^{-1}}c_s$ lies on a geodesic, which is ipso facto horizontal, and which passes through the identity element of the tangent cone. Let $V = \pi^{-1}L$ where L is a one dimensional vector subspace of $\mathbf{R}^h$. Then V is a normal connected graded Lie subgroup of the graded nilpotent group $\bar{M}_{x_0}$.

The tangent subbundle of V intersects the restriction to V of the horizontal subbundle of $\bar{M}_{x_0}$ in a right invariant rank one bundle on V, denoted H_V. There is one and only one oriented maximal trajectory to H_V passing through any point of V and parametrized by arc length; namely, right translates of the one parameter subgroup $\{\exp \mu X; \mu \in \mathbf{R}\}$ with X a unit horizontal tangent vector to V at x_0.

In particular any convergent subsequence of $\{h_{s^{-1}}c_s; 0 < s \leq a\}$ with $s \to 0$ converges to the geodesic segment $\{\exp \mu X; -1 \leq \mu \leq 1\}$; that is, has a single limit as asserted in 9.4. By 9.5.2, this geodesic segment is $\lim_{s\to 0} h_{s^{-1}}c_s$. By (9.5.0), this implies Lemma 9.5.

10. The Differential of a Quasi-conformal Map

10.1. In a *cc* space, a horizontal path $c(t)$ $(-a \leq t \leq a)$ which is *cc* differentiable at $t = 0$ induces a map $\dot{c}$ of the tangent vector $\frac{d}{dt}\big|_{t=0}$ to $\mathbf{R}$ at 0 to the tangent cone of M at $c(0)$; the image horizontal vector in the tangent cone is denoted $\dot{c}(0)$ or $\frac{dc}{dt}\big|_{t=0}$.

Let $\varphi : \mu \to \mu'$ be a quasi-conformal map of *cc* spaces. Set $M_{a,I} = \{x \in M \mid L_\varphi(x, r) \leq Ir \text{ for } 0 \leq r \leq a\}$ and let $M^0_{a,I}$ denote the set of density points in $M_{a,I}$. Set

$$M_I = \bigcup_{a\downarrow 0} M^0_{a,I}, \quad M^0 = \bigcup_{I\uparrow\infty} M_I .$$

Then $x \in M^0$ implies $x \in M^0_{a,I}$ for some $a > 0$, $I < \infty$. By the results in §6 and §7, the Lipshitz bound I_φ is finite almost everywhere on M. This and the density point theorem imply that M^0 is a subset of full measure in M.

The map $\varphi : M \to M'$ induces a continuous map $\varphi_t : (M, td) \to (M', td')$ for each $t \in \mathbf{R}^+$.

DEFINITION 10.1.1: We say that the family $\{\varphi_t\}$, $t \geq 1$, is *equicontinuous at* $x_0 \in M$ if this family is uniformly continuous on balls with center x_0 of radius R in (M, td) for every $R > 0$. In other words, $\{\varphi_t\}$ is equicontinuous at $x_0 \in M$ if for every $R > 0$ the following condition is satisfied: for every $\varepsilon > 0$ there exists $\delta > 0$ such that for all $t \geq 1$, whenever $y, z \in M$, $d(x_0, y) < \frac{R}{t}$, $d(x_0, z) < \frac{R}{t}$ and $d(y, z) < \frac{\delta}{t}$ we have $d(\varphi(y), \varphi(z)) < \frac{\varepsilon}{t}$.

LEMMA 10.2. *If $x_0 \in M^0$ then $\{\varphi_t\}$ is equicontinuous at x_0.*

Proof: We have $x_0 \in M^0_{a,I}$ for some $a > 0$, $I < \infty$. Since x_0 is a density point of $M_{a,I}$, it follows from Proposition 2.2(b) that

$$\lim_{x \to x_0} \frac{\operatorname{dist}(x, M_{a,I})}{d(x, x_0)} = 0 . \tag{$*$}$$

Fix $\delta > 0$ and $\eta > 0$. Let $y, z \in M$, $d(y, z) < \frac{\delta}{t}$ and $d(x_0, y) < \frac{R}{t}$. It follows from $(*)$ that, for sufficiently large t, there exists $y' \in M^0_{a,I}$ with $d(y', y) < \eta \frac{R}{t}$. We can, by taking $\eta R + \delta < a$, also assume that $d(y', y) < a$ and $d(y', z) \leq d(y', y) + d(y, z) < a$. Then

$$\begin{aligned} d'\big(\varphi(y), \varphi(z)\big) &\leq d'\big(\varphi(y'), \varphi(y)\big) + d'\big(\varphi(y'), \varphi(z)\big) \\ &\leq I\big(d(y', y) + d(y', z)\big) \leq I\left(\eta\frac{R}{t} + \eta\frac{R}{t} + \frac{\delta}{t}\right) \\ &= \frac{I}{t}(2\eta R + \delta) . \end{aligned}$$

Given therefore any $\varepsilon > 0$, for sufficiently small η and δ, $d'(\varphi(y), \varphi(z)) < \frac{\varepsilon}{t}$. This proves the equicontinuity of $\{\varphi_t\}$ at x_0.

10.3. Given any point $x \in M$ and any horizontal smooth vector field X defined in a neighborhood of x with the path $\varphi((\exp tX)(x))$ absolutely continuous and *cc* differentiable at $t = 0$, we set

$$\dot{\varphi}_x\big(X(x)\big) = \frac{d}{dt}_{|_{t=0}} \varphi\big((\exp tX)(x)\big) .$$

By Corollary 6.5, we can assert:

Given any smooth horizontal vector fields $Y_1, \ldots, Y_n$ on M, there is a set of full measure M^* on M such that

$$\dot{\varphi}_x\big(Y_i(x)\big) \text{ exists} , \quad i = 1, \ldots, m \quad \text{for all} \quad x \in M^* .$$

DEFINITION 10.3.1: Let $\{\varphi_t\}$ denote the family of (10.1.1). We say that $\varphi : M \to M'$ is *cc differentiable* at $x_0 \in M$ if and only if φ_t converges uniformly on compact sets as $t \to \infty$ to a map of the tangent cone to M at x_0 to the tangent cone of M' at $\varphi(x_0)$.

DEFINITION: Let M be a manifold and X a measurable vector field on M. Let U be an open set of M. We say that X is *integrable on* U if there exists an interval $(-a,a)$ in $\mathbf{R}$ and continuous map $\rho : (-a,a) \times U \to M$ such that for almost all $x \in U$ and $-a < t < a$, $t \to \rho(t,x)$ is an absolutely continuous curve with $\frac{d}{dt}\rho(t,x) = X(\rho(t,x))$ for almost all t and $\rho(0,x) = x$.

In this circumstance, we denote $\rho(t,x)$ by $(\exp tX)(x)$. We shall not assume that ρ is unique but whenever the notation $\exp tX$ is used, the ρ remains fixed throughout the discussion. If, in addition $\frac{d\rho}{dt}(0,x) = X(x)$, we say that X is *proper* at x.

EXAMPLE 10.3.2: Let $Y_i'(\varphi(x))$ denote $\varphi_x(Y_i(x))$ for all $x \in M^*$, $i = 1,\ldots,m$. Then, by Corollary 6.5 and Theorem 7.1, $Y_1',\ldots,Y_m'$ are examples of integrable vector fields on M. The vector field Y_i' is called the *image of* Y_1 *under* φ, $i = 1,\ldots,m$.

Let X be an integrable vector field on a *cc* space (M,d), and let X_t denote the same vector field on the *cc* space (M,td) centered at a point $x_0 \in M$. Let $\bar{X}$ denote a smooth vector field on the tangent cone to (M,d) at the point x_0. Set $\rho_\infty : t \to \exp t\bar{X}$.

For an integrable vector field X on U and the map ρ as above, let us consider the set

$$\Omega_X = \left\{ x \in U \;\middle|\; \begin{array}{l} x \text{ is a point of density for} \\ \text{a set } A \text{ where} \\ \frac{1}{s} d(\rho(s,y), \in_y (sX(y)) \\ \text{tends to 0 uniformly when} \\ s \to 0 \text{ , for all } y \text{ in } A \end{array} \right\}.$$

PROPOSITION 10.4. *Let $t_n \to \infty$ and let us assume that $X_{t_n} \to \bar{X}$ uniformly where X_{t_n} denotes the same vector field X on the space $(M, t_n d)$ centered at $x \in \Omega_X$. Then $\bar{X}$ is a right invariant horizontal vector field on the tangent cone $(\bar{M}_x, \bar{d})$ with $\bar{X}(\bar{1}_x) = X(x)$.*

Proof: Let z be a point in $(\bar{M}_x, \bar{d})$. We have that x is a point of density of a set A where $\frac{1}{s}d(\rho(s,y), \in_y (sX(y))$ tends to 0 uniformly when $s \to 0$. Then it follows from Proposition 2.2 (b) that we can find a sequence y_n in A such that $\alpha_{t_n}(y_n) \to z$. Then in view of Proposition 8.12 and the definition of the set A, $\rho_{t_n}(s, y_n)$ tends to a right coset relative to a one-parameter subgroup in $(\bar{M}_x, \bar{d})$ corresponding to $X(x)$. (Here we use the fact that since ρ is continuous, the restriction of X to A is continuous.)

THEOREM 10.5. *Let $\varphi : M \to M'$ be a quasi-conformal map of cc spaces and let $\varphi_t : (M, td) \to (M', td')$ denote a continuous map induced by φ. Then for almost all $x \in M$, φ is cc differentiable at x; the maps φ_t converge uniformly on compact sets as $t \to \infty$ to a continuous map $d\varphi_x : (\bar{M}_x, \bar{d}) \to (\bar{M}'_{\varphi(x)}, \bar{d}')$, and $d\varphi_x$ is a group isomorphism of the tangent cones equivariant with respect to their homotheties.*

Proof: Let $Y_1, \ldots, Y_h$ be horizontal smooth vector field on M such that, for every $x \in M$, $\{Y_1(x), \ldots, Y_h(x)\}$ form a basis in the horizontal space $T^1(x)$. In view of 10.3.2, the images $Y'_1, \ldots, Y'_k$ of $Y_1, \ldots, Y_h$ under φ are integrable vector fields on M'. Using Remark 9.4.1 we can also deduce from Theorem 7.1 and Lemma 9.5 that the sets $\Omega_{Y'_1}, \ldots, \Omega_{Y'_h}$ have full measure on M'. Let us consider a point x from the intersection $M^0 \cap \Omega_{Y'_1} \cap \cdots \cap \Omega_{Y'_h}$. By Lemma 10.2 the family $\{\varphi_t\}$ is equicontinuous at x_0. But in view of Proposition 10.4, each limit map ψ for the family $\{\varphi_t\}$ is equivariant with respect to the one-parameter subgroup $u_i \in (\bar{M}_x, \bar{d})$ corresponding to $Y_i(x)$, $1 \le i \le h$. Clearly $\psi(\bar{1}_x) = \bar{1}_{\varphi(x)}$. We have $\psi(\exp s\bar{Y}_i g) = \exp s\bar{Y}'_i \cdot \psi(g)$ where the $\bar{Y}'_i$ is determined from Y'_i as in Proposition 10.4. By induction, for any $i_1, \ldots, i_m$ in $\{1, \ldots, h\}$,

$$\psi(\exp s_1 \bar{Y}_{i_1} \cdots \exp s_m \bar{Y}_{i_m}) = \exp s_1 \bar{Y}'_{i_1} \cdots \exp s_m \bar{Y}'_{i_m}.$$

But $\{u_1, \ldots, u_h\}$ generate $(\bar{M}_x, \bar{d})$. Therefore each limit map of the family $\{\varphi_t\}$ is the same. Consequently $\{\varphi_t\}$ converges to a group isomorphism $d\varphi_x : (\bar{M}_x, \bar{d}) \to (\bar{M}_{\varphi(x)}, d(x))$ which maps $\exp s\bar{Y}_i$ to $\exp s\bar{Y}'_i$, $i = 1, \ldots, h$. The equivariance of $d\varphi_x$ follows from the fact that it takes horizontal subgroups to horizontal subgroups by Corollary 6.5 and Remark 8.4.3.

References

[Go] R. Goodman, Nilpotent Lie Groups: Structure and Applications to Analysis, Lecture Notes in Math, Springer, Berlin, 562, 1970.

[Gr] M. Gromov, Structures métriques pour les variétés Riemannienes, CEDIC, Paris, 1981.

[KR] A. Koranyi, M. Reimann, Foundations for the theory of quasi-conformal mappings of the Heisenberg group, Adv. in Math. III:1 (1995), 1-87.

[M] G. Metivier, Fonction spectrale et valeurs propres d'une classe d'operateurs non elliptiques, Comm. Partial Differential Equations 1 (1976), 479-519.

[Mi] J. Mitchell, On Carnot-Carathéodory metrics, J. Differential Geometry 21 (1985), 35-45.

[Mo1] G.D. Mostow, Quasi-conformal mappings in n-spaces and the rigidity of hyperbolic space forms, Publ. Math. IHES 34 (1968), 53-104.

[Mo2] G.D. Mostow, Strong Rigidity of Locally Symmetric Spaces, Annals of Math Studies, Princeton Univ. Press, Princeton, NJ, 1973.

[Mo3] G.D. Mostow, A remark on quasi-conformal mappings on Carnot groups, Michigan Math. J. 41 (1994), 31-37.

[P] P. Pansu, Métriques de Carnot-Carathéodory, et quasi-isométries des espaces symmétriques de rang un, Ann. of Math. 2:129 (1989), 1-60.

G.A. Margulis and G.D. Mostow
Dept. of Math.
Yale University
New Haven, CT 06520
USA

Submitted: August 1994

Geometric And Functional Analysis
Vol. 5, No. 2 (1995)
1016-443X/95/0200434-11$1.50+0.20/0

THE SEMICLASSICAL ELECTRON IN A MAGNETIC FIELD AND LATTICE

SOME PROBLEMS OF LOW DIMENSIONAL "PERIODIC" TOPOLOGY

S.P. Novikov

1. Hamiltonian Systems and Foliations

For any *Phase space*, i.e. a manifold M with the *Poisson structure*, we have a scew-symmetric Poisson Tensor h^{ij} with 2 upper indices in any local coordinates (x^i), such that the *Poisson Bracket* $\{f,g\} = h^{ij} f_i g_j$ satisfies the Jacoby identity. Here f_i means a partial derivative of the function $f(x)$ on the manifold and the standard summation rule is used for tensor indices. The Poisson bracket is well-defined for multivalued functions as well (i.e. closed 1–forms $df = f_i dx^i, dg = g_j dx^j$).

An Annihilator (Casimir) for the Poisson bracket contains all (perhaps well-defined only locally) C^∞-functions on M such that $\{f,g\} = 0$ for any function g. So the annihilator is a sheaf on M. It determines some integrable foliation A in any domain where the rank of the matrix h is constant. This matrix may be reduced to a constant in such a domain.

A Hamiltonian system determined by the Hamiltonian H is such that $f_t = \{f,H\}$ for any (perhaps locally well-defined) function f. Here H may not be a function, but a multivalued function, i.e. dH may be a closed 1–form on the manifold M.

For obvious reasons any trajectory $x(t)$ belongs to the intersection of 2 foliations:

$$x(t) \subset (dH = 0) \cap A .$$

In particular, in the case of the so-called *Symplectic manifolds* a Poisson Tensor is nondegenerate $det(h^{ij}) \neq 0$ and foliation A is trivial. We are coming to the codimension 1 foliations with Morse-type singularities determined by the closed 1-form $dH = 0$. Such foliations were studied by this author in early 80s (see [N1,2]) and later by the author's pupils in Moscow (see [N3]). An analogue of Morse theory was constructed for the critical points of multivalued functions H on finite-dimensional manifolds (Novikov, Farber, Sikorav, Pazitnov). A nice conjecture by the author about the "quasiperiodic structure" of this foliation was proved (see the papers of Zorich, Le Tu and Alaniya in [N3]; this problem is still open for the analogous foliations given by 2 and more equations). This quasiperiodic structure may be nontrivial even in the case of the closed 2-dimensional manifolds M.

Topological theory of generic Hamiltonian systems on surfaces was constructed by Katok, Hubbard and others in 70s. Topological invariants of such foliations come from Poincaré maps along trajectories crossing any closed transversal curve. These maps preserve some length element on the circle and have a finite number of the 1-st kind of discontinuity points. In particular, Hubbard studied foliations determined by the real part of the square root from the generic holomorphic quadratic differential on the Riemann surface with some complex structure. Such foliations present, more or less, all generic topological types.

However, nobody has studied a special class of foliations determined by the real part of the meromorphic or even by the holomorphic (1-st kind) 1-form for the genus $g \geq 2$.

In the case $g = 2$ any algebraic curve is hyperelliptic

$$M : \left(y^2 = P_5(x)\right) .$$

Our foliations present the simplest natural generalization of the straight-line flows on the torus. They are given by the effective algebraic formula:

$$Re\left[\frac{(ax+b)dx}{\sqrt{P_5(x)}}\right] = 0 .$$

In some special nongeneric cases (for example, if all coefficients are real in the hyperelliptic case) topological invariants of these foliations may probably be calculated analytically. To study their small perturbations we have to use analytical perturbation theory. In general we have to use a computer to find them. Until now almost nothing has been done for this very concrete class.

Recently Zorich found some nonstandard ergodic properties of the generic Hamiltonian foliations on surfaces.

We shall see later in section 3 that for our goals some very nongeneric class is important:

Let some closed 2-dimensional submanifold in the 3-torus be given (the so-called *Fermi surface* M^2 *in the space of quasimomenta* T^3); a closed form ω on M^2 is by definition a restriction of the constant 1-form on T^3, determined by an *external magnetic field*.

Our leaves will coincide with the semiclassical (or adiabatic) trajectories of the quantum electron in the Lattice and the external magnetic field in the space of quasimomenta. Some beautiful problems of *Periodic Topology* appear here (see section 3 below).

2. Semiclassics for the Linear ODEs and Foliations

Let me demonstrate here how the foliations determined by the real part of the meromorphic 1-form may appear in the problems of semiclassical analysis for linear ODE systems.

Consider some linear ODE system

$$\epsilon\Psi_t = \Lambda(t,\epsilon)\Psi \ , \qquad \epsilon \to 0$$

Here Λ is an $(n \times n)$-matrix, whose elements depend on the variable t as a rational function:

$$\Lambda = \Lambda_0(t) + \Lambda_1(t)\epsilon + \dots$$

We may start from the case when all $\Lambda_i = 0$ for $i \geq 1$ (as in the papers of the author and P. Grinevich in the theory of the so-called "String equation"; see [N4],[GN]).

How is a semiclassical approximation constructed?

Semiclassical Formulas make sense only for systems which are diagonal in the zero order:

$$\Lambda_0(t) = \operatorname{diag}\left[\lambda_1(t), \dots, \lambda_n(t)\right]$$

Therefore the first step is:

STEP 1. Diagonalization of the system in the zero order.

After the substitution $\Psi = U(t)\Phi$ we have a gauge transformation for the system:

$$\epsilon\Phi_t = \bar{\Lambda}\Phi = [U^{-1}\Lambda U - \epsilon U^{-1}U_t]\Phi$$

Taking $U(t)$ such that the matrix $U^{-1}\Lambda_0 U$ is diagonal we are approaching the desired form. However *our matrices* $U(t), \bar{\Lambda}(t)$ *live on the Riemann surface* $\Gamma \det[\Lambda_0(t) - y] = 0$.

For traceless 2×2-matrices Λ this Riemann surface has the form $y^2 = \det[\Lambda_0(t)]$.

STEP 2. Formal semiclassical series on the Riemann surface.

We may choose the following expression such that it formally solves the ODE system above:

$$\Phi_{sc} = \left(1 + \sum_{i\geq 1} \epsilon^i A_i\right) \exp\left\{\epsilon^{-1}B_{-1} + B_0 + \sum_{i\geq 1}\epsilon^i B_i\right\}$$

Here all A_i are offdiagonal matrices whose entries are algebraic functions on the surface Γ, all B_i are the diagonal matrices whose entries are integrals from the algebraic (meromorphic) 1-forms on Γ,

$$B_{-1} = \operatorname{diag}\left\{ \pm \int_{t_0}^{t} \sqrt{det[\Lambda_0(t)]}dt \right\}$$

Our formal series lives in fact on the surface which is a branching covering over the surface Γ. Only the first 2 terms (corresponding to the powers $\epsilon^{-1}, \epsilon^0$ in the exponent) are really important.

STEP 3. Semiclassical asymptotics for the exact solutions.

Consider the oriented foliation

$$Re[\Omega] = 0 , \qquad \Omega = \sqrt{\det[\Lambda_0(t)]}dt$$

on the surface Γ.

The semiclassical formula above gives right asymptotics for the exact solutions of the linear ODE system for $\epsilon \to 0$ along the path of integration for B_0 if this path is transversal to our foliation in the positive direction where the real part of the integral increases.

For the linear ODE systems of order n we use the analogous foliations $Re[\lambda_i(t)] = 0$ for all eigenvalues of the matrix $\Lambda_0(t)$.

3. Semiclassical (or Adiabatic) Motion of the Quantum Electron in the Periodic Lattice and Weak Magnetic Field

In the absence of a magnetic field, quantum states of an electron in the lattice Γ correspond to "Bloch waves" $\psi_n(x,p)$. These functions satisfy to the stationary Schroedinger equation in the variable x (we ignore all dimensional units here)

$$H\psi_n = E_n\psi , \quad H = -\Delta + V(x) , \quad V(x+\Gamma) = V(x) .$$

Let the lattice Γ be generated by the basic vectors $(1,00), (0,1,0), (0,0,1)$ in the x-space R^3. It is important for our goals to emphasize that this basis is orthonormal in the euclidean metric in 3-space and therefore it is "better" than any other basis obtained from it by the group $SL_3(Z)$. Only the action of the group $SL_3(Z) \cap O_3(R)$ leads to the equivalent bases in our problem. Corresponding translations we denote by T_1, T_2, T_3. We have

$$T_i H = H T_i .$$

We may choose the basic eigenfunctions $\psi_n(x,p)$ such that

$$T_j\psi_n = \exp\{2\pi i p_j\}\psi_n , \qquad j = 1,2,3 ,$$

for the real vector p. By definition, this vector is well-defined modulo a dual lattice in p-space: (p_1,p_2,p_3) is equivalent to $(p_1+n_1, p_2+n_2, p_3+n_3)$ for the integers n_j.

Therefore p is a point of the torus T^3. People call p a "*Quasimomentum*". The *dispersion relation* for the energy $E_n(p)$, which is an eigenvalue of the Hamiltonian operator with $B = 0$, is well-defined as a real function on the torus T^3. For the generic operators in the 3-dimensional case we may have an equality $E_n(p) = E_m(p)$, $n \neq m$, for the isolated points p only. Therefore, in the generic case there are no such points on the important chosen surface – the so-called *Fermi surface* $M^2 : \{E = E_0\}$. This energy level E_0 (*Fermi Level*) is an intrinsic invariant of a metal, depending on the number of free electrons in it. It may have a very complicated topology, for example, for some "noble metals" such as gold and others (see in the book of A. Abrikosov [A], for example).

Now consider this system in an external relatively "weak" homogeneous and constant (i.e. constant in space and time) magnetic field B, which does not deform the lattice itself.

It is very difficult to study this system (see, for example, my survey article [N5] for the Schroedinger operator in the magnetic field and lattice, translated into English by AMS in 1985). In the case where all magnetic fluxes are irrational numbers, no exact magnetic analogue of Bloch waves exists. Nobody knows whether any convenient picture exists or not, which might be built on the base of the exact eigenfunctions here. Chern classes for the dispersion relations based on magnetic Bloch functions for the generic 2-dimensional Schroedinger operator in the external magnetic field with rational flux through an elementary cell in the lattice were studied for the first time in the work [N6]. It was rediscovered later by physicists in connection with the Integral Quantum Hall effect.

Remark: Alan Connes conjectured after the author's talk at the Geometry Conference (Tel-Aviv University, December 1993) that all topological invariants of the adiabatic electron in the magnetic field (see below) can be expressed naturally through the language of noncommutative C^*-algebras. It should be point out that in fact the language of von Neumann algebras looks very poor for our problems; probably, only a few of the interesting physical quantities could be treated naturally through them. In the cases where the author knew such treatment was possible, it was only a hard mathematical foundation and extension of the understanding, already reached, using standard geometry and topology in the more special cases (for example, in the case of the integral Quantum Hall effect). For the problems below, not one of our "adiabatic" topological quantities has yet been explained through quantum C^*-algebras. It would be interesting to do this, but probably very difficult.

Many years ago quantum solid state physicists started to use the semiclassical or adiabatic one-zone picture in practical studies. In particular, physicists from the Kharkov-Moscow school (for example, I. Lifshitz, M. Azbel, M. Kaganov) actively used it. This picture leads to interesting topological problems. The author began thinking about it in the early 80s (see, for example [N1]). Some topological problems about this picture, posed in the early 80s, have now been solved by the author's pupils (see below). Let us describe this picture.

We shall now work with one value of n only and forget this index. In the semiclassical one-zone adiabatic approach people use the dispersion relation $E(p)$ as a classical Hamiltonian function in the phase space $R^3 \times T^3$ with the canonical coordinates (x, p) such that $\{p_i, p_j\} = 0$, in particular. In the external magnetic field $B(x)$ we use the same Hamiltonian $E(p)$, but change the Poisson brackets

$$\{p_i, p_j\} = eB_{ij}(x) .$$

In a case of interest to physicists, we have a field B independent of x and t. Therefore, the components of the quasimomenta give us a closed subalgebra of Poisson brackets. In this small phase space T^3 we have a Poisson bracket with nontrivial annihilator A and a Hamiltonian $E(p)$. "*Symplectic leaves*" $A =$ const are the planes, orthogonal to the magnetic field $B = (b_{23}, b_{31}, b_{12})$ as a vector in the euclidean 3-space R^3, which is a universal covering over the torus T^3. We have, in fact, a multivalued annihilator function on the torus T^3. The levels of the Hamiltonian are Fermi surfaces $M^2 \subset T^3$ in the space of the quasimomenta. Therefore, the trajectories in p-space are the sections of Fermi surfaces by the planes orthogonal to the magnetic field.

We may consider this picture as a Hamiltonian foliation of the Fermi surface. This foliation is obviously very special and nongeneric between the Hamiltonian foliations of the surfaces. We shall use the word "generic" now inside this special subclass only.

Geometrically, this foliation is obtained by the family of parallel plane sections of the periodic surface in the universal covering euclidean p-space R^3. The *periodic surface* is given by the equation $E(p) = 0$. The function $E(p)$ here is periodic in all 3 variables p_1, p_2, p_3. Its analytical structure is unknown. There are many different solid media with complicated Fermi surfaces. So we may consider all this class as well: nobody knows any additional physical restrictions.

Which nontrivial topological properties may have a family of parallel plane sections of the generic periodic surface in R^3?

As possibilities we may have the following types of trajectories:

1. *isolated point* (a critical point of our foliation),
2. *closed curve in R^3* (a curve homotopic to zero in T^3),

3. *periodic nonclosed curve in* R^3 (a curve, closed in T^3, but nonhomotopic to zero),
4. *nonperiodic nonclosed curve in* R^3 (such a curve in T^3 has a closure which is generically 2-dimensional; physicists call it an "*Open Trajectory*"). It definitely exists for some media.

In case 4, we may ask about the asymptotics of this trajectory in the space R^3 for the time $t \to \pm\infty$. The author's very first conjecture (see [N1]) was that there exists generically an asymptotic direction, which is the same for both signs $+\infty$ and $-\infty$ in time. In fact, the topology of this picture proved to be much more interesting than he expected originally.

The *Topological closure* of an open trajectory we define as a minimal compact 2-manifold, with possibly several components of the boundary (which is a part of the Fermi surface), containing our trajectory, such that each component of the boundary is a closed trajectory, homotopic to zero in the torus T^3. A genus of this surface is by definition a *genus of an open trajectory*.

For any trajectory with genus equal to g there exist a closed 2-manifold N^2 of the genus $g(N^2) = g$ with the curve $y(t) \subset N^2$ and a continuous mapping (singular bordism, which is an immersion, not the imbedding in general) $f : N^2 \to T^3$ such that $f(y(t)) = x(t)$. Obviously we have $g(N^2) \leq g(M^2)$ where M^2 is the Fermi surface.

In the most interesting case of **rank 3** the image of the map $\pi_1(M^2) \to \pi_1(T^3) = Z^3$ has a rank equal to 3. A family of 1-forms with constant coefficients $\omega_0 + \epsilon\omega_1$ we call a *perturbation* of the form ω_0 for all small enough values of the parameter ϵ.

The following results were obtained in fact by Zorich in 1984 and by Dynnikov in 1993:

THEOREM 1. *For any Fermi surface $M^2 \subset T^3$ of rank equal to 3 and any small perturbation of the rational 1-form with constant coefficients with Morse type singularities only on the surface M^2, the genus of the open trajectory is less than or equal to 1 (see [Z]).*

THEOREM 2. *Let some generic periodic function–dispersion relation E on the space of Quasimomenta be given, which determines nondegenerate Fermi-surfaces for $E_0 \leq E \leq E_1$. Let also the constant magnetic field be given with Morse singularities only on the Fermi surface $E = E_0$, such that there exists an open trajectory in it with genus equal to $g \geq 2$ (and this number is maximal for all trajectories on this Fermi surface).*

Any Fermi surface $E = E_1$ close enough to the original surface $E = E_0$ may have open trajectories with genus no more than $g - 1$ (see [D]).

(Let me to point out that there are no such theorems in the papers by Zorich and Dynnikov. They formulated as a main theorem only the Corollary below. The author extracted these theorems from their proofs.)

The proof of the Theorem 1 (Zorich) is purely topological (see below).

The proof of Theorem 2 (Dynnikov) also uses some metric arguments. It is much more complicated then the proof of Theorem 1. Let me present here the idea of the proof of Theorem 1:

We may think, without any loss of generality, that $\omega_0 = dp_3$ because this form is rational. Let also dp_3 be a Morse form on the Fermi surface $E = E_0$. All the leaves of the foliation $dp_3 = 0$ of the Fermi surface are compact. Nonsingular leaves $p_3 = \text{const}$ are equal to the disjoint union of the circles in the tori $[(E = E_0) \cap T^2] \subset T^3$. We take a minimal number of the values $p_3 = c_i$, $i = 1, 2, \ldots$ such that all these leaves are nonsingular and divide the Fermi surface into cylinder like pieces or elementary "pants" W_k, whose 2 boundary leaves ∂W_k contain exactly 3 circles $\partial W_k = S_1 \cup S_2 \cup S_3$ (there is only 1 critical point inside). So we conclude that 2 of them (say S_1, S_2) belong to the same torus T^2, $p_3 = \text{const}$. They are nonselfintersecting and pairwise nonintersecting closed curves in this torus.

Therefore, only 2 cases are possible: the domain between them in the torus T^2 is equal to the cylinder or one of them bounds a disk in T^2. In both cases, we conclude that at least one of the curves S_1, S_2, S_3 is homotopic to zero in T^2 (in the first case it will be the third curve S_3, whose homology class in the torus is equal to the difference $S_1 - S_2$).

Now we cut the Fermi surface along all these curves (one for each of the elementary pants) homotopic to zero in the torus. It is easy to prove that all connected pieces of the Fermi surface will have a genus equal to 1 after cutting.

Adding any closed small perturbation to the form $\omega_0 = dp_3$, we observe that for any compact leaf homotopic to zero in T^3 there exists a compact leaf of the perturbed foliation homotopic to zero and close to the original one. Therefore, cutting a Fermi surface along this perturbed leaf, we get its decomposition on the genus 1 pieces bounded by the circles homotopic to zero. Any noncompact leaf does not cross these closed curves (because they are also leaves) and belongs to the required piece with genus equal to 1). Theorem 1 is proved.

As already mentioned, the proof of Theorem 2 is much more complicated. Dynnikov and Tsarev constructed examples of open orbits with any genus (to appear). The author does not know whether these examples are completely nongeneric or not (i.e. do they appear in any generic finite dimensional family of Fermi surfaces for a fixed magnetic field or not). For a fixed Fermi surface their dependence on a magnetic field might be point set theoretically complicated. It would be natural to study this numerically in order to formulate good conjectures.

As a *corollary* we conclude that any generic open trajectory belongs to some strip of finite width in the universal covering space of quasimomenta R^3. The proof of this is the following:

For any map $f : T^2 \to T^3$, of the 2-torus in the 3-torus monomorphic on the fundamental group $\pi_1(T^2)$, the covering map $R^2 \to R^3$ belongs completely to the domain containing points whose distance from a fixed 2-plane is bounded. This plane in R^3 is determined by the image of the fundamental group of the 2-torus in the 3-lattice. Therefore this plane is based on the integral sublattice

$$L_f = f_*(\pi_1(T^2)) = Z^2 \subset Z^3 .$$

The section of this domain by the plane orthogonal to the magnetic field (i.e. by the plane $d\omega = 0$) is a strip in the plane with finite width. This proves our corollary.

It is not difficult to prove that this trajectory passes this strip in one (oriented) direction.

The same is true for the both asymptotics of time $t \to \pm\infty$.

The basis in the 2-sublattice L_f is well-defined modulo $SL_2(Z)$-transformations. It means in fact that a basic element in the second exterior power is fixed

$$l_f \in \Lambda^2(Z^2) \subset \Lambda^2(Z^3) = H_2(T^3) , \quad l_f = f_*(\mu) .$$

The element μ here is an oriented fundamental homology class of the 2-torus. A full description of the element l_f needs 3 integers (its coordinates in $H_2(T^3)$ in the original basis of the second exterior power of 3-lattice). These 3 integers are relatively prime (if nonzero), because it is easy to prove that we have in fact an imbedding of the 2-torus in the 3-torus. Any connected oriented codimension 1 submanifold has a homology class which is trivial or indivisible in the group $H_2(T^3, Z) = Z^3$. If the image of the fundamental group is Z^3, we have at least one such torus. Their number is always no more then the genus of Fermi surface.

Two different 2-tori in T^3, constructed by Zorich and Dynnikov, do not intersect each other (it is not difficult to extract this from their constructions). We conclude from this that they determine the same element in the group $H_2(T^3)$, because otherwise their intersection would be nontrivial homologically. *All open trajectories determine the same indivisible element in the group $H_2(T^3) = Z^3$ depending on the Fermi surface and generic magnetic field only. This is the most interesting topological invariant in this picture.*

It looks like the plane $L_f \subset Z^3$ should be "more or less" parallel to the original plane given by the magnetic field or close to it. However, for irrational magnetic fields it cannot be the same plane. We are not able to estimate how far it can be from this original plane.

An additional topological invariant of the open trajectory is its so-called "rotation number" on the 2-torus, but this invariant depends on the magnetic field as a continuous (locally nonconstant) function. The topology of the torical splitting of the Fermi surface and its integer-valued invariants are well-defined in the generic case and ae locally constant. Therefore we may ask the following fundamental question:

Which kind of observable quantities in the conductivity of normal metals may correspond to the integer-valued topological invariants of generic open orbits in the external homogeneous, not very strong, magnetic field?

The author would like to point out that each trajectory determines some "*adiabatic wave function*"

$$\phi(x,t) = \psi(x, p(t))$$

associated with the 2-torus found by the theorems above. The observable quantities should be extracted from this wave function. However, a very deep understanding of solid state physics of concrete media is necessary to discover which kind of quantity might directly correspond to the second homology class of 2-torus in T^3 – the "topological closure" of the trajectory (or of the adiabatic wave function).

Another problem is to analyze the quasiperiodic structure of the leaves in the 2-plane orthogonal to the magnetic field, which definitely exists as a corollary from the very general results on such foliations. What kind of "quasicrystals" do we get in this plane? Is it possible to extract something from it for the physical quantities?

Remark (added in December 1994): After some analysis and discussions of this problem in the author's seminar in Moscow, his student Malcev (a physicist) pointed out that in the presence of an open orbit with average direction η in the space of quasimomenta, we have some special properties for conductivity in a relatively strong magnetic field:

In the plane, orthogonal to this field, conductivity has a nonzero limit for $B \to \infty$ only in the direction orthogonal to the vector η.

This fact may be extracted from the material of the textbook by L. Landau and E. Lifshitz, vol 10, where calculations were performed using the semiclassical (adiabatic) picture, described above.

Discussing this remark on the basis of Theorems 1 and 2, we take into account the following properties of the direction η:

it generically exists.

What is more important, *it belongs to some integral plane in the lattice $Z^3 = H_1(T^3, Z) \subset R^3$ in the space of quasimomenta. The last plane is locally rigid under variation of the direction of the magnetic field.*

Therefore, we come to the following

CONCLUSION. *All (small enough) variations of the magnetic field lead to the vectors η in the same integral 2-plane in the 3-lattice. This property may be tested directly by experiment under appropriate conditions, depending on the actual metal with complicated Fermi surface.*

For example, some "noble metals" (such as gold) have a very complicated Fermi surface. However, some serious work should be done to understanding to which real physical parameters (temperature, magnetic field and so on) this effect may correspond.

I thank my friend and colleague L. Falkovski for consultations in solid state physics.

References

[D] I. DYNNIKOV, Proof of Novikov's Conjecture on the semiclassical motion of electron, Math. Zametki 53:5 (1993), 57-68.

[GN] P.G. GRINEVICH, S.P. NOVIKOV, String equation – 2. Physical solution, Algebra and Analysis (1994), (dedicated to the 60th birthday of L.D. Faddeev).

[A] A.A. ABRIKOSOV, Introduction to the Theory of Metals, Moscow, Nauka (1987).

[N1] S.P. NOVIKOV, The Hamiltonian formalism and a multivalued analog of Morse theory, Uspekhi Math. Nauk (Russian Math Surveys) 37:5 (227) (1982), 3–49.

[N2] S.P. NOVIKOV, Critical points and level surfaces of multivalued functions. Proceedings of the Steklov Institute of Mathematics, 1986, AMS, iss 1, 223-232.

[N3] S.P. NOVIKOV, Quasiperiodic structures in topology. In the proceedings of the Conference "Topological methods in Modern Mathematics", Stonybrook University, June 1991 (dedicated to the 60th birthday of John Milnor). Stonybrook, 1993.

[N4] S.P. NOVIKOV, Quantization of the finite gap potentials and string equation, Functional Analysis and its Applications 24:4 (1990), 196-206.

[N5] S.P.NOVIKOV, Two dimensional Schroedinger Operator in the periodic fields. Current Problems in Mathematics, VINITI, 1983, v 23, 3-22. (Translated by AMS in the January 1985).

[N6] S.P. NOVIKOV, Bloch functions in a magnetic field and vector bundles. Typical dispersion relations and their quantum numbers, Doklady AN SSSR (Russian Math Dokl) 257:3 (1981), 538-543.

[Z] A.V. ZORICH, Novikov's problem on the semiclassical motion of electron in the homogeneous magnetic field. Uspekhi Math Nauk (RMS) 39:5 (1984), 235-236.

S.P. Novikov
Department of Mathematics
University of Maryland
College Park, MD 20742
USA

Submitted: October 1994

Geometric And Functional Analysis
Vol. 5, No. 2 (1995)
1016-443X/95/0200445-19$1.50+0.20/0

WIDTHS OF NONNEGATIVELY CURVED SPACES

G. PERELMAN

0. Introduction and Outline of Proof

0.1. The purpose of this paper is to present a proof of the following conjecture, formulated by M. Gromov ([G2]).

The volume and the widths of a closed Riemannian manifold M^n with nonnegative sectional curvatures satisfy

$$c^{-1}\operatorname{Vol}(M^n) \le \prod_{k=0}^{n-1} w_k(M^n) \le c\operatorname{Vol}(M^n)\ , \tag{$*$}$$

where $c = c(n)$ is a constant.

(Recall that the k-dimensional (Uryson) width $w_k(X)$ of a metric space X is defined as the exact lower bound of those $\delta > 0$ for which there exists a k-dimensional space P and a continuous map $f : X \to P$ all of whose inverse images have diameters at most δ.)

The relation $(*)$ is well known for convex sets in euclidean spaces. It was verified by Gromov for almost flat manifolds. He also suggested that to prove $(*)$ for nonnegatively curved manifolds it may be helpful to use a collapsing technique. To illustrate this approach we indicate an argument that proves something weaker than $(*)$.

0.2. Consider a sequence M_i^n of closed nonnegatively curved Riemannian manifolds of diameter 1. We will show that either both $\operatorname{Vol}(M_i^n)$ and $\prod_{k=0}^{n-1} w_k(M_i^n)$ stay bounded away from zero or they both tend to zero. Indeed, passing to a subsequence, we may assume that M_i^n converge in Gromov-Hausdorff sense to some metric space M. The space M is not necessarily a Riemannian manifold, but it is a nonnegatively curved Alexandrov space. Two possibilities may occur: either M has dimension n or M has lower dimension m, m is an integer. In the first case, it is known that M contains a bilipschitz copy of a euclidean ball, B^n, and moreover, for sufficiently large i, each M_i^n contains such a copy, with lipschitz constants bounded away from zero and infinity; it follows that volumes and widths of M_i^n are bounded away from zero. In the second case, it is easy to show that $\operatorname{Vol}(M_i^n) \to 0$, because for each fixed $\delta > 0$ the number of balls of radius δ needed to cover M_i^n is, for sufficiently large i, approximately the same as for the limit space M, and therefore can be estimated from above by $c\delta^{-m}$. On the other hand, the widths $w_k(M_i^n)$ with $k \geq m$ also tend to zero, because it is known that M is locally contractible, and therefore M_i^n can be continuously mapped into M with small inverse images.

The actual proof of $(*)$ is more involved, and its logical structure is different. It consists of two steps. First we introduce a different collection of measurements, called packing widths, and prove $(*)$ for those. Then we establish essential equivalence between widths and packing widths. The following subsections contain an outline of the proof.

CONVENTION. In this paper we denote by c various positive constants, which may depend only on dimension; several different constants may be denoted by c in the same formula. The equalities of the form $A = cB$ mean the same as $cA \leq B \leq cA$, that is the ratio A/B is bounded away from zero and infinity.

0.3. We define the ν-packing number $N(X, \nu)$ of a metric space X as the maximal number of disjoint open metric balls of radius ν contained in X. Of course, $N(X, \nu)$ is decreasing in ν, and for a general X one cannot say much more. However, if X is a rectangular solid $[0, \ell_0] \times \ldots \times [0, \ell_{n-1}]$ in $\mathbf{R}^n$, with markedly different $\ell_0 > \ell_1 > \ldots > \ell_{n-1}$, then $N(X, \nu)$ behaves like a polynomial in ν^{-1} of degree k when ν is varying between ℓ_k and ℓ_{k-1}. It turns out that the packing function of a nonnegatively curved manifold behaves in a similar fashion; the borderlines can be called the packing widths.

Notice that, unlike the widths, which show the size of the thickest part of the space, and are obviously monotone w.r.t. inclusion, the packing widths are average characteristics, and need not be monotone. Therefore, one should expect that the relation $(*)$ is easier to prove for packing widths, and that equivalence between widths and packing widths is not trivial.

0.4. As we have seen in our model argument 0.2, the constant in $(*)$ may deteriorate for a sequence M_i^n only if the limit space M has dimension $m < n$. In this case M contains a bilipschitz copy of euclidean ball B^m, and one may expect that, for i large enough, M_i^n contains a bilipschitz copy of the product of B^m and an $(n-m)$-dimensional space of small diameter. However, such a result is not known. Still, it can be shown that M_i^n contains a subset, which fibers over B^m, with fibres of essentially constant small diameter, and moreover, the packing function of this fibered subset behaves as though it was a product.

A positive conclusion, which can be drawn from this argument "by contradiction", is that any nonnegatively curved manifold M^n contains a subset, of size comparable with the size of M, which looks like a product of euclidean ball of some dimension $m \leq n$ and an $(n-m)$-dimensional "fibre" of small diameter. The maximal such m can be called the virtual dimension of M^n.

0.5. Now we can indicate how $(*)$ can be proved for packing widths pw_k. We use reverse induction on the virtual dimension. The base ($v\dim(M^n) = n$) is clear, because in this case each packing width is comparable with the diameter, while the volume is comparable with its n-th power.

Assume that $v\dim(M^n) = m < n$. Still, both the volume and the product of packing widths of M^n are comparable to the corresponding values for its fibered subset W. Consider a ball $B \subset W$ whose radius is equal to the diameter d of a typical fibre of W. Since, with regard to the packing numbers, the subset W is similar to a product, it is not hard to show that

$$\mathrm{Vol}(W) \approx (\mathrm{diam}\, W/d)^m \,\mathrm{Vol}(B) ,$$

$$pw_k(W) \approx pw_k(B) \qquad \text{if } k \geq m ,$$

$$pw_k(W) \approx (\mathrm{diam}\, W/d) pw_k(B) \qquad \text{if } k < m .$$

Therefore, the verification of $(*)$ for M can be reduced to that for B. It remains to observe that $v\dim(B) > m$, and use the assumption of induction.

0.6. The proof of equivalence between widths and packing widths can be split into several implications.

(a) If $pw_{k-1}(B) \geq w$ for some ball $B \subset M^n$, then there exists a continuous map of k-dimensional cube I^k into B of size $\geq cw$. (The size of the map is the minimal distance between the images of opposite faces.)

(b) If $pw_k(B) \leq w$ for each ball $B \subset M^n$, then $w_k(M^n) \leq cw$.

(c) If $pw_k(M^n) \leq w$ then $pw_k(B) \leq cw$ for each ball $B \subset M^n$.

Indeed, the inequality $w_k(M^n) \geq cpw_k(M^n)$ follows from (a) and Lebesgue lemma, while the opposite inequality follows from (b) and (c).

The assertion (a) can be easily proved by contradiction, as in 0.2. To prove (b) map M^n to the nerve of an appropriate covering and project to its k-skeleton. The proof of (c), outlined in the next subsection, uses (a),(b) and the following implication.

(d) Let W be the fibered subset of M, as in 0.4, 0.5. If there is a continuous map of I^k into M^n of size $\geq w$ then there is a similar map into W of size $\geq cw$.

A natural way to deform a map keeping its size bounded away from zero is to move each point along a shortest geodesic connecting it with some fixed point in W. Unfortunately, the shortest geodesics need not be unique, so the actual proof is more involved.

0.7. We conclude our outline by indicating how (a),(b),(d) can be used to prove (c). Again we use reverse induction on $v\dim(M^n)$. The base ($v\dim(M^n) = n$) is clear, because in this case we have $pw_k(M^n) \geq c\,\mathrm{diam}(M^n) \geq c\, pw_k(B)$.

Assume that $v\dim(M^n) = m < n$, and let $pw_k(B) = w' \gg w$. (Notice that we may assume $k \geq m$ since otherwise $pw_k(M^n) \geq c\; pw_k(W) \geq c\,\mathrm{diam}(W) \geq c\,\mathrm{diam}(M^n)$.) According to (a), we can find a continuous map of I^{k+1} into B of size $\geq cw'$; applying (d) find a similar map into W; it follows that $w_k(W) \geq cw'$, therefore according to (b), we can find a ball $B' \subset W$ with $pw_k(B') \geq cw'$. Let B'' be a ball, concentric with B', with radius equal to the diameter of a typical fibre of W. Then in case $B' \supset B''$, B' is itself "almost" a fibered subset, and arguing as in 0.5, we get $pw_k(B') \leq c\; pw_k(B'')$ $(k \geq m)$. On the other hand, if $B' \subset B''$ then we obtain the same conclusion from the inductional assumption applied to B''. Taking into account that (as in 0.5), $pw_k(B'') = c\; pw_k(M)$ $(k \geq m)$, we conclude that $w' \leq c\; pw_k(M) = cw$. □

0.8 Structure of the paper, prerequisites and notation. In §1 we present a formal proof of the relation (∗) modulo three assertions of technical nature, marked by Roman figures I, II, III. Then we prove each of them separately in §§2–4.

We are working in the category of Alexandrov spaces. The reader unfamiliar with Alexandrov spaces may almost always assume that the space in question is Riemannian. The only exception is when we are considering limit spaces; the only information we need about them is that any such space has integer dimension and contains an almost isometric copy of a (small) cross of euclidean coordinate axes; this very elementary result is contained in §§5,6 of [BGP], which in fact form the basis of most of our arguments. Familiarity with [BGP, §10] and [P, §2] is helpful in §2 and §3 respectively. The proof of Assertion III for Alexandrov spaces uses the technique of gradient curves, developed in [PPe]; we included an alternative proof for Riemannian manifolds.

NOTATION.

$B_p(R)$ means an open metric ball of radius R centered at p.

$I^k_v(R)$ means a euclidean cube $\{u \in \mathbf{R}^k : |u_i - v_i| \leq R,\ 1 \leq i \leq k\}$.

$\overline{pq}$ means a shortest geodesic between p and q.

$\tilde{\angle} PrQ$ is the comparison angle, that is, the angle at $\tilde{r}$ in the triangle $\tilde{p}\tilde{q}\tilde{r}$ on the euclidean plane, such that $|\tilde{p}\tilde{q}| = |PQ|$, $|\tilde{p}\tilde{r}| = |Pr|$, $|\tilde{q}\tilde{r}| = |Qr|$; here P and Q are compact sets not containing the point r.

Σ_p means the space of directions at p (the unit sphere of the tangent space); if A is a compact set not containing p, then $A' \subset \Sigma_p$ is the (compact) set of directions of all shortest geodesics $\overline{pq}$ such that $q \in A$ and $|pq| = |pA|$.

Vol means the Hausdorff measure of the appropriate dimension.

1. The Formal Proof

1.1 The class of objects. We define $\mathcal{M}_n$ as the collection of all triplets $(o, M, \tilde{M})$, where $\tilde{M}$ is a not necessarily complete nonnegatively curved Alexandrov space of dimension $\leq n$, $o \in M \subset \tilde{M}$, $M = B_o(1)$, diam $M \geq 1$, $\tilde{M} = B_o(10)$, and each ball $B_o(R)$ with $R < 10$ is relatively compact in $\tilde{M}$. It is not hard to see that $\mathcal{M}_n$ is compact in Gromov-Hausdorff topology. (Indeed, it can be shown, by an argument similar to [BGP, §3] that the Toponogov angle comparison holds for triangles poq provided that $\angle poq$ is small in comparison with the $10 - |op|$ and $10 - |oq|$; this allows one to estimate the packing numbers $N(\tilde{M}, \nu)$ and prove compactness as in [BGP, §8].

All of our constructions will take place in the spaces M, and we will write $M \in \mathcal{M}_n$ ignoring $\tilde{M}$ and o. Notice that if $M \in \mathcal{M}_n$ and B is a ball in M then there is a unique way to rescale the metric of B to make it an element of $\mathcal{M}_n$.

Throughout the paper the value of n will remain fixed.

1.2 Packing numbers. Recall that the ν-packing number $N(X, \nu)$ of a metric space X was defined as the maximal number of disjoint open balls of radius ν contained in X; in other words, $N(X, \nu)$ is the maximal number of points in a 2ν-discrete net in X. In an m-dimensional nonnegatively curved space, the packing numbers satisfy

$$N\big(B_p(R), \nu R\big) \leq c\nu^{-m} \quad \text{and}$$

$$N\big(B_p(\nu R), \nu\nu'\big) \geq N\big(B_p(R), \nu'\big) \quad \text{when } 0 < \nu \leq 1\,.$$

Therefore, for any m-dimensional $M \in \mathcal{M}_n$, we have

$$N(X, \nu) \leq N(X, \nu') \leq cN(X, \nu)(\nu'/\nu)^{-m} \text{ if } \nu' \leq \nu, \text{ for any } X \subset M, \text{ and}$$

$$N(M, \nu\nu') \geq cN(M, \nu)N(M, \nu')\,, \text{ if } \nu, \nu' \leq 1\,, \text{ in particular}$$

$$N(M, \nu') \geq \big(cN(M, \nu)\big)^{[-\log(\nu'/\nu)]}\,.$$

It is also true that $\lim_{\nu \to 0} N(M, \nu)\nu^{-m} = c_0 \operatorname{Vol}(M)$, where c_0 is the same for all m-dimensional M. (This is well known in the Riemannian case; for general Alexandrov spaces the additional ingredient in the proof is [BGP, 10.9].

1.3 Packing widths. Fix a small number $\lambda = \lambda(n) > 0$. Define the k-th packing width $pw_k(X)$ of a metric space X as the exact lower bound of those $\nu > 0$ which satisfy

$$N(X,\nu)/N(X,\lambda\nu) \geq \lambda^{k+\frac{1}{2}} .$$

The "correctness" of this definition for elements of $\mathcal{M}_n$ is ensured by the following assertion, proved in §2.

ASSERTION I. *For sufficiently small positive λ (i.e. $0 < \lambda \leq c$) there exists $c(\lambda) > 0$ such that*

$$\text{If } M \in \mathcal{M}_n,\ \nu > 0,\ 0 < \nu' < c(\lambda)\nu \ \text{ and } \ N(M,\nu)/N(M,\lambda\nu) \leq \lambda^{k+\frac{1}{4}}$$
$$\text{then } \ N(M,\nu')/N(M,\lambda\nu') \leq \lambda^{k+\frac{3}{4}} .$$

This means, in particular, that

$$\begin{aligned}&\text{if } \ N(M,\nu)/N(M,\lambda\nu) \geq (\leq) c\lambda^{k+\frac{1}{2}}\\ &\text{then } \ \nu \geq (\leq) c(\lambda) pw_k(M) \ \text{ and}\\ &\text{if } \ N(M_1,\nu) = cN(M_2,\nu) \ \text{ for all } \nu\\ &\text{then } \ pw_k(M_1) = c(\lambda) pw_k(M_2) .\end{aligned}$$

Of course, the last statements depend on the fact that λ is small in comparison with the constants c; we will always assume this is the case for constants denoted by c, and use $c(\lambda)$ otherwise. Formally speaking, the actual value of the parameter λ is to be determined in the end of the proof, when the values of all c are known.

1.4 Fibered subsets. Fix a small positive number $\delta = \delta(n) < c$. Let $M \in \mathcal{M}_n$, $p, a_i, b_i \in M$, $1 \leq i \leq k$. Suppose that p, a_i, b_i satisfy the following set of inequalities

$$\begin{aligned}&\tilde\angle a_i p b_i > \pi - \delta \ , \ \ \tilde\angle a_i p b_j > \pi/2 - \delta \ , \ \ \tilde\angle a_i p a_j > \pi/2 - \delta \ ,\\ &\tilde\angle b_i p b_j > \pi/2 - \delta \ , \ \ |pa_i| > \delta^{-1}R \ , \ \ |pb_i| > \delta^{-1}R \ , \ \ (1 \leq i \neq j \leq k) \ .\end{aligned} \tag{1}$$

Then $\{a_i, b_i\}$ is called a (k,δ,R)-strainer at p. A strainer determines an associated map $f : M \to \mathbf{R}^k$, $f(x) = \{|xa_1|, \ldots, |xa_k|\}$. This map is obviously c-lipschitz; it is also c-open in $B_p(\delta^{-\frac{1}{2}}R)$, that is if $f(q_1) = v_1$ and v_2 is close to v_1 then there exists q_2 such that $f(q_2) = v_2$ and $|q_1q_2| \leq c|v_1v_2|$; if $k = \dim M$ then f is a bilipschitz homeomorphism (see [BGP, 5.4]).

The inverse image $W = f^{-1}(I^k_{f(p)}(R))$ is called a (k,δ,R)-fibered subset of M (or simply a rank k fibered subset), if $\operatorname{diam}(f^{-1}(v)) \leq \delta R$ for every $v \in I^k_{f(p)}(R)$. It is not hard to show that $f : W \to I^k_{f(p)}(R)$ is indeed a (trivial) fibration, at least if M is Riemannian. Indeed, the inequalities (1) continue to hold, with slightly bigger δ, for any point $q \in W$ in place of p.

Therefore the directions of $\overline{qa_i}$, $\overline{qb_i}$ are contained in a $c\delta$-neighborhood of some k-dimensional plane in the tangent space at q, and it is elementary to check that f is a submersion near q. If M is a general Alexandrov space, then one can modify the strainer to make all the angles of the type $\tilde{\angle} a_i q a_j$ strictly bigger than $\pi/2$, and then use [P, 1.4(B)].

We will prove the following in §3.

ASSERTION II. *If $W = f^{-1}(I^k_{f(p)}(R))$ is a rank k fibered subset of $M \in \mathcal{M}_n$ then*

$$\max\left(\operatorname{diam} f^{-1}(v)\right) \le c \min\left(\operatorname{diam} f^{-1}(v)\right)$$
$$\max\left(N(f^{-1}(v),\nu)\right) \le c \min\left(N(f^{-1}(v),\nu)\right) \quad \textit{for any} \ \ \nu > 0\,,$$

where max *and* min *are taken over all $v \in I^k_{f(p)}(R)$.*

In fact, the first inequality follows from the second one because the fibres $f^{-1}(v)$ are connected (see [BGP, 11.11]).

1.5 Virtual dimension.

1.5.1 LEMMA. *For any small $\delta > 0$ there exists $R(\delta) > 0$ such that*

(a) *Any $M \in \mathcal{M}_n$ contains a (k, δ, R)-fibered subset for some $1 \le k \le n$, $R \ge R(\delta)$.*

(b) *Moreover, if M admits a $(m, 10\delta, \delta^2)$-strainer at some point, then* (a) *holds with some $k \ge m$.*

Proof: (a) We argue by contradiction. Suppose M_j is a sequence of elements of $\mathcal{M}_n$ which violates our assertion for a given $\delta > 0$; we may assume that M_j converge in Gromov-Hausdorff topology to some $M \in \mathcal{M}_n$, $\dim M = k \le n$. According to [BGP, 6.7], one can find a $(k, \delta/2, 2R)$-strainer $\{\overline{a}_i, \overline{b}_i\}$ at some point $\overline{p} \in M$, for some $R > 0$. It is clear that, for j large enough, the space M_j will admit a (k, δ, R)-strainer $\{a_i, b_i\}$ at some point p, such that a_i, b_i, p are close to $\overline{a}_i, \overline{b}_i, \overline{p}$, respectively, and therefore will contain a (k, δ, R)-fibered subset because each fibre is close to some point of $\overline{M}$. This is a contradiction.

(b) The $(m, 10\delta, \delta^2)$-strainers in M_j converge to a strainer in M; the existence of such a strainer in M implies $k = \dim M \ge m$. □

1.5.2. At this point we fix (until the end of §1) the value of δ consistent with all the statements in 1.4, 1.5.1. From now on, a rank k fibered subset will mean $(k, \delta, R(\delta))$-fibered subset.

Define the virtual dimension $v\dim(M)$ of a space $M \in \mathcal{M}_n$ as the largest k such that M contains a rank k fibered subset. The virtual dimension of a ball in M is defined as virtual dimension of its rescaling that belongs to $\mathcal{M}_n$.

1.5.3 LEMMA. *Let $M \in \mathcal{M}_n$, $v\dim(M) = k < \dim M$. Let $W \subset M$ be its rank k fibered subset, with associated map f, $x \in W$, $F = f^{-1}(f(x))$, $d = \operatorname{diam} F$. Then $v\dim(B_x(2d)) > k$.*

Proof: Let a_{k+1}, b_{k+1} be the endpoints of a diameter of F, and let q be the midpoint of $\overline{a_{k+1}b_{k+1}}$. Clearly a_{k+1}, b_{k+1}, $q \in B_x(2d)$. It is easy to show that $\{a_i, b_i\}$, $1 \le i \le k+1$ form a $(k+1, 10\delta, \delta d/100)$-strainer at q, and this property is obviously retained if a_i, b_i are replaced by the corresponding points on $\overline{qa_i}, \overline{qb_i}$ lying in $B_x(2d)$. Therefore our assertion follows from 1.5.1(b). □

1.6 Proof of (*) for packing widths. In fact, we are going to prove that

$$\prod_{k=0}^{n-1} pw_k(M) = c(\lambda)\operatorname{Vol}(M) \quad \text{for all} \quad M \in \mathcal{M}_n\,. \tag{**}$$

We use reverse induction on virtual dimension. If $v\dim(M) = n$ then its rank n fibered subset is a c-bilipschitz copy of the cube $I^n(1)$. Therefore $N(M,\nu) = cN(W,\nu) = c\nu^{-n}$ for all $0 < \nu \le 1$, $\operatorname{Vol}(M) = c$, and according to 1.3, $pw_k(M) = c(\lambda)$, $0 \le k \le n-1$. Thus (**) holds in this case.

Now assume that $v\dim(M) = m < n$, and let W be its rank m fibered subset, with associated map f, $x \in W$, $F = f^{-1}(f(x))$, $d = \operatorname{diam} F$. Of course, we still have $N(M,\nu) = cN(W,\nu)$ for all $\nu > 0$. Moreover, it is easy to see from Assertion II and c-openness of f, that for any ball $B_x(R) \subset W$ with $R \ge d$, the packing function satisfy

(1) $N(B_x(R),\nu) = \max\{1, c(R/\nu)^m\, N(F,\nu)\}$.

In particular

(2) $N(M,\nu) = cN(W,\nu) = \max\{1, c\nu^{-m}N(F,\nu)\}$,

(3) $N(B,\nu) = \max\{1, c(d/\nu)^m N(F,\nu)\}$ for $B = B_x(2d)$.

Therefore

(4) $N(M,\lambda^{-1})/N(M,1) = c$, $N(M,1)/N(M,\lambda) \le c\lambda^m$,

(5) $N(M,\lambda^{-1}d)/N(M,d) \ge c\lambda^m$,

(6) $N(B,\lambda^{-1}d)/N(B,d) = c$, $N(B,d)/N(B,\lambda d) \le c\lambda^m$,

(7) $N(M,\nu)/N(M,\lambda\nu) = cN(B,\nu)/N(B,\lambda\nu)$ for $0 < \nu \le d$.

Using Assertion I and its corollaries in 1.3 we can estimate

(8) $pw_k(M) = c(\lambda)$, $pw_k(B) = c(\lambda)d$ for $0 \le k \le m-1$ (from (4),(6))

(9) $pw_k(M) \le c(\lambda)d$, $pw_k(B) \le c(\lambda)d$ for $k \ge m$ (from (5),(6))

(10) $pw_k(M) = c(\lambda)pw_k B$ for $k \ge m$ (from (7),(9)).

Thus $c(\lambda)d^m \prod_{k=0}^{n-1} pw_k(M) = \prod_{k=0}^{n-1} pw_k(B)$. On the other hand, $\operatorname{Vol}(B)/\operatorname{Vol}(M) = \lim_{\nu\to 0} N(B,\nu)/N(M,\nu) = cd^m$ (from (2),(3)). Therefore, (**) for M follows from (**) for B. It remains to observe that

$v\dim(B) > m$ by 1.5.3, and therefore B satisfies (scale invariant) (**) by the assumption of induction. □

1.7 Widths and packing widths.

1.7.1 LEMMA. *There exists $c(\lambda)>0$ such that if $M \in \mathcal{M}_n$ satisfies $N(M,\lambda) \geq \lambda^{-k-\frac{1}{4}}$ then there exists a continuous map $\phi : I^{k+1} \to M$ of size $s(\phi) \geq c(\lambda)$.*

(Recall that $s(\phi)$ is the minimal distance between the ϕ-images of opposite faces of I^{k+1}.)

The proof is by contradiction and very similar to 1.5.1. The condition on the packing number guarantees that the limit space has dimension $\geq k+1$, and therefore the spaces M_j (with large j) contain rank $\geq k+1$ fibered subsets of uniform size, whence the result. □

1.7.2 COROLLARY. $w_k(M) \geq c(\lambda) pw_k(M)$.

Indeed, let $w = pw_k(M)$. Then $N(M,w)/N(M,\lambda w) = \lambda^{k+\frac{1}{2}}$. Therefore there exists a ball $B(w) \subset M$ such that $N(B(w),\lambda w) \geq c\lambda^{-k-\frac{1}{2}} \geq \lambda^{-k-\frac{1}{4}}$. It remains to scale this ball to the unit size, apply 1.7.1, scale back, and use the Lebesgue lemma.

1.7.3 LEMMA. *Let $w > 0$ and assume that every ball $B(w) \subset M$ satisfies $N(B(w),\lambda w) \leq \lambda^{-k-\frac{3}{4}}$. Then $w_k(M) \leq w$.*

Proof (cf. [G1, p.52]): Consider a covering of M by balls of radius $w/2$ with multiplicity at most c, a subordinate partition of unity, and a corresponding map of M to the nerve of this covering. We want to project the image of M to the k-skeleton of this nerve; the inverse images of the resulting map will obviously have diameters $\leq w$. The required projection is constructed inductively; at each step we project the current image of M in simplices of some dimension $i+1$ to their boundaries. Such a map is lipschitz, with lipschitz constant $c\lambda_i^{-1}$, if in each simplex we project from a point which is at distance $\geq \lambda_i w$ from the current image of M and not too close to the boundary of the simplex. (We assume that the simplices of the nerve have size w.) Thus the lipschitz constant Λ_i of the map from M to the i-skeleton can be estimated as $\Lambda_i \leq c\prod_{j\geq i}\lambda_j^{-1}$. We have to check that it is possible to choose (small) parameters $\lambda_i > 0$, $k \leq i \leq c$, so that for each i in each $i+1$-simplex a point at distance $\geq \lambda_i w$ from the current image of M can be found.

Our assumption on the packing number implies that $N(B(w),\nu w) \leq c\nu^{-k-\frac{7}{8}}$ if $\nu^{10k} \geq \lambda$ (cf. 1.2).

It follows that $N(B(w),\Lambda_{i+1}^{-1}\lambda_i w) \leq c\lambda_i^{-k-\frac{15}{16}}$ if $\lambda_i \leq \Lambda_{i+1}^{-100k}$, $\lambda_i^{100k} \geq \lambda$, and therefore, in this case, the current image of M leaves enough empty space in each $(i+1)$-simplex, $(i \geq k)$. The conditions $\lambda_i^{100k} \geq \lambda$ and $\lambda_i \leq (c\prod_{j>i}\lambda_j)^{100k}$ for our small parameters λ_i, $k \leq i \leq c$, can be easily satisfied if λ is small enough, $\lambda < c$. □

1.7.4 Corollary. *$w_k(M) \leq c(\lambda)\sup pw_k(B)$, where sup is taken over all balls $B \subset M$.*

1.8 Proposition. *$pw_k(B) \leq c(\lambda)pw_k(M)$ for any $M \in \mathcal{M}_n$ and any ball $B \subset M$.*

The proof is based on the following statement, proved in §4.

Assertion III. *For any $0 < R < 1$ there exists $\gamma = \gamma(R) > 0$ with the following property. Let $M \in \mathcal{M}_n$, $p, q, r \in M$, $q \in \overline{pr}$. Then any continuous map $\phi : I^k \to B_q(\gamma|qr|)$ can be deformed to a map $\tilde{\phi} : I^k \to B_p(R)$ such that $s(\tilde{\phi}) \geq cR^2 s(\phi)$.*

We prove our proposition using reverse induction on $v\dim(M)$. If $v\dim(M) = n$, then M contains a c-bilipschitz copy of $I^n(1)$, therefore $pw_k(M) \geq c(\lambda) \geq c(\lambda)pw_k(B)$ for any ball $B \subset M$.

Now assume $v\dim M = m < n$, let W be its rank m fibered subset, with associated map f. Notice that we may assume $k \geq m$ since otherwise $pw_k(M) = c(\lambda)$, see 1.6(8).

Take any ball $B \subset M$, let $w = pw_k(B)$. Obviously we can find a ball $B_1 = B_r(\lambda w) \subset B$, such that $N(B_1, \lambda^2 w) \geq c\lambda^{-k-\frac{1}{2}}$. We want to show that a ball with similar properties can be found in W.

Take a point p such that $B_p(R(\delta)) \subset W$, and let q be a point on $\overline{pr}$ such that $|qr| = \lambda w$. Consider a ball $B_2 = B_q(\gamma(\frac{1}{2}R(\delta))|qr|)$. Projecting B_1 into B_2 from q we easily see that $N(B_2, \lambda^2 w) \geq c\lambda^{-k-\frac{1}{2}}$. Therefore, according to 1.7.1 we can find a continuous map $\phi : I^{k+1} \to B_2$ of size $s(\phi) \geq c(\lambda)w$. Applying Assertion III we can deform it to a map of size $\geq c(\lambda)w$ into $B_p(\frac{1}{2}R(\delta))$. Hence, according to Lebesgue lemma and 1.7.3, there is a ball $B_3 = B_x(w') \subset W$ such that $w' \geq c(\lambda)w$ and $N(B_3, \lambda w') \geq \lambda^{-k-\frac{3}{4}}$. In particular, $pw_k(B_3) \geq c(\lambda)w$.

Let $d = \operatorname{diam} f^{-1}(f(x))$, $B_4 = B_x(2d)$. We claim that $pw_k(B_3) \leq c(\lambda)pw_k(B_4)$. Indeed, if $w' \geq d$, then we can apply 1.6(1) to B_3 and, arguing as in 1.6, obtain $pw_k(B_3) = c(\lambda)pw_k(B_4)$ for $k \geq m$. Otherwise, $B_3 \subset B_4$, and we justify our claim using the inductional assumption for appropriately rescaled B_4.

Finally we get $w \leq c(\lambda)pw_k(B_3) \leq c(\lambda)pw_k(B_4) \leq c(\lambda)pw_k(M)$, where the last inequality is a part of 1.6(10), since $k \geq m$. □

2. Proof of Assertion I

The goal of this section is to prove the following.

ASSERTION I. *For sufficiently small* $\lambda > 0$ *there exists a positive number* $c(\lambda)$ *such that if* $M \in \mathcal{M}_n$, $\nu > 0$, $0 < \nu' < c(\lambda)\nu$ *and* $N(M,\nu)/N(M,\lambda\nu) < \lambda^{k+\frac{1}{4}}$, *then* $N(M,\nu')/N(M,\lambda\nu') < \lambda^{k+\frac{3}{4}}$.

At first we establish (a slightly stronger form of) a particular case of our assertion, namely the case $\nu = 1$, and then deduce the general case from this particular one.

PARTICULAR CASE. *For sufficiently small* $\lambda > 0$ *there exists a positive number* $c(\lambda)$, *such that if* $M \in \mathcal{M}_n$, $0 < \nu < c(\lambda)$, $N(M,\lambda) > \lambda^{-k-\frac{1}{16}}$, *then* $N(M,\nu)/N(M,\lambda\nu) < \lambda^{k+\frac{15}{16}}$.

Proof of the Particular Case: Argue by contradiction. Assume that there exist sequences $M_j \in \mathcal{M}_n$, $\nu_j \to 0$, such that $N(M_j,\lambda) > \lambda^{-k-\frac{1}{16}}$ and $N(M_j,\nu_j)/N(M_j,\lambda\nu_j) \geq \lambda^{k+\frac{15}{16}}$. We may assume that M_j converges in the Gromov-Hausdorff sense to some $M \in \mathcal{M}_n$; clearly $\dim M \geq k+1$ if λ is small enough. This contradicts the following.

PROPOSITION. *Given* $M \in \mathcal{M}_n$, $\dim M \geq m$, $\mu > 0$, $\delta > 0$ *there exists* $R > 0$ *and* $\overline{\nu} > 0$ *such that if* $\overline{M} \in \mathcal{M}_n$, $d_H(\overline{M},M) < \overline{\nu}$ *then* $N(\overline{M}\backslash\overline{M}(m,\delta,R),\nu) \leq \mu N(\overline{M},\nu)$ *for all* $\nu < \overline{\nu}$.

(Here d_H is the Gromov-Hausdorff distance, and $\overline{M}(m,\delta,R)$ denotes the set of all points $p \in \overline{M}$ which have (m,δ,R)-strainers.

Indeed, apply the proposition with $\overline{M} = M_j$, for j large enough, $\mu = \frac{1}{2}$, $m = k+1$, and δ sufficiently small, ($\delta < c$). Then for every $p \in M_j(k+1,\delta,R)$ we easily obtain $N(B_p(\nu_j),\lambda\nu_j) \geq c\lambda^{-k-1}$, whence $N(M_j,\nu_j)/N(M_j,\lambda\nu_j) \leq c\lambda^{k+1}$.

Proof of the Proposition: We use induction on m (cf. [BGP, 10.9]). The base $m = 0$ is trivial. The step of induction can be reduced (using covering arguments, taking advantage of μ, δ, and rescaling) to the second statement of the following.

LEMMA 1. *Given* $M \in \mathcal{M}_n$, $\dim M \geq m+1$, $\delta > 0$ *small enough* ($\delta < c(\operatorname{Vol} M)$), $\mu > 0$, *there exist* $R > 0$, $\overline{\nu} > 0$ *such that if* $\overline{M} \in \mathcal{M}_n$, $d_H(\overline{M},M) < \overline{\nu}$, $\{a_i,b_i\}$ *is an* $(m,\delta,\delta/10)$*-strainer at* $p \in \overline{M}$, *with associated map* f, *and* $U(\rho)$ *denotes a "cylinder"* $B_p(\rho) \cap f^{-1}(I^m_{f(p)}(\rho^2))$, *then*

(a) $\operatorname{diam} U_p(\delta) > \delta/10$

(b) $N(U_p(\delta)\backslash\overline{M}(m+1,100\delta,\delta R),\nu) < \mu \cdot N(U_p(10\delta),\nu)$ *for all* $\nu < \overline{\nu}$.

Proof of (a): The assertion can be easily reduced to the case $\overline{M} = M$. Let q_1, q_2 be the endpoints of a diameter of $U_p(\delta)$, and assume that $|q_1q_2| \leq \delta/10$.

Using the volume estimates (cf. [BGP, 8.6]), we can find either a direction $\xi_1 \in \Sigma_{q_1}$, such that $|q_2'\xi_1| > \pi/2 + c(\mathrm{Vol}\, M)$, or a direction $\xi_2 \in \Sigma_{q_2}$, such that $|q_1'\xi_2| > \pi/2 + c(\mathrm{Vol}\, M)$. Assuming the former, move q_1 a little bit in the direction ξ_1 and then apply the procedure of consecutive approximations, as in [BGP, 5.8], to get a point $q_3 \in f^{-1}(q_1)$. Since for all x near q_1 the distances in Σ_x satisfy $|a_i'q_2'| > \pi/2 - 10\delta$, $|b_i'q_2'| > \pi/2 - 10\delta$, the above mentioned consecutive approximations cannot reduce the distance from q_2 considerably. Since the first move substantially increased this distance, we conclude that $|q_2q_3| > |q_2q_1|$ – a contradiction.

Proof of (b): Let $V_q = (U_p(\delta)\backslash B_q(\delta/20))\backslash \overline{M}(m+1, 100\delta, \delta R)$. In fact, we are going to show that

$$N(V_q, \nu) \leq (\mu/2)N(U_p(10\delta), \nu) \quad \text{for all } q \in U_p(\delta)\ , \tag{1}$$

if $2\nu < \delta R$ and R is small enough. Applying (1) to points $q_1, q_2 \in U_p(\delta)$ at distance $\geq \delta/10$, we get (b).

Let $\{x_\alpha\}$ be a maximal 2ν-discrete net in V_q. For each α fix a shortest geodesic $\overline{qx}_\alpha$ and divide it by points $x_{\alpha j}$ into $[\delta/50R]$ equal parts; clearly, each part has length $> 2R$. It is easy to see that $x_{\alpha j} \in U_p(10\delta)$. Furthermore, the pairwise distances between those of $x_{\alpha j}$, which lie outside $B_q(\delta/50)$, are at least $\min\{2\nu, \delta R\}/100$. (Indeed, if $|x_{\alpha j_\alpha}x_{\beta j_\beta}| < \min\{2\nu, \delta R\}/100$ then either $|qx_{\alpha j_\alpha}|/|qx_\alpha| = |qx_{\beta j_\beta}|/|qx_\beta|$ and hence $|x_\alpha x_\beta| < 2\nu$ – a contradiction, or, say, $|qx_{\alpha j_\alpha}|/|qx_\alpha| < |qx_{\beta j_\beta}|/|qx_\beta|$ and therefore $|\overline{qx}_\alpha, x_\beta| < \delta R$ and $|qx_\alpha| > |qx_\beta| + R$ – a contradiction with the assumption $x_\beta \notin \overline{M}(m+1, 100\delta, \delta R)$.) The number of $x_{\alpha j}$ outside $B_q(\delta/50)$ is at least $(\delta/100R)N(V_q, \nu)$; therefore we conclude that $N(U_p(10\delta), \nu/100) \geq (\delta/100R)N(V_q, \nu)$ when $2\nu < \delta R$. Since $N(U_p(10\delta), \nu) \geq cN(U_p(10\delta), \nu/100)$, we see that (1) is satisfied for sufficiently small R. □

In order to prove the general case of Assertion I we need the following.

LEMMA 2. *For sufficiently small $\lambda > 0$ we have*

$$N(M, \lambda\nu)/N(M, \nu) \leq cN(M, \lambda^{m+1}\nu)/N(M, \lambda^m\nu)$$

for all $M \in \mathcal{M}_n$, positive integers m and $0 < \nu < 1$.

Proof: Let L_i denote a maximal $2\lambda^i\nu$-discrete set in M, $i = 0, \ldots, m+1$. We say that a point $x \in L_i$ is an immediate ancestor of a point $y \in L_{i+1}$, $x \to y$, if $|xy| \leq 2\lambda^i\nu$; in this case y is an immediate descendant of x. The relation "x is an ancestor of y", $x \Rightarrow y$ is defined by transitivity.

It is clear that a point can have at most c immediate ancestors. Moreover, if $x \Rightarrow y$ and x has N immediate descendants, then y must have at least cN immediate descendants, cf. 1.2

Let $2A = N(M,\lambda\nu)/N(M,\nu)$, and let B be a constant which is large in comparison with all c appearing in the argument. Since λ is small, we may assume $A \gg B$. Denote by L_0' the set of all points in L_0 which have at least A immediate descendants, and let $L_i' = \{y \in L_i : x \Rightarrow y \text{ for some } x \in L_0'\}$. Similarly, let L_m'' be the set of all points in L_m which have at most AB^{-1} immediate descendants, and let $L_i'' = \{x \in L_i : x \Rightarrow y \text{ for some } y \in L_m''\}$.

It is clear that

$$|L_i'| \geq (cA)^{i-j}|L_j'| , \quad |L_j''| \geq (cAB^{-1})^{j-i}|L_i''| , \quad i \geq j . \tag{2}$$

Moreover, $L_i' \cap L_i'' = \emptyset$, $i = 0,\dots,m$.

The definition of A implies $|L_1'|/|L_0| \geq A$. Therefore, using (2) we obtain $|L_m''| \leq (cB^{-1})^m A^m |L_0''| \leq (cB^{-1})^m |L_m'| < |L_m \backslash L_m''|$. Since each $x \in L_m \backslash L_m''$ has at least AB^{-1} descendants, we get $N(M,\lambda^{m+1}\nu)/N(M,\lambda^m\nu) \geq cB^{-1}A$. □

Now we can finish the proof of Assertion I. Assume that $N(M,\nu')/N(M,\lambda\nu') \geq \lambda^{k+\frac{3}{4}}$. Then, according to Lemma 2, $N(M,\lambda^{-m-1}\nu')/N(M,\lambda^{-m}\nu') \geq c\lambda^{k+\frac{3}{4}}$ whenever $\lambda^{-m-1}\nu' < 1$. Multiplying these inequalities, we easily get

$$N(M,\nu') \leq N(M,\nu)(\nu/\nu')^{k+\frac{13}{16}} \tag{3}$$

if ν'/ν is small enough ($\nu'/\nu < c(\lambda)$).

On the other hand, since $N(M,\nu)/N(M,\lambda\nu) \leq \lambda^{k+\frac{1}{4}}$, there exist at least $cN(M,\nu)$ points of the maximal 2ν-discrete net L_0, having $\geq \lambda^{-k-\frac{1}{8}}$ immediate descendants each. Let p be one of these. Applying the statement of the Particular Case, proved above, to the rescaled ball $B_p(\nu)$, we see that $N(B_p(\nu),\nu'')/N(B_p(\nu),\lambda\nu'') < \lambda^{k+\frac{15}{16}}$ if $\nu'' < c(\lambda)\nu$. Multiplying such inequalities, we get $N(B_p(\nu),\nu') > (\nu/\nu')^{k+\frac{7}{8}}$ if $\nu'/\nu < c(\lambda)$. Hence $N(M,\nu') \geq cN(M,\nu)(\nu/\nu')^{k+\frac{7}{8}} > N(M,\nu)(\nu/\nu')^{k+\frac{13}{16}}$ – a contradiction to (3). □

3. Proof of Assertion II

The proof of the Assertion is based on the following.

LEMMA. *Let Σ^n be a complete Alexandrov space with curvature ≥ 1, $1 \leq k \leq n$, $\varepsilon > 0$. Let $v_1,\dots,v_k$ satisfy $\sum_{i=1}^k v_i^2 = 1$ and $v_i \geq \varepsilon$ for all $i = 1,\dots,k$. Let A_i, B_i be compact subsets of Σ^n, such that*

$$|A_iB_i| > \pi - \delta , \qquad |A_iB_j| > \pi/2 - \delta ,$$
$$|A_iA_j| > \pi/2 - \delta , \quad |B_iB_j| > \pi/2 - \delta , \quad (1 \leq i \neq j \leq k), \qquad \text{where } \delta > 0$$

is sufficiently small ($\delta < c(\varepsilon)$). Let $D \subset \Sigma^n$ be a compact subset, such that $|A_iD| > \pi/2 - \rho$, $1 \leq i \leq k$, where $0 < \rho < \delta$.

Then one can find a point $p \in \Sigma^n$ *and a number* μ, $|1-\mu| < \varepsilon$, *such that*

(1) $\cos|A_i p| = \mu v_i$, $i = 1, \dots, k$

(2) $|Dp| > \pi/2 - c(\varepsilon)\rho$.

Proof: At first we construct a point $q \in \Sigma^n$, which satisfies (2) and

$$\cos|A_i q| = v_i + \kappa(\delta) \ , \qquad i = 1, \dots, k \tag{1'}$$

where κ denotes functions which tend to 0 when their arguments tend to 0.

Consider k pairs of points $(\tilde{A}_i \tilde{B}_i)$ on the standard sphere S^{k-1}, such that $|\tilde{A}_i \tilde{A}_j| = \pi/2$, $|\tilde{B}_i \tilde{B}_j| = \pi/2$, $|\tilde{A}_i \tilde{B}_j| = \pi/2$, $|\tilde{A}_i \tilde{B}_i| = \pi$, $(1 \le i \ne j \le k)$, and let $\tilde{q}_1, \dots \tilde{q}_k \in S^{k-1}$ be constructed according to the following rule:

$$\tilde{q}_1 = \tilde{A}_1 \ , \quad \tilde{q}_{j+1} \in \overline{\tilde{q}_j \tilde{A}_{j+1}} \ , \quad \sin|\tilde{q}_j \tilde{q}_{j+1}| = v_{j+1}(v_1^2 + \dots + v_{j+1}^2)^{-\frac{1}{2}} \ .$$

It is clear that $\cos|\tilde{A}_i \tilde{q}_k| = v_i$ for all $i = 1, \dots, k$. Now consider a corresponding sequence $q_1, \dots, q_k \in \Sigma^n$, such that $q_1 \in A_1$, q_{j+1} lies on some shortest geodesic between q_j and A_{j+1} and $\sin|q_j q_{j+1}| = v_{j+1}(v_1^2 + \dots + v_{j+1}^2)^{-\frac{1}{2}}$. It is easy to prove by induction on j that $||A_i q_j| - |\tilde{A}_i \tilde{q}_j|| < \kappa(\delta)$, $||B_i q_j| - |\tilde{B}_i \tilde{q}_j|| < \kappa(\delta)$, and $|Dq_j| > \pi/2 - c(\varepsilon)\rho$ (cf. [BGP, 12.2]). Hence $q = q_k$ satisfies (1′) and (2).

Observe that for any point r in a $\kappa(\delta)$-neighborhood of q we have $\tilde{\angle} A_i r A_j > \pi/2 + c(\varepsilon)$, $1 \le i \ne j \le k$. Moreover, we have $\tilde{\angle} A_i r D > \pi/2$ whenever $|Dr| \le \pi/2 - c(\varepsilon)\rho$ for an appropriate $c(\varepsilon)$. Applying [P, Lemma 1], we obtain a set of directions $\xi_{\alpha\beta} \in \Sigma_r$, $1 \le \alpha \ne \beta \le k$, such that $|A_i' \xi_{\alpha\beta}| = \pi/2$ if $i \ne \alpha$, $i \ne \beta$, $|A_\alpha' \xi_{\alpha\beta}| > \pi/2 + c(\varepsilon)$, $|A_\beta' \xi_{\alpha\beta}| < \pi/2 - c(\varepsilon)$, and moreover, an analysis of the proof of that lemma shows that we can in addition ensure $|D' \xi_{\alpha\beta}| = \pi/2$ whenever $|Dr| \le \pi/2 - c(\varepsilon)\rho$. Now we can use these directions to carry out the procedure of consecutive approximations, starting from q. The result of this procedure is a point p, which satisfies (2) and (1) with some μ, $|\mu - 1| < \kappa(\delta) < \varepsilon$. □

Now fix some small ε (say, $\varepsilon = 1/100k$), and choose a small $\delta > 0$ to satisfy the conditions of the lemma. Let $v \in \mathbf{R}^k$, $|v| = 1$, $v_i > \varepsilon$ for all $i = 1, \dots, k$. Let $u_1, u_2 \in I^k_{f(p)}(20R)$ be such that $u_2 = u_1 + \omega v$, $\omega > 0$.

CLAIM. $N(f^{-1}(u_1), \nu/2) \ge N(f^{-1}(u_2), \nu)$ *for all* $\nu < \delta R$.

Indeed, let $\{z_\alpha\}$ be a maximal 2ν-discrete net in $f^{-1}(u_2)$. Apply our lemma to each Σ_{z_α}, with $A_i = a_i'$, $B_i = b_i'$, $D = \cup_\beta \{z_\beta' : |z_\alpha z_\beta| \le 4\nu\}$, $\rho = c\nu\delta R^{-1}$. We obtain a direction $\xi_\alpha \in \Sigma_{z_\alpha}$, such that $\cos|a_i' \xi_\alpha| = \mu v_i$ for some μ, $|1-\mu| < \varepsilon$, and $|z_\beta' \xi_\alpha| > \pi/2 - c\nu\delta R^{-1}$ if $0 < |z_\alpha z_\beta| \le 4\nu$. Move each of z_α a little bit in the direction ξ_α and apply consecutive approximations to obtain $\bar{z}_\alpha \in f^{-1}(u_2 - \bar{\omega} v)$, with some very small $\bar{\omega} > 0$. Clearly $|\bar{z}_\alpha \bar{z}_\beta| \geq 2\nu - c\nu\delta R^{-1}\bar{\omega}$. Our claim follows from this construction repeated sufficiently many times.

It follows immediately from the claim above that

$$N\big(f^{-1}(u_1), \nu\big) \geq cN\big(f^{-1}(u_2), \nu\big) \ . \tag{3}$$

To prove the similar inequality for arbitrary u_1, u_2, we replace our strainer by the opposite one, that is $\bar{a}_i = b_i$, $\bar{b}_i = a_i$; let $\bar{f}$ denote its associated map. It is clear that

$$N\big(f^{-1}(I^k_{f(x)}(2R)), \nu\big) = cN\big(\bar{f}^{-1}(I^k_{\bar{f}(x)}(2R)), \nu\big) \quad \text{for any } x \in B_p(100kR) \ .$$

It follows easily from (3) and c-openness that

$$N\big(f^{-1}(I^k_{f(x)}(2R)), \nu\big) = c \max\big\{N(f^{-1}(u), \nu) \ , \ u \in I^k_{f(x)}(R)\big\}(\nu/R)^{-k} \ ,$$

and a similar estimate holds for $\bar{f}$. Therefore

$$\max\big\{N(f^{-1}(u), \nu), u \in I^k_{f(x)}(R)\big\} = c \max\big\{N(\bar{f}^{-1}(u), \nu), u \in I^k_{\bar{f}(x)}(R)\big\} \ . \tag{4}$$

Let $p_+, p_- \in B_p(100kR)$ be such that $f(p_\pm) = f(p) \pm 10R \cdot (1, \ldots, 1)$, and let $u_1, u_2 \in I^k_{f(p)}(R)$. Then (3) implies that

$$\begin{aligned} &N\big(f^{-1}(u_2), \nu\big) \leq cN\big(f^{-1}(u), \nu\big) \quad \text{for all } u \in I^k_{f(p_-)}(R) \ , \\ &N\big(f^{-1}(u_1), \nu\big) \geq cN\big(f^{-1}(u), \nu\big) \quad \text{for all } u \in I^k_{f(p_+)}(R) \ , \quad \text{and} \\ &N\big(\bar{f}^{-1}(\bar{u}_+), \nu\big) \geq cN\big(\bar{f}^{-1}(\bar{u}_-), \nu\big) \end{aligned}$$

for all pairs $(\bar{u}_+, \bar{u}_-)$ such that $\bar{u}_\pm \in I^k_{\bar{f}(p_\pm)}(R)$. Hence applying (4) we obtain the desired inequality $N(f^{-1}(u_1), \nu) \geq cN(f^{-1}(u_2), \nu)$.

4. Proof of Assertion III

The crucial step in the proof of Assertion III is the following

PROPOSITION. *Let $p, r \in M$, let $\phi : I^k \to M$ be a continuous map such that $\tilde{\angle} pxr > \pi/2$ for each $x \in \phi(I^k)$. Then for $\delta > 0$ small enough there exists a continuous map $\phi_\delta : I^k \to M$ with the following properties*

(1) $s(\phi_\delta) \geq (1 - \delta)s(\phi) - c\delta^2$

(2) $|p\phi_\delta(u)| \leq |p\phi(u)| - \delta \cos^2 \tilde{\angle} p\phi(u)r + c\delta^2$ *for each* $u \in I^k$

(3) $\tilde{\angle} pr\phi_\delta(u) \leq \tilde{\angle} pr\phi(u) + c\delta^2$ *for each* $u \in I^k$,

where constants c, as well as the upper bound for δ may depend on $|p\phi(I^k)|$, $|r\phi(I^k)|$, $s(\phi)$, $\inf\{\tilde{\angle} pxr - \pi/2, x \in \phi(I^k)\}$, and do not deteriorate when those quantities are bounded away from zero.

It is not hard to see that the proposition implies Assertion III. Indeed, given R, choose $\gamma > 0$ so small that $\sin 2\gamma < R/10$. Now given $p, r \in M$ (where we may assume $|pr| > R$), $q \in \overline{pr}$, $\phi : I^k \to B_q(\gamma|qr|)$, consider a sequence of maps $\phi_0 = \phi$, $\phi_1 = \phi_\delta, \dots, \phi_i = (\phi_{i-1})_\delta, \dots$, for appropriately chosen very small $\delta > 0$. Condition (3) guarantees that for all $x \in \phi_i(I^k)$, we have $\tilde{\angle} prx < 2\gamma$ (since this is obviously true for ϕ_0); therefore $\tilde{\angle} pxr > \pi - 5R^{-1}\gamma > 5\pi/6$ for $x \in \phi_i(I^k)$ whenever $|px| \geq R/2$; hence it follows easily from (2) that the image of some ϕ_i will be contained in $B_p(R)\backslash B_p(R/2)$; finally, the conditions (1),(2) imply that the size of our maps, considered as a function of the distance of their image from p, satisfies a discretized version of a differential inequality $\frac{ds}{dt} \leq \frac{2s}{t}$, which implies the required estimate for $s(\phi_i)$.

We present two proofs of the proposition. The first one works only in the Riemannian case, and requires c to depend on the parameters of the manifold M. The second one is more involved, but it works for general (even infinite dimensional) Alexandrov spaces.

First proof: Consider a fine triangulation T of I^k such that $\mathrm{diam}(\phi(\Delta)) < \delta^2$ for each simplex $\Delta \in T$. For each vertex $v \in \mathrm{skel}_0(T)$ let $\phi_\delta(v)$ be a point on some shortest geodesic $\overline{p\phi(v)}$ such that $|\phi_\delta(v)\phi(v)| = \delta$. Now we can construct ϕ_δ inductively on the skeleta of I^k: assign a number to each $v \in \mathrm{skel}_0(T)$, and for any simplex $\Delta \in T$, whose boundary has already been mapped, span $\phi_\delta(\Delta)$ using geodesics connecting points of $\phi_\delta(\partial\Delta)$ with the ϕ_δ-image of its vertex having the smallest number. The condition (1) is now clear, since it is immediate for $\mathrm{skel}_0(T)$, and because if x, y_1, y_2 form a triangle in M with $|y_1y_2| \approx \delta$, $|xy_1|, |xy_2| > c$, and $y \in \overline{y_1y_2}$, then $|xy| \geq \min\{|xy_1|, |xy_2|\} - c\delta^2$.

To check (2) and (3) for $u \in I^k$, consider the simplex $\Delta = \mathrm{conv}\{v_0, \dots, v_\ell\}$ containing u. Let $x = \phi(u)$, $y = \phi_\delta(u)$, $y_i = \phi_\delta(v_i)$. Since $\mathrm{diam}(\phi(\Delta)) < \delta^2$, we can easily see that $\tilde{\angle} rxy_i \geq \tilde{\angle} rxp - c\delta$. On the other hand, it is clear from our construction that the direction y' of shortest geodesic $\overline{xy}$ is contained in a $c\delta$-neighborhood of the convex hull of directions y'_i in Σ_x. Since Σ_x is the standard unit sphere, a simple geometric argument shows that $\min_{0\leq i\leq \ell} |y'y'_i| \leq \pi - \min_{0\leq i\leq \ell} |r'y'_i| + c\delta$; it follows that $\tilde{\angle} y_ixy \leq \pi - \tilde{\angle} rxp + c\delta$ for some i, and therefore, $\tilde{\angle} pxy \leq \pi - \tilde{\angle} rxp + c\delta$. The last inequality implies (3). Finally, the estimate (2) follows from (3) and the estimate $|ry| \geq |rx| - \delta\cos\tilde{\angle} rxp + c\delta^2$, which is immediate for y_i in place of y, and can be proved for y by the same argument that we used to prove (1). □

Second proof: Again take a fine triangulation T of I^k and for each $v \in \mathrm{skel}_0(T)$ let $\phi_\delta(v)$ be a point on $\overline{p\phi(v)}$ at distance δ from $\phi(v)$. Introduce some notation: for $v_i \in \mathrm{skel}_0(T)$ let $x_i = \phi(v_i)$, $y_i = \phi_\delta(v_i)$, $\alpha_i = \pi - \tilde{\angle} p x_i r$; note that $|\alpha_i \alpha_j| \le c\delta$ if v_i, v_j are vertices of the same simplex.

The map ϕ_δ will be constructed in two steps. First we construct a map $\psi_\delta : I^k \to M$, such that $\psi_\delta(v_i) = y_i$, and for any $\Delta \in T$, say $\Delta = \mathrm{conv}\{v_0, \dots, v_\ell\}$, and any $u \in \Delta$ there exists $0 \le j \le \ell$ satisfying

$$\left|\psi_\delta(u)\psi_\delta(v_j)\right| \le \delta \sin\alpha_j + c\delta^2 \ . \tag{4}$$

Then we deform ψ_δ to a map ϕ_δ which still satisfies (4) and, in addition, the requirement that $\phi_\delta(u)$ is, in some sense, contained in a convex hull of $\phi_\delta(v_j)$, $0 \le j \le \ell$.

To construct ψ_δ we need the following

LEMMA 1. (a) *Let $x \in M$ satisfy $\tilde{\angle} pxr = \pi - \alpha > \pi/2$, $y \in \overline{px}$, $|xy| = \delta$. Let γ be a dist_r-gradient curve, extending a shortest geodesic $\overline{rx}$, and let $z = \gamma \circ \rho^{-1}(|rx| + \delta\cos\alpha)$, where ρ denotes the standard ρ-parametrization of gradient curves, see* [PPe, 3.3]. *Then $|yz| \le \delta \sin\alpha$.*

(b) *Let $x_1 \in \overline{xy}$, $\alpha_1 = \pi - \angle p x_1 r$, γ_1 a dist_r-gradient curve extending $\overline{rx}_1$, $z_1 = \gamma_1 \circ \rho_1^{-1}(|rx_1| + |x_1 y|\cos\alpha_1)$. Then $|yz_1| \le \delta\sin\alpha$.*

Proof: (a) This is immediate from [PPe, 3.3(b)].

(b) Applying (a) to x_1 in place of x we get $|yz_1| \le |x_1 y| \sin\alpha_1$, and clearly $|x_1 y|\sin\alpha_1 \le \delta\sin\alpha$. □

Let K denote I^k with segments $v_i \times [0,1]$ attached at $v_i \times \{0\}$, $v_i \in \mathrm{skel}_0(T)$; let L denote $K \times [0,1]$ with $v_i \times \{1\} \times [0,1]$ shrunk to points. We have a map $\phi_K : K \to M$ which coincides with ϕ on I^k and maps any segment $v_i \times [0,1]$ to $\overline{x_i y_i}$. The map ϕ_K can be extended to L as follows. Given $u \in K$ construct a dist_r-gradient curve γ_u extending $\overline{r\phi_K(u)}$ and map $u \times [0,1]$ to the arc of $\gamma_u \circ \rho_u^{-1}$ between $|r\phi_K(u)|$ and $|r\phi_K(u)| - |\phi_K(u) y_i| \cos \tilde{\angle} p \phi_K(u) r$ in case $u \in v_i \times [0,1]$, and between $|r\phi(u)|$ and $|r\phi(u)| - \delta\cos\tilde{\angle} p\phi(u) r$ in case $u \in I^k$. (When we say that a segment is mapped to an arc of a curve, we always mean that as a point travels along the segment with constant speed, its image also travels along the arc with constant speed, measured w.r.t. the given parameter on the curve.)

This construction defines a continuous map $\phi_L : L \to M$, extending ϕ_K. On the other hand, the inclusion $K \hookrightarrow L$ can be extended, inductively on the skeleta of T, to a map $e : I^k \times [0,1] \to L$, such that for each simplex $\Delta \in T$, say $\Delta = \mathrm{conv}\{v_0, \dots, v_\ell\}$ we have

$$e(\Delta \times \{1\}) \subset \left((\cup_{0 \le j \le \ell} v_j \times [0,1]) \cup \Delta\right) \times \{1\} \subset L \ .$$

Therefore ψ_δ can be defined as the restriction of $\phi_L \circ e$ to $I^k \times \{1\}$, with condition (4) following from Lemma 1. The first step of the construction is complete.

To carry out the second step we need the following

LEMMA 2. *Let $\psi : \Delta \to M$ be a continuous map,* $\operatorname{diam}(\psi(\Delta)) \leq 2\delta$. *Let $q \in M$ satisfy $|q\psi(u)| \geq s \gg \delta$ for each $u \in \Delta = \operatorname{conv}\{v_0, \ldots, v_\ell\}$. Let $R_0, \ldots, R_\ell > 0$ be given, such that $\psi(u) \in \cup_{0 \leq i \leq \ell} B_{v_i}(R_i)$ for each $u \in \Delta$. Then there exists a deformation Ψ of ψ into another map $\overline{\psi}$ such that*

$$\Psi\big((u,t)\big) \in \cup_{0 \leq i \leq \ell} B_{v_i}(R_i) \quad \textit{for all} \quad (u,t) \in \Delta \times [0,1] \,,$$

Ψ is trivial on $\partial\Delta$,

$$\big|q\overline{\psi}(u)\big| \geq a = \min_{v \in \partial\Delta} \big|q\psi(v)\big| - 10\delta^2/s \quad \textit{for all} \quad u \in \Delta \,,$$

$$\big|\psi(u)\overline{\psi}(u)\big| \leq 10\delta^{-1}s\big(|\psi(v_0)\psi(u)| - |\psi(v_0)\overline{\psi}(u)|\big) \quad \textit{for all} \quad u \in \Delta \,,$$

and

$$\big|\overline{\psi}(u_1)\overline{\psi}(u_2)\big| / \big|\psi(u_1)\psi(u_2)\big| \leq 1 + 100\delta^{-2}s^2\big(|\psi(u_1)\overline{\psi}(u_1)| + |\psi(u_2)\overline{\psi}(u_2)|\big) \,, \quad \textit{for all} \quad u_1, u_2 \in \Delta \,.$$

Proof: For each u let γ_u be a dist_q-gradient curve starting at $\psi(u)$. We let Ψ map $u \times [0,1]$ to the arc of γ_u between $|q\psi(u)|$ and a, in case $|q\psi(u)| < a$, and to the point $\psi(u)$ otherwise. For any $x \in \gamma_u([|q\psi(u)|, a])$ we clearly have $\tilde{\angle} qxv_i > \pi/2 + 2\delta/s$, hence $|\nabla \operatorname{dist}_q|(x) > \delta/s$. Now the properties of $\overline{\psi}$ and Ψ listed above can be easily proved by an argument similar to [PPe, 3.3]. □

Now we can improve ψ_δ inductively on skeleta, using deformations provided by Lemma 2 for an infinite sequence of points playing the role of q. The estimates of Lemma 2 guarantee that there will be a limit map provided that we keep s bounded away from zero. It is easy to choose the sequence in such a way that the limit map $\phi_\delta : I^k \to M$ satisfies

(5) $|\phi_\delta(u)\phi_\delta(u')| \geq \min_{v \in \partial\Delta} |\phi_\delta(v)\phi_\delta(u')| - 100\delta^2/s(\phi)$ whenever $u \in \Delta$ and u' belongs to the opposite face of I^k,

(6) $|\phi_\delta(u)r| \geq \min_{v \in \partial\Delta} |\phi_\delta(v)r| - 100\delta^2/|r\phi(I^k)|$ whenever $u \in \Delta$.

It remains to observe that (1) follows easily from (5), (3) follows easily from (4) (which is preserved by the deformations according to Lemma 2), and (2) follows from (3) and (6). □

Remark: We suspect that the sequence of maps $\phi_0 = \phi, \phi_1 = \phi_\delta, \ldots, \phi_i = (\phi_{i-1})_\delta, \ldots$ will converge, as $\delta \to 0$, to a continuous deformation, under which every point travels along some dist_p-gradient curve.

Acknowledgments. I am grateful to M. Gromov for introducing me to the problem and for helpful discussions. A considerable part of this work was done in January-March 1991, when I enjoyed the hospitality of IHES.

References

[BGP] Yu. Burago, M. Gromov, G. Perelman, Alexandrov spaces with curvatures bounded below, Uspehi matem. nauk. 47:2 (1992), 3–51.

[G1] M. Gromov, Volume and bounded cohomology, Publ. Math. IHES 56 (1982), 5–100.

[G2] M. Gromov, Widths and related invariants of Riemannian manifolds, Asterisque 163/164 (1988), 93–109.

[P] G. Perelman, Elements of Morse Theory in Alexandrov spaces, St. Petersburg Math. Journal 5:1 (1994), 205–213.

[PPe] G. Perelman, A. Petrunin, Quasigeodesics and gradient curves in Alexandrov spaces, to appear.

G. Perelman
Steklov Institute and Department of Mathematics
St.Petersburg, University of California
Russia Berkeley, CA 94720
USA

Submitted: August 1994

Geometric And Functional Analysis
Vol. 5, No. 2 (1995)
1016-443X/95/0200464-18$1.50+0.20/0

ON THE CONFORMAL AND CR AUTOMORPHISM GROUPS

R. Schoen

Introduction

This paper concerns the behavior of conformal diffeomorphisms between Riemannian manifolds, and that of CR diffeomorphisms between strictly pseudoconvex CR manifolds. We develop a new approach to the subject which involves the scalar curvature theory, and the conformally invariant Laplace operator, and its subelliptic analogue in the CR case. We first describe the results in the conformal case. Our main quantitative result (Proposition 2.1) concerns a conformal embedding of a ball with a well controlled Riemannian metric into a Riemannian manifold with zero scalar curvature. It says that if the derivative of the transformation is large at the origin, then it is uniformly large, so that the map is uniformly expanding. Furthermore, it shows that the full curvature tensor on the image of the ball must be very small. This result can be immediately applied to show that the conformal automorphism group G of a scalar flat manifold (not necessarily complete) acts properly unless the manifold is isometric to $\mathbf{R}^n$. Recall that a group G acts properly on a manifold M if for any compact subset K of M, the set

$$G_K = \{F \in G : F(K) \cap K \neq \phi\}$$

is compact in G. Note that if M is compact, then G acts properly on M only if G is compact. We then remove the condition that M have vanishing scalar curvature, and show that the conformal automorphism group of any Riemannian manifold (M not conformally equivalent to $\mathbf{R}^n$ or S^n) acts properly (Theorems 3.3 and 3.4). The idea of our proof for noncompact M is to show that any such M either has a scalar flat conformal metric, or one of scalar curvature -1. This dichotomy is distinguished by the sign of the lowest eigenvalue of the conformal Laplacian. Note that we do not require our metrics to be complete; in fact, our procedure would choose a scalar flat metric on the standard unit ball rather than the hyperbolic metric. We stress that this dichotomy is elementary from an analytic point of view, and follows from maximum and Harnack principles together with linear elliptic theory. This paper does not use anything difficult concerning the Yamabe problem, but is an application of the general theory. Another issue which we consider here is the topology on the conformal automorphism group. A classical theorem asserts that this group is a Lie group with Lie topology coinciding with the C^k topology for k sufficiently large. Initially, the properness works only for the compact open topology, so it is useful to compare these topologies. Our methods also show directly that these topologies are the same (Proposition 1.1).

This problem has an interesting history. The compactness of the identity component of the conformal automorphism group of a compact manifold $\neq S^n$ was shown by Obata ([O]), and compactness of the full group by Ferrand ([F1]) more than 20 years ago. Around the same time Alekseevskii ([A1,2]) announced the properness of the action of the conformal automorphism group for noncompact M. (See also Yoshimatsu ([Y]) and Ferrand ([F2]).) In the past few years, Alekseevskii's argument has met with serious criticism, and been found to be incomplete (see the discussion of Gutschera ([G])). Recently Ferrand ([F3]) has given a proof of the theorem asserted by Alekseevskii. The contribution of this paper is to provide an alternate approach to these issues based on the conformal scalar curvature theory (as well as to prove the quantitative estimates on conformal diffeomorphisms described above).

Our methods also can be extended to obtain analogous results for the CR automorphism group of a strictly pseudoconvex CR manifold. This is done in section 4 of this paper. There is a connection and curvature theory in this setting which was developed by Tanaka ([T]) and Webster ([W]), as well as a Carnot metric theory. The corresponding operator is now subelliptic, so our methods exploit estimates for such operators. Most of the analytic estimates which we need in our setting were obtained by Jerison and Lee ([JL1]). We prove a similar result for a CR embedding of a ball into a scalar flat pseudohermitian manifold, showing that the map is uniformly expanding for Carnot distance if its derivative is large at the center. We are able to show that the CR automorphism group acts properly unless the manifold is CR equivalent to the Heisenberg group or the standard CR sphere. We also show that the compact open topology on the CR automorphism group is equivalent to the C^k topology for any $k \geq 1$. This is a somewhat deeper result than the corresponding conformal result, since it shows that all derivatives of a CR diffeomorphism can be controlled in terms of a C^0 bound.

In the CR case, the compactness of the automorphism group was proved for compact M of dimension greater than 3 by Burns (see [L]). The method of Ferrand was extended to obtain compactness of the automorphism group in the C^0 topology for compact M of any dimension by Pansu ([P]). Lee ([L]) proved the compactness of the identity component for compact M of any dimension. His proof involves an analysis of CR Killing vector fields which is of independent interest. It seems that neither the properness of the action of the CR automorphism group for noncompact M, nor the equivalence of the C^0 and C^k topologies was previously known.

We are pleased to express our appreciation to Jack Lee for his detailed comments on a preliminary version of this paper, and for providing a number of references for results on CR geometry.

1. Conformal Preliminaries

In this section we introduce that part of the scalar curvature theory for conformal metrics which will be needed. We then use this theory to prove that the C^0 and C^k topologies (any $k \geq 1$) coincide for the conformal automorphism group of a smooth conformal structure. Let M^n be a smooth n-dimensional manifold, and suppose g_0 is a Riemannian metric on M. We denote by $[g_0]$ the conformal equivalence class of g_0; that is, $[g_0]$ is the set of all Riemannian metrics of the form $e^{2\lambda}g_0$ where $\lambda \in C^\infty(M)$. Note that we do not require any completeness hypothesis on our metrics. A conformal diffeomorphism $F : (M, [g_0]) \to (N, [h_0])$ is then a diffeomorphism such that $F^*h_0 \in [g_0]$.

If we let R_{g_0} denote the scalar curvature function for the metric g_0, and if $g = u^{\frac{4}{n-2}}g_0$ (assume $n \geq 3$) for a positive smooth function

$$R_g = -c(n)^{-1}u^{-\frac{n+2}{n-2}}L_{g_0}u\ , \qquad L_{g_0}u = \Delta_{g_0}u - c(n)R_{g_0}u \tag{1.1}$$

where $c(n) = \frac{n-2}{4(n-1)}$. Implicit in this formula is the conformal invariance property of the linear operator L which may be expressed

$$L_{g_0}\varphi = v^{-\frac{n+2}{n-2}}L_{g_1}(v\varphi) \tag{1.2}$$

where $g_0 = v^{\frac{4}{n-2}}g_1$. This operator, which we refer to as the conformal Laplace operator, will play a fundamental role in this work.

We first prove the equivalence of the C^0 and C^k topologies for conformal diffeomorphisms.

PROPOSITION 1.1. *Let $F_i : (M, [g_0]) \to (N, [h_0])$, $i = 1, 2, \ldots$, be a sequence of conformal diffeomorphisms such that both $\{F_i\}, \{F_i^{-1}\}$ converge uniformly on compact subsets. For any $k \geq 1$, both sequences $\{F_i\}$, $\{F_i^{-1}\}$ converge in the C^k topology on compact subsets. Moreover, there is a (smooth) conformal diffeomorphism F such that $F_i \to F$, $F_i^{-1} \to F^{-1}$.*

Proof: By hypothesis, there are continuous maps $F : M \to N, G : N \to M$ such that $F_i \to F$, $F_i^{-1} \to G$. We see easily that F is a homeomorphism, and $G = F^{-1}$. It suffices to prove that all derivatives of $\{F_i\}$ and $\{F_i^{-1}\}$ are uniformly bounded on compact subsets. Suppose $x_0 \in M$, and $y_0 = F(x_0) \in N$. For any $r_0 > 0$, there exists $\delta_0 > 0$ such that $F_i(B_{\delta_0}(x_0)) \subseteq B_{r_0}(y_0)$ and $F_i^{-1}(B_{\delta_0}(y_0)) \subseteq B_{r_0}(x_0)$ for all i sufficiently large. If r_0 is small enough, then the lowest Dirichlet eigenvalues of the operators $-L_{g_0}$ on $B_{r_0}(x_0)$ and $-L_{h_0}$ on $B_{r_0}(y_0)$ are positive, so there is a unique positive solution of the boundary value problems

$$L_{g_0}u = 0 \text{ in } B_{r_0}(x_0)\ , \quad u = 1 \text{ on } \partial B_{r_0}(x_0)$$
$$L_{h_0}v = 0 \text{ in } B_{r_0}(y_0)\ , \quad v = 1 \text{ on } \partial B_{r_0}(y_0)\ .$$

Now let $g = u^{\frac{4}{n-2}} g_0$ and $h = v^{\frac{4}{n-2}} h_0$ in $B_{r_0}(x_0), B_{r_0}(y_0)$ respectively. The metrics g, h then have vanishing scalar curvature. For i sufficiently large we then consider $F_i^* h = u_i^{\frac{4}{n-2}} g_0$ where $u_i = (v \circ F_i)|F_i'|^{\frac{n-2}{2}}$ where $|F_i'|$ is the linear stretch factor for F_i taken with respect to g_0, h_0. The function u_i is then a solution of $L_{g_0}u_i = 0$ in $B_{\delta_0}(x_0)$. Now the volume of $(B_{r_0}(y_0), h)$ is bounded, and hence

$$\int_{B_{\delta_0}(x_0)} u_i^{\frac{2n}{n-2}} d\mu_{g_0} \le c\ .$$

Therefore, by elliptic estimates we have a uniform bound on u_i in $B_{\delta/2}(x_0)$. This implies a bound on $|F_i'|$ in $B_{\delta/2}(x_0)$. We conclude that for every compact subset $K \subseteq M$ we have $\sup_K |F_i'|$ is bounded. Similarly, we have $\sup_{K_1} |(F_i^{-1})'| \le c$ for every compact subset $K_1 \subseteq N$. This implies by the chain rule that for all $x \in K$

$$c^{-1} \le |F_i'(x)| \le c$$

for a constant $c = c(K)$ and any K compact in M. To get higher derivative bounds, we return to the metric $F_i^* h = u_i^{\frac{4}{n-2}} g_0$. We now have shown for $x \in B_\delta(x_0)$ we have $c^{-1} \le u_i(x) \le c$. By elliptic theory we then get uniform bounds on all derivatives of u_i. We now have uniform bounds up to any order on the pullback metrics $F_i^* h$ with respect to the fixed background metric g_0. Since $F_i : (B_{\delta/2}(x_0), F_i^* h) \to (B_{r_0}(y_0), h)$ is an isometry onto its image, we then have uniform bounds on the C^k norm of F_i in a neighborhood of x_0. (To see this one can observe that F_i takes normal coordinates to normal coordinates. Since both metrics are controlled up to any order in terms of the background metrics g_0, h_0, the normal coordinate systems are uniformly controlled up to any order.) This gives bounds on the C^k norm of $\{F_i\}$ uniformly on compact subsets of M, and we similarly get C^k bounds uniformly on compact subsets of N for the $\{F_i^{-1}\}$. This completes the proof of Proposition 1.1.

2. The Scalar Flat Case

We prove here a basic result which will be applied to establish the main theorem under the assumption that (M, g_0) admits a scalar flat metric of the form $g = e^{2\lambda} g_0$. This result concerns the image of a small ball under a conformal transformation whose derivative is large at the center. Let B^n denote the unit ball in $\mathbf{R}^n$, and let h be a Riemannian metric on B^n satisfying the bounds

$$\Lambda^{-1}(\delta_{ij}) \leq (h_{ij}(x)) \leq \Lambda(\delta_{ij}) \qquad \forall x \in B^n \tag{2.1}$$
$$\|h_{ij}\|_{2,\alpha} \leq \Lambda$$

for some $\Lambda > 0$, $\alpha \in (0,1)$ where the first inequality indicates bounds on the eigenvalues of the matrix (h_{ij}), and $\|\cdot\|_{2,\alpha}$ denotes the $C^{2,\alpha}(B^n)$ norm of a function. If $F : B^n \to M^n$ is a conformal diffeomorphism, then we have $F^*g = |F'(x)|^2 h$ where the conformal stretch factor is denoted $|F'(x)|$.

PROPOSITION 2.1. *If $F : (B^n, h) \to (M^n, g)$ is a conformal diffeomorphism with $\lambda = |F'(0)|$, and if the scalar curvature of g vanishes then we have:*

i) *For a constant c depending only on Λ, n we have*

$$c^{-1}\lambda \leq |F'(x)| \leq c\lambda \qquad \forall x \in B_{1/2}(0)\ ,$$

and $F(B^n_{1/2})$ satisfies

$$B_{\lambda/(2c)}(F(0)) \subseteq F(B^n_{1/2}) \subseteq B_{c\lambda/2}(F(0))$$

where $B_r(P)$ for $P \in M$ denotes the geodesic ball of radius r centered at P.

ii) *The curvature tensor Riem(g) satisfies*

$$\sup_{B_{c^{-1}\lambda}(F(0))} \|Riem(g)\| \leq c\lambda^{-2}$$

for a constant c depending only on Λ, n.

Proof: Part i) is a consequence of the Harnack inequality applied as follows: Since g has vanishing scalar curvature, so does F^*g, so this implies that the function $u = |F'|^{\frac{n-2}{2}}$ satisfies $Lu = 0$ in B^n where L is the conformal Laplacian for h; i.e.

$$Lu = \Delta_h u - \frac{n-2}{4(n-1)} R(h) \cdot u\ .$$

The standard Harnack inequality which holds under hypothesis (2.1) then proves the first inequality of i). The condition on $F(B^n_{1/2})$ follows from this since for any curve $\gamma \subseteq B^n_{1/2}$ we have

$$c^{-1}\lambda \; Length(F(\gamma)) \leq Length(\gamma) \leq c\lambda \; Length(F(\gamma)) \; .$$

To prove ii), we consider the metric on B^n given by $g_1 = \lambda^{-2}F^*g = v^{\frac{4}{n-2}}h$ where

$$v = \left(\lambda^{-1}|F'|\right)^{\frac{n-2}{2}} \; .$$

This metric is scalar flat since g is, and thus $Lv = 0$. From i), v is bounded above and below by positive constants. Standard elliptic theory using (2.1) and Schauder estimates then gives a bound on $\|v\|_{C^{2,\alpha}(B^n_{1/2})}$. This implies

$$\sup_{B^n_{1/2}} \|Riem(g_1)\| \leq c \; ,$$

and since $\|Riem(g_1)\| = \lambda^2\|Riem(g)\|$, we have established part ii). This completes the proof of Proposition 2.1.

As a corollary of the above proposition, we prove the main theorem for scalar flat manifolds. Recall that the group G of conformal transformations of (M,g) is given the compact-open topology; i.e. the topology of uniform convergence on compact sets. We showed in Proposition 1.1 that this topology is equivalent to the topology of C^k convergence on compact subsets for any $k \geq 1$. We say that the group G acts *properly* on M if for every compact subset $K \subseteq M$, the set

$$G_K = \{F \in G : F(K) \cap K \neq \emptyset\}$$

is a compact subset of G.

THEOREM 2.2. *If (M,g) is scalar flat and not isometric to $\mathbf{R}^n$, then G acts properly.*

Proof: We claim that for any $K \subseteq M$ compact, and any $F \in G_K$ we have

$$\sup_{K_1} |F'| \leq c(K, K_1) \tag{2.2}$$

for any compact $K_1 \subseteq M$ unless (M,g) is isometric to $\mathbf{R}^n$. Inequality (2.2) implies the conclusion of the theorem by the Arzela-Ascoli theorem. (Note that $F^{-1} \in G_K$ if $F \in G_K$, and so the same bound holds for F^{-1}.) Since $u = |F'|^{\frac{n-2}{2}}$ is a solution of $Lu = 0$, (2.2) follows from a bound on $|F'|$ at any chosen point of K together with the Harnack inequality. If no such bound exists, then there is a sequence $\{F_i\} \subseteq G_K$ so that

$$\min_K |F_i'| \to \infty \; .$$

Let $x_i \in K$ be a point such that $F_i(x_i) \in K$, and apply Proposition 2.1 in a small coordinate chart centered at x_i. We conclude that (M, g) is complete, flat, and simply connected (since it is an increasing union of diffeomorphic images of the ball). This proves Theorem 2.2.

3. The Main Theorem in the Conformal Case

In this section we reduce the proof of the main theorem to the scalar flat case. The main point in this reduction is to show that for a "negative" conformal class the conformal group is proper. Recall the Yamabe invariant

$$Q(M) = \inf\left\{ -\int_M \varphi L_0 \varphi d\mu_{g_0} : \int_M \varphi^{\frac{2n}{n-2}} d\mu_{g_0} = 1 \ , \ \varphi \in C_c^\infty(M) \right\}$$

where L_0 denotes the conformal Laplacian of g_0. This number (possibly $Q(M) = -\infty$) is an invariant of the conformal class of the metric g_0. Recall that if M is compact without boundary, then $Q(M) > 0$ $(= 0)(< 0)$ depending on whether there is a metric in the conformal class of g_0 with positive (zero) (negative) scalar curvature respectively. These are in turn equivalent to the lowest eigenvalue of $-L$, $\lambda_1(-L)$, being positive (zero) (negative). The proof of the following result is well known in PDE circles.

PROPOSITION 3.1. *If (M, g) has scalar curvature -1, then the conformal group acts properly. If M is compact and $R(g) = -1$, then the conformal group is the isometry group, and in particular is compact.*

Proof: In this case we have stronger bounds on conformal transformations. If $F : M \to M$ is a conformal transformation, then for any $K \subseteq M$ compact we have

$$\sup_K |F'| \leq c(n, K) \ . \tag{3.1}$$

To see this, let $u = |F'|^{\frac{n-2}{2}}$ so that u satisfies

$$Lu - c(n) u^{\frac{n+2}{n-2}} = 0 \tag{3.2}$$

where $c(n) = \frac{n-2}{4(n-1)}$, and L is the conformal Laplacian of g. Let $\varphi \in C_c^\infty(M)$ be a nonnegative function and multiply the equation by $\varphi^{\frac{n+2}{2}}$. Integrating by parts we have

$$c(n) \int_M (\varphi^{\frac{n-2}{2}} u)^{\frac{n+2}{n-2}} d\mu = \int_M \left[c(n) \varphi^{\frac{n+2}{2}} u + u \Delta \varphi^{\frac{n+2}{2}}\right] d\mu \ .$$

Expanding we have

$$\Delta \varphi^{\frac{n+2}{2}} = \frac{n+2}{2} \varphi^{\frac{n}{2}} \Delta \varphi + \left(\frac{n+2}{2}\right)\left(\frac{n}{2}\right) \varphi^{\frac{n-2}{2}} |\nabla \varphi|^2 \ .$$

A standard use of Young's inequality then gives

$$\int_M (\varphi^{\frac{n-2}{2}} u)^{\frac{n+2}{n-2}} d\mu \le c \int_M [\varphi^{\frac{n+2}{2}} + (\varphi|\Delta\varphi|)^{\frac{n+2}{4}} + |\nabla\varphi|^{\frac{n+2}{2}}] d\mu .$$

This implies local bounds on $\int u^{\frac{n+2}{n-2}} d\mu$, and since $Lu \ge 0$, standard mean value inequalities imply (3.1). If M is compact without boundary, the maximum principle applied to (3.2) implies $u \equiv 1$, and hence F is an isometry. This completes the proof of Proposition 3.1.

We note the following result for noncompact manifolds.

LEMMA 3.2. *If (M, g_0) is noncompact, then $Q(M) \ge 0$ if and only if there is a scalar flat* (possibly incomplete) *metric in the conformal class of g_0.*

Proof: The condition $Q(M) \ge 0$ is equivalent to the condition $\lambda_1(-L) \ge 0$ since both simply are equivalent to the statement that

$$-\int_M \varphi L\varphi d\mu \ge 0$$

for all $\varphi \in C_c^\infty(M)$. This is in turn equivalent (cf [F-CS, Theorem 1]) to the existence of a positive solution u of $Lu = 0$. The metric $g = u^{\frac{4}{n-2}} g_0$ then becomes the scalar flat metric in the conformal class.

We now prove the main theorem in the noncompact case.

THEOREM 3.3. *If (M, g_0) is any noncompact manifold which is not conformally diffeomorphic to $\mathbf{R}^n$, then the conformal group acts properly.*

Proof: We should perhaps remark that this result is very easy in case $n = dim M = 2$, since in this case, the conformal surface either has a unique complete hyperbolic metric, or a complete flat metric. In the former case, the conformal group acts isometrically for this hyperbolic metric, and hence acts properly. In the flat case, M is either conformally equivalent to the flat $\mathbf{R}^2$, or to the cylinder $S^1 \times \mathbf{R}$. In the former case we are done, and in the latter, the conformal group preserves the flat metric, and hence acts properly. Thus we assume that $n \ge 3$

If $Q(M, g_0) \ge 0$, then the previous lemma combined with Theorem 2.2 establishes the result. Thus we assume $Q(M, g_0) < 0$. Let $\Omega_i \subset M$ be smooth compact domains which exhaust M; i.e. $\Omega_i \subset \Omega_{i+1}$, $M = \cup_{i=1}^\infty \Omega_i$. Since $Q(M, g_0) = \lim_{i\to\infty} Q(\Omega_i, g_0)$, it follows that for i large we have $Q(\Omega_i, g_0) < 0$. Standard existence theory then gives a solution u_i of

$$Lu_i - c(n) u_i^{\frac{n+2}{n-2}} = 0 \text{ in } \Omega_i , \quad u_i > 0 ,$$
$$u_i = 0 \text{ on } \partial\Omega_i .$$

As in the proof of Proposition 3.1, we have for any compact $K \subseteq M$

$$\sup_K u_i \le c(n, K)\ .$$

Elliptic theory then gives local estimates on all derivatives of the u_i, so there is a subsequence, also denoted $\{u_i\}$, which converges uniformly along with all derivatives on compact subsets to a limit u which is a solution of $Lu - c(n)u^{\frac{n+2}{n-2}} = 0$. The Harnack principle implies that either $u \equiv 0$ or $u > 0$ everywhere on M. If $u > 0$, then $g = u^{\frac{4}{n-2}} g_0$ has scalar curvature identically equal to -1, and the theorem follows from Proposition 3.1. If $u \equiv 0$, then let $x_0 \in M$ be any chosen point, and set $v_i = u_i(x_0)^{-1}u_i$. The function v_i then satisfies

$$Lv_i - c(n)\big(u_i(x_0)\big)^{\frac{4}{n-2}} v_i^{\frac{n+2}{n-2}} = 0\ .$$

Since $u_i(x_0) \to 0$, and v_i satisfy the Harnack inequality with $v_i(x_0) = 1$, we see that v_i is locally bounded along with all of its derivatives. Passing to a new subsequence, we get $v_i \to v$ where $v > 0$ everywhere on M is a solution of $Lv = 0$. By Lemma 3.2 this implies $Q(M, g_0) \ge 0$ contrary to our assumption. This completes the proof of Theorem 3.3.

THEOREM 3.4. *Suppose (M, g_0) is compact and closed. The conformal group of (M, g_0) is compact unless (M, g_0) is conformally diffeomorphic to the standard n-sphere.*

Proof: If $n = dim M = 2$, then M has a constant curvature conformal metric, and this has nonpositive curvature unless M is conformally equivalent to S^2 or $\mathbf{RP}^2$. In all cases except S^2, the conformal group agrees with the isometry group, and hence is compact. Thus we may assume $n \ge 3$.

If $Q(M, g_0) \le 0$, then the conformal group agrees with the isometry group. Thus we assume $Q(M, g_0) > 0$. We show that if (M, g_0) is not conformally diffeomorphic to the standard n-sphere, then we have a constant $c = c(M)$

$$\sup_M |F'| \le c$$

for all conformal diffeomorphisms F. Suppose on the contrary that there is a sequence $\{F_i\}$ with $\sup_M |F_i'| \to \infty$. Let $x_i \in M$ be a maximum point of $|F_i'|$. Consider small coordinate balls $B_{r_0}(x_i)$ for a fixed $r_0 > 0$ so that hypothesis (2.1) is satisfied for the metric g_0 in each of these balls (in suitable coordinates). Since M is compact this can clearly be done. Choose a point $y_i \in M$ such that $y_i \notin F(B_{r_0}(x_i))$. Since $Q(M, g_0) > 0$, the lowest eigenvalue of $-L_0$ is positive, and thus there is a unique positive solution G_i of $L_0 G_i = 0$ on $M - \{y_i\}$ normalized so that $\min_M G_i = 1$. Consider now the metric g_i on $M - \{y_i\}$ given by

$$g_i = \left(\frac{G_i}{G_i(F_i(x_i))}\right)^{\frac{4}{n-2}} g_0 \ .$$

This is a scalar flat metric, and we have

$$F^* g_i = u_i^{\frac{4}{n-2}} g_0 \ , \quad u_i = \frac{G_i \circ F_i}{G_i(F_i(x_i))} |F_i'|^{\frac{n-2}{2}} \ .$$

Denoting by λ_i the maximum derivative of F_i, we then know from Proposition 2.1 that $F_i(B_{r_0}(x_i))$ contains a ball in the g_i metric of radius a multiple of λ_i. If $G_i(F_i(x_i))$ is large, then such a ball would contain the full sublevel set $\{y : G_i(y) \leq G_i(F_i(x_i))\}$ since $g_i \leq g_0$ on this set, and the set is approximately the complement of a small ball centered at y_i in M (from the local expansion of the Green's function). Since the Harnack inequality holds for u_i in $B_{r_o/2}(x_i)$, and x_i is a maximum point for $|F_i'|$, it follows that

$$G_i(F_i(x)) \geq c^{-1} G_i(F_i(x_i))$$

for $x \in B_{r_0/2}(x_i)$. Since there is a point $x \in B_{r_0/2}$ for which $F_i(x)$ is a minimum point for G_i, we conclude $G_i(F_i(x_i)) \leq c$. (We have shown that $F_i(x_i)$ and y_i are separated by a fixed positive distance.) Proposition 2.1 now implies that the metrics $\bar{g}_i = G_i^{\frac{4}{n-2}} g_0$ have small curvature on large balls. Choosing a subsequence so that $y_i \to y$ it follows that $G_i \to G$ uniformly along with derivatives on compact subsets of $M - \{y\}$, and that the metric $G^{\frac{4}{n-2}} g_0$ is flat. Thus $(M - \{y\}, g_0)$ is conformally diffeomorphic to $\mathbf{R}^n = S^n - \{\infty\}$.

To see that M is conformally diffeomorphic to S^n, suppose $F : M - \{y\} \to S^n - \{\infty\}$ is a conformal diffeomorphism, and choose a flat conformal metric g_1 on a neighborhood of $\infty \in S^n$. We then have a conformal diffeomorphism of punctured balls

$$F : (B_0 - \{y\}, g_0) \to (B_1 - \{\infty\}, g_1) \ .$$

Since g_1 is scalar flat, the function $u = |F'|^{\frac{n-2}{2}}$ is a positive solution of $L_0 u = 0$ with an isolated singularity at y. Since the volume of the image

$$Vol(F(B_0 - \{y\})) = \int_{B_0 - \{y\}} |F'|^n d\mu_{g_0} = \int_{B_0 - \{y\}} u^{\frac{2n}{n-2}} d\mu_{g_0} < \infty \ ,$$

we see easily that u has a removable singularity at y, and extends as a smooth positive solution of $Lu = 0$ on B_0. The map F then extends as an isometry of $(B_0, u^{\frac{4}{n-2}} g_0) \to (B_1, g_1)$ both of which are smooth metrics. Therefore F defines a smooth conformal diffeomorphism of (M, g_0) with S^n. This completes the proof of Theorem 3.4.

4. Theorems on CR Automorphisms

In this section we prove the CR analogues of our results of the previous sections. Let M^{2n+1} be a strictly pseudoconvex CR manifold ($n \geq 1$). This structure is determined by an n-dimensional subbundle of the complexified tangent bundle of M. We denote this subbundle $T_{1,0}M$, and we require $T_{1,0} \cap \overline{T_{1,0}} = 0$, as well as the integrability condition $[T_{1,0}, T_{1,0}] \subseteq T_{1,0}$. Suppose θ is a real 1-form on M which vanishes on the codimension one subspace $H = Re(T_{1,0} \oplus \overline{T_{1,0}}) \subseteq TM$. Following Webster ([W]), we refer to the choice of such θ as a pseudohermitian structure on M. Such a choice of θ determines a Levi form L_θ defined by

$$L_\theta(X,Y) = -2id\theta(X \wedge \overline{Y})$$

for $X, Y \in T_{1,0}$. The CR manifold is called strictly pseudoconvex if the Levi form is positive definite on $T_{1,0}$. Note that L_θ is a hermitian form, and if $\theta' = u\theta$ for u a positive smooth function, then $L_{\theta'} = uL_\theta$, so the notion of pseudoconvexity does not depend on the choice of θ. Given (M, θ), we can define a Riemannian metric g on H by defining $g(X,Y) = 1/2Re(L_\theta(X + iJX, Y - iJY))$ for $X, Y \in H$, where $J : H \to H$ is the complex structure on the horizontal space H given by $J(V + \overline{V}) = i(V - \overline{V})$ for $V \in \Gamma(T_{1,0})$.

A pseudohermitian manifold (M, θ) induces a Carnot distance $d(\cdot,\cdot)$ given by

$$d(x,y) = \inf \{length(\gamma) : \gamma \text{ horizontal from } x \text{ to } y\}$$

for $x, y \in M$, where a horizontal curve is one whose tangent vector at each point lies in H. Chow's theorem guarantees that any two points can be joined by a horizontal curve, and it is known that the Carnot distance induces the manifold topology on M. The reader may see Strichartz ([S]) for a discussion of these issues. We will denote by $B_r(x)$ the metric ball of radius r centered at x. These will play the role analogous to the Riemannian balls in the conformal case.

There is a natural transversal vector field T on a pseudohermitian manifold (assuming both M and H are orientable) determined by the conditions $\theta(T) = 1$ and $d\theta(T, V) = 0 \ \forall \ V \in TM$. We may then extend the metric g to all of TM by requiring that T be a unit vector orthogonal to H. There is a pseudohermitian connection ∇ which has been introduced and studied by Tanaka ([T]) and Webster ([W]). It may be characterized by the conditions that g, T, J are all parallel, as well as certain conditions on the torsion of this connection which we describe: Let $\mathbf{T}(X,Y) = \nabla_X Y - \nabla_Y X - [X,Y]$ be the torsion; then we require

$$\mathbf{T}(X,Y) = -d\theta(X,Y)T \ , \quad \mathbf{T}(T, J(X)) = -J(\mathbf{T}(T,X)) \ , \text{ for } X, Y \in H \ .$$

The pseudohermitian torsion is then given by

$$\tau(X,Y) = g(\mathbf{T}(T,X),Y) \text{ for } X,Y \in H$$

where τ is a section of $H^* \otimes H^*$. There is a pseudohermitian curvature tensor **Riem** which is a section of $H \otimes H^* \otimes H^* \otimes H^*$. The traces of this tensor are the Ricci tensor **Ric**, and R, the scalar curvature. The scalar curvature $R = R_\theta$ is of primary interest for us here. We will need to use the general form of the transformation formula for **Riem** when the contact form θ is changed, and will need the precise transformation formula for the scalar curvature. These formulae may be found in [JL2]. If we define a new contact form θ_1 by $\theta_1 = u^{2/n}\theta$ where u is a positive smooth function, then we have the general form

$$\mathbf{Riem}_{\theta_1} = \mathbf{Riem}_\theta + E_1(u,\nabla u,\nabla\nabla u)\ , \quad \tau_{\theta_1} = \tau_\theta + E_2(u,\nabla u,\nabla\nabla u) \tag{4.1}$$

where the dependence on u includes u^{-1}.

Finally the Yamabe-type theory involves the transformation formula for the scalar curvature. If θ_1 is as above, then this transformation law is related to a pseudoconformal invariant Laplace operator which is a subelliptic operator. For a smooth function φ on M, the trace taken on H of its second covariant derivatives with respect to ∇ is denoted $\Delta_\theta\varphi$. We then have the formula (see [JL1])

$$R_{\theta_1} = -b(n)^{-1}u^{-\frac{n+2}{n}}L_\theta u \tag{4.2}$$

where $L_\theta u = \Delta_\theta u - b(n)R_\theta u$, and $b(n) = \frac{n+1}{2(2n+1)}$. The pseudoconformal invariance property is then

$$L_{\theta_1}\varphi = u^{-\frac{n+2}{n}}L_\theta(u\varphi)\ . \tag{4.3}$$

Finally we note that the volume form of a pseudohermitian manifold (M,θ) is $d\mu_\theta = \theta\wedge d\theta^n = d\mu_g$, and we see that it transforms as $d\mu_{\theta_1} = u^{2(n+1)/n}d\mu_\theta$.

A diffeomorphism $F : (M,\theta) \to (N,\sigma)$ between pseudohermitian manifolds is a CR diffeomorphism if it is contact, meaning that its differential at any point maps horizontal vectors to horizontal vectors, and its differential defines a complex linear transformation between the horizontal spaces. We first prove the CR analogue of Proposition 1.1.

PROPOSITION 1.1′. *Let $F_i : (M,\theta) \to (N,\sigma)$, $i = 1,2,\dots$ be a sequence of CR diffeomorphisms such that both $\{F_i\},\{F_i^{-1}\}$ converge uniformly on compact subsets. For any $k \geq 1$, both sequences $\{F_i\},\{F_i^{-1}\}$ converge in the C^k topology on compact subsets. Moreover, there is a (smooth) CR diffeomorphism F such that $F_i \to F$, $F_i^{-1} \to F^{-1}$.*

Proof: We indicate the necessary modifications to the proof of Proposition 1.1. There is a limiting homeomorphism F as in that proof. We let g_0 and h_0 denote the Riemannian metrics associated to the contact forms

θ and σ respectively. Given $x_0 \in M$, and $y_0 = F(x_0) \in N$, we may choose δ_0, r_0 as above, where the balls are now taken to be balls in the Carnot distance. Using subelliptic theory which is summarized in [JL1, Section 5], we may produce u, v as above, and we consider the new contact forms $\theta_1 = u^{2/n}\theta$, $\sigma_1 = v^{2/n}\sigma$, and their corresponding Riemannian metrics g, h. These then have vanishing Webster (pseudohermitian) scalar curvature, and hence if we write $F_i^*\sigma = |F_i'|\theta$, we then have $F_i^*\sigma_1 = u_i^{2/n}\theta$ where $u_i = (v \circ F_i)|F_i'|^{n/2}$. The function u_i is then in the kernel of L_θ. The volume bound gives

$$\int_{B_{\delta_0}} u_i^{\frac{2(n+1)}{n}} \, d\mu_\theta \leq c \, .$$

The C^k bounds for any k are then a consequence of subelliptic theory (see [JL1, Proposition 5.7]). We may now complete the proof by using the pseudohermitian normal coordinates of Jerison and Lee ([JL2]) in place of the Riemannian normal coordinate argument used above. This establishes Proposition 1.1′.

We now establish the CR analogue of the material of section 2. We first need a geometric characterization of the Heisenberg group as a pseudohermitian manifold. Let $\mathbf{H}^{2n+1}$ denote the CR manifold which is $\mathbf{C}^n \times \mathbf{R}$ with the subspace $T_{0,1}$ being the space spanned by the vector fields $Z_j = \partial/\partial z_j + i\bar{z}_j\partial/\partial t$ where t denotes the $\mathbf{R}$ coordinate, and $j = 1, \ldots, n$. There is a natural flat pseudohermitian structure on $\mathbf{H}$ given by the contact form

$$\theta_0 = dt + \Sigma_{j=1}^n (iz_j d\bar{z}_j - i\bar{z}_j dz_j) \, .$$

The metric g_0 associated with θ_0 has vanishing pseudohermitian torsion and curvature. We need the following global characterization of $(\mathbf{H}, \theta_0)$.

LEMMA 4.1. *Suppose (M, θ) is a pseudohermitian manifold which is simply connected, and has zero pseudohermitian torsion and curvature. If closed Carnot balls of bounded radius are compact in M, then there is a CR diffeomorphism $F : \mathbf{H} \to M$ such that $F^*(\theta) = \theta_0$. Thus (M, θ) and $(\mathbf{H}, \theta_0)$ are equivalent as pseudohermitian manifolds.*

Proof: It is known that the vanishing of torsion and curvature imply that (M, θ) is locally equivalent to $(\mathbf{H}, \theta_0)$. Since the local equivalence is unique up to composition with a global automorphism of $\mathbf{H}$, we can use the fact that M is simply connected to define a global "developing map" $G : M \to \mathbf{H}$ which is a local pseudohermitian equivalence near each point of M. (The developing map in this setting was first defined by Burns and Shnider ([BSh]), and extended by Z. Li ([Li]).) Without loss of generality we may assume that $0 \in \mathbf{H}$ lies in the image of G. The condition that the balls of finite radius are compact in M enables us to lift curves from $\mathbf{H}$ to M, since it guarantees that lifts remain in a compact set (lengths of curves are preserved by G). Since $\mathbf{H}$ is simply connected, this enables us to define an inverse of G, and hence G is a diffeomorphism. Taking $F = G^{-1}$ then completes the proof of Lemma 4.1.

We would now like to define a condition analogous to (2.1) which guarantees that suitable estimates hold for solutions of $Lu = 0$. The difficulty is that such a sharp condition seems to be hard to find in the subelliptic literature. To prove Theorem 2.2 we actually don't need the full force of Proposition 2.1, but it suffices to consider the domain of F to be a compact subset of M. It is not important (for the proof of Theorem 2.2) to quantify the dependence of these constants on the order of differentiability of the CR structure. To deal with this problem, we replace (2.1) with the estimates which are needed in the proof of Proposition 2.1. Let D_r denote the Euclidean ball of radius r centered at 0 in $\mathbf{R}^{2n+1}$, and suppose we have a pseudohermitian structure on D_1 with contact form σ. Let h be the Riemannian metric associated with σ, and suppose the following bounds are satisfied

$$\sup_{D_1} \left\{\|\mathbf{Riem}_\sigma\|_h + \|\tau_\sigma\|_h\right\} \leq \Lambda\,, \quad \Lambda^{-1} \leq d(0,x) \leq \Lambda \;\forall x \in \partial D_{1/2}\,. \tag{4.4}$$

Assume further that for any positive solution u of $L_\sigma u = 0$ in D_1 we have the bounds

$$\sup_{D_{3/4}} u \leq \Lambda \inf_{D_{3/4}} u\,, \quad \|u\|_{C^2(D_{1/2})} \leq \Lambda \sup_{D_{3/4}} u\,. \tag{4.5}$$

We now prove an analogue to Proposition 2.1 under the hypotheses (4.4),(4.5).

PROPOSITION 2.1$'$. *If $F : (D_1, \sigma) \to (M, \theta)$ is a CR diffeomorphism onto its image with $\lambda = |F'(0)|$, and if the Webster scalar curvature of θ vanishes, then we have:*

i) *For a constant c depending only on Λ we have*

$$c^{-1}\lambda \leq \left|F'(x)\right| \leq c\lambda$$

and $F(D_{1/2})$ satisfies

$$B_{\lambda/(2c)}(F(0)) \subseteq F(D_{1/2}) \subseteq B_{c\lambda/2}(F(0))$$

where $B_r(P)$ denotes the Carnot ball of radius r and center P.

ii) *The pseudohermitian curvature $\mathbf{Riem}_\theta$ satisfies*

$$\sup_{B_{c^{-1}\lambda}(F(0))} (\|\mathbf{Riem}_\theta\|_g + \|\tau_\theta\|g) \leq c\lambda^{-2}$$

for a constant c depending only on Λ.

Proof: The proof is almost identical to that of Theorem 2.2 since we have hypothesized the relevant estimates in (4.4) and (4.5). We write $F^*\theta = |F'|\sigma$, and then $u = |F'|^{n/2}$ is a solution of $L_\sigma u = 0$. Statement i) then follows from the Harnack inequality together with the distance bound given in (4.4).

To prove ii), we let $\sigma_1 = \lambda^{-1}F^*\theta = v^{2/n}\sigma$, so that $v = (\lambda^{-1}|F'|)^{n/2}$. From (4.5), we have a C^2 bound on v, and from i) we have upper and lower bounds on v, so we may apply (4.1) to see that the pseudohermitian curvature and torsion are bounded in $D_{1/2}$. This implies that the curvature and torsion of $F^*\theta$ are bounded by a constant times λ^{-2}, and in light of i), completes the proof of ii) and of Proposition 2.1$'$.

THEOREM 2.2$'$. *If (M,θ) is scalar flat and not equivalent as a pseudohermitian manifold to $(\mathbf{H},\theta_0)$, then the CR automorphism group of M acts properly.*

Proof: The proof proceeds in complete analogy with the proof of Theorem 2.2. The domain in the application of Proposition 2.1$'$ is a small ball in M. The local validity of (4.4) is clear, and (4.5) follows from [JL1, Propositions 5.9 and 5.12]. The argument shows that if the CR automorphism group acts improperly, then the pseudohermitian torsion and curvature of (M,θ) both vanish. The proof also shows that the Carnot distance function on M is proper, and hence by Lemma 4.1, (M,θ) is equivalent to $(\mathbf{H}^{2n+1},\theta_0)$. This completes the proof of Theorem 2.2$'$.

We can define a CR Yamabe invariant by setting

$$Q(M) = \inf\left\{ - \int_M \varphi L_\theta \varphi d\mu_\theta : \int_M \varphi^{\frac{2(n+1)}{n}} d\mu_\theta = 1\ , \quad \varphi \in C_c^\infty(M)\right\}$$

where M is a strictly pseudoconvex CR manifold, and θ is a choice of contact form. The invariant is clearly independent of the choice of θ. The operator L_θ has discrete spectrum for the Dirichlet boundary condition on any compact smooth subdomain of M, and we define $\lambda_1(M)$ to be the infimum of the first Dirichlet eigenvalue of $-L_\theta$ taken over compact subdomains of M. It is then clear that $Q(M) \geq 0$ if and only if $\lambda_1(-L_\theta) \geq 0$. We can now prove.

PROPOSITION 3.1$'$. *If (M,θ) has scalar curvature -1, then the CR automorphism group acts properly. If M is compact and $R_\theta = -1$, then the CR automorphism group is the isometry group, and in particular is compact.*

Proof: Let F be a CR automorphism of (M,θ). We prove

$$\sup_K |F'| \leq c(n,K)$$

for compact subsets $K \subseteq M$. If we set $u = |F'|^{n/2}$, then u satisfies the equation

$$L_\theta u - b(n) u^{\frac{n+2}{n}} = 0 \, .$$

By a similar integration by parts argument as given in the proof of Proposition 3.1, we can show that $\int u^{\frac{n+2}{n}}$ is locally bounded. To get a pointwise bound on u we may proceed by first bounding L^p norms locally for any $p > 0$. To do this, observe that if k is a positive integer, and we set $p_k = 1 + k/n$, then we may compute

$$L_\theta u^{p_k} \geq p_k b(n) u^{p_{k+1}} - (p_k - 1) b(n) u^{p_k} \, .$$

Integrating against a smooth function of compact support, we see that it is possible to bound the integral of $u^{p_{k+1}}$ in terms of the integral of u^{p_k} on a slightly larger set. Since $p_1 = (n+1)/n$, we have a bound on the integral of u^{p_1} locally, so by induction, we have a bound on $\int u^p$ for any $p > 0$. The pointwise bound on u may now be obtained by applying [JL1, Theorem 5.15]. The result for compact manifolds follows easily from the maximum principle. This completes the proof of Proposition 3.1′.

LEMMA 3.2′. *If (M, θ) is a pseudohermitian manifold, then $Q(M) \geq 0$ if and only if there is a contact form on M with vanishing scalar curvature.*

Proof: As pointed out above, the condition that $Q(M) \geq 0$ is equivalent to $\lambda_1(-L_\theta) \geq 0$. If we assume this condition, then the argument of [F-CS, Theorem 1] can be applied, since it only uses local solvability of the Dirichlet problem

$$L_\theta u = 0 \text{ in } \Omega \, , \quad u = 1 \text{ on } \partial\Omega$$

for a compact smooth region Ω in M, where the solution u is positive and smooth. The proof then uses only a local Harnack principle together with local interior derivative estimates on bounded solutions of $L_\theta u = 0$.

Conversely, if u is a positive solution of $L_\theta u = 0$ on M, and Ω is any smooth compact domain in M, then we may see that $\lambda_1(\Omega) \geq 0$, by considering a (positive) first Dirichlet eigenfunction v of L_θ on Ω, thus

$$L_\theta v + \lambda_1(\Omega) v = 0 \, , \quad v = 0 \text{ on } \partial\Omega \, .$$

Now there is a positive multiple av such that $av \leq u$ in Ω, and such that there is a point $x_0 \in \Omega$ with $av(x_0) = u(x_0)$. Elementary calculus then implies that $L(av - u) \geq 0$ at x_0. Therefore we see that $\lambda_1(\Omega) \geq 0$. This completes the proof of Lemma 3.2′.

THEOREM 3.3′. *Let M^{2n+1} $n \geq 1$ be a strongly pseudoconvex CR-manifold which is noncompact. The CR-automorphism group acts properly on M unless M is CR diffeomorphic to the Heisenberg group $\mathbf{H}^{2n+1}$.*

Proof: This theorem can be proven exactly analogously with the corresponding conformal proof. The existence of the scalar curvature -1 metric which is used there follows in the CR case from the work of Jerison and Lee ([JL1]). It can also be done by a version of the method of sub and supersolutions which was extended to subelliptic problems by Minicozzi in his PhD thesis ([M]). The remainder of the proof works as above, so we omit the details.

Finally we handle the compact case and prove the analogue of Theorem 3.4.

THEOREM 3.4′. *Let M^{2n+1} $(n \geq 1)$ be a compact closed, strongly pseudoconvex CR manifold. The CR-automorphism group is compact unless M is CR diffeomorphic to S^{2n+1} with its standard CR structure (as the boundary of the unit ball in $\mathbf{C}^{n+1}$).*

Proof: We refer to the necessary modifications to extend the conformal proof. First we need a fundamental solution with a leading order asymptotic expansion (we only need that it go uniformly to infinity at its singularity). This analysis has been done by Z. Li ([Li]). Using this together with the argument of Theorem 3.4, we can show that if the CR automorphism group of M is noncompact, then there is a point $y \in M$ such that $(M - \{y\}, \theta)$ is CR equivalent to the Heisenberg group. To prove that (M, θ) is CR equivalent to the sphere, we may then apply a similar extension argument as in the conformal case. The issue is to prove smooth extension across a point of a positive solution of $L_\theta u = 0$ with $u \in L^{2(n+1)/n}$. This follows from [JL1, Proposition 5.10, Theorem 5.15]. This completes the proof of Theorem 3.4′.

References

[A1] D.V. Alekseevskii, Groups of transformations of Riemannian spaces, Mat. Sbornik 89:131 (1972), 280-296 and Math. USSR Sbornik 18 (1972), 285-301.

[A2] D.V. Alekseevskii, Uspehi Mat. Nauk 28 (1973), 225-226 (Russian).

[BSh] D. Burns, S. Shnider, Spherical hypersurfaces in complex manifolds, Invent. Math. 33 (1976), 223-246.

[F1] J. Lelong-Ferrand, Transformations conformes et quasi-conformes des variétés riemanniennes compactes, Memoires Academie Royale de Belgique, classe des sciences 39 (1971), 1-44.

[F2] J. Ferrand, Sur un lemma d'Alekseevskii relatif aux transformations conformes, C.R.A.S. Paris 284 (1977), 121-123.

[F3] J. Ferrand, The action of conformal transformations on a Riemannian manifold, to appear.

[F-CS] D. Fischer-Colbrie, R. Schoen, The structure of complete stable minimal surfaces in 3-manifolds of non-negative scalar curvature, Comm. Pure Appl. Math. XXXIII (1980), 199-211.

[G] K.R. GUTSCHERA, Invariant metrics for groups of conformal transformations, to appear.

[JL1] D. JERISON, J. LEE, The Yamabe problem on CR manifolds, J. Diff. Geom. 25 (1987), 167-197.

[JL2] D. JERISON, J. LEE, Intrinsic CR normal coordinates and the CR Yamabe problem, J. Diff. Geom. 29(1989), 303-343.

[L] J.M. LEE, CR manifolds with noncompact connected automorphism groups, to appear.

[Li] Z. LI, On spherical CR manifolds with positive Webster scalar curvature, to appear.

[M] W. MINICOZZI, Geometric variational problems related to symplectic geometry, Stanford PhD thesis, 1994.

[O] M. OBATA, The conjectures on conformal transformations of Riemannian manifolds, J. Diff. Geom. 6 (1971), 247-258.

[P] P. PANSU, Distances Conformes et Cohomologie L^n, Publ. Univ. Pierre et Marie Curie 92 (1990).

[S] R. STRICHARTZ, Sub-Riemannian geometry, J. Diff. Geom. 24 (1986), 221-263; corrections, JDG 30 (1989), 595-596.

[T] N. TANAKA, A Differential Geometric Study on Strongly Pseudo-Convex Manifolds, Kinokuniya Co. Ltd., Tokyo, 1975.

[W] S. WEBSTER, Pseudo-hermitian structures on a real hypersurface, J. Diff. Geom. 13 (1978), 25-41.

[Y] Y. YOSHIMATSU, On a theorem of Alekseevskii concerning conformal transformations, J. Math. Soc. Japan 28 (1976), 278-289.

Richard Schoen
Mathematics Department
Stanford University
Stanford, CA 94305
USA
e-mail: schoen@gauss.stanford.edu

Submitted: November 1994

Geometric And Functional Analysis
Vol. 5, No. 2 (1995)
1016-443X/95/0200482-46$1.50+0.20/0

L^2 RIEMANN–ROCH THEOREM FOR ELLIPTIC OPERATORS

M.A. Shubin

Introduction

1. Let X be a compact riemannian surface (i.e. a compact complex manifold with $\dim_{\mathbf{C}} X = 1$) or, in other words, a non-singular complex algebraic curve. Denote by g the genus of X, i.e. a non-negative integer such that X is homeomorphic to the sphere with g handles. In particular, if $g = 0$ then X is just a riemannian sphere, and if $g = 1$ then topologically X is a 2-torus (in this case X is called an elliptic curve). Consider a *divisor* μ on X, i.e. an element of the free abelian group generated by the points in X. So μ is in fact a finite collection of points $x_1, \dots, x_k$ in X with multiplicities $p_1, \dots, p_k$ which are arbitrary integers. Let us consider the space $\mathcal{O}(\mu)$ of meromorphic functions on X which are *subordinated* to μ. This means that $\mathcal{O}(\mu)$ consists of the meromorphic functions u which satisfy the following conditions:

1°. u is holomorphic in X except possibly at the points x_j with $p_j > 0$.
2°. At any point x_j with $p_j > 0$ the function u is allowed to have a pole of order $\leq p_j$.
3°. At any point x_j with $p_j < 0$ the function u is required to have a zero of multiplicity at least $|p_j|$.

It is easy to see that $\mathcal{O}(\mu)$ is a linear space. Denote $r(\mu) = \dim_{\mathbf{C}} \mathcal{O}(\mu)$.

Similarly we can consider the space $\mathcal{O}'(\mu)$ which is defined in exactly the same way but instead of meromorphic functions we should take meromorphic differentials (meromorphic forms), which locally have the form $f(z)dz$ where z is a local complex parameter and f is a meromorphic function. Denote $r'(\mu) = \dim_{\mathbf{C}} \mathcal{O}'(\mu)$.

Define the inverse divisor μ^{-1} by keeping the same points x_j but changing signs of all integers p_j.

Define also the degree of the divisor μ as $d(\mu) = \sum p_j$.

Then the classical Riemann-Roch theorem is the following equality:

$$r(\mu) = 1 - g + d(\mu) + r'(\mu^{-1}) \; .$$

Partially supported by NSF grant DMS-9222491.

The simplest corollary of this theorem is the inequality

$$r(\mu) \geq 1 - g + d(\mu) .$$

It implies the existence of non-trivial meromorphic functions subordinated to μ, i.e. functions with permitted poles at some points and prescribed zeros at other points, provided $1 - g + d(\mu) > 0$.

The restrictions on the meromorphic functions in the definition of $\mathcal{O}(\mu)$ can be easily described in terms of the growth (or vanishing) conditions near the points x_j:

$$u(x) = o\big(|x - x_j|^{-p_j-1}\big) \quad \text{as} \quad x \to x_j .$$

But there is also another possibility: they can be described in terms of distribution theory with the help of the $\bar{\partial}$ operator. Namely, the function u has a pole of order $\leq p > 0$ at x_j if and only if it can be extended to a distribution $\hat{u}$ in a neighbourhood of x_j so that

$$\bar{\partial}\hat{u} = \sum_{|\alpha| \leq p-1} c_\alpha \partial^\alpha \delta(x - x_j) ,$$

where δ is the (real) Dirac delta-function, α is a multiindex (so ∂^α is a mixed derivative of order $|\alpha|$) and c_α are constants.

Similarly, the fact that u has a zero of order at least $p > 0$ at x_j can be described as the fact that u is orthogonal to all distributions which are linear combinations of the derivatives of the Dirac δ-function of the same form as above.

All this means that we can describe $\mathcal{O}(\mu)$ as the space of all solutions of $\bar{\partial}u = 0$ defined outside of the points x_j with $p_j > 0$ such that, first, they can be extended to distributions on X so that applying $\bar{\partial}$ to the extension we get to a specific finite-dimensional space of distributions, and second, that they are orthogonal to another finite-dimensional space of distributions.

All this can be done in a much more general context. First of all, instead of $\bar{\partial}$ we can consider a general elliptic operator A on a real compact manifold X. The poles of the solutions of $Au = 0$ (point singularities with restrictions of growth) can be described in the same way with the help of the Dirac δ-function; also vanishing conditions at a finite subset in X can be written with the help of the δ-functions as well. But then we can pass to considering much more general singularities and vanishing conditions by allowing arbitrary finite-dimensional spaces of distributions. In fact we need two such spaces, which are supported on two disjoint nowhere dense closed subsets, one of them used to describe singularities and the other to define vanishing conditions. All these data are conveniently combined in the notion of a rigged divisor which we will describe in more detail later.

N. Nadirashvili was first to produce a Riemann-Roch theorem of such a kind for the Laplacian on a riemannian manifold, with point singularities and vanishing conditions. This allowed him to prove an estimate of multiplicity of zero of the Coulomb potential of k point charges: he proved that if the potential does not vanish identically, then the multiplicity of its zeros is $\leq k-1$.

The Riemann-Roch type theorems are closely connected with some special duality theorems which provide necessary and sufficient conditions for an elliptic equation to have a solution with permitted singularities and required vanishing conditions. These duality theorems are proved simultaneously with the corresponding Riemann-Roch theorems, though such a connection was not explicit in the classical case where the duality theorem is equivalent to the Serre duality. It is possible to deduce some local solvability results from such duality theorems. They claim solvability for an elliptic equation near a compact nowhere dense set provided any finite number of orthogonality conditions is imposed upon the solution and obvious necessary conditions for the right-hand side are satisfied. For example the Riemann-Roch theorem with distributed singularities as described above allows us to prove an approximate local solvability of the Cauchy problem for the Laplace operator in $\mathbf{R}^n$ with the data on a compact set D of Lebesgue measure 0:

$$\Delta u = f \text{ near } D\ ; \quad u \approx 0 \text{ and } \nabla u \approx 0 \text{ on } D\ .$$

Here the approximate equality might mean, e.g. equality of any number of Fourier coefficients with respect to any orthonormal system in $L^2(D, d\nu)$ where $d\nu$ is a finite Borel measure on D. Note that D might be a complicated set (e.g. a Sierpiński carpet) and not just a hypersurface as for the classical Cauchy problem.

Another generalization, which is the main goal of this paper, is the L^2 Riemann-Roch theorem on regular coverings of compact manifolds, i.e. manifolds with a free action of a discrete group Γ such that X/Γ is a compact manifold without boundary. Since X is allowed to be non-compact (this is the case when Γ is infinite), some conditions at infinity are needed. They are provided by imposing the condition of finiteness of the L^2-norm (or an appropriate uniform Sobolev norm). Still the corresponding spaces of solutions can be infinite-dimensional but they prove to be finite-dimensional in the von Neumann sense. This means that they have finite Γ-dimensions. Here the Γ-dimension is a function with values in $[0, \infty]$ which is defined on so called Hilbert Γ-modules which are just Hilbert spaces with an unitary action of Γ such that they are Γ-invariant subspaces in the Hilbert tensor products $L^2\Gamma \otimes \mathcal{H}$, where $L^2\Gamma$ is the Hilbert space of all square-integrable functions on Γ (with respect to the canonical discrete measure) and $\mathcal{H}$ is an arbitrary Hilbert space. We shall explain the precise definition of the Γ-dimension in section 1. Now we shall just note that it has all the properties of the usual dimension and $\dim_\Gamma L^2\Gamma = 1$.

We shall provide the precise formulation of the L^2 Riemann-Roch theorem later. It has the usual existence results as corollaries. Let us give an example of such a result.

Consider the standard Laplacian Δ in $\mathbf{R}^3$. Denote $\Gamma = \mathbf{Z}^3$ and let Γ act on $\mathbf{R}^3$ by translations. Let us suppose that D^+ and D^- are disjoint Γ-invariant discrete subsets in $\mathbf{R}^3$. Let *density*$(D^\pm)$ mean the number of the points of the corresponding set $D^\pm$ in the fundamental domain of Γ. Let us assume that

$$density(D^+) > density(D^-)\ .$$

Then there exists a non-trivial function $u \in L^2(\mathbf{R}^3)$ such that $\Delta u = 0$ on $\mathbf{R}^3 - D^+$ and $u = 0$ on D^-. We shall return to this example at the end of the next subsection.

2. Now let us give the precise formulation of the L^2 Riemann-Roch theorem.

Note that the number $1 - g$ in the classical Riemann-Roch theorem is equal to the index of $\bar\partial : C^\infty(X) \to \Lambda^{0,1}(X)$. Here the index of an operator A is given by

$$\operatorname{ind} A = \dim \operatorname{Ker} A - \dim \operatorname{Coker} A = \dim \operatorname{Ker} A - \dim \operatorname{Ker} A^*\ ,$$

where A^* is the adjoint operator to A. We shall need this definition for the case when A is an elliptic differential operator on a compact manifold (acting on smooth sections of vector bundles). Then both $\dim \operatorname{Ker} A$ and $\dim \operatorname{Coker} A$ are finite, so the index $\operatorname{ind} A$ is well defined. It can be calculated in terms of the principal symbol of A by the Atiyah-Singer index formula.

In the case of Γ-invariant operators on a regular covering X the Γ-index of Atiyah

$$\operatorname{ind}_\Gamma A = \dim_\Gamma \operatorname{Ker} A - \dim_\Gamma \operatorname{Ker} A^*$$

is used. It is proved by Atiyah that this index coincides with the usual index of A on the quotient manifold X/Γ.

Suppose that an elliptic Γ-invariant differential operator A on X is given. The permitted singularities can be situated on a Γ-invariant closed nowhere dense set $D^+ \subset X$ and are described in terms of a given Γ-invariant distribution space L^+ such that $\operatorname{supp} f \subset D^+$ for all $f \in L^+$, L^+ is a subspace in a uniform negative Sobolev space and $\dim_\Gamma L^+ < \infty$ where $\dim_\Gamma$ is the von Neumann Γ-dimension. Namely, the permitted singularities are singularities of solutions u of the equation $Au = 0$ defined on $X - D^+$ such that u can be extended to a distribution $\hat{u}$ on X with $A\hat{u} \in L^+$. Also u should be in L^2 in a generalized sense which can be formulated, e.g. by saying that the extension $\hat{u}$ is in a uniform negative Sobolev space.

Similarly the required orthogonality conditions say that u should be orthogonal to a Γ-invariant distribution space L^- such that L^- belongs to a uniform negative Sobolev space, $\dim_\Gamma L^- < \infty$ and all elements of L^- are supported in a Γ-invariant closed nowhere dense set D^- such that $D^+ \cap D^- = \emptyset$.

All these data are encoded into a notion of a *rigged divisor* which is a tuple

$$\mu = (D^+, L^+; D^-, L^-) .$$

Denote the corresponding space of solutions with permitted singularities and required orthogonality conditions by $L(\mu, A)$; denote also $r(\mu, A) = \dim_\Gamma L(\mu, A)$.

The symmetry of the description of singularities and orthogonality conditions allows us to interchange the pairs (D^+, L^+) and (D^-, L^-) to form the inverse divisor

$$\mu^{-1} = (D^-, L^-; D^+, L^+)$$

which is naturally associated with the adjoint operator A^*.

Then the main result of this paper is the following Riemann–Roch type formula

$$r(\mu, A) = \operatorname{ind}_\Gamma A + \deg_A(\mu) + r(\mu^{-1}, A^*) . \tag{0.1}$$

Here $\deg_A(\mu)$ is a number which is expressed in terms of Γ-dimensions $\ell^\pm = \dim_\Gamma L^\pm$ and Γ-dimensions $\tilde{\ell}^\pm = \dim_\Gamma \tilde{L}^\pm$ of "secondary spaces" $\tilde{L}^\pm$ as follows

$$\deg_A(\mu) = (\ell^+ - \tilde{\ell}^+) - (\ell^- - \tilde{\ell}^-) ;$$

the spaces $\tilde{L}^\pm$ are defined as follows

$$\tilde{L}^+ = \{u \mid \operatorname{supp} u \subset D^+ \ , \ u \in W^{-\infty} \ , \ Au \in L^+\} ,$$
$$\tilde{L}^- = \{v \mid \operatorname{supp} v \subset D^- \ , \ v \in W^{-\infty} \ , \ A^*v \in L^-\} ,$$

where $W^{-\infty} = \cup_{s \in Z} W^s$, W^s is the uniform Sobolev space on X.

The formula (0.1) can be used to establish the existence of non-trivial solutions with permitted singularities and prescribed orthogonality conditions. Namely, it follows from (0.1) that

$$r(\mu, A) \geq \operatorname{ind}_\Gamma A + \deg_A(\mu) . \tag{0.2}$$

Therefore the inequality $\mathrm{ind}_\Gamma A + \deg_A(\mu) > 0$ implies that $r(\mu, A) > 0$, hence the space $L(\mu, A)$ is nontrivial. In fact it is infinite-dimensional in the usual sense.

The simplest example of the application of this argument is the existence of a non-trivial solution u of the equation $Au = 0$ such that u is in L^2 in a natural generalized sense (i.e. belongs to a uniform negative Sobolev space), u is allowed to have poles (of a bounded order) on a discrete Γ-invariant set and required to vanish on another discrete Γ-invariant set. One concrete example of this sort was given above at the end of the previous subsection. There we had $X = \mathbf{R}^3$, $A = \Delta$, $\Gamma = \mathbf{Z}^3$, $L^{\pm}$ are spaces of L^2 linear combinations of δ-functions supported at the points of $D^{\pm}$ respectively. Then $\mathrm{ind}_\Gamma A = 0$, $\tilde{L}^{\pm} = 0$, $\dim_\Gamma L^{\pm} = density(D^{\pm})$ and $\deg_A(\mu) = density(D^+) - density(D^-)$. The fundamental solution of Δ (which equals $(4\pi|x|)^{-1}$) is locally in L^2 so the requirement $u \in L^2$ for a distribution u with $\Delta u \in L^+$ is equivalent to the requirement $u \in W^{-N}(\mathbf{R}^3)$ for any $N < 0$. Therefore we obtain the desired existence result.

3. Let us give some (minimal) history, connections and references.

The standard proof of the classical Riemann-Roch theorem can be found, e.g. in the book [GH]. Another approach is to note that meromorphic functions and meromorphic $(1,0)$-forms can be identified with holomorphic sections of holomorphic line bundles on the riemannian surface, therefore the Riemann–Roch theorem is a particular case of the Atiyah–Singer index theorem for elliptic operators (see, e.g. [P]). A generalization to many-dimensional complex manifolds and holomorphic vector bundles which is due to F. Hirzebruch (see, e.g. [Hi]) is also a particular case of the Atiyah–Singer index theorem.

Our generalization goes in a different direction. This direction was first indicated by N. Nadirashvili ([N]) for the Laplacian on a riemannian manifold. The case of general elliptic operators and point singularities was first considered in [GrS1], and the case of distributed singularities is described in [GrS2]. These results do not follow from the Atiyah–Singer index theorem but supplement it, allowing rather general singularities of solutions and vanishing requirements. Note that the generality of the setting leads in fact to simpler proofs and gives a new point of view, especially in duality theorems.

This paper is a continuation and extension of earlier papers [GrS1] and [GrS2] where only compact (or similar to compact) situations were considered. As a starting point of this paper we take a generalization of the Atiyah–Singer index theorem to regular coverings which is due to M. Atiyah ([A2]). It has important applications, e.g. the existence of non-trivial L^2-

solutions of the Dirac equation and similar equations which provide spaces where discrete series of representations of semi-simple Lie groups can be constructed. There are further generalizations of the Atiyah theorem, e.g. the ones by I.M. Singer ([Si]), A. Connes ([Con1,2]) and J. Roe ([R]). All of them can be applied to elliptic operators on non-compact manifolds without the action of any discrete group, though with some requirements that might replace necessary elements which appear in [A2] from the group action. A. Connes ([Con2]) and J. Roe ([R]) gave also examples of existence results for L^2 (in the generalized sense) meromorphic functions on $\mathbf{C}$ without any action of a discrete group.

4. Let us describe the structure of this paper.

It is written independent of the previous papers [GrS1] and [GrS2].

In section 1 we recall some important facts about Hilbert Γ-modules and their morphisms, and prove some results in preparation for the next sections. The most important notion here is the notion of a Γ-Fredholm operator in Hilbert Γ-modules. This notion in a slightly different context but much broader generality appeared in papers of M. Breuer ([B]). A Γ-Fredholm operator is a bounded linear operator $A : L' \to L''$, where L', L'' are Hilbert Γ-modules, such that A commutes with the action of Γ in L' and L'', $\dim_\Gamma \operatorname{Ker} A < \infty$ and $\operatorname{Im} A$ contains a closed Γ-invariant subspace of finite Γ-codimension. An important tool we use is the fact, noticed by M. Breuer ([B]), that a much stronger statement about $\operatorname{Im} A$ follows: it is almost closed, i.e. Γ-dense in its closure $\overline{\operatorname{Im} A}$. This means that for every $\varepsilon > 0$ there exists a Γ-invariant subspace $L_\varepsilon \subset \operatorname{Im} A$ such that it is closed in L'' and $\dim_\Gamma(\overline{\operatorname{Im} A} \ominus L_\varepsilon) < \varepsilon$. The importance of this statement is clear from the fact that the intersection of an almost closed subspace M with any closed Γ-invariant subspace L is again almost closed with the closure $\bar{M} \cap L$ which implies in particular that $M \cap L$ is non-trivial if $\bar{M} \cap L$ is non-trivial. Also if $B : L' \to L''$ is any bounded Γ-invariant operator and $M \subset L''$ is almost closed then $B^{-1}(M)$ is Γ-dense in $B^{-1}(\bar{M})$ and in particular almost closed.

We also discuss in section 1 the duality in Hilbert Γ-modules.

Note that a notion similar to Γ-density in the context of random elliptic operators was introduced in [FS] though it was not actually used there.

Section 2 is devoted to complexes of Hilbert Γ-modules. An important invariant of such a complex is its reduced cohomology. We introduce a class of Γ-Fredholm complexes, i.e. complexes such that their Laplacians are Γ-Fredholm. For a short exact sequence of complexes of Hilbert Γ-modules a standard construction provides a sequence of reduced cohomologies. We prove that if all complexes in the sequence are Γ-Fredholm then the sequence of reduced cohomologies is almost exact, i.e. the image of any map in this sequence is dense (and even Γ-dense) in the kernel of the next map. This important fact is due to J. Cheeger and M. Gromov ([CGr]) who proved it in a slightly different context. The most important corollary of this almost exact sequence is the additivity property of the Euler Γ-characteristic. We also prove some results that simplify checking that a complex is Γ-Fredholm.

Sections 1 and 2 actually contain many more facts and details than are really necessary for the L^2 Riemann–Roch theorem. This is done because it seemed to the author that these facts might be important in other analytic contexts.

Section 3 contains the formulation of the L^2 Riemann–Roch theorem for a rigged Γ-divisor. We also provide some examples here.

Section 4 contains the proof of the main theorem. It also contains a closely connected duality result which describes the closure of the image of the given operator A on spaces of sections with permitted singularities and required orthogonality conditions. Though the result seems much weaker than in the compact case where the image was closed, in fact it is a generalization of the corresponding compact result because the corresponding operator is Γ-Fredholm and the image is almost closed.

5. This paper was written during the author's visit to the Forschungsinstitut für Mathematik, ETH Zürich. I am very grateful to many people at the Institute for their hospitality. In particular, I am very grateful to Rahel Boller for her careful work in typing the manuscript of this paper.

1. Preliminaries on Hilbert Γ-modules

In this section we collected some abstract definitions and facts about Hilbert Γ-modules and Γ-Fredholm operators. Most of these facts are well known ([B],[Co],[D]) though their importance is sometimes underestimated. Together they show that it is almost as easy to work with Hilbert Γ-modules of finite Γ-dimension and their morphisms as with the usual finite-dimensional linear spaces and their linear maps.

A. Let Γ be a discrete group with the neutral element e. Denote

$$L^2\Gamma = \left\{ f \;\middle|\; f : \Gamma \longrightarrow \mathbf{C} \,, \sum_{\gamma\in\Gamma} |f(\gamma)|^2 < \infty \right\} .$$

This is a Hilbert space with the scalar product

$$(f,g) = \sum_{\gamma\in\Gamma} f(\gamma)\overline{g(\gamma)} \,, \quad f,g \in L^2\Gamma \,.$$

It has a natural orthonormal basis of δ-functions $\{\delta_\gamma \mid \gamma \in \Gamma\}$, where $\delta_\gamma(x) = 1$ if $\gamma = x$ and 0 otherwise.

There are two natural unitary representations of Γ in $L^2\Gamma$: left regular and right regular representations $\gamma \mapsto L_\gamma$, $\gamma \mapsto R_\gamma$. They are homomorphisms $\Gamma \to U(L^2\Gamma)$ (here $U(\mathcal{H})$ for any Hilbert space $\mathcal{H}$ denotes the set of all unitary operators in $\mathcal{H}$), where L_γ, R_γ are given by the formulas:

$$(L_\gamma f)(x) = f(\gamma^{-1}x) , \qquad x \in \Gamma ,$$
$$(R_\gamma f)(x) = f(x\gamma) , \qquad x \in \Gamma .$$

Denote $\mathfrak{A}_\ell$ (resp. $\mathfrak{A}_r$) the von Neumann algebra in $L^2\Gamma$ generated by $\{L_\gamma \mid \gamma \in \Gamma\}$ (resp. $\{R_\gamma \mid \gamma \in \Gamma\}$). This is simply a weak closure of the set of all finite linear combinations of the operators L_γ (resp. R_γ). For any Hilbert space $\mathcal{H}$ denote $\mathcal{B}(\mathcal{H})$ the algebra of all bounded linear operators in $\mathcal{H}$. For any subset $M \subset \mathcal{B}(\mathcal{H})$ its commutant is defined as

$$M' = \{B \mid B \in \mathcal{B}(\mathcal{H}) , \ BA = AB \quad \text{for any} \quad A \in M\} .$$

Then $\mathfrak{A}'_\ell = \mathfrak{A}_r$, $\mathfrak{A}'_r = \mathfrak{A}_\ell$ (see, e.g. [D, part I, ch. 9]).

Let us consider a linear map

$$\tau : \mathcal{B}(L^2\Gamma) \longrightarrow \mathbf{C} , \quad \tau(A) = (A\delta_e, \delta_e) .$$

Its restriction to $\mathfrak{A}_\ell$ or $\mathfrak{A}_r$ is a natural faithful finite trace on these algebras, i.e. for $\mathfrak{A} = \mathfrak{A}_\ell$ or $\mathfrak{A}_r$

(i) $\tau : \mathfrak{A} \to \mathbf{C}$ is linear;

(ii) $\tau(AB) = \tau(BA)$, $A, B \in \mathfrak{A}$;

(iii) $A \in \mathfrak{A}$, $A \geq 0$ implies $\tau(A) \geq 0$ with the equality for $A = 0$ only.

Obviously τ is also weakly continuous. In particular it is normal (i.e. if $A_\alpha \in \mathfrak{A}$ and $A_\alpha \nearrow A$ then $\tau(A_\alpha) \to \tau(A)$).

Let $\mathcal{H}$ be a complex Hilbert space. Consider the Hilbert tensor product $L^2\Gamma \otimes \mathcal{H}$ which is a Hilbert space with the orthonormal basis $\{\delta_\gamma \otimes \ell_j \mid \gamma \in \Gamma, j \in J\}$ where $\{\ell_j \mid j \in J\}$ is an orthonormal basis in $\mathcal{H}$. There are two natural actions of Γ given there by $\gamma \mapsto L_\gamma \otimes I$ (resp. $\gamma \mapsto R_\gamma \otimes I$). They generate von Neumann algebras $\mathfrak{A}_\ell \otimes I$ (resp. $\mathfrak{A}_r \otimes I$) with the commutants $\mathfrak{A}_r \otimes \mathcal{B}(\mathcal{H})$ (resp. $\mathfrak{A}_\ell \otimes \mathcal{B}(\mathcal{H})$).

Both algebras $\mathfrak{A}_\ell \otimes \mathcal{B}(\mathcal{H})$ and $\mathfrak{A}_r \otimes \mathcal{B}(\mathcal{H})$ have a natural faithful normal semifinite ([D]) trace $\mathrm{Tr}_\Gamma = \tau \otimes \mathrm{Tr}$, where Tr is the usual trace on $\mathcal{B}(\mathcal{H})$. We shall call it the Γ-trace. This trace induces a dimension function $\dim_\Gamma$ on Γ-invariant subspaces $L \subset L^2\Gamma \otimes \mathcal{H}$. More specifically, we can take a subspace L which is invariant under all operators $L_\gamma \otimes I$ (or $R_\gamma \otimes I$) which is equivalent to saying that $P_L \in \mathfrak{A}_r \otimes \mathcal{B}(\mathcal{H})$ (resp. $P_L \in \mathfrak{A}_\ell \otimes \mathcal{B}(\mathcal{H})$) where P_L is the orthogonal projection in $L^2\Gamma \otimes \mathcal{H}$ with the image L. Then by definition

$$\dim_\Gamma L = \mathrm{Tr}_\Gamma P_L . \tag{1.1}$$

Obviously $\dim_\Gamma(L^2\Gamma \otimes \mathcal{H}) = \dim_{\mathbf{C}} \mathcal{H}$.

DEFINITION 1.1: The Hilbert space $L^2\Gamma \otimes \mathcal{H}$ with the unitary action of Γ given by $\gamma \mapsto L_\gamma \otimes I$ is called a *free Hilbert* Γ*-module* of rank $k = \dim_{\mathbf{C}} \mathcal{H}$.

For simplicity of notation we shall write L_γ instead of $L_\gamma \otimes I$ on $L^2\Gamma \otimes \mathcal{H}$ and on all Γ-invariant subspaces of $L^2\Gamma \otimes \mathcal{H}$.

DEFINITION 1.2: (i) A *(projective) Hilbert* Γ*-module* L is a Hilbert space with a unitary action of Γ such that it can be isometrically imbedded as a Γ-invariant subspace of a free Hilbert Γ-module (with the induced action).

We shall omit the word "projective" because we will not need any Hilbert Γ-modules except projective ones.

(ii) The Γ*-dimension* of a Hilbert Γ-module imbedded into a free Hilbert Γ-module is defined by (1.1).

An important fact is that the Γ-dimension $\dim_\Gamma L$ does not in fact depend on the choice of the imbedding. It has the following properties:

(a) $0 \le \dim_\Gamma L \le \infty$ for any Hilbert Γ-module L; if $L \neq \{0\}$ then $\dim_\Gamma L > 0$;
(b) $\dim_\Gamma(L_1 \oplus L_2) = \dim_\Gamma L_1 + \dim_\Gamma L_2$;
(c) if $L_{j+1} \subset L_j$, $j = 1, 2, \ldots$, are Hilbert Γ-modules and $\dim_\Gamma L_1 < \infty$ then $\dim_\Gamma \cap L_j = \lim_{j\to\infty} \dim_\Gamma L_j$.

DEFINITION 1.3: A *morphism* of Hilbert Γ-modules is a bounded linear operator $A : L_1 \to L_2$ (where L_1, L_2 are Hilbert Γ-modules) such that A commutes with the action of Γ in L_1 and L_2, i.e. $AL_\gamma = L_\gamma A$ for every $\gamma \in \Gamma$.

Sometimes it is useful to accept unbounded operators as morphisms of Hilbert Γ-modules (see, e.g. [CGr]), but this causes many complications. We shall not need unbounded morphisms for our purpose.

DEFINITION 1.4: For any Hilbert Γ-module L denote $\mathfrak{A}_\Gamma$ (or $\mathfrak{A}_\Gamma(L)$ if it is necessary to specify L) the von Neumann algebra of all Γ-endomorphisms of L (i.e. morphisms of Hilbert Γ-modules $A : L \to L$). We shall say that $\mathfrak{A}_\Gamma$ is the *von Neumann algebra associated with the Hilbert* Γ*-module* L.

There is a natural (faithful, normal, semifinite) trace on $\mathfrak{A}_\Gamma$ which we shall denote Tr_Γ. It is obtained by embedding L into a free Hilbert Γ-module and taking the restriction of the Γ-trace given on the free module. This trace does not depend on the choice of the embedding.

DEFINITION 1.5: A morphism of Hilbert Γ-modules $A : L_1 \to L_2$ is called an *almost isomorphism* if $\mathrm{Ker}\, A = 0$ and $\mathrm{Im}\, A$ is dense in L_2.

LEMMA 1.6. *If L_1, L_2 are Hilbert Γ-modules such that there exists an almost isomorphism $A : L_1 \to L_2$ then $\dim_\Gamma L_1 = \dim_\Gamma L_2$.*

Proof: Let us take the polar decomposition $A = US$ where $S \geq 0$ and $U : L_1 \to L_2$ a partial isometry with $\operatorname{Ker} U = \operatorname{Ker} S$ (in fact then $S = \sqrt{A^*A}$ and S is usually denoted $|A|$). In our case $\operatorname{Ker} S = \operatorname{Ker} A = 0$ and U will be a unitary isomorphism of Γ-modules. Hence $\dim_\Gamma L_1 = \dim_\Gamma L_2$. □

DEFINITION 1.7: (i) A linear (not necessarily closed) subspace $M \subset L$ in a Hilbert Γ-module L is called Γ-*dense* in L if for every $\varepsilon > 0$ there exists a closed (in L) Γ-invariant subspace $M_\varepsilon \subset M$ such that $\dim_\Gamma(L \ominus M_\varepsilon) < \varepsilon$. (Here $L \ominus M_\varepsilon = L \cap M_\varepsilon^\perp$, so that $L = M_\varepsilon \oplus (L \ominus M_\varepsilon)$.)

(ii) A linear Γ-invariant subspace M in a Hilbert Γ-module L is called *almost closed* if M is Γ-dense in its closure $\bar{M}$.

LEMMA 1.8. *If $M \subset L$ is Γ-dense in L then M is dense in L.*

Proof: If M is not dense in L, then for any closed Γ-invariant subspace $N \subset M$ we have

$$\dim_\Gamma(L \ominus N) \geq \dim_\Gamma(L \ominus \bar{M}) > 0$$

which contradicts the hypothesis that M is Γ-dense in L. □

Remark: It might happen that a Γ-invariant dense subspace $M \subset L$ in a Hilbert Γ-module L is not Γ-dense. For example if Γ is countable, then the space M of all finite linear combinations of δ-functions is not Γ-dense in $L^2\Gamma$. Indeed, a closed subspace in M is then necessarily finite-dimensional in the usual sense whereas a non-trivial closed Γ-invariant subspace in $L^2\Gamma$ is necessarily infinite-dimensional.

DEFINITION 1.9: A morphism of Hilbert Γ-modules $A : L_1 \to L_2$ is called a Γ-*Fredholm* operator if the following two conditions are satisfied

(i) $\dim_\Gamma \operatorname{Ker} A < \infty$;

(ii) there exists a closed Γ-invariant subspace $M \subset L_2$, such that $M \subset \operatorname{Im} A$ and $\dim_\Gamma(L_2 \ominus M) < \infty$.

For a Hilbert Γ-module L and its Hilbert Γ-submodule M we shall use the notation

$$\operatorname{codim}_\Gamma M = \dim_\Gamma(L \ominus M)\ .$$

The Γ-*index* of a Γ-Fredholm operator $A : L_1 \to L_2$ is

$$\operatorname{ind}_\Gamma A = \dim_\Gamma \operatorname{Ker} A - \dim_\Gamma \operatorname{Ker} A^* = \dim_\Gamma \operatorname{Ker} A - \operatorname{codim}_\Gamma \overline{\operatorname{Im} A}\ .$$

Though $\operatorname{ind}_\Gamma A$ is real-valued, it is stable under norm continuous deformations of Γ-Fredholm operators. More exactly, for any Γ-Fredholm operator $A : L_1 \to L_2$ there exists $\varepsilon > 0$, such that if $B : L_1 \to L_2$ is a morphism of Hilbert modules with $\|B\| < \varepsilon$ then $A + B$ is a Γ-Fredholm operator and $\operatorname{ind}_\Gamma(A + B) = \operatorname{ind}_\Gamma A$.

Let L be a Hilbert Γ-module, $\mathfrak{A}_\Gamma$ the von Neumann algebra associated with L. We shall introduce some important ideals in $\mathfrak{A}$.

DEFINITION 1.10: The ideal of all Γ-*trace class* operators is the set of all finite linear combinations of operators $A \in \mathfrak{A}_\Gamma$ such that $A \geq 0$ and $\mathrm{Tr}_\Gamma A < \infty$. Denote this set by $J_1(\mathfrak{A}_\Gamma)$. It is really a two-sided ideal in $\mathfrak{A}_\Gamma$.

The Γ-trace can be uniquely extended to a linear map $\mathrm{Tr}_\Gamma : J_1(\mathfrak{A}_\Gamma) \to \mathbf{C}$. If $A \in J_1(\mathfrak{A}_\Gamma)$ and $B \in \mathfrak{A}_\Gamma$, then

$$\mathrm{Tr}_\Gamma(AB) = \mathrm{Tr}_\Gamma(BA)\ .$$

If $A \in \mathfrak{A}_\Gamma$ and $A = US$ its polar decomposition then $A \in J_1(\mathfrak{A}_\Gamma)$ if and only if $S \in J_1(\mathfrak{A}_\Gamma)$ (or $\mathrm{Tr}_\Gamma S < \infty$).

DEFINITION 1.11: An operator $B \in \mathfrak{A}_\Gamma$ is called a Γ-*Hilbert–Schmidt* operator if $B^*B \in J_1(\mathfrak{A}_\Gamma)$ (or $\mathrm{Tr}_\Gamma(B^*B) < \infty$). Denote $J_2(\mathfrak{A}_\Gamma)$ the set of all Γ-Hilbert–Schmidt operators. It is in fact also a two-sided ideal in $\mathfrak{A}_\Gamma$ and $(J_2(\mathfrak{A}_\Gamma))^2 = J_1(\mathfrak{A}_\Gamma)$. If $B \in \mathfrak{A}_\Gamma$ and $B = US$ its polar decomposition then $B \in J_2(\mathfrak{A}_\Gamma)$ if and only if $S \in J_2(\mathfrak{A}_\Gamma)$.

DEFINITION 1.12: The operators in the norm closure of $J_1(\mathfrak{A}_\Gamma)$ (or $J_2(\mathfrak{A}_\Gamma)$) are called Γ-*compact*. They form a two-sided ideal in $\mathfrak{A}_\Gamma$ which is denoted $J_\infty(\mathfrak{A}_\Gamma)$.

The following important proposition is due to M. Breuer ([B]) though it is proved there in a slightly different but essentially more general context.

PROPOSITION 1.13. *Suppose that $A : L_1 \to L_2$ is a morphism of Hilbert Γ-modules. Then the following conditions are equivalent:*

(i) *A is Γ-Fredholm;*

(ii) *there exists a morphism of Hilbert Γ-modules $B : L_2 \to L_1$ such that both operators $AB - I$ and $BA - I$ are in the Γ-trace class;*

(iii) *the same as* (ii) *except Γ-trace class is replaced by the class of all Γ-Hilbert–Schmidt operators;*

(iv) *the same as* (ii) *except Γ-trace class is replaced by the class of all Γ-compact operators;*

(v) *$\dim_\Gamma \mathrm{Ker}\, A^* < \infty$ and if $A^*A = \int \lambda dE_\lambda$ is the spectral decomposition of A^*A then there exists $\lambda > 0$ such that $\mathrm{Tr}_\Gamma E_\lambda < \infty$.*

COROLLARY 1.14. (i) *A morphism $A : L_1 \to L_2$ of Hilbert Γ-modules is Γ-Fredholm if and only if the adjoint morphism $A^* : L_2 \to L_1$ is Γ-Fredholm.*

(ii) *If $A_1 : L_1 \to L_2$ and $A_2 : L_2 \to L_3$ are Γ-Fredholm morphisms of Hilbert Γ-modules then $A_2A_1 : L_1 \to L_3$ is also Γ-Fredholm.*

Proof: (i) If B has the property described in (ii) of Proposition 1.13, then B^* will have the same property with respect to A^*.

(ii) If $B_1 : L_2 \to L_1$, $B_2 : L_3 \to L_2$ are such as in (ii) of Proposition 1.13 with respect to A_1, A_2, then B_1B_2 has the same property with respect to A_2A_1. □

LEMMA 1.15. *If* $A : L_1 \to L_2$ *is* Γ*-Fredholm then* $\operatorname{Im} A$ *is almost closed.*

Proof: Let us consider $T = \sqrt{AA^*}$; then $\operatorname{Im} T = \operatorname{Im} A$ because $A = TV$, where V is a partial isometry with $\operatorname{Im} V = (\operatorname{Ker} T)^\perp$. It follows that $T = AV^*$ is also Γ-Fredholm. If $T = \int \lambda dE_\lambda$ is the spectral decomposition of T, then there exists $\lambda > 0$ such that $\operatorname{Tr}_\Gamma E_\lambda < \infty$. It follows from the normality of the trace that $\operatorname{Tr}_\Gamma E_\lambda \to \dim_\Gamma \operatorname{Ker} T$ as $\lambda \downarrow 0$. Hence $\operatorname{Im}(I - E_\lambda)$ is a closed Γ-invariant subspace in $\operatorname{Im} T$, such that

$$\operatorname{codim}_\Gamma \operatorname{Im}(I - E_\lambda) = \dim_\Gamma \operatorname{Im} E_\lambda = \operatorname{Tr}_\Gamma E_\lambda$$

$$\begin{aligned}\dim_\Gamma(\overline{\operatorname{Im} T} \ominus \operatorname{Im}(I - E_\lambda)) &= \operatorname{codim}_\Gamma \operatorname{Im}(I - E_\lambda) - \operatorname{codim}_\Gamma \overline{\operatorname{Im} T} \\ &= \dim_\Gamma \operatorname{Im} E_\lambda - \dim_\Gamma \operatorname{Ker} T \longrightarrow 0 \quad \text{as} \quad \lambda \downarrow 0\end{aligned}$$

which proves that $\operatorname{Im} T = \operatorname{Im} A$ is Γ-dense in its closure. □

Remark: In a slightly different context this lemma is proved in [B]. A similar lemma for random elliptic operators in $\mathbf{R}^n$ was proved in [FS].

COROLLARY 1.16. *If* $A : L_1 \to L_2$ *is a morphism of Hilbert* Γ*-modules and* $\dim_\Gamma L_2 < \infty$ *then* $\operatorname{Im} A$ *is almost closed.*

Proof: The induced operator $L_1/\operatorname{Ker} A \to L_2$ is obviously Γ-Fredholm and has the same image. □

LEMMA 1.17. *Let* N *be a closed* Γ*-invariant subspace of a Hilbert* Γ*-module* L*. Suppose that* M *is an almost closed* Γ*-invariant subspace in* L*. Then* $M \cap N$ *is* Γ*-dense in* $\bar{M} \cap N$*. In particular* $M \cap N$ *is almost closed and its closure equals* $\bar{M} \cap N$*.*

Proof: It is sufficient to prove that $M \cap N$ is Γ-dense in $\bar{M} \cap N$. Suppose that M_ε is a closed Γ-invariant subspace in M such that $\dim_\Gamma(\bar{M} \ominus M_\varepsilon) < \varepsilon$. It is sufficient to prove that then $\dim_\Gamma[(\bar{M} \cap N) \ominus (M_\varepsilon \cap N)] < \varepsilon$. But identifying $\bar{M} \ominus M_\varepsilon \cong \bar{M}/M_\varepsilon$ and $(\bar{M} \cap N) \ominus (M_\varepsilon \cap N) \cong \bar{M} \cap N/M_\varepsilon \cap N$ we obtain a natural injective morphism of Hilbert Γ-modules $\bar{M} \cap N/M_\varepsilon \cap N \to \bar{M}/M_\varepsilon$. Therefore $\dim_\Gamma[(\bar{M} \cap N) \ominus (M_\varepsilon \cap N)] = \dim_\Gamma(\bar{M} \cap N/M_\varepsilon \cap N) \le \dim_\Gamma \bar{M}/M_\varepsilon = \dim_\Gamma(\bar{M} \ominus M_\varepsilon)$ and the desired inequality follows. □

Remark: Generally $M \cap N$ is not necessarily dense in $\bar{M} \cap N$. For example, it might happen that N is finite dimensional in the usual sense and $N \subset \bar{M}$ but $N \cap M = 0$. So the action of Γ and the requirement of the almost closedness of M are essential.

Sometimes the following generalization of Lemma 1.17 is useful.

LEMMA 1.18. *If* M_1, M_2 *are two almost closed subspaces in a Hilbert* Γ*-module* L*, then* $M_1 \cap M_2$ *is* Γ*-dense in* $\bar{M}_1 \cap \bar{M}_2$*. In particular,* $M_1 \cap M_2$ *is almost closed and its closure equals* $\bar{M}_1 \cap \bar{M}_2$*.*

Proof: For any $\varepsilon > 0$ we can find Γ-invariant subspaces $M_i^{(\varepsilon)} \subset M_i$, $i = 1, 2$, such that $M_i^{(\varepsilon)}$ is closed in L and $\dim_\Gamma(\bar{M}_i \ominus M_i^{(\varepsilon)}) < \varepsilon$, $i = 1, 2$. It follows that $\dim_\Gamma[(\bar{M}_1 \cap \bar{M}_2) \ominus (M_1^{(\varepsilon)} \cap M_2^{(\varepsilon)})] < 2\varepsilon$ which proves the desired statement. □

LEMMA 1.19. *Let $B : L_1 \to L_2$ be a morphism of Hilbert Γ-modules, $M \subset L_2$ be a Γ-invariant subspace which is almost closed. Then $B^{-1}(M)$ is Γ-dense (and, in particular, dense) in $B^{-1}(\bar{M})$. In particular $B^{-1}(M)$ is almost closed and its closure is equal to $B^{-1}(\bar{M})$.*

Proof: Let $L \subset M$ be a closed Γ-invariant subspace such that $\dim_\Gamma(\bar{M} \ominus L) < \varepsilon$. Let us prove that $\dim_\Gamma(B^{-1}(\bar{M}) \ominus B^{-1}(L)) < \varepsilon$. Identifying $\bar{M} \ominus L \cong \bar{M}/L$, $B^{-1}(\bar{M}) \ominus B^{-1}(L) \cong B^{-1}(\bar{M})/B^{-1}(L)$ we see that B induces an injective morphism of Hilbert Γ-modules $B^{-1}(\bar{M})/B^{-1}(L) \to \bar{M}/L$. Therefore

$$\begin{aligned}\dim_\Gamma (B^{-1}(\bar{M}) \ominus B^{-1}(L)) &= \dim_\Gamma (B^{-1}(\bar{M})/B^{-1}(L)) \\ &\leq \dim_\Gamma(\bar{M}/L) = \dim_\Gamma(\bar{M} \ominus L) < \varepsilon\end{aligned}$$

which ends the proof. □

Remark: Generally $B^{-1}(M)$ is not necessarily dense in $B^{-1}(\bar{M})$. For example let B be a finite rank operator such that $B \neq 0$, $\operatorname{Im} B \subset \bar{M}$ but $\operatorname{Im} B \cap M = 0$. Then $B^{-1}(M) = \operatorname{Ker} B$ and $B^{-1}(\bar{M}) = L_1$. So the action of Γ and the almost closedness of M are essential here.

The following lemma gives a very useful description of almost isomorphisms for Hilbert Γ-modules with finite Γ-dimension.

LEMMA 1.20. *Let $A : L_1 \to L_2$ be a morphism of Hilbert Γ-modules, $\dim_\Gamma L_1 < \infty$ or $\dim_\Gamma L_2 < \infty$. Then A is an almost isomorphism if and only for every $\varepsilon > 0$ if there exist closed Γ-invariant subspaces $L_i^{(\varepsilon)} \subset L_i$, $i = 1, 2$, such that $\operatorname{codim}_\Gamma L_i^{(\varepsilon)} < \varepsilon$, $i = 1, 2$, $A(L_1^{(\varepsilon)}) = L_2^{(\varepsilon)}$ and the restriction of A to $L_1^{(\varepsilon)}$ defines a topological isomorphism of Hilbert Γ-modules $L_1^{(\varepsilon)}$ and $L_2^{(\varepsilon)}$.*

Proof: (i) Suppose that for every $\varepsilon > 0$ there exist $L_i^{(\varepsilon)}$ with the desired properties. Note first that then $\operatorname{Ker} A = 0$. Indeed $\operatorname{Ker} A \cap L_1^{(\varepsilon)} = 0$ for all ε and $\operatorname{Ker} A \neq 0$ would imply that $\operatorname{codim}_\Gamma L_1^{(\varepsilon)} \geq \dim_\Gamma \operatorname{Ker} A$. Similarly if $\operatorname{Im} A$ is not dense then we would have $\operatorname{codim}_\Gamma L_2^{(\varepsilon)} \geq \dim_\Gamma \operatorname{Ker} A^*$. Hence A is an almost isomorphism. In particular we have then $\dim_\Gamma L_1 = \dim_\Gamma L_2 < \infty$.

(ii) Vice versa, suppose that A is an almost isomorphism. It follows that $\dim_\Gamma L_1 = \dim_\Gamma L_2 < \infty$. Consider the polar decomposition $A = US$, $S = |A| = \sqrt{A^*A}$; then U is a unitary isomorphism of Hilbert Γ-modules L_1 and L_2, $S : L_1 \to L_1$ is an almost isomorphism. Let $S = \int \lambda dE_\lambda$ be the spectral decomposition of S. Then A defines a topological isomorphism of $\operatorname{Ker} E_\lambda = \operatorname{Im}(I - E_\lambda)$ to $A(\operatorname{Ker} E_\lambda)$. It remains to note that $\operatorname{codim}_\Gamma \operatorname{Ker} E_\lambda = \dim_\Gamma \operatorname{Im} E_\lambda = \operatorname{Tr}_\Gamma E_\lambda \to 0$ as $\lambda \downarrow 0$, so we can take $L_1^{(\varepsilon)} = \operatorname{Ker} E_\lambda$ where $\lambda = \lambda(\varepsilon)$ is sufficiently small. □

LEMMA 1.21. *Let $A : L_1 \to L_2$ be a morphism of Hilbert Γ-modules. Suppose that $M_i \subset L_i$, $i = 1, 2$, are closed Γ-submodules, such that* $\dim_\Gamma L_i/M_i < \infty$, $i = 1, 2$, *and* $A(M_1) \subset M_2$. *Denote $A_M : M_1 \to M_2$ the restriction of A. Then A is Γ-Fredholm if and only if A_M is Γ-Fredholm.*

Proof: Denote P_1 the orthogonal projection on M_i in L_i, $i = 1, 2$. Obviously P_1, P_2 are Γ-Fredholm and $I - P_1$, $I - P_2$ are in the Γ-trace class. Hence $A(I - P_1)$ is in the Γ-trace class also and the operators A and AP_1 are Γ-Fredholm simultaneously because $A = AP_1 + A(I - P_1)$. Obviously $\operatorname{Ker} A_M \subset \operatorname{Ker} A$ and $\operatorname{Im} A_M = \operatorname{Im} AP_1$. Hence if A is Γ-Fredholm, then AM is Γ-Fredholm too.

Now $P_2AP_1 = P_2A_MP_1$, hence

$$A = P_2A_MP_1 + (I - P_2)AP_1 + P_2A(I - P_1) + (I - P_2)A(I - P_1) .$$

All the terms on the right-hand side except possibly the first one are in Γ-trace class. Hence, if A_M is Γ-Fredholm then A is also Γ-Fredholm. □

B. We shall also need some facts about duality between Hilbert Γ-modules.

DEFINITION 1.22: Suppose that L, L' are Hilbert Γ-modules. A Γ-*duality* or Γ-*pairing* between them is a bilinear or sesquilinear continuous map $(\cdot,\cdot) : L \times L' \to \mathbf{C}$ which is Γ-invariant in the following sense:

$$(\gamma u, \gamma v) = (u, v) , \qquad u \in L , \ v \in L' , \ \gamma \in \Gamma .$$

It is called *non-degenerate* if for every $u \in L - \{0\}$ there exists $v \in L'$ such that $(u, v) \neq 0$ and for every $v \in L' - \{0\}$ there exists $u \in L$ such that $(u, v) \neq 0$.

For any linear subspace $M \subset L$ denote M° its *annihilator* in L':

$$M^\circ = \{v \mid v \in L', (u, v) = 0 \quad \text{for every} \quad u \in L\} .$$

It is a closed subspace in L'. If M is Γ-invariant then M° is Γ-invariant too. Similarly, if N is a linear subspace in L' we denote N° its *annihilator* in L:

$$N^\circ = \{u \mid u \in L , \ (u, v) = 0 \quad \text{for every} \quad v \in N\} .$$

It is a closed subspace in L, which is Γ-invariant if N itself is Γ-invariant.

EXAMPLE: The inner product in a Hilbert Γ-module L provides an example of a non-degenerate Γ-pairing in L (i.e. between L and L). The annihilator then becomes the orthogonal complement.

LEMMA 1.23. *Let $(\cdot,\cdot) : L \times L' \to \mathbf{C}$ be a non-degenerate pairing between Hilbert Γ-modules L and L'. Let M be a Γ-invariant subspace in L. Then*
(i) $M \subset (M^\circ)^\circ$;
(ii) $\operatorname{codim}_\Gamma \bar{M} \geq \dim_\Gamma M^\circ$;
(iii) $\operatorname{codim}_\Gamma M^\circ = \dim_\Gamma \bar{M}$.

Proof: 1) (i) is obvious and it implies that

$$\operatorname{codim}_\Gamma \bar{M} \geq \operatorname{codim}_\Gamma (M^\circ)^\circ .$$

Hence (iii) implies (ii) if we apply (iii) to M^0 instead of M. So we have only to prove (iii).

2) For any Hilbert Γ-module M denote M^* the linear space of all bounded linear (or antilinear, depending on whether the given duality is bilinear or sesquilinear) maps $M \to \mathbf{C}$. Then M^* is again a Hilbert Γ-module (which can be identified with M): the action of Γ on M^* is defined by the formula

$$(\gamma f)(u) = f(\gamma^{-1}u) , \qquad f \in M^* , \ u \in M , \ \gamma \in \Gamma .$$

It is easy to check that the natural pairing

$$M^* \times M \to \mathbf{C}$$
$$\{f,u\} \mapsto (f,u) = f(u)$$

is Γ-invariant, so it becomes a Γ-pairing in the sense of Definition 1.22. Obviously $\dim_\Gamma M^* = \dim_\Gamma M$. Let us consider the following exact sequence

$$0 \longrightarrow M^\circ \xrightarrow{i} L' \xrightarrow{p} \bar{M}^*$$

where i is the natural inclusion, p is defined as follows:

$$(ph)(u) = (h,u) , \qquad h \in L' , \ u \in \bar{M} .$$

Obviously i and p are morphisms of Hilbert Γ-modules. Therefore p induces an injective morphism of Hilbert Γ-modules $\hat{p} : L'/M^\circ \to \bar{M}^*$. It follows that

$$\operatorname{codim}_\Gamma M^\circ = \dim_\Gamma L'/M^\circ \leq \dim_\Gamma \bar{M}^* = \dim_\Gamma \bar{M} .$$

3) Consider a natural linear map $j : \bar{M} \to (L')^*$ which maps $u \in \bar{M}$ to $ju = (u,\cdot) \in (L')^*$. This map is a morphism of Hilbert Γ-modules and it is injective due to the non-degeneracy of the duality. Since all elements

in $j(\bar{M})$ vanish on M° it obviously defines also an injective morphism of Hilbert Γ-modules $\hat{j} : \bar{M} \to (L'/M^0)^*$ (which is dual to $\hat{p}$ is an obvious sense). It follows that

$$\dim_\Gamma \bar{M} \le \dim_\Gamma (L'/M^\circ)^* = \dim_\Gamma L'/M^\circ = \operatorname{codim}_\Gamma M^\circ$$

which proves (iii). □

2. Complexes of Hilbert Γ-modules

A. Definition 2.1: A sequence

$$L : 0 \longrightarrow L_0 \xrightarrow{d_0} L_1 \xrightarrow{d_1} \ldots \longrightarrow L_k \xrightarrow{d_k} L_{k+1} \longrightarrow \ldots \xrightarrow{d_{N-1}} L_N \longrightarrow 0 \quad (2.1)$$

of Hilbert Γ-modules and their morphisms is called a *complex of Hilbert Γ-modules* if $d_{k+1} \circ d_k = 0$, i.e. $\operatorname{Im} d_k \subset \operatorname{Ker} d_{k+1}$. It is called an *almost exact sequence* of Hilbert Γ-modules if $\overline{\operatorname{Im} d_k} = \operatorname{Ker} d_{k+1}$ (i.e. $\operatorname{Im} d_k$ is dense in $\operatorname{Ker} d_{k+1}$) for all $k = -1, 0, \ldots, N-1$. (Here by definition $L_{-1} = L_{N+1} = 0$, $d_{-1} = d_N = 0$.)

Definition 2.2: The *reduced cohomologies* of a complex of Hilbert Γ-modules (2.1) are defined as Hilbert Γ-modules

$$\bar{H}^k(L) = \operatorname{Ker} d_k / \overline{\operatorname{Im} d_{k-1}} \ .$$

Lemma 2.3. *If L is a complex of Hilbert Γ-modules and $\dim_\Gamma L_k < \infty$ for all $k = 0, 1, \ldots, N$, then*

$$\sum_{j=0}^{N} (-1)^j \dim_\Gamma L_j = \sum_{j=0}^{N} (-1)^j \dim_\Gamma \bar{H}^j(L) \ .$$

We skip the proof which does not differ from the proof of the corresponding statement for the trivial group Γ (and for the spaces L_j which are finite-dimensional in the usual sense) except almost isomorphisms should be used instead of usual isomorphisms and Lemma 1.6 should be applied.

Corollary 2.4. *If (2.1) is an almost exact sequence of Hilbert Γ-modules with finite Γ-dimensions then*

$$\sum_{j=0}^{N} (-1)^j \dim_\Gamma L_j = 0 \ .$$

Definition 2.5. *Laplacians of a Hilbert complex of Γ-modules (2.1) are operators*

$$\Delta_k = d_k^* d_k + d_{k-1} d_{k-1}^* \ ,$$

$k = 0, 1, \ldots, N$.

Obviously Δ_k is a selfadjoint operator in L_k. It also commutes with the action of Γ (so it is again a morphism of Hilbert Γ-modules). It follows that all its spectral projections commute with the action of Γ, hence the Γ-dimension of their images is well defined. Denote $\mathcal{H}^k = \operatorname{Ker}\Delta_k$ (the space of all "harmonic forms"). Obviously $\mathcal{H}^k = \operatorname{Ker} d_k \cap \operatorname{Ker} d^*_{k-1}$.

LEMMA 2.6 (Kodaira decomposition). *There is the following decomposition of L_k into an orthogonal sum of Hilbert Γ-modules*

$$L_k = \overline{\operatorname{Im} d_{k-1}} \oplus \mathcal{H}^k \oplus \overline{\operatorname{Im} d^*_k} \ . \tag{2.2}$$

In this decomposition

$$\operatorname{Ker} d_k = \overline{\operatorname{Im} d_{k-1}} \oplus \mathcal{H}^k \ , \quad \operatorname{Ker} d^*_{k-1} = \mathcal{H}^k \oplus \overline{\operatorname{Im} d^*_k} \ . \tag{2.3}$$

In particular there is a natural isomorphism of Hilbert Γ-modules $\mathcal{H}^k \cong \bar{H}^k(L)$.

We omit the proof that actually does not use the action of Γ, and the Γ-invariance of all the subspaces is obvious.

Suppose that we have two complexes of Hilbert Γ-modules: $L = \{L_j, d_j, j = 1, \ldots, N\}$ (see (2.1)) and $L' = \{L'_j, d'_j, j = 1, \ldots, N\}$.

DEFINITION 2.7: A *morphism* $f : L' \to L$ of complexes of Hilbert Γ-modules is a set of morphisms of Hilbert Γ-modules $f = \{f_j : L'_j \to L_j, j = 1, \ldots, N\}$, such that they commute with the differentials: $f_{j+1}d'_j = d_j f_j$ for all j.

The composition $f \circ g : L'' \to L$ of two morphisms of complexes of Hilbert Γ-modules $f : L' \to L$ and $g : L'' \to L'$ is defined by the maps $(f \circ g)_j = f_j \circ g_j$. It is again a morphism of complexes of Hilbert Γ-modules.

Let us consider a sequence

$$\cdots \longrightarrow L^{(k-1)} \xrightarrow{f^{(k-1)}} L^{(k)} \xrightarrow{f^{(k)}} L^{(k+1)} \longrightarrow \cdots \tag{2.4}$$

where all $L^{(k)}$ are complexes of Hilbert Γ-modules, $f^{(k)}$ their morphisms.

DEFINITION 2.8: The sequence (2.4) is called *exact* if it is exact in the usual algebraic sense in every term, i.e. $\operatorname{Im} f_j^{(k-1)} = \operatorname{Ker} f_j^{(k)}$ for all k and j. In particular $\operatorname{Im} f_j^{(k-1)}$ has to be a closed subspace in $L_j^{(k)}$.

Now suppose that we have a short exact sequence

$$0 \longrightarrow L' \xrightarrow{i} L \xrightarrow{r} L'' \longrightarrow 0 \tag{2.5}$$

of complexes of Hilbert Γ-modules. The differentials in L', L, L'' will be denoted d', d, d'' respectively. From this sequence we shall produce a sequence of reduced cohomologies

$$\cdots \longrightarrow \bar{H}^k(L') \xrightarrow{i^*_k} \bar{H}^k(L) \xrightarrow{r^*_k} \bar{H}^k(L'') \xrightarrow{\delta_k} \bar{H}^{k+1}(L') \xrightarrow{i^*_{k+1}} \cdots \ . \tag{2.6}$$

Let us describe the maps in this sequence. First of all i_k^* and r_k^* are induced by i_k and r_k and they are well defined. Indeed, $i_k(\operatorname{Ker} d'_k) \subset \operatorname{Ker} d_k$ and $i_k\left(\overline{\operatorname{Im} d'_{k-1}}\right) \subset \overline{\operatorname{Im} d_{k-1}}$ because i_k is bounded; hence i_k defines a bounded linear map $i_k^* : \bar{H}^k(L') \to \bar{H}^k(L)$ which is obviously a morphism of Hilbert Γ-modules. Similarly r_k^* is defined as a morphism of Hilbert Γ-modules.

To describe δ_k let us first introduce a bounded morphism of Hilbert Γ-modules $r_k^{(-1)} : L''_k \to L_k$, such that $r_k r_k^{(-1)} = \mathrm{Id}$ on L''_k. The simplest way to do this is to take $r_k^{(-1)}$ as the inverse operator to the restriction of r_k to $(\operatorname{Ker} r_k)^\perp = \operatorname{Im} r_k^* = \operatorname{Ker} i_k^*$. The latter subspace (which is a Hilbert Γ-module) will be then equal to $\operatorname{Im} r_k^{(-1)}$.

Let us identify $\bar{H}^k$ with the space $\mathcal{H}^k$ of "harmonic forms" – see Lemma 2.6 (and we shall denote by $\mathcal{H}^k(L)$ the corresponding space in the complex L if there is a need to specify L). Denote by P'_k the orthogonal projection on $\mathcal{H}^k(L')$ in L'_k.

Now we can define δ_k by the formula

$$\delta_k = P'_{k+1} i_{k+1}^{-1} d_k r_k^{(-1)} : \mathcal{H}^k(L'') \longrightarrow \mathcal{H}^k(L') \,. \tag{2.7}$$

This linear operator obviously commutes with the action of Γ. It is easy to see that it is everywhere defined on $\mathcal{H}^k(L'')$. Indeed, if $\omega \in \mathcal{H}^k(L'')$, then $d_k r_k^{(-1)}\omega \in \operatorname{Im} i_{k+1} = \operatorname{Ker} r_{k+1}$ because $r_{k+1} d_k r_k^{(-1)}\omega = d_k r_k r_k^{(-1)}\omega = d_k\omega = 0$. Now it is obvious that δ_k is bounded, hence is a morphism of Hilbert Γ-modules.

LEMMA 2.9. *The sequence* (2.6) *is a complex, i.e.* (i) $r_k^* i_k^* = 0$, (ii) $\delta_k r_k^* = 0$ *and* (iii) $i_{k+1}^* \delta_k = 0$, *for all* k.

We shall postpone the proof until shorter notation will be introduced.

Unfortunately the sequence (2.6) is not always exact or even almost exact (see an example in [CGr]). An additional condition for the complexes is required to obtain the almost exactness.

DEFINITION 2.10: A complex L of Hilbert Γ-modules is called Γ-*Fredholm* if all its Laplacians Δ_k are Γ-Fredholm operators.

Note that if L is Γ-Fredholm then $\dim_\Gamma \bar{H}^k(L) = \dim_\Gamma \mathcal{H}^k(L) < \infty$.

THEOREM 2.11 ([CGr]). *If in the exact sequence* (2.5) *all the complexes* L', L, L'' *are* Γ-*Fredholm, then the sequence of reduced cohomologies* (2.6) *is almost exact. Moreover the image of every map in* (2.6) *is* Γ-*dense in the kernel of the next map.*

Remark: The last statement follows from the first one because all the spaces $\bar{H}^k(L)$ have finite Γ-dimensions (see Corollary 1.16). But there is no way here to prove density without proving Γ-density, so we shall actually prove Γ-density and then use Lemma 1.8.

Now we shall introduce shorter notation which allows us to avoid using too many sub- and superscripts. A proof of Theorem 2.11 (different from the one in [CGr]) will be given later. We shall identify the complex (2.1) with the (graded) Hilbert Γ-module

$$L = \bigoplus_{k=0}^{N} L_k$$

which we shall denote by the same letter. Then we can introduce a differential $d : L \to L$ which is a morphism of Hilbert Γ-modules, such that $d|L_k = d_k$. It has degree 1 and $d^2 = 0$. Instead of the system of reduced cohomologies $\{\bar{H}^k(L) \mid k = 0, 1, \ldots, N\}$ a single reduced cohomology Γ-module

$$\bar{H}(L) = \bigoplus_{k=0}^{N} \bar{H}^k(L)$$

can be considered. Obviously

$$\bar{H}(L) = \operatorname{Ker} d / \overline{\operatorname{Im} d} \ .$$

Instead of the Laplacians Δ_k we can consider the Laplacian $\Delta = \bigoplus_{k=0}^{N} \Delta_k : L \to L$ which is a morphism of Hilbert Γ-modules; obviously

$$\Delta = dd^* + d^*d \ .$$

The Hodge decomposition (2.2) becomes

$$L = \overline{\operatorname{Im} d} \oplus \mathcal{H} \oplus \overline{\operatorname{Im} d^*} \ , \qquad (2.2')$$

where

$$\mathcal{H} = \mathcal{H}(L) = \operatorname{Ker} \Delta = \bigoplus_{k=0}^{N} \mathcal{H}^k \ .$$

Also (2.3) can be rewritten as

$$\operatorname{Ker} d = \overline{\operatorname{Im} d} \oplus \mathcal{H} \ , \quad \operatorname{Ker} d^* = \mathcal{H} \oplus \overline{\operatorname{Im} d^*} \qquad (2.3')$$

and there is a natural isomorphism of Hilbert Γ-modules $\mathcal{H} \cong \bar{H}(L)$. A morphism of complexes of Hilbert Γ-modules $f : L' \to L$ is just a morphism of the corresponding Γ-modules of degree 0, such that $fd' = df$ (here d', d are the differentials in L', L respectively). It induces a morphism of reduced cohomologies $f^* : \bar{H}(L') \to \bar{H}(L)$.

Suppose that we have a short exact sequence of complexes (2.5). Then the sequence (2.6) becomes a triangle

$$\begin{array}{ccc} \bar{H}(L') & \xrightarrow{i^*} & \bar{H}(L) \\ \delta \nwarrow & & \swarrow r^* \\ & \bar{H}(L'') & \end{array} \tag{2.6'}$$

Here δ has the degree 1 and its definition (2.7) can be rewritten as follows:

$$\delta = P'i^{-1}dr^{(-1)} : \mathcal{H}(L'') \longrightarrow \mathcal{H}(L') \,, \tag{2.7'}$$

where $r^{(-1)} = \bigoplus_k r_k^{(-1)} : L'' \to L$, $i^{-1} = \bigoplus_k i_k^{-1}$, $P' = \bigoplus_k P'_k$, so P' is the orthogonal projection on $\mathcal{H}(L')$. Denote also P, P'' the orthogonal projections to $\mathcal{H}(L), \mathcal{H}(L'')$ in L, L'' respectively.

Proof of Lemma 2.9: Let us consider all the maps in (2.6) or (2.6′) as maps in the spaces of "harmonic forms". Then in fact $r^* = P''r : \mathcal{H}(L) \to \mathcal{H}(L'')$, $i^* = Pi : \mathcal{H}(L') \to \mathcal{H}(L)$.

Note first that

$$r^*i^* = P''rPi = P''ri - P''r(I-P)i = -P''r(I-P)i \,.$$

Now $i(\mathcal{H}(L')) \subset \operatorname{Ker} d$, hence $(I-P)i(\mathcal{H}(L')) \subset \overline{\operatorname{Im} d}$ due to (2.3). Therefore $r(I-P)i(\mathcal{H}(L')) \subset \overline{\operatorname{Im} d''}$ and $P''r(I-P)i = 0$ on $\mathcal{H}(L')$ which proves (i).

Let us consider the map δr^* on $\mathcal{H}(L)$. We have

$$\delta r^* = P'i^{-1}dr^{(-1)}P''r : \mathcal{H}(L) \longrightarrow \mathcal{H}(L') \,.$$

For any $\omega \in \mathcal{H}(L)$ we have $d\omega = 0$, hence $r\omega \in \operatorname{Ker} d''$ and

$$P''(r\omega) = r\omega + \alpha \,, \quad \alpha \in \overline{\operatorname{Im} d''} \,.$$

We shall prove now that application of $P'i^{-1}dr^{(-1)}$ to both $r\omega$ and α gives 0.

Note that $r^{(-1)}r\omega = \omega + i\lambda$, $\lambda \in L'$. Since $d\omega = 0$ we have

$$P'i^{-1}dr^{(-1)}r\omega = P'i^{-1}di\lambda = P'i^{-1}id\lambda = P'd\lambda = 0 \,.$$

Now instead of taking all $\alpha \in \overline{\operatorname{Im} d''}$ it is sufficient to take only $\alpha \in \operatorname{Im} d''$ because the operator $P'i^{-1}dr^{(-1)}$ is continuous. So suppose that $\alpha = d''\beta$, $\beta \in L''$. Then

$$r^{(-1)}\alpha = r^{(-1)}d''\beta = r^{(-1)}d''rr^{(-1)}\beta = r^{(-1)}rdr^{(-1)}\beta = dr^{(-1)}\beta + i\gamma \,,$$

where $\gamma \in L'$. Here we should apply $P'i^{-1}d$ to both terms in the right-hand side. The first term then disappears and we get

$$P'i^{-1}dr^{(-1)}\alpha = P'i^{-1}di\gamma = P'i^{-1}id'\gamma = P'd'\gamma = 0$$

which ends the proof of (ii).

At last we consider the map $i^*\delta$ on $\mathcal{H}(L'')$. We have

$$i^*\delta = PiP'i^{-1}dr^{(-1)} : \mathcal{H}(L'') \longrightarrow \mathcal{H}(L) \,.$$

Note that obviously $Pii^{-1}dr^{(-1)} = Pdr^{(-1)} = 0$ on $\mathcal{H}(L'')$. Hence it is sufficient to prove that $Pi(I-P')i^{-1}dr^{(-1)}$ vanishes on $\mathcal{H}(L'')$. Consider $\omega \in \mathcal{H}(L'')$. Then $i^{-1}dr^{(-1)}\omega \in \operatorname{Ker} d'$ because $id'i^{-1}dr^{(-1)}\omega = dii^{-1}dr^{(-1)}\omega = d^2r^{(-1)}\omega = 0$. Therefore $(I-P')i^{-1}dr^{(-1)}\omega \in \overline{\operatorname{Im} d'}$. So we have to prove that Pi vanishes on $\overline{\operatorname{Im} d'}$. But again we can replace $\overline{\operatorname{Im} d'}$ by $\operatorname{Im} d'$ and the conclusion becomes obvious because $Pid' = Pdi = 0$. □

LEMMA 2.12. *Let L be a complex of Hilbert Γ-modules with the differential d. Then the following conditions are equivalent:*

(i) *L is Γ-Fredholm;*

(ii) *the operator $\tilde{d} : L/\operatorname{Ker} d \to \operatorname{Ker} d$ induced by d is Γ-Fredholm;*

(iii) *$\dim_\Gamma \bar{H}(L) < \infty$ and the operator $d : \overline{\operatorname{Im} d^*} \to \overline{\operatorname{Im} d}$ is Γ-Fredholm;*

(iv) *the adjoint complex L^* with the same total space $L^* = \bigoplus_k L_k$ as L but with the differential d^* is Γ-Fredholm;*

(v) *the operator $\tilde{d}^* : L^*/\operatorname{Ker} d^* \to \operatorname{Ker} d^*$ induced by d^* is Γ-Fredholm;*

(vi) *$\dim_\Gamma \bar{H}(L^*) < \infty$ and the operator $d^* : \overline{\operatorname{Im} d} \to \overline{\operatorname{Im} d^*}$ is Γ-Fredholm;*

(vii) *$\dim_\Gamma \bar{H}(L) < \infty$ and the operator $d^* : \overline{\operatorname{Im} d} \to \overline{\operatorname{Im} d^*}$ is Γ-Fredholm.*

Proof: The Kodaira decomposition (2.2) obviously implies the equivalence of (ii) and (iii), as well as the equivalence of (v) and (vi). Note also that the space of "harmonic forms" $\mathcal{H}(L) = \operatorname{Ker} d \cap \operatorname{Ker} d^*$ is the same for L and L^*, therefore (vi) and (vii) are equivalent. Operators $d : \overline{\operatorname{Im} d^*} \to \overline{\operatorname{Im} d}$ and $d^* : \overline{\operatorname{Im} d} \to \overline{\operatorname{Im} d^*}$ are adjoint to each other, so Corollary 1.14 implies the equivalence of (iii) and (vii). Therefore all the statements (ii),(iii),(v),(vi) and (vii) are equivalent. It remains to prove the equivalence of (i) to one of them (then the equivalence of (iv) to the rest of the statements will follow because we can replace L by L^* and apply the same arguments).

Now in the Kodaira decomposition (2.2′) all the spaces are invariant with respect to Δ which there becomes the direct sum

$$\Delta = dd^* \oplus 0 \oplus d^*d\ .$$

Here d^*d can be considered as the composition of the operators d^*, d restricted as in (vi),(iii), so they become almost isomorphisms. But then Proposition 1.13 implies that d^*d is Γ-Fredholm if and only if d is Γ-Fredholm. This immediately gives us the equivalence of (i) and (iii). □

Proof of Theorem 2.11: (i) First let us prove the almost exactness in the term $\bar{H}(L)$ in (2.6′). We shall identify the reduced cohomology spaces with the corresponding spaces of "harmonic forms", as in the proof of Lemma 2.9.

Suppose that $\omega \in \mathcal{H}(L)$. Then $r^*\omega = 0$ means that $P''r\omega = 0$ or $r\omega \in \overline{\operatorname{Im} d''}$ (because $r\omega \in \operatorname{Ker} d''$ and due to (2.3′)). Therefore

$$\operatorname{Ker} r^* = r^{-1}\left(\overline{\operatorname{Im} d''}\right) \cap \mathcal{H}(L)\ .$$

The almost exactness in the term $\bar{H}(L)$ will follow if we prove the following two statements:

(A) $r^{-1}(\operatorname{Im} d'') \cap \mathcal{H}(L)$ is Γ-dense in $r^{-1}(\overline{\operatorname{Im} d''}) \cap \mathcal{H}(L)$;

(B) $r^{-1}(\operatorname{Im} d'') \cap \mathcal{H}(L) \subset \operatorname{Im} i^*$.

To prove (A) note that $\operatorname{Im} d''$ is Γ-dense in $\overline{\operatorname{Im} d''}$ due to Lemmas 2.12 and 1.15. Therefore $r^{-1}(\operatorname{Im} d'')$ is Γ-dense in $r^{-1}(\overline{\operatorname{Im} d''})$ due to Lemma 1.19. Then (A) follows due to Lemma 1.17.

The statement (B) is proved more or less algebraically and is true just because it is true in the corresponding purely algebraic situation. Suppose that $\omega \in r^{-1}(\operatorname{Im} d'') \cap \mathcal{H}(L)$, i.e. $\omega \in \mathcal{H}(L)$ and $r\omega = d''\alpha$, $\alpha \in L''$. We have

$$d''\alpha = d''rr^{(-1)}\alpha = rdr^{(-1)}\alpha ,$$

so $r(\omega - dr^{(-1)}\alpha) = 0$ and $\omega - dr^{(-1)}\alpha = i\beta$. Applying the projection P to both sides here we get $\omega = Pi\beta$. It remains to replace β by a harmonic element.

Note first that $d'\beta = 0$ because $id'\beta = di\beta = d(\omega - dr^{(-1)}\alpha) = 0$. Using (2.3′) we get $\beta = P'\beta + \gamma$, $\gamma \in \overline{\operatorname{Im} d'}$. So

$$\omega = PiP'\beta + Pi\gamma = i^*P'\beta + Pi\gamma .$$

It remains to prove that $Pi\gamma = 0$. But this immediately follows from the fact that $i(\overline{\operatorname{Im} d'}) \subset \overline{\operatorname{Im} d}$ and P vanishes on $\overline{\operatorname{Im} d}$.

(ii) Let us check the almost exactness in the term $\bar{H}(L'')$ in (2.6′).

Suppose that $\omega \in \mathcal{H}(L'')$ and $\delta\omega = P'i^{-1}dr^{(-1)}\omega = 0$. Note that $i^{-1}dr^{(-1)}\omega \in \operatorname{Ker} d'$ because

$$id'i^{-1}dr^{(-1)}\omega = dii^{-1}dr^{(-1)}\omega = d^2r^{(-1)}\omega = 0 .$$

Hence due to (2.3′) the equality $\delta\omega = 0$ is equivalent to the inclusion $i^{-1}dr^{(-1)}\omega \in \overline{\operatorname{Im} d'}$ or $dr^{(-1)}\omega \in i(\overline{\operatorname{Im} d'})$. So we have

$$\operatorname{Ker} \delta = (dr^{(-1)})^{-1} i(\overline{\operatorname{Im} d'}) \cap \mathcal{H}(L'')$$

and it is sufficient to prove that

(A) $(dr^{(-1)})^{-1} i(\operatorname{Im} d') \cap \mathcal{H}(L'')$ is Γ-dense in $(dr^{(-1)})^{-1} i(\overline{\operatorname{Im} d'}) \cap \mathcal{H}(L'')$;

(B) $(dr^{(-1)})^{-1} i(\operatorname{Im} d') \cap \mathcal{H}(L'') \subset \operatorname{Im} r^*$.

Here (A) is proved exactly as in (i) if we notice that $i(\operatorname{Im} d')$ is Γ-dense in $i(\overline{\operatorname{Im} d'})$ because $\operatorname{Im} d'$ is Γ-dense in $\overline{\operatorname{Im} d'}$ by Lemmas 2.12 and 1.15 and $i : L' \to i(L')$ is a topological isomorphism of Γ-modules.

Let us prove (B). Suppose that $\omega \in \mathcal{H}(L'')$ and $dr^{(-1)}\omega \in i(\operatorname{Im} d')$. This means that $dr^{(-1)}\omega = id'\alpha = di\alpha$, hence $d(r^{(-1)}\omega - i\alpha) = 0$ or $r^{(-1)}\omega - i\alpha \in \operatorname{Ker} d$ where $\alpha \in L'$. Note that $\omega = r(r^{(-1)}\omega - i\alpha)$, hence

$$\omega = P''r(r^{(-1)}\omega - i\alpha) .$$

Again it remains to replace $r^{(-1)}\omega - i\alpha$ by a harmonic element. We have due to (2.3′)

$$r^{(-1)}\omega - i\alpha = P(r^{(-1)}\omega - i\alpha) + \beta ,$$

where $\beta \in \overline{\operatorname{Im} d}$. Hence

$$\omega = P''rP(r^{(-1)}\omega - i\alpha) + P''r\beta .$$

But $r\beta \in \overline{\operatorname{Im} d''}$, hence $P''r\beta = 0$, so

$$\omega = P''rP(r^{(-1)}\omega - i\alpha) = r^*P(r^{(-1)}\omega - i\alpha)$$

which proves (B).

(iii) It remains to prove almost exactness in the term $\bar{H}(L')$ in (2.6′).

Suppose that $\omega \in \mathcal{H}(L')$ and $i^*\omega = Pi\omega = 0$. This means that $i\omega \in \overline{\operatorname{Im} d}$, so we have

$$\operatorname{Ker} i^* = i^{-1}(\overline{\operatorname{Im} d}) \cap \mathcal{H}(L') .$$

Again it is sufficient to prove the following two statements:

(A) $i^{-1}(\operatorname{Im} d) \cap \mathcal{H}(L')$ is Γ-dense in $i^{-1}(\overline{\operatorname{Im} d}) \cap \mathcal{H}(L')$;

(B) $i^{-1}(\operatorname{Im} d) \cap \mathcal{H}(L') \subset \operatorname{Im} \delta$.

Proof of (A) is identical with the proof of (A) in (i). To prove (B) suppose that $\omega \in i^{-1}(\operatorname{Im} d) \cap \mathcal{H}(L')$, i.e. $\omega \in \mathcal{H}(L')$ and $i\omega = d\alpha$, $\alpha \in L$. Then $\omega = i^{-1}d\alpha$ and applying P' we get $\omega = P'i^{-1}d\alpha$. Now denote $Q = r^{(-1)}r$, so $Q^2 = Q$, $\operatorname{Ker} Q = \operatorname{Im}(I - Q) = \operatorname{Ker} r = \operatorname{Im} i$. We can write then

$$\alpha = Q\alpha + (I - Q)\alpha = r^{(-1)}r\alpha + i\beta , \quad \beta \in L' .$$

Notice that we can apply $i^{-1}d$ to $i\beta$ because $di\beta = id'\beta \in \operatorname{Im} i$. Since $i^{-1}d$ can be applied to α as well, it can be applied to $r^{(-1)}r\alpha$. Therefore we can write

$$\omega = P'i^{-1}dr^{(-1)}(r\alpha) + P'i^{-1}di\beta .$$

But $i^{-1}di\beta = i^{-1}id'\beta = d'\beta \in \operatorname{Im} d'$, so $P'i^{-1}di\beta = 0$. Hence

$$\omega = P'i^{-1}dr^{(-1)}(r\alpha) .$$

It remains to replace $r\alpha$ by a "harmonic form" (an element of $\mathcal{H}(L'')$). Note first that $r\alpha \in \operatorname{Ker} d''$. Indeed, $d''r\alpha = rd\alpha = ri\omega = 0$. Therefore $r\alpha = P''r\alpha + \gamma$ with $\gamma \in \overline{\operatorname{Im} d''}$. Denoting $\kappa = P''r\alpha$ we obtain

$$\omega = \delta\kappa + P'i^{-1}dr^{(-1)}\gamma$$

and it is sufficient to prove that the last term vanishes. Again it is sufficient to consider the case when $\gamma \in \operatorname{Im} d''$, i.e. $\gamma = d''\lambda$, $\lambda \in L''$.

Denote $\xi = r^{(-1)}\gamma = r^{(-1)}d''\lambda$. Then $r\xi = d''\lambda = d''rr^{(-1)}\lambda = rdr^{(-1)}\lambda$. Hence $r(\xi - dr^{(-1)}\lambda) = 0$ and $\xi = dr^{(-1)}\lambda + i\eta$, $\eta \in L'$. It follows that $d\xi = di\eta = id'\eta$ and

$$P'i^{-1}dr^{(-1)}\gamma = P'i^{-1}d\xi = P'i^{-1}id'\eta = P'd'\eta = 0$$

which proves that $\omega = \delta\kappa$. □

DEFINITION 2.13: Suppose that L is a Γ-Fredholm complex of Hilbert Γ-modules (2.1). Its *Euler Γ-characteristic* is defined as

$$\chi_\Gamma(L) = \sum_{j=0}^{N}(-1)^j \dim_\Gamma \bar{H}^j(L)\,.$$

COROLLARY 2.14. *Suppose that in the short exact sequence* (2.5) *all complexes* L', L, L'' *are Γ-Fredholm. Then*

$$\chi_\Gamma(L) = \chi_\Gamma(L') + \chi_\Gamma(L'')\,.$$

Proof: The result immediately follows if we apply Corollary 2.4 to the almost exact sequence (2.6). □

LEMMA 2.15. *A short complex of Hilbert Γ-modules*

$$L : 0 \longrightarrow L_0 \xrightarrow{d_0} L_1 \longrightarrow 0 \tag{2.8}$$

is Γ-Fredholm if and only if the operator d_0 *is Γ-Fredholm. If this is the case, then*

$$\operatorname{ind}_\Gamma d_0 = \chi_\Gamma(L)\,. \tag{2.9}$$

Proof: The equivalence of the conditions of being Γ-Fredholm for L and d_0 follows immediately from Lemma 2.12. The formula (2.9) is obvious. □

Let us mention that the following well known statement can be easily deduced from Corollary 2.14:

PROPOSITION 2.16. *Let* $A : E \to F$ *and* $B : F \to G$ *are morphisms of Hilbert Γ-modules which are both Γ-Fredholm. Then* $BA : E \to G$ *is also Γ-Fredholm and*

$$\operatorname{ind}_\Gamma BA = \operatorname{ind}_\Gamma A + \operatorname{ind}_\Gamma B\,.$$

Proof: The operator BA is Γ-Fredholm by Corollary 1.14. Now consider the following commutative diagram of morphisms of Hilbert Γ-modules:

$$\begin{array}{ccccccccc}
0 & \longrightarrow & E & \xrightarrow{1_E \oplus A} & E \oplus F & \xrightarrow{A+(-1_F)} & F & \longrightarrow & 0 \\
 & & \Big\downarrow{\scriptstyle A} & & \Big\downarrow{\scriptstyle (BA)\oplus 1_F} & & \Big\downarrow{\scriptstyle B} & & \\
0 & \longrightarrow & F & \xrightarrow{B\oplus 1_F} & G \oplus F & \xrightarrow{1_G - B} & G & \longrightarrow & 0
\end{array}$$

where $1_E, 1_F, 1_G$ are the identity morphisms of the corresponding Hilbert Γ-modules. Note that $BA \oplus 1_F$ is Γ-Fredholm and $\operatorname{ind}_\Gamma((BA) \oplus 1_F) = \operatorname{ind}_\Gamma(BA)$. Now in this diagram both rows are algebraically exact and all three columns can be considered as short Γ-Fredholm complexes due to Lemma 2.15. Hence we can apply Corollary 2.14 and the result immediately follows. □

B. Now we shall provide a useful tool to prove that a complex is Γ-Fredholm.

LEMMA 2.17. *Consider a short exact sequence* (2.5) *of complexes of Hilbert Γ-modules. Suppose that among 3 complexes L', L, L'' one is Γ-Fredholm and one of the other two has finite Γ-dimension. Then the third complex is Γ-Fredholm.*

Proof: 1) The case when $\dim_\Gamma L < \infty$ is obvious because then $\dim_\Gamma L' < \infty$ and $\dim_\Gamma L'' < \infty$. The remaining possibilities are as follows:

(i) $\dim_\Gamma L' < \infty$, L is Γ-Fredholm;
(ii) $\dim_\Gamma L' < \infty$, L'' is Γ-Fredholm;
(iii) L' is Γ-Fredholm, $\dim_\Gamma L'' < \infty$;
(iv) L is Γ-Fredholm, $\dim_\Gamma L'' < \infty$.

Note that (iii) will be reduced to (ii) and (iv) to (i) if we replace all the complexes L', L, L'' by the adjoint complexes (see (iv) in Lemma 2.12) and the morphisms i, r by the adjoint operators. So we have to consider the cases (i) and (ii) only. Therefore we will suppose that $\dim_\Gamma L' < \infty$.

2) Let us consider the following commutative diagram of morphisms of Hilbert Γ-modules

$$\begin{array}{ccccccccc} & & L'/\operatorname{Ker} d' & \xrightarrow{\tilde{i}} & L/\operatorname{Ker} d & \xrightarrow{\tilde{r}} & L''/\operatorname{Ker} d'' & \longrightarrow & 0 \\ & & \downarrow{\scriptstyle \tilde{d}'} & & \downarrow{\scriptstyle \tilde{d}} & & \downarrow{\scriptstyle \tilde{d}''} & & \\ 0 & \longrightarrow & \operatorname{Ker} d' & \xrightarrow{i_0} & \operatorname{Ker} d & \xrightarrow{r_0} & \operatorname{Ker} d'' & & \end{array}$$

where $\tilde{f}$ is induced by f for $f = i, r, d', d$ or d''. The morphisms i_0, r_0 are restrictions of i, r respectively. In this diagram the second row is exact and $\tilde{r}$ is surjective, so the first row is exact in the term $L''/\operatorname{Ker} d''$.

Let us prove that r_0 is Γ-Fredholm. To do this replace $\operatorname{Ker} d$ by $r^{-1}(\operatorname{Ker} d'')$. Then we have an exact sequence of Hilbert Γ-modules

$$0 \longrightarrow L' \xrightarrow{i} r^{-1}(\operatorname{Ker} d'') \xrightarrow{r_1} \operatorname{Ker} d'' \longrightarrow 0 ,$$

where r_1 is induced by r. Since $\dim_\Gamma L' < \infty$, the operator r_1 in this sequence is Γ-Fredholm. Let us compare r_1 and r_0. We have

$$r^{-1}(\operatorname{Ker} d'') = \operatorname{Ker}(d''r) = \operatorname{Ker}(rd) = d^{-1}(\operatorname{Ker} r) = d^{-1}(\operatorname{Im} i) \supset \operatorname{Ker} d .$$

But $\dim_\Gamma d^{-1}(\operatorname{Im} i)/\operatorname{Ker} d < \infty$ because the operator $d^{-1}(\operatorname{Im} i)/\operatorname{Ker} d \to \operatorname{Im} i$ induced by d, is injective. Therefore $\dim_\Gamma r^{-1}(\operatorname{Ker} d'')/\operatorname{Ker} d < \infty$ and r_0 is a restriction of r_1 to a closed subspace of finite Γ-codimension, so r_0 is Γ-Fredholm by Lemma 1.21.

Now it is obvious that $\tilde{r}$ is also Γ-Fredholm because $\operatorname{Ker} \tilde{r} = r^{-1}(\operatorname{Ker} d'')/\operatorname{Ker} d$ and $\tilde{r}$ is surjective.

3) Let us consider the equality

$$\tilde{d}''\tilde{r} = r_0\tilde{d}$$

where both operators $\tilde{r}, r_0$ are Γ-Fredholm. Since the Γ-Fredholm property means invertibility modulo Γ-trace class operators, the operator $\tilde{d}$ is Γ-Fredholm if and only if the operator $\tilde{d}''$ is Γ-Fredholm. This proves that L'' is Γ-Fredholm in the case (i) and that L is Γ-Fredholm in the case (ii) as required. □

C. Our next goal is to relax requirements leading to the additivity of the Euler characteristic in Corollary 2.14 in a particular case that we shall need later.

LEMMA 2.18. *Suppose that* (2.5) *is a short sequence of complexes of Hilbert Γ-modules and their morphisms such that it is exact in the terms L', L and almost exact in the term L'' (i.e. $\operatorname{Im} r$ is dense in L''). Then we can replace d'' and r by new maps so that* (2.5) *will become an exact (in all terms) sequence of complexes of Hilbert Γ-modules.*

Proof: Let us consider the polar decomposition $r = TV$ of the morphism r; here V is a partial isometry from L to L'' with $\operatorname{Ker} V = \operatorname{Ker} r = \operatorname{Im} i$, $\operatorname{Im} V = \overline{\operatorname{Im} r} = L''$; $T = \sqrt{rr^*}$ is a self-adjoint (bounded) operator in L'' with $\operatorname{Ker} T = \operatorname{Ker} r^* = 0$, $\operatorname{Im} T = \operatorname{Im} r$, so in fact T is an almost isomorphism of the Hilbert Γ-module L'' to itself.

We shall replace r by V and d'' by $b'' = VdV^*$. Then we obtain the exact sequence of Hilbert Γ-modules

$$0 \longrightarrow L' \stackrel{i}{\longrightarrow} L \stackrel{V}{\longrightarrow} L'' \longrightarrow 0$$

and it remains to prove that V becomes a morphism of complexes, i.e. that $b''V = Vd$. This means that $VdV^*V = Vd$ or $Vd(I-P) = 0$ where $P = V^*V$ is an orthogonal projection with $\operatorname{Ker} P = \operatorname{Ker} V = \operatorname{Im} i$. Hence $I - P$ is an orthogonal projection with $\operatorname{Im}(I-P) = \operatorname{Im} i$. Hence we have to prove that $Vdi = 0$ which is obvious because $Vdi = Vid' = 0$ because $Vi = 0$. □

Remark: In fact the following diagram

$$\begin{array}{ccccc} L & \xrightarrow{V} & L'' & \xrightarrow{T} & L'' \\ \downarrow{\scriptstyle d} & & \downarrow{\scriptstyle b''} & & \downarrow{\scriptstyle d''} \\ L & \xrightarrow{V} & L'' & \xrightarrow{T} & L'' \end{array}$$

is commutative. We already proved the commutativity of the left square. Now using this commutativity we get

$$d''TV = d''r = rd = TVd = Tb''V$$

and we can remove V from the equality $d''TV = Tb''V$ because V is surjective. This gives the desired commutativity of the right square.

COROLLARY 2.19. *Suppose that* (2.5) *is a short sequence of complexes of Hilbert* Γ*-modules such that it is exact in the terms* L', L *and almost exact in the term* L''. *Suppose also that* $\dim_\Gamma L'' < \infty$. *Then* L' *is* Γ*-Fredholm if and only if* L *is* Γ*-Fredholm, and if it is true then*

$$\chi_\Gamma(L) = \chi_\Gamma(L') + \chi_\Gamma(L'') .$$

Proof: Using Lemmas 2.18 and 2.17 we should only notice that $\chi_\Gamma(L'')$ does not depend on the choice of the differential in L'' due to Lemma 2.3. □

3. Main Theorem and Examples

Let X be a C^∞-manifold (without boundary), with a free action of a discrete group Γ such that $\bar{X} = X/\Gamma$ is compact, $\dim_{\mathbf{R}} X = n$. Suppose that E is a complex vector Γ-bundle over X, i.e. a C^∞ complex vector bundle over X with a free C^∞ action of Γ on the space of this bundle, such that this action covers the action of Γ on X and is linear on fibers (i.e. the action of $\gamma \in \Gamma$ on E maps linearly the fiber E_x of E over $x \in X$ to the fiber $E_{\gamma x}$ over γx).

Denoting $\bar{E} = E/\Gamma$ we see that $\bar{E}$ is a complex C^∞ vector bundle on $\bar{X}$ and $E = \pi^*\bar{E}$ where $\pi : X \to \bar{X}$ is the canonical projection.

Let us take a Γ-invariant positive smooth density $d\mu(x)$ on X which defines a Γ-invariant measure on X. For any complex vector Γ-bundle E let us take a Γ-invariant C^∞ hermitian positive definite scalar product in fibers of E and denote it h_E. Then we can define the Hilbert space $L^2(X, E)$ as the space of (classes of) measurable sections u of E such that

$$\|u\|^2_{L^2(X,E)} = \int_X h_E\big(u(x), u(x)\big)\, d\mu(x) < \infty . \tag{3.1}$$

It is easy to see that the space $L^2(X, E)$ does not depend on the choice of $d\mu$ and h_E: the change of these arbitrary elements leads to replacing of the norm (3.1) by an equivalent norm.

If $E = \mathbf{C}_X$ is the trivial vector bundle over X with the fiber $\mathbf{C}$ then we obtain $L^2(X, \mathbf{C}_X) = L^2(X)$.

If u is a section of a vector Γ-bundle E, then the action of $\gamma \in \Gamma$ on u is defined as follows:

$$(R_\gamma u)(x) = \hat{\gamma}_{\gamma^{-1}x}\left[u(\gamma^{-1}x)\right] , \tag{3.2}$$

where $\hat{\gamma}_y : E_y \to E_{\gamma y}$ for any $y \in X$ denotes the action induced by γ on the fiber E_y. Then $\{R_\gamma \mid \gamma \in \Gamma\}$ define a unitary representation of Γ in $L^2(X, E)$ (with the norm (3.1) induced by a Γ-invariant density on X and a Γ-invariant hermitian scalar product in E). Moreover then $L^2(X, E)$ becomes a (projective) Hilbert Γ-module. Indeed, we can always find a complex C^∞ vector bundle $\bar{F}$ on $\bar{X}$ such that $\bar{E} \oplus \bar{F} = \mathbf{C}^N_{\bar{X}}$ where $\mathbf{C}^N_{\bar{X}}$ is the trivial vector bundle with the fiber $\mathbf{C}^N$ over $\bar{X}$ (see, e.g. [A1]). But then $F = \pi^* \bar{F}$ is a complex vector Γ-bundle over X and $E \oplus F$ is canonically isomorphic to $\mathbf{C}^N_X$, so

$$L^2(X, E) \oplus L^2(X, F) = L^2(X, \mathbf{C}^N_X) = L^2(X) \otimes \mathbf{C}^N . \tag{3.3}$$

This gives an imbedding of $L^2(X, E)$ into a Hilbert space $L^2(X) \otimes \mathbf{C}^N$ where Γ acts on the first factor only by the formula

$$(R_\gamma u)(x) = u(\gamma^{-1}x) , \quad u \in L^2(X) . \tag{3.4}$$

Now $L^2(X)$ can be identified with the free Hilbert Γ-module $L^2\Gamma \otimes L^2(F)$, where F is a fundamental domain of the action of Γ on X and the action of $\gamma \in \Gamma$ on $L^2\Gamma \otimes L^2(F)$ is given by the operator $R_\gamma \otimes I$ (see notation in §1). To do this we should identify $f \in L^2(X)$ with $\sum_{g \in \Gamma} \delta_g \otimes f_g$, where $f_g(x) = f(g^{-1}x) \mid F \in L^2(F)$. It is easy to check then that the operator (3.4) corresponds to $R_\gamma \otimes I$ in $L^2\Gamma \otimes L^2(F)$. Therefore (3.3) gives an imbedding of $L^2(X, E)$ into a free Hilbert Γ-module; hence $L^2(X, E)$ is a Hilbert Γ-module.

We shall also denote $C^\infty(X, E)$ the space of all C^∞-sections of E over X; $C^\infty_c(X, E)$ will be the space of all such sections with a compact support. The formula (3.2) defines an action of Γ on these spaces as well.

For any complex C^∞ vector Γ-bundle E over X denote by E^* any complex vector Γ-bundle which is equipped with a C^∞ bilinear or sesquilinear non-degenerate Γ-invariant duality of bundles $E \times E^* \to \Omega(X)$ where $\Omega(X)$ is the density bundle over X (which is obviously a Γ-bundle also). Then we have bilinear or sesquilinear duality on sections

$$(\cdot, \cdot) : C^\infty_c(X, E) \times C^\infty(X, E^*) \longrightarrow \mathbf{C} , \quad (u, v) = \int_X (u(x), v(x))_x , \tag{3.5}$$

where $(\cdot, \cdot)_x$ denotes the given duality in the fibers over the point $x \in X$. This duality has the following Γ-invariance property

$$(R_\gamma u, R_\gamma v) = (u, v)\ , \qquad u \in C_c^\infty(X, E)\ ,\ v \in C^\infty(X, E^*)\ . \tag{3.6}$$

Replacing $C^\infty(X, E^*)$ by $C_c^\infty(X, E^*)$ in (3.5) we can then extend the duality by continuity to a Γ-invariant (linear or sesquilinear) non-degenerate duality

$$(\cdot,\cdot) : L^2(X, E) \times L^2(X, E^*) \longrightarrow \mathbf{C}\ , \tag{3.7}$$

There is a canonical choice of E^*: we can take

$$E^* = \mathrm{Hom}_{\mathbf{C}}(E, \Omega(X)) \quad \text{or} \quad E^* = \overline{\mathrm{Hom}}_{\mathbf{C}}(E, \Omega(X))\ ,$$

where Hom, $\overline{\mathrm{Hom}}$ denote linear and antilinear homomorphisms of bundles respectively. But sometimes more general choice of E^* is convenient. For example, if we are given a C^∞ hermitian positive definite Γ-invariant scalar product h_E in fibers of E and a positive smooth Γ-invariant density $d\mu(x)$ on X, then we can identify E^* with E using the sesquilinear duality

$$E_x \times E_x \longrightarrow \Omega(X)_x\ ,$$
$$u, v \longmapsto h_E(u, v)d\mu(x)\ .$$

Then (3.7) becomes the scalar product in $L^2(X, E)$ with the corresponding norm given by (3.1).

Denote by $\mathcal{D}'(X, E)$ the space of all distributional sections of E over X (it can be defined as the dual space to $C_c^\infty(X, E^*)$). If E is a Γ-bundle then the action of Γ on $\mathcal{D}'(X, E)$ can be defined as the extension by continuity of the action given by (3.2).

Consider a Γ-invariant connection ∇ in E. Then for any $s \in \mathbf{Z}_+$ (i.e. a non-negative integer s) we can define the Sobolev space $W^s(X, E)$ as the space of $u \in \mathcal{D}'(X, E)$ such that $\nabla^k u \in L^2(X, E)$ for all $k = 0, 1, \ldots, s$, so that

$$\|u\|^2_{W^s(X,E)} = \sum_{k=0}^{s} \|\nabla^k u\|^2_{L^2(X,E)} < \infty\ . \tag{3.8}$$

Here the derivatives are of course taken in the sense of distributions.

The spaces $W^s(X, E)$ are uniform Sobolev spaces (they can be defined on any manifold of bounded geometry – see, e.g. [R],[S]). Equality (3.3) will still hold if we replace L^2 by W^s, so $W^s(X, E)$ is a Hilbert Γ-module. Note that $W^0(X, E) = L^2(X, E)$.

For a negative $s \in \mathbf{Z}$ the space $W^s(X, E)$ can be defined as a dual space (of distributional sections of E) to $W^{-s}(X, E)$ with respect to the (sesquilinear) duality given by the scalar product in $L^2(X, E)$ (see (3.1)). It is again a Hilbert Γ-module. Denote also

$$W^\infty(X, E) = \bigcap_{s \in \mathbf{Z}} W^s(X, E)\ , \quad W^{-\infty}(X, E) = \bigcup_{s \in \mathbf{Z}} W^s(X, E)\ .$$

By $W^s(X), W^\infty(X), W^{-\infty}(X)$ we shall denote the corresponding spaces with $E = \mathbf{C}_X$.

Replacing $C^\infty(X, E^*)$ by $C_c^\infty(X, E^*)$ in (3.5) we can then extend the duality by continuity to a Γ-invariant (linear or sesquilinear) duality

$$(\cdot,\cdot) : W^s(X, E) \times W^{s'}(X, E^*) \longrightarrow \mathbf{C} , \quad s + s' \geq 0 \qquad (3.9)$$

with the same invariance property (3.6).

For any section space $\mathcal{F} \subset \mathcal{D}'(X, E)$ and any closed subset $D \subset X$ denote

$$\mathcal{F}_D = \{u \mid u \in \mathcal{F} , \ \operatorname{supp} u \subset D\} .$$

If D and $\mathcal{F}$ are Γ-invariant (and E is a Γ-bundle) then $\mathcal{F}_D$ is Γ-invariant either. We shall write $\mathcal{D}'_D(X, E), W^s_D(X, E), W^{-\infty}_D(X, E), \ldots$ instead of $\mathcal{D}'(X, E)_D, W^s(X, E)_D, W^{-\infty}(X, E)_D, \ldots$. If D is Γ-invariant and E is a Γ-bundle then $W^s_D(X, E)$ is a Hilbert Γ-module.

Suppose that E, F are vector Γ-bundles over X,

$$A : C^\infty(X, E) \longrightarrow C^\infty(X, F) \qquad (3.10)$$

a differential operator of order d which commutes with the action of Γ in $C^\infty(X, E)$. Then A naturally defines a differential operator $\bar{A}$ on $\bar{X} = X/\Gamma$

$$\bar{A} : C^\infty(\bar{X}, \bar{E}) \longrightarrow C^\infty(\bar{X}, \bar{F}) , \qquad (3.11)$$

where $\bar{E} = E/\Gamma$, $\bar{F} = F/\Gamma$. Also A can be extended by continuity to a morphism of Hilbert Γ-modules

$$A : W^s(X, E) \longrightarrow W^{s-d}(X, F) \qquad (3.12)$$

for any $s \in \mathbf{Z}$. Now suppose that A is elliptic (or, equivalently, $\bar{A}$ is elliptic). Then M. Atiyah ([A2]) proved that the morphism (3.12) is Γ-Fredholm and its Γ-index $\operatorname{ind}_\Gamma A$ is given by the formula

$$\operatorname{ind}_\Gamma A = \operatorname{ind} \bar{A} . \qquad (3.13)$$

If A is a Γ-invariant differential operator (3.10) then the adjoint operator

$$A^* : C^\infty(X, F^*) \longrightarrow C^\infty(X, E^*) \qquad (3.14)$$

is again a Γ-invariant differential operator, such that

$$(Au, v) = (u, A^* v) , \quad u \in C_c^\infty(X, E) , \quad v \in C^\infty(X, F^*) . \qquad (3.15)$$

This will also be true with the extended duality (like in (3.9)) for all $u \in W^{s+d}(X, E)$, $v \in W^{s'}(X, E^*)$, $s + s' \geq 0$.

Note also that the following elliptic regularity result is true provided A is a Γ-invariant elliptic differential operator of order d

$$u \in W^{-\infty}(X, E) , \quad Au \in W^s(X, F) \quad \text{imply} \quad u \in W^{s+d}(X, E) \qquad (3.16)$$

for any $s \in \mathbf{Z}$ (see, e.g. [S] for a more general statement on manifolds of bounded geometry).

DEFINITION 3.1: *A rigged* Γ*-divisor* (associated with a Γ-invariant elliptic differential operator A acting as in (3.10)) is a tuple

$$\mu = (D^+, L^+; D^-, L^-)$$

where D^+, D^- are disjoint closed nowhere dense Γ-invariant subsets in X, $L^\pm$ are Γ-invariant linear spaces of distributional sections

$$L^+ \subset \mathcal{D}'_{D^+}(X, F) , \quad L^- \subset \mathcal{D}'_{D^-}(X, E^*) ,$$

such that the following conditions are fulfilled

1) there exist $N_1, N_2 \in \mathbf{Z}_+$ such that

$$L^+ \subset W^{-N_1}(X, F) , \quad L^- \subset W^{-N_2}(X, E^*) \tag{3.17}$$

and $L^\pm$ are closed subspaces in the corresponding Hilbert spaces W^{-N_j};

2) $\ell^\pm = \dim_\Gamma L^\pm < \infty$.

Remark: The fact that L is closed in $W^p(X, F)$ and has a finite Γ-dimension does not generally imply that L is closed in $W^q(X, F)$ for any other number $q \in \mathbf{R}$. However if this happens, then the Γ-dimension of L will not depend on the choice of the Sobolev space. More generally, if $q < p$ then the closure $\bar{L}$ of L in $W^q(X, F)$ with the induced Hilbert structure has the same Γ-dimension as L (with the Hilbert structure induced from $W^p(X, F)$) because then the inclusion $L \subset \bar{L}$ is an injective continuous morphism of Hilbert Γ-modules with a dense image, i.e. almost isomorphism (see Lemma 1.6). Therefore the numbers $\ell^\pm$ do not depend on the choice of N_1 and N_2.

We shall also need "*secondary*" spaces of distributional sections which are defined as follows

$$\tilde{L}^+ = \{u \mid u \in W^{-\infty}_{D^+}(X, E) , \ Au \in L^+\} ,$$
$$\tilde{L}^- = \{v \mid v \in W^{-\infty}_{D^-}(X, F^*) , \ A^*v \in L^-\} .$$

They are also obviously Γ-invariant. The inclusions (3.17) and the elliptic regularity in the uniform spaces W^s (see, e.g. [S]) imply that

$$\tilde{L}^+ \subset W^{-N_1+d}(X, E) , \quad \tilde{L}^- \subset W^{-N_2+d}(X, F^*) ,$$

obviously $\tilde{L}^\pm$ are closed subspaces there. We have morphisms of Hilbert Γ-modules

$$A : \tilde{L}^+ \longrightarrow L^+ , \quad A^* : \tilde{L}^- \longrightarrow L^- ,$$

which are injective because $D^\pm$ are nowhere dense, hence the equations $Au = 0$, $A^*v = 0$ cannot have solutions supported on these sets. Therefore

$$\tilde{\ell}^\pm = \dim_\Gamma \tilde{L}^\pm < \infty .$$

DEFINITION 3.2: The Γ-*degree* of the given rigged Γ-divisor μ (associated with A) is

$$\deg_A(\mu) = (\ell^+ - \tilde{\ell}^+) - (\ell^- - \tilde{\ell}^-) .$$

Of course the numbers $\ell^\pm, \tilde{\ell}^\pm, \deg_A(\mu)$ depend on the choice of Γ. We do not indicate this dependence directly to simplify notations.

DEFINITION 3.3: The *inverse divisor* to a given rigged Γ-divisor $\mu = (D^+, L^+; D^-, L^-)$ associated with A, is the rigged Γ-divisor

$$\mu^{-1} = (D^-, L^-; D^+, L^+) ,$$

associated with the adjoint operator A^*.

Note that

$$\deg_{A^*}(\mu^{-1}) = -\deg_A(\mu) .$$

Now we shall introduce the main space of L^2-solutions with permitted singularities on D^+ and vanishing conditions on D^-.

DEFINITION 3.4: Denote

$$L(\mu, A) = \{u \mid u \in C^\infty(X - D^+, E) , \ \exists \tilde{u} \in W^{-\infty}(X, E) , \\ \tilde{u} = u \ \text{ on } \ X - D^+ , \ A\tilde{u} \in L^+ \text{ and } (u, L^-) = 0\} ;$$

$r(\mu, A) = \dim_\Gamma L(\mu, A)$.

Here $(u, L^-) = \{(u, g) \mid g \in L^-\} \subset \mathbf{C}$ and we write 0 instead of $\{0\}$, so the equality $(u, L^-) = 0$ means that u is orthogonal to L with respect to the given duality. This makes sense (in spite of the fact that u is defined on $X - D^+$ only) because all distribution sections from L^- are supported on D^-. Indeed, we can extend u from a Γ-invariant neighbourhood of D^- to a section $\hat{u} \in W^\infty(X, E)$; if we then take $(\hat{u}, g)$ for $g \in L^-$, then the result will not depend on the extension (and we can denote it (u, g)).

We will show now that $L(\mu, A)$ has a natural structure of a Hilbert Γ-module. Obviously $L(\mu, A)$ is a Γ-invariant subspace in $C^\infty(X - D^+, E)$. Let us consider the following space

$$\tilde{L}(\mu, A) = \{\tilde{u} \mid \tilde{u} \in W^{-\infty}(X, E) , \ A\tilde{u} \in L^+ \ \text{ and } \ (u, L^-) = 0\} .$$

Obviously $\tilde{u} \in \tilde{L}(\mu, A)$ implies that $u = \tilde{u} \mid (X - D^+) \in L(\mu, A)$, so $\tilde{L}(\mu, A)$ is exactly the set of all "regularizations" $\tilde{u}$ of elements $u \in L(\mu, A)$, i.e. the set of all $\tilde{u}$ that appear in the definition of $L(\mu, A)$. Now actually $\tilde{L}(\mu, A) \subset W^{-N_1+d}(X, E)$ and $\tilde{L}(\mu, A)$ is a closed Γ-invariant subspace in $W^{-N_1+d}(X, E)$, where N_1 is the number from (3.17) in Definition 3.1. Therefore $\tilde{L}(\mu, A)$ is a Hilbert Γ-module. Note now that there is an exact sequence

$$0 \longrightarrow \tilde{L}^+ \xrightarrow{\ i\ } \tilde{L}(\mu, A) \xrightarrow{\ r\ } L(\mu, A) \longrightarrow 0$$

where i and r are natural inclusion and restriction maps. Therefore we can supply $L(\mu, A) \cong \tilde{L}(\mu, A)/\tilde{L}^+$ by a structure of a Hilbert Γ-module.

Our first main result is

THEOREM 3.5 (The Riemann–Roch Theorem for the rigged Γ-divisor μ).

$$r(\mu, A) = \mathrm{ind}_\Gamma A + \deg_A(\mu) + r(\mu^{-1}, A^*) \; . \tag{3.18}$$

Note that $\mathrm{ind}_\Gamma\, A$ can be calculated by the Atiyah theorem (3.13) and the Atiyah–Singer index formula. Later we shall show how to calculate $\deg_A(\mu)$ in an important particular case of point divisors.

COROLLARY 3.6. $r(\mu, A) \geq \mathrm{ind}_\Gamma A + \deg_A(\mu)$.

In particular, $\mathrm{ind}_\Gamma A + \deg_A(\mu) > 0$ implies that $r(\mu, A) > 0$, i.e. $L(\mu, A) \neq \{0\}$. In fact $\dim_\Gamma L(\mu, A) > 0$ which in the case of infinite Γ implies that $L(\mu, A)$ is in fact infinite-dimensional in the usual sense. So this corollary provides an existence result for L^2-solutions of $Au = 0$ with singularities on D^+ and orthogonality conditions on D^-.

EXAMPLE 3.7: Suppose that

$$|\mu| = \{x_1, x_2, \dots\} \subset X$$

is a discrete Γ-invariant subset and

$$p : |\mu| \longrightarrow \mathbf{Z} - \{0\}$$

$$x_i \mapsto p_i$$

is a Γ-invariant function. These data define a rigged Γ-divisor (associated with any elliptic operator A of order d acting as in (3.10)) which we will write as

$$\mu = \prod_{i=1}^{\infty} x_i^{p_i} \; . \tag{3.19}$$

Namely, denote $D^\pm = \{x_i \mid \pm p_i > 0\}$. Suppose that $U^\pm \supset D^\pm$ are Γ-invariant neighbourhoods of $D^\pm$, such that Γ-invariant sets of local coordinates and trivializations of bundles E, F, E^*, F^* are given in $U^\pm$. Introduce then the distribution spaces $L^\pm$ which are locally represented in these coordinates and trivializations as

$$L^\pm = \left\{ f \mid f(x) = \sum_{\pm p_i > 0} \; \sum_{|\alpha| \leq |p_i| - 1} c_{i\alpha} \delta^{(\alpha)}(x - x_i) \; , \; \sum_{i,\alpha} |c_{i\alpha}|^2 < \infty \right\} , \tag{3.20}$$

where δ means the Dirac measure, $\delta^{(\alpha)}$ denotes its derivative corresponding to the multiindex α, $c_{i\alpha}$ are vector coefficients from $\mathbf{C}^q$ (identified with the fiber of F or E^* at the point x_i) where q is the dimension of the fibers of the bundles E, F (they are equal due to the ellipticity of A), $|c_{i\alpha}|$ is the standard euclidean norm of $c_{i\alpha}$ in $\mathbf{C}^q$.

The rigged divisors of the form (3.19) will be called *point* Γ*-divisors*.

Since A is elliptic (of order d) it is easy to check that the "secondary" spaces are locally represented in a similar form

$$\tilde{L}^{\pm}=\Big\{v \mid v(x)=\sum_{\pm p_i>0}\ \sum_{|\alpha|\le|p_i|-1-d} c_{i\alpha}\delta^{(\alpha)}(x-x_i), \sum_{i,\alpha}|c_{i\alpha}|^2<\infty\Big\}. \quad (3.21)$$

Obviously $L^{\pm}$ are Γ-invariant and by the Sobolev imbedding theorem

$$L^+ \subset W^{-N}(X,F)\ ,\quad L^- \subset W^{-N}(X,E^*)$$

provided $N > \max_i(|p_i|-1)+n/2$, where $n=\dim_{\mathbf{R}} X$. They are obviously closed subspaces in W^{-N} for any such N. Choosing a fundamental domain $F\subset X$ of Γ so that $\partial F \cap |\mu| = \emptyset$, it is easy to see that

$$L^{\pm} = L^2\Gamma \otimes L_0^{\pm}\ ,\quad \tilde{L}^{\pm} = L^2\Gamma\otimes\tilde{L}_0^{\pm}$$

where $L_0^{\pm}, \tilde{L}_0^{\pm}$ are finite-dimensional complex vector spaces of distributions which are defined exactly as $L^{\pm}, \tilde{L}^{\pm}$ but with summation restricted to the points $x_i \in F$ only. Therefore we see that

$$\ell^{\pm} = \dim_\Gamma L^{\pm} = \dim L_0^{\pm} = q\sum_{\substack{\pm p_i>0\\ x_i\in F}} \binom{n+|p_i|-1}{n}\ ,$$

$$\tilde{\ell}^{\pm} = \dim_\Gamma \tilde{L}^{\pm} = \dim \tilde{L}_0^{\pm} = q\sum_{\substack{\pm p_i>0\\ x_i\in F}} \binom{n+|p_i|-1-d}{n}\ ,$$

where $\binom{N}{n} = \frac{N!}{n!(N-n)!}$ if $N\ge n$ and 0 otherwise. It follows that

$$\deg_A(\mu) = q\sum_{x_i\in F} \operatorname{sign} p_i\left[\binom{n+|p_i|-1}{n} - \binom{n+|p_i|-1-d}{n}\right] \quad (3.22)$$

(See more details on these numbers in [GrS1]).

The space $L(\mu, A)$ in this case consists of solutions of the equation $Au = 0$ that are defined on $X - |\mu|^+$, where $|\mu|^+ = \{x_i \mid p_i > 0\}$ and may have isolated singularities at the points x_i with $p_i > 0$ (poles of order $\le p_i$); these solutions should vanish at all points x_i with $p_i < 0$ with the multiplicity of zeros at least $|p_i|$; these solutions should also be L^2-solutions in a natural generalized sense.

Corollary 3.6 implies in particular the existence of non-trivial generalized L^2-solutions of the equation $Au = 0$ which are allowed to have isolated poles in a given Γ-invariant discrete set and are required to vanish at all the points in another Γ-invariant discrete set, which is disjoint with the first set. The inequality $\mathrm{ind}_\Gamma A + \deg_A(\mu) > 0$ will be fulfilled if we allow poles of sufficiently high order at the first set.

EXAMPLE 3.8: This example is a particular case of the previous one.

Let X be a (non-compact) Riemann surface with a free action of a discrete group Γ so that $\bar{X} = X/\Gamma$ is a compact Riemann surface. Let g denote the genus of $\bar{X}$.

For example we can take $X = \mathbf{C}$ with $\Gamma = \mathbf{Z}^2$ acting by translations (then $g = 0$). Also we can take $X = \{z \mid z \in \mathbf{C}, \operatorname{Im} z > 0\}$ and Γ a Fuchsian group (then g can be any integer with $g \geq 2$ depending on Γ).

Consider the operator

$$A = \bar{\partial} : C^\infty(X) \longrightarrow \Lambda^{0,1}(X) ,$$

then

$$A^* = \bar{\partial} : \Lambda^{1,0}(X) \longrightarrow \Lambda^{1,1}(X) = \Lambda^2(X) ,$$

if the duality between $C_c^\infty(X)$ and $\Lambda^2(X)$ is given by multiplication and integration and the duality between $\Lambda_c^{0,1}(X)$ (the forms in $\Lambda^{0,1}(X)$ with a compact support) and $\Lambda^{1,0}(X)$ is given by external multiplication and integration. Then $\operatorname{Ind}_\Gamma A = 1 - g$.

Consider the divisor of the form (3.19). Then (3.22) implies that

$$\deg_A(\mu) = \sum_{x_i \in F} p_i$$

because here we have $n = 2$, $q = 1$ and $d = 1$. So Theorem 3.5 in this case means

$$r(\mu, A) = 1 - g + \sum_{x_i \in F} p_i + r(\mu^{-1}, A^*) \tag{3.23}$$

and Corollary 3.6 gives the inequality

$$1 - g + \sum_{x_i \in F} p_i > 0 \tag{3.24}$$

as a sufficient condition for the space $L(\mu, A)$ to be non-trivial.

Let us describe the space $L(\mu, A)$ in this case.

Let us choose a local parameter z near every point x_i in a Γ-invariant way, so that z_i is the value of the parameter that corresponds to x_i. Then $L(\mu, A)$ consists of meromorphic functions f on X that satisfy the following conditions:

(i) near x_i we should have $f(z) = (z - z_i)^{-p_i} g(z)$, where g is holomorphic near z_i;

(ii) let U be a Γ-invariant neighbourhood of the set $\{x_i \mid p_i > 0\}$ where the above-mentioned local parameter is defined, let

$$f(z) = \sum_{k=1}^{p_i} \frac{c_{ik}}{(z - z_i)^k} + h(z)$$

be the Laurent decomposition of f near x_i up to a regular part h which is holomorphic in U; then

$$\sum_{i,k} |c_{ik}|^2 + \|h\|^2_{L^2(U)} + \|f\|^2_{L^2(X-U)} < \infty$$

where the spaces L^2 are taken with respect to a measure induced by a positive Γ-invariant density on X.

The space $L(\mu, A^*)$ is described similarly: the only difference is that meromorphic functions should be replaced by meromorphic $(1,0)$-forms.

Note however that in this particular example the formula (3.23) can be deduced from the M. Atiyah index theorem ([A2]) (formula (3.13)) if instead of scalar meromorphic functions we consider holomorphic sections of the holomorphic line bundle associated with the divisor μ.

Note also that existence results for meromorphic functions of more general type (with non-periodic configuration of poles and zeros) which are in L^2 in the generalized sense, were given by A. Connes ([Con2]) and J. Roe ([R]), who extended the M. Atiyah theorem to more general situations (foliations and manifolds of bounded geometry).

EXAMPLE 3.9: This will be again a particular case of Example 3.7. Let $X = \mathbf{R}^n$, and $\Gamma = \mathbf{Z}^n$ act by translations. Consider the standard flat Laplacian Δ on $\mathbf{R}^n$. Consider a point Γ-divisor with

$$|\mu| = \bigcup_{1\le i\le m} (\mathbf{Z}^n + \tau_i) \ , \quad \tau_i \in \mathbf{R}^n \ , \ \tau_i \not\equiv \tau_j \operatorname{mod} \mathbf{Z}^n \text{ if } i \ne j \ ; \ p(\tau_i) = p_i \ .$$

We have then $\operatorname{ind}_\Gamma \Delta = 0$ and

$$\deg_A(\mu) = \sum_{i=1}^m \operatorname{sign} p_i \left[\binom{n + |p_i| - 1}{n} - \binom{n + |p_i| - 3}{n} \right] .$$

The space $L(\mu, \Delta)$ consists of harmonic functions, which are defined outside the set $\bigcup_{p_i>0}(\mathbf{Z}^n + \tau_i)$, near any point in $\mathbf{Z}^n + \tau_i$ with $p_i > 0$ can be represented as a sum of derivatives of the order $\le p_i$ of the fundamental solution of Δ and a harmonic function near this point, vanish at all points in $\mathbf{Z}^n + \tau_i$ if $p_i < 0$, with the zeros of the multiplicity $\ge |p_i|$, are in L^2 in the generalized sense (which implies in particular that they really are in $L^2(\mathbf{R}^n)$ if $n \le 3$ and $p_i \le 1$ for all i). Theorem 3.5 then becomes

$$r(\mu, \Delta) = \deg_A(\mu) + r(\mu^{-1}, \Delta)$$

and the Corollary 3.6 gives

$$r(\mu, \Delta) \ge \deg_A(\mu)$$

which implies the existence of non-trivial harmonic functions with a permitted periodic configuration of poles and required periodic configuration of zeros according to μ, provided $\deg_A(\mu) > 0$.

EXAMPLE 3.10: For the same X, Γ, A, A^* as in the Example 3.8 consider another rigged divisor μ which is defined as follows. Consider discrete Γ-invariant sets $D^{\pm} \subset X$ such that

$$D^+ \cap F = \{\bar{x}_1, \ldots, \bar{x}_k\} , \quad D^- \cap F = \{\bar{y}_1, \ldots, \bar{y}_\ell\} ,$$

where all the points $\bar{x}_1, \ldots, \bar{x}_k, \bar{y}_1, \ldots, \bar{y}_\ell$ are distinct. Define in Γ-invariant real local coordinates x_1, x_2 with $z = x_1 + ix_2$

$$L^+ = \Big\{ \sum_{x_i \in D^+} c_i \delta(x - x_i) d\bar{z} \,\Big|\, \sum_i |c_i|^2 < \infty \Big\} ,$$

$$L^- = \Big\{ \sum_{y_j \in D^-} \sum_{\alpha=1}^{2} c_{j\alpha} \frac{\partial}{\partial x_\alpha} \delta(x - y_j) dx_1 \wedge dx_2 \,\Big|\, \sum_{j,\alpha} |c_{j\alpha}|^2 < \infty \Big\} .$$

Then for the secondary spaces we will have

$$\tilde{L}^+ = \{0\} , \quad \tilde{L}^- = \Big\{ \sum_{y_j \in D^-} b_j \delta(x - y_j) dz \,\Big|\, \sum_j |b_j|^2 < \infty \Big\} .$$

It is easy to see that $\ell^+ = k$, $\ell^- = 2\ell$, $\tilde{\ell}^+ = 0$, $\tilde{\ell}^- = \ell$, so

$$\deg_A(\mu) = k - \ell .$$

The space $L(\mu, A)$ (with $A = \bar{\partial} : C^\infty(X) \to \Lambda^{0,1}(X)$) is the space of all meromorphic functions on X which are in L^2 in the same generalized sense as in the Example 3.8, with possible simple poles at the points in D^+ and with critical points at the points in D^-. The space $L(\mu^{-1}, A^*)$ is the space of all meromorphic (1,0) forms (in L^2 in the generalized sense) which might have poles of order 2 with vanishing residues at the points in D^- and should vanish at all points in D^+.

Theorem 3.5 becomes then

$$r(\mu, A) = 1 - g + k - \ell + r(\mu^{-1}, A^*)$$

with the Corollary 3.6 giving inequalities

$$r(\mu, A) \geq 1 - g + k - \ell , \quad r(\mu^{-1}, A^*) \geq g - 1 + \ell - k$$

and providing corresponding existence results.

EXAMPLE 3.11: Suppose that $X = \mathbf{R}^3$, $\Gamma = \mathbf{Z}^3$ act by translations, $A = -\Delta$. Suppose that

$$D^+ = \bigcup_{i=1}^{k} (\mathbf{Z}^3 + \bar{x}_i) , \quad D^- = \bigcup_{j=1}^{\ell} (\mathbf{Z}^3 + \bar{y}_j) ,$$

where all the points $\bar{x}_1,\ldots,\bar{x}_k,\bar{y}_1,\ldots,\bar{y}_\ell$ are distinct mod$\mathbf{Z}^3$ (i.e. define distinct images in $\mathbf{R}^3/\mathbf{Z}^3$). Define

$$L^+ = \Big\{ \sum_{x_i \in D^+} q_i \delta(x - x_i) \,\Big|\, \sum_i |q_i|^2 < \infty \Big\} ,$$

$$L^- = \Big\{ \sum_{y_j \in D^-} \sum_{\alpha=1}^{3} c_{j\alpha} \frac{\partial}{\partial x_\alpha} \delta(x - y_j) \,\Big|\, \sum_{j,\alpha} |c_{j\alpha}|^2 < \infty \Big\} .$$

Then $\dim_\Gamma L^+ = k$, $\dim_\Gamma L^- = 3\ell$. Obviously $\tilde{L}^+ = \tilde{L}^- = \{0\}$, so for the divisor $\mu = (D^+, L^+; D^-, L^-)$ we have $\deg_A(\mu) = k - 3\ell$. Therefore

$$r(\mu, \Delta) = k - 3\ell + r(\mu^{-1}, \Delta) .$$

The space $L(\mu, \Delta)$ is the space of Coulomb potentials u of point charges q_i with a periodic configuration D^+ such that $u \in L^2(\mathbf{R}^3)$ and u has all points in D^- as their critical points (or equilibrium positions of the corresponding system of charges).

The space $L(\mu^{-1}, \Delta)$ will consist of the dipole potentials v of dipoles at all points $y_j \in D^-$ with the vector (c_{j1}, c_{j2}, c_{j3}) defining the orientation and the dipole momentum, such that v vanishes at all points in D^+ and is in L^2 in a natural generalized sense.

It follows that $L(\mu, \Delta) \neq \{0\}$ if $k > 3\ell$. Similarly $L(\mu^{-1}, \Delta) \neq \{0\}$ if $k < 3\ell$.

4. Localization, Duality and Proofs

A. The first idea of the proof of Theorem 3.5 is a localization which begins with the introduction of the following space

$$\begin{aligned} W^s(\mu, A) = \{u \mid u \in \mathcal{D}'(X - D^+, E) \,,\ \exists \tilde{u} \in W^{-\infty}(X, E), \\ \tilde{u} = u \text{ on } X - D^+ \,,\ A\tilde{u} \in L^+ + W^{s-d}(X, F) \,, (u, L^-) = 0\} \,, \end{aligned} \tag{4.1}$$

where $s \gg 0$, i.e. s is sufficiently large. At the moment it is sufficient to take $s > \max(N_2, d + n/2)$ where N_2 is the number from (3.17) in Definition 3.1; this is needed to ensure first that the condition $(u, L^-) = 0$ makes sense and second that $L^+ \cap W^{s-d}(X, F) = \{0\}$. Consider also the space

$$\tilde{W}^s(\mu, A) = \{\tilde{u} \mid \tilde{u} {\in} W^{-\infty}(X, E) \,,\ A\tilde{u} {\in} L^+ + W^{s-d}(X, F), (u, L^-){=}0\} \,, \tag{4.2}$$

this is the space of all "regularizations" of elements $u \in W^s(\mu, A)$, i.e. the space of all $\tilde{u}$ that might appear in (4.1). For any $f \in L^+ + W^{s-d}(X, F)$ denote f_s and f_r its "singular" and "regular" parts, i.e. $f_s \in L^+$, $f_r \in W^{s-d}(X, F)$ and $f = f_s + f_r$. Then we can define the following norm on

$\tilde{W}^s(\mu, A)$;

$$\|\tilde{u}\|^2_{\tilde{W}^s(\mu,A)} = \|\tilde{u}\|^2_{W^{-N_1+d}(X,E)} + \|(A\tilde{u})_s\|^2_{W^{-N_1}(X,F)} + \|(A\tilde{u})_r\|^2_{W^{s-d}(X,F)} \,. \tag{4.3}$$

With this norm $\tilde{W}^s(\mu, A)$ becomes a Hilbert space, hence a Hilbert Γ-module. Obviously $\tilde{L}^+$ is its closed subspace (and Γ-submodule). Moreover we have an exact sequence

$$0 \longrightarrow \tilde{L}^+ \xrightarrow{i} \tilde{W}^s(\mu, A) \xrightarrow{r} W^s(\mu, A) \longrightarrow 0\,, \tag{4.4}$$

where i and r are natural inclusion and restriction maps. We shall use this sequence to define on $W^s(\mu, A)$ a structure of a Hilbert Γ-module $\tilde{W}^s(\mu, A)/\tilde{L}^+$.

Note that the space $W^s(\mu, A)$ is already sufficiently large: it includes all functions $u \in W^s(X, E)$ such that $\operatorname{supp} u \cap (D^+ \cup D^-) = \emptyset$; in particular it includes $C_c^\infty(X - (D^+ \cup D^-), E)$.

Suppose that $u \in W^s(\mu, A)$. Consider Au which is defined on $X - D^+$. It is obvious from (4.1) that Au can be extended to a section in $W^{s-d}(X, F)$. Denote this extension by $\tilde{A}u$. We have

$$(\tilde{A}u, \tilde{L}^-) = (Au, \tilde{L}^-) = (u, A^*\tilde{L}^-) \subset (u, L^-) = \{0\}\,.$$

This motivates the introduction of the following space.

$$\tilde{W}^{s-d}_\mu(A) = \{f \mid f \in W^{s-d}(X, F), (f, \tilde{L}^-) = 0\}\,. \tag{4.5}$$

It is a closed subspace in $W^{s-d}(X, F)$, hence a Hilbert Γ-module. So $\tilde{A}$ defines a linear map

$$\tilde{A} : W^s(\mu, A) \longrightarrow \tilde{W}^{s-d}_\mu(A)\,. \tag{4.6}$$

In terms of the extension $\tilde{u}$ in (4.1) we can write that $\tilde{A}u = (A\tilde{u})_r$, i.e. $\tilde{A}u$ is the regular part of $A\tilde{u}$ (obviously $(A\tilde{u})_r$ does not depend on the choice of the extension $\tilde{u}$ of the given section u). The operator $\tilde{A}$ commutes with the action of Γ.

Let us consider the following commutative diagram

$$\begin{array}{ccccccccc} 0 & \longrightarrow & \tilde{L}^+ & \xrightarrow{i} & \tilde{W}^s(\mu, A) & \xrightarrow{r} & W^s(\mu, A) & \longrightarrow & 0 \\ & & \big\downarrow{\scriptstyle A_s} & & \big\downarrow{\scriptstyle \hat{A}} & & \big\downarrow{\scriptstyle \tilde{A}} & & \\ 0 & \longrightarrow & L^+ & \xrightarrow{i_1} & \tilde{W}^{s-d}_\mu(A) \oplus L^+ & \xrightarrow{r_1} & \tilde{W}^{s-d}_\mu(A) & \longrightarrow & 0 \end{array} \tag{4.7}$$

where the first line coincides with the exact sequence (4.4), $A_s, \hat{A}$ are restrictions of A, and the operators i_1, r_1 are the natural inclusion and restriction maps. It follows from the definition (4.3) of the norm in $\tilde{W}^s(\mu, A)$ that the operator $\hat{A}$ here is bounded. Hence, the operator $r_1\hat{A} = \tilde{A}r$ is bounded too and $\tilde{L}^+ \subset \operatorname{Ker}(\tilde{A}r)$. It follows that $\tilde{A}$ is a bounded linear operator, hence it is a morphism of Hilbert Γ-modules.

LEMMA 4.1. *$\tilde{A}$ is Γ-Fredholm and*

$$\mathrm{ind}_\Gamma \tilde{A} = \mathrm{ind}_\Gamma A + \deg_A(\mu) . \tag{4.8}$$

Proof: 1) Let us consider the diagram (4.7) as a short exact sequence of short complexes of Hilbert Γ-modules which are columns of this diagram. Since $\dim_\Gamma L^+ < \infty$ and $\dim_\Gamma \tilde{L}^+ < \infty$, we can apply Lemma 2.17 which shows that $\tilde{A}$ and $\hat{A}$ are Γ-Fredholm simultaneously, and if it is the case then Corollary 2.14 and Lemma 2.15 imply

$$\mathrm{ind}_\Gamma \tilde{A} = \mathrm{ind}_\Gamma \hat{A} - \mathrm{ind}_\Gamma A_s .$$

But

$$\mathrm{ind}_\Gamma A_s = \dim_\Gamma \tilde{L}^+ - \dim_\Gamma L^+ = \tilde{\ell}^+ - \ell^+$$

by Lemma 2.3. Hence

$$\mathrm{ind}_\Gamma \tilde{A} = \mathrm{ind}_\Gamma \hat{A} + (\ell^+ - \tilde{\ell}^+) . \tag{4.9}$$

2) Let us consider the following space

$$W^s_\mu(A) = \{u \mid u \in W^s(X,E), (u, L^-) = 0\} , \quad s \gg 0 .$$

Obviously it is a closed Γ-invariant subspace in $W^s(X,E)$, hence a Hilbert Γ-module. Now consider the commutative diagram

$$\begin{array}{ccccccccc} 0 & \longrightarrow & W^s_\mu(A) & \longrightarrow & \tilde{W}^s(\mu, A) & \xrightarrow{r_2\hat{A}} & L^+ & \longrightarrow & 0 \\ & & \Big\downarrow{\scriptstyle A_\mu} & & \Big\downarrow{\scriptstyle \hat{A}} & & \Big\downarrow{\scriptstyle \mathrm{Id}} & & \\ 0 & \longrightarrow & \tilde{W}^{s-d}_\mu(A) & \longrightarrow & \tilde{W}^{s-d}_\mu(A) \oplus L^+ & \xrightarrow{r_2} & L^+ & \longrightarrow & 0 \end{array} \tag{4.10}$$

where r_2 is the natural projection, A_μ is the restriction of A. We want to show that the rows here are exact. It is not obvious in the term L^+ of the first row only. Let us prove that the operator $r_2\hat{A}$ (which maps u to the singular part of Au) is surjective. Consider a parametrix of A: a pseudodifferential Γ-invariant operator $B : C^\infty_c(X,F) \to C^\infty_c(X,E)$ of order $-d$ such that B is almost local, i.e. the Schwartz kernel $K_B(x,y)$ of B vanishes if $\mathrm{dist}(x,y) \geq \varepsilon > 0$ (here the distance function should be induced by a Γ-invariant riemannian metric), and $AB = I - T$, where T is also almost local (with the same ε) and has a C^∞ Schwartz kernel (see [A2] or [S]). For any $f \in L^+$ we have then $Tf \in W^\infty(X,F)$. Obviously $Bf = 0$ in a neighbourhood of D^- if ε is sufficiently small. Therefore for $\tilde{u} = Bf$ we shall have $\tilde{u} \in \tilde{W}^s(\mu, A)$ for any s because $Bf \in W^{-N_1+d}(X,E)$, and $\tilde{u} = ABf = f - Tf$; it follows also that $r_2\hat{A}\tilde{u} = f$ which proves the surjectivity of $r_2\hat{A}$.

Now we can consider the diagram (4.10) as a short exact sequence of complexes of Hilbert Γ-modules. Applying Lemma 2.17 we see that $\hat{A}$ and A_μ are Γ-Fredholm simultaneously and

$$\mathrm{ind}_\Gamma \hat{A} = \mathrm{ind}_\Gamma A_\mu \ . \tag{4.11}$$

3) Let us consider the commutative diagram

$$\begin{array}{ccccccccc} 0 & \longrightarrow & W^s_\mu(A) & \xrightarrow{i_\mu} & W^s(X,E) & \xrightarrow{p_\mu} & (L^-)' & \longrightarrow & 0 \\ & & \big\downarrow{\scriptstyle A_\mu} & & \big\downarrow{\scriptstyle A} & & \big\downarrow{\scriptstyle (A^*)'} & & \\ 0 & \longrightarrow & \tilde{W}^{s-d}_\mu(A) & \xrightarrow{\tilde{i}_\mu} & W^{s-d}(X,F) & \xrightarrow{\tilde{p}_\mu} & (\tilde{L}^-)' & \longrightarrow & 0 \end{array} \tag{4.12}$$

Here for $\tilde{L} = L^-$ or $\tilde{L}^-$ we denote the space of all continuous linear (or antilinear) maps $L \to \mathbf{C}$ by L'. Of course L' may be identified with L but it is not convenient now. The maps $i_\mu, \tilde{i}_\mu$ are natural inclusion maps. The maps $p_\mu, \tilde{p}_\mu$ are defined as follows:

$$(p_\mu u)(g) = (u,g) \ , \quad u \in W^s(X,E) \ , \ g \in L^- \ ,$$
$$(\tilde{p}_\mu f)(h) = (f,h) \ , \quad f \in W^{s-d}(X,F) \ , \ h \in \tilde{L}^- \ .$$

The map $A^* : \tilde{L}^- \to L^-$ induces the dual map $(A^*)' : (L^-)' \to (\tilde{L}^-)'$ defined by the formula:

$$((A^*)'v, u) = (v, A^*u) \ , \quad u \in \tilde{L}^- \ , \ v \in (L^-)' \ ,$$

where $(\cdot,\cdot)$ denotes natural bilinear (or sesquilinear) pairing between $(\tilde{L}^-)'$ and $\tilde{L}^-$ or between $(L^-)'$ and L^-.

The rows of the diagram (4.12) are obviously exact in all terms except $(L^-)'$ and $(\tilde{L}^-)'$. Let us show that they are almost exact in these terms. This follows from the fact that the adjoint operators to p_μ and $\tilde{p}_\mu$ are just canonical embeddings $L^- \subset W^{-s}(X,E^*)$, $\tilde{L}^- \subset W^{-s+d}(X,F^*)$, hence $p_\mu, \tilde{p}_\mu$ have dense images.

Now we can apply Corollary 2.19 to conclude that A_μ is Γ-Fredholm (because A is Γ-Fredholm – see, e.g. [A2]) and besides

$$\begin{aligned} \mathrm{ind}_\Gamma A_\mu = \mathrm{ind}_\Gamma A - \mathrm{ind}_\Gamma (A^*)' &= \mathrm{ind}_\Gamma A - \left(\dim_\Gamma (L^-)' - \dim_\Gamma (\tilde{L}^-)'\right) \\ &= \mathrm{ind}_\Gamma A - (\ell^- - \tilde{\ell}^-) \ . \end{aligned}$$

Combining this with (4.11) and (4.9) we conclude that $\tilde{A}$ is Γ-Fredholm and (4.8) is true. □

Remark 4.2: Note that $\mathrm{Ker}\,\tilde{A} = L(\mu, A)$ for every integer $s \gg 0$. Therefore Lemma 4.1 means that

$$r(\mu, A) = \mathrm{ind}_\Gamma A + \deg_A(\mu) + \dim_\Gamma \mathrm{Coker}\,\tilde{A} \tag{4.13}$$

where $\mathrm{Coker}\,\tilde{A} = \tilde{W}^{s-d}_\mu(A)/\overline{\mathrm{Im}\,\tilde{A}}$. Therefore for the proof of Theorem 3.5 it remains to prove that

$$\dim_\Gamma \mathrm{Coker}\,\tilde{A} = \dim_\Gamma \mathrm{Ker}\,\tilde{A}^* \ , \tag{4.14}$$

because $\dim_\Gamma \operatorname{Ker} \tilde{A}^* = r(\mu^{-1}, A^*)$. Here we denote by $\tilde{A}^*$ the result of applying the "tilde" operation to A^* and *not* the adjoint operator to $\tilde{A}$. The equality (4.14) is not obvious. To prove it we shall need a refined use of duality which gives interesting analytic results by itself.

B. We shall introduce now an important pairing or duality between spaces that we described.

DEFINITION 4.3: The duality

$$W^s(\mu, A) \times \tilde{W}^{s'-d}_{\mu^{-1}}(A^*) \longrightarrow \mathbf{C}\,, \quad s \gg 0\,, s' \gg 0\,, \tag{4.15}$$

is defined as follows:

$$(u, f) = (r^{-1}u, f)\,, \quad u \in W^s(\mu, A)\,,\ f \in \tilde{W}^{s'-d}_{\mu^{-1}}(A^*) \tag{4.16}$$

where r is taken from the exact sequence (4.4), i.e. $r^{-1}u = \tilde{u}$ is any element in $\tilde{W}^s(\mu, A)$ such that $r\tilde{u} = u$. Here $(r^{-1}u, f)$ in the right-hand side of (4.16) is taken in the sense of the natural duality

$$W^{-N_1+d}(X, E) \times W^{N_1-d}(X, E^*) \longrightarrow \mathbf{C}$$

and it makes sense because

$$\tilde{W}^s(\mu, A) \subset W^{-N_1+d}(X, E)\,, \quad \tilde{W}^{s'-d}_{\mu^{-1}}(A^*) \subset W^{N_1-d}(X, E^*)$$

if $s > \max(N_2, d + n/2)$, $s' \geq N_1$. Since the inclusion maps are continuous, we actually get the duality of Hilbert Γ-modules

$$\tilde{W}^s(\mu, A) \times \tilde{W}^{s'-d}_{\mu^{-1}}(A^*) \longrightarrow \mathbf{C} \tag{4.17}$$

but this duality is degenerated if $\tilde{L}^+ \neq \{0\}$ because $(\tilde{L}^+, f) = 0$ for any $f \in \tilde{W}^{s'-d}_{\mu^{-1}}(A^*)$. But this exactly shows that the right-hand side in (4.16) does not depend on the choice of $\tilde{u} = r^{-1}u$, so the formula (4.16) really defines the duality (4.15) as the duality of Hilbert Γ-modules.

Let us for any open set $U \subset X$ denote by $W^s_{\mathrm{loc}}(U, E)$ the set of all section u of E over U, such that $\varphi u \in W^s(X, E)$ for every $\varphi \in C^\infty_c(U)$.

Note that the duality (4.15) is already non-degenerate because

$$C^\infty_c(X - (D^+ \cup D^-), E) \subset W^s(\mu, A) \subset W^s_{\mathrm{loc}}(X - D^+, E)$$

and

$$C^\infty_c(X - (D^+ \cup D^-), E^*) \subset \tilde{W}^{s'-d}_{\mu^{-1}}(A^*) \subset W^{s'-d}(X, E^*)\,.$$

Similarly (by replacing A by A^* and μ by μ^{-1}) we define the duality of Hilbert Γ-modules

$$\tilde{W}^{s-d}_{\mu}(A) \times W^{s'}(\mu^{-1}, A^*) \longrightarrow \mathbf{C}\,, \quad s \gg 0\,, s' \gg 0\,, \tag{4.18}$$

which is also non-degenerate.

LEMMA 4.4. *We have*

$$(\tilde{A}u, v) = (u, \tilde{A}^* v) , \quad u \in W^s(\mu, A) , \ v \in W^{s'}(\mu^{-1}, A^*) ,$$

where the dualities on the left and right-hand sides are the dualities (4.18) *and* (4.15) *respectively.*

Proof: Taking $\tilde{u} = r^{-1}u$ and using the duality (4.17) (and similar duality for (4.18)) we get

$$(\tilde{A}u, v) = (A\tilde{u}, \tilde{v}) = (\tilde{u}, A^*\tilde{v}) = (u, \tilde{A}^* v)$$

because the singular supports of $\tilde{u}$ and $\tilde{v}$ do not intersect. □

Now we can formulate the main duality result.

THEOREM 4.5 (Duality Theorem). *For any $s \gg 0$ and $s' \gg 0$*

(i) $\overline{\operatorname{Im} \tilde{A}} = (\operatorname{Ker} \tilde{A}^*)^\circ$, *i.e. $f \in \overline{\operatorname{Im} \tilde{A}}$ if and only if $f \in \tilde{W}^{s-d}_\mu(A)$ and $(f, \operatorname{Ker} \tilde{A}^*) = 0$ in the duality* (4.18);

(ii) $\dim_\Gamma \operatorname{Coker} \tilde{A} = \dim_\Gamma \operatorname{Ker} \tilde{A}^*$, *where* $\operatorname{Coker} \tilde{A} = \tilde{W}^{s-d}_\mu(A)/\overline{\operatorname{Im} \tilde{A}}$.

We shall need the following

LEMMA 4.6. *In the pairing* (4.18)

$$(\operatorname{Im} \tilde{A})^\circ = \operatorname{Ker} \tilde{A}^* .$$

Proof: Clearly

$$\operatorname{Ker} \tilde{A}^* = \left\{ v \mid v \in W^{s'}(\mu^{-1}, A^*) , \ A^* v = 0 \text{ on } X - (D^+ \cup D^-) \right\} .$$

Lemma 4.6 follows because $\operatorname{Im} \tilde{A}$ contains all sections Au with $u \in C_c^\infty(X - (D^+ \cup D^-), E)$. □

Proof of Theorems 3.5 and 4.5: Due to Lemmas 1.23 and 4.6 we have in the pairing (4.18)

$$\overline{\operatorname{Im} \tilde{A}} \subset (\operatorname{Ker} \tilde{A}^*)^\circ , \tag{4.19}$$

$$\operatorname{codim}_\Gamma \overline{\operatorname{Im} \tilde{A}} \geq \dim_\Gamma \operatorname{Ker} \tilde{A}^* , \tag{4.20}$$

and we have to prove that both these inclusion and inequality are in fact equalities. Actually the equality in (4.20) (which is the same as the equality (4.14)) will obviously imply the equality in (4.19), hence Theorem 4.5 will immediately follow and so will Theorem 3.5 due to Remark 4.2.

Note that (4.13) and (4.20) imply

$$r(\mu, A) = \operatorname{ind}_\Gamma A + \deg_A(\mu) + \dim_\Gamma \operatorname{Coker} \tilde{A} \geq \operatorname{ind}_\Gamma A + \deg_A(\mu) + r(\mu^{-1}, A^*) .$$

Let us replace here μ by μ^{-1} and A by A^*. Since $\operatorname{ind}_\Gamma A^* = -\operatorname{ind}_\Gamma A$,

$\deg_{A^*}(\mu^{-1}) = -\deg_A(\mu)$, we obtain then the opposite inequality. Therefore we obtain the equality

$$\dim_\Gamma \operatorname{Coker} \tilde{A} = r(\mu^{-1}, A^*)$$

which is the same as the equality in (4.20). This completes the proof of both Theorems 3.5 and 4.5. □

The Duality Theorem 4.5 actually means the following. Suppose that we are given $f \in \tilde{W}_\mu^{s-d}(A)$ and want to find approximate solutions $u \in W^s(\mu, A)$ of the equation $Au = f$, i.e. a sequence of functions $u_k \in W^s(\mu, A)$, $k = 1, 2, \dots$, such that

$$\|Au_k - f\|_{W^{s-d}(X,F)} \longrightarrow 0 \quad \text{as} \quad k \longrightarrow \infty .$$

This can be done if and only if $(f, \operatorname{Ker} \tilde{A}^*) = 0$.

This is of course a very weak solvability statement. But in fact a much stronger statement can be made about the image $\operatorname{Im} \tilde{A}$.

COROLLARY 4.7. *$\operatorname{Im} \tilde{A}$ is Γ-dense in $(\operatorname{Ker} \tilde{A}^*)^\circ$, i.e. for any $\varepsilon > 0$ there exists a closed Γ-invariant subspace $L_\varepsilon \subset (\operatorname{Ker} \tilde{A}^*)^\circ$ such that $L_\varepsilon \subset \operatorname{Im} \tilde{A}$ and $\dim_\Gamma[(\operatorname{Ker} \tilde{A}^*)^\circ \ominus L_\varepsilon] < \varepsilon$.*

Proof: We proved that $\tilde{A}$ is a Γ-Fredholm operator with $\overline{\operatorname{Im} \tilde{A}} = (\operatorname{Ker} \tilde{A}^*)^\circ$. Hence $\operatorname{Im} \tilde{A}$ is almost closed and Γ-dense in $(\operatorname{Ker} \tilde{A}^*)^\circ$ (see Definition 1.7 and Lemma 1.15). □

References

[A1] M. ATIYAH, K-theory, Addison-Wesley, Amsterdam, 1988.

[A2] M. ATIYAH, Elliptic operators, discrete groups and von Neumann algebras, Astérisque 32–33 (1976), 43–72.

[B] M. BREUER, Fredholm theories in von Neumann algebras I, II, Math. Ann. 178 (1968), 243–254; 180 (1969), 313–325.

[CGr] J. CHEEGER, M. GROMOV, Bounds on the von Neumann dimension of L^2-cohomology and the Gauss–Bonnet theorem for open manifolds, J. Differential Geometry 21 (1985), 1–34.

[Co] J. COHEN, Von Neumann dimension and the homology of covering spaces, Quart. J. Math. Oxford 30 (1979), 133–142.

[Con1] A. CONNES, Sur la théorie non commutative de l'intégration, Springer Lecture Notes Math. 725 (1979), 19-143.

[Con2] A. CONNES, A survey of foliations and operator algebras, Proc. Symp. Pure Math. 38 (1982), Part I, 521–628.

[D] J. DIXMIER, Von Neumann algebras, North-Holland, Amsterdam, 1981.

[FS] B.V. FEDOSOV, M.A. SHUBIN, The index or random elliptic operators, I, II, Matem. Sbornik 106:1 and 3 (1978), 108–140 and 455–483.

[GH] P. Griffiths, J. Harris, Principles of algebraic geometry, John Wiley & Sons, New York, 1978.

[GrS1] M. Gromov, M. Shubin, The Riemann–Roch theorem for elliptic operators, in "I.M. Gelfand Seminar, part 1", American Math. Soc., Providence, R.I., 1993.

[GrS2] M. Gromov, M. Shubin, The Riemann–Roch theorem for elliptic operators and solvability of elliptic equations with additional conditions on compact subsets, Preprint ETH Zürich, 1993.

[Hi] F. Hirzebruch, Topological Methods in Algebraic Geometry, Springer-Verlag, Berlin, 1966.

[N] N.S. Nadirashvili, Harmonic functions with a given set of singularities, Functional Anal. and Appl. 22:1 (1988), 64–66.

[P] R. Palais, Seminar on the Atiyah–Singer index theorem, Princeton Univ. Press, Princeton, 1965.

[R] J. Roe, An index theorem on open manifolds, I, II, J. Differential Geometry 27 (1988), 87-113, 115–136.

[S] M.A. Shubin, Spectral theory of elliptic operators on non-compact manifolds, Astérisque 207 (1992), 35–108.

[Si] I.M. Singer, Some remarks on operator theory and index theory, Lecture Notes Math., 575 (1977), 128–138.

Mikhail A. Shubin
Department of Mathematics
Northeastern University
Boston, MA 02115, USA
E-mail: shubin@neu.edu

Submitted: August 1994